Microorganisms for Sustainability

Volume 16

Series editor

Naveen Kumar Arora, Environmental Microbiology, School for Environmental Science, Babasaheb Bhimrao Ambedkar University, Lucknow, Uttar Pradesh, India

More information about this series at http://www.springer.com/series/14379

Ashok Kumar • Swati Sharma
Editors

Microbes and Enzymes in Soil Health and Bioremediation

Springer

Editors
Ashok Kumar
Department of Biotechnology and Bioinformatics
Jaypee University of Information Technology
Waknaghat, Himachal Pradesh, India

Swati Sharma
University Institute of Biotechnology (UIBT)
Chandigarh University
Mohali, Punjab, India

ISSN 2512-1901 ISSN 2512-1898 (electronic)
Microorganisms for Sustainability
ISBN 978-981-13-9116-3 ISBN 978-981-13-9117-0 (eBook)
https://doi.org/10.1007/978-981-13-9117-0

This Springer imprint is published by the registered company Springer Nature Singapore Pte Ltd.
The registered company address is: 152 Beach Road, #21-01/04 Gateway East, Singapore 189721, Singapore

Preface

This book was encouraged by the aspiration that daily human activities in the modern world add to the huge amount of waste into the environment. In order to meet the food demand and requirement, chemical fertilizers in excess have been added to agricultural land and harmful pesticides sprayed onto the crops and vegetables which altered the normal soil microflora and fauna. Plastic pollution becomes the biggest threat in the twenty-first century. Microplastic particles and fibers, crude oil, paints, varnishes, and other daily used stuff in modern life have created the threat to human life. This book had been written to provide a framework for the role of microbes and enzymes to maintain the health of the soil. This book mainly emphasizes the interaction of various pollutants with the soil and water on the Earth's surface.

Researchers across the globe have been trying to generate alternative ways from bio-based products which are more eco-friendly and biodegradable. The chemical products, such as oil, grease, adhesives, and detergents, could be replaced with bio-oil, bio-lubricants, bioadhesives, and biosurfactants which are sensitive to microbial attack and eco-friendly. But there are various challenges for the complete replacement with bio-based products which make it not feasible. A remarkable development in the genomic technologies in the last three decades has enabled the development of various engineered microbial strains and recombinant enzymes which are capable to attack the highly stable bond network in the polymers, oxides, and various complex organic compounds.

We compiled the chapters written by various experienced researchers working in the environmental biotechnology and relevant areas. This book will be helpful to people working in industry and academia, young professionals, and students. The first four chapters will introduce the various pollutants entering in nature and how the microbial population is getting affected. Various microbes important for the growth and development of the plant have also been discussed.

In fundamental nature, the rationale of this book is to provide a toolbox from which researchers, students, and environmentalists working in earth science will be benefited. Another major reason for editing the book was the topic of the research area of our interest. Generally, we spend many hours to collect the information on a wide range of topic and able to get little information or puzzling results. Thus, in the book, we compiled the chapters on all the important issues which need to be solved urgently. Chapter 1 describes the various environmental challenges and implications. Chapter 2 deals with microbes and processes in the bioremediation of soil and provides basic information about the bioremediation processes. Chapters 3, 4 and 5 deal with the effect of pollution on the physical and chemical properties of soil and the roles of microbes in plant growth and development. Chapters 6, 7 and 8 described the role of microbial enzymes in the degradation and removal of various pollutants. These chapters are emphasizing on the hydrolases, laccases, esterases, and other various microbial enzymes. Chapters 9 and 10 describe the microbial interventions for sustainable bioremediation strategies and degradation of chemical pesticides, while Chaps. 11, 12 and 13 discuss about phenolic compounds. Chapters 14 and 16 discuss on soil that has also been contaminated with heavy metals and pharmaceutical products. Chapter 15 describes the role of biosurfactants in soil health and bioremediation. We definitely hope that the present book will be beneficial for all the early-stage researchers and industrialists.

Waknaghat, Himachal Pradesh, India — Ashok Kumar
Mohali, Punjab, India — Swati Sharma

Contents

About the Series Editor

Naveen Kumar Arora, Ph.D., microbiology, Professor in the Department of Environmental Science, Babasaheb Bhimrao Ambedkar University (A Central University), Lucknow, Uttar Pradesh, India, is a renowned researcher in the field of environmental microbiology and biotechnology. His specific area of research is rhizosphere biology and PGPRs. He has more than 50 research papers published in premium international journals and several articles published in magazines and dailies. He is editor of 10 books, published by Springer. He is a member of several national and international societies, Secretary General of Society for Environmental Sustainability, in editorial board of four journals, and reviewer of several international journals. He is also the Editor in Chief of the journal *Environmental Sustainability* published by Springer Nature. He has delivered lectures in conferences and seminars around the globe. He has a long-standing interest in teaching at the PG level and is involved in taking courses in bacteriology, microbial physiology, environmental microbiology, agriculture microbiology, and industrial microbiology. He has been an advisor to 118 postgraduate and 8 doctoral students. Recently, he was awarded for excellence in research by the Honorable Governor of Uttar Pradesh. Although an academician and researcher by profession, he has a huge obsession for the wildlife and its conservation and has authored a book, *Splendid Wilds*. He is the President of Society for Conservation of Wildlife and has a dedicated website www.naveenarora.co.in for the cause of wildlife and environment conservation.

About the Editors

Dr. Ashok Kumar is an Assistant Professor at the Department of Biotechnology and Bioinformatics, Jaypee University of Information Technology, India. He completed his Ph.D. in Biotechnology at the Department of Biotechnology, Himachal Pradesh University, India, and worked as a Postdoctoral Fellow at the State Key Laboratory of Agricultural Microbiology, Huazhong Agricultural University, China. He also worked as a Brain Pool Researcher/Assistant Professor in the Department of Chemical Engineering, Konkuk University, Seoul, South Korea.. He has a keen interest in enzyme immobilization, biomaterials, biopolymers, nanobiotechnology, biocatalysis, waste management, biomass degradation, biofuel synthesis, and biotransformation. He has published 47 research papers in various internationally respected journals, 1 book, and 15 book chapters.

Dr. Swati Sharma is working as an Assistant Professor at University Institute of Biotechnology (UIBT), Chandigarh University, Mohali, Punjab, India. She has completed her Ph.D. from the Department of Chemical and Natural Resources Engineering, Universiti Malaysia Pahang , Malaysia. She worked as a Visiting Researcher at the College of Life and Environmental Sciences at Konkuk University, South Korea. She has completed her master's degree (M.Sc.) at Dr. Yashwant Singh Parmar University of Horticulture and Forestry, India, and also worked as a Program Co-coordinator at the Himalayan Action Research Center, Dehradun, and Senior Research

Fellow at the Indian Agricultural Research Institute. Her current research is in the field of bioplastics, hydrogels, keratin nano-fibers and nanoparticles, biodegradable polymers, and polymers with antioxidant and anticancerous activities and sponges. She has published 16 research papers in various respected international journals, 1 book, and a number of book chapters.

Chapter 1
Let's Protect Our Earth: Environmental Challenges and Implications

Ashok Kumar, Tanvi Sharma, Sikandar I. Mulla, Hesam Kamyab, Deepak Pant, and Swati Sharma

Abstract Microbial enzymes play a vital role to maintain the soil health and removal of pollutants from the contaminated land. Soil microflora is closely associated to maintain the fertility of the soil. Use of chemical pesticides, fertilizers and other volatile sprays in the agricultural practices threatening the healthy microbial population in the soil. Every single particle of healthy soil is loaded with millions of bacteria which interact with the nutrients available in the surrounding and sustain the nutrient cycle, and this microflora is an essential component of life on earth. The rapid increase in the industrialization and urbanization polluted the water and air heavily which affected the microbial populations and their existence too. Some microbes have been evolved to breakdown the complex toxic pollutants entering the soil into non-harmful components and helping to maintain the soil fertility. Thus, it is urgently needed to identify these microorganisms and enzymes which are involved in restoring the remediation of toxic substances and restoration of microflora required for a normal life.

A. Kumar · T. Sharma
Department of Biotechnology and Bioinformatics, Jaypee University of Information Technology, Waknaghat, Solan, Himachal Pradesh, India

S. I. Mulla
Department of Biochemistry, School of Applied Sciences, Reva University, Bangalore, Karnataka, India

Key Laboratory of Urban Pollutant Conversion, Institute of Urban Environment, Chinese Academy of Sciences, Xiamen, China

H. Kamyab
Department of Engineering, UTM Razak School of Engineering and Advanced Technology, Universiti Teknologi Malaysia, Kuala Lumpur, Malaysia

D. Pant
School of Chemical Sciences, Department of Chemistry, Central University of Haryana, Mahendragarh, Haryana, India

S. Sharma (✉)
University Institute of Biotechnology (UIBT), Chandigarh University, Mohali, Punjab, India

A. Kumar, S. Sharma (eds.), *Microbes and Enzymes in Soil Health and Bioremediation*, Microorganisms for Sustainability 16,
https://doi.org/10.1007/978-981-13-9117-0_1

Keywords Enzymes · Soil health · Pesticides · Microflora · Industrialization · Pollutants

1.1 Introduction

In recent years, an increase in population growth and rapid industrialization not only ameliorate the standard of life but also affected the quality of the environment. Due to the release of harmful pollutants into the ecosystem such as plastics, pharmaceutical ingredients, greenhouse gases, pesticides, and synthetic dyestuffs, every part of the earth has severely affected. These pollutants not merely caused the teratogenic, carcinogenic, mutagenic, and toxic effects on humans beings or organisms but also created a serious risk to the environment (Jacob et al. 2018; Liu et al. 2019). The heavy metal ions from the contaminated land enter in the crop products and edible vegetable fruits or in the fish or aquatic organism from contaminated water which ultimately reaches in the human body. The continuous flow of these poisonous substances or metal impurities allows them to accumulate inside the human body and alter the normal microflora. All microbial genera, bacteria, fungi, algae, nematode, and protozoa play a significant role in bioremediations and maintaining soil health. To meet the energy crisis and food demand with the growing population, it is very important to save the agricultural land from contamination and maintain productivity.

These pollutants are broadly dispersed everywhere in the environment because of various human activities and industrial processes (Bilal et al. 2019b; Rasheed et al. 2019). Numerous methods such as filtration, reverse osmosis, incineration, lagooning treatment, landfill deposition, and bioremediation using microbes and their enzymes have been applied to treat harmful pollutants (Bilal et al. 2019a; Kuppusamy et al. 2017). The major advantages of using microbial enzymes for the degradation of environmental pollutants are high efficiency, minimum by-products, no secondary pollution, economic feasibility, and environmentally safe (Garcia-Garcia et al. 2016).

Various bacterial genera were used in bioremediation that can convert pollutant into less toxic compounds: *Pseudomonas* sp., *Achromobacter* sp., *Burkholderia* sp., *Rhodococcus* sp., *Ralstonia* sp., *Alcaligenes* sp., *Sphingomonas* sp., *Dehalococcoides* sp., and *Comamonas* sp. that ultimately reduce pollutants (Lloyd et al. 2003). However, highly diverse and specific microorganisms present in nature efficiently remove the several pollutants. But microbial-based remediation is usually slow, as compared to the daily production of a huge amount of waste which causes the pollutants to accumulate in the environment. Nevertheless, molecular biology allows producing novel strains of the microorganism with desirable features for the bioremediation process, thus considerably improving the degradation capability of pollutants (Zhao et al. 2017). Microbes play a very important role in nutrient cycles by intracellular digestion of complex macromolecules and converting these into smaller units in their metabolic activities. Secondly, the enzyme secreted in the extracellular environment facilitates the conversion of complex macromolecules into micromolecules which can be easily absorbed by other living species.

1.2 Major Environmental Challenges

1.2.1 Global Warming and Climate Change

Global warming is defined as the rise in earth temperature due to the increased level of carbon dioxide (CO_2) and other greenhouse gases (GHG). It is directly connected to the percentage of CO_2 present in the earth's atmosphere. The consequences of global warming are a rise in sea level, acidic oceans, increased air pollution, deviations in the cropping, and disease patterns. CO_2 is considered as the primary GHG that imparts to climate change, produced by the burning of fossil fuels such as coal, natural gas, and oil (He et al. 2018). A biological method like photosynthesis occurring in the plants convert CO_2 and water into organic compounds and maintains the equilibrium by fixing atmospheric CO_2 on the earth (Mondal et al. 2016). In the literature, many examples are already described for CO_2 conversion using microbes and their enzymes. Various algal species such as *Chlorella vulgaris*, *Nannochloropsis* sp., *Scenedesmus quadricauda*, *Chlamydomonas reinhardtii*, and *Nannochloris* sp. have been studied to sequester CO_2 (Eloka-Eboka and Inambao 2017). During photosynthesis, the RuBisCO enzyme in a photosynthetic organism is responsible for converting CO_2 into inorganic carbon. The major limitation of RuBisCO is a low affinity for CO_2 (Pavlik et al. 2017). However, CO_2 removal using biological methods is not suitable for region specific large-scale CO_2 sequestration such as industries outlets and polluted cities.

In a recent study, *Methylobacterium extorquens* formate dehydrogenase was reported for the conversion of CO_2 to formate (Jang et al. 2018). Carbonic anhydrase is an enzyme that is mostly used for conversion of CO_2 to bicarbonate (Sharma et al. 2018). Many bacterial species having CA enzyme are *Aeribacillus pallidus, Lactobacillus delbrueckii, Bacillus* sp., *Pseudomonas fragi, Serratia* sp., have been studied for conversion of CO_2 into calcium carbonate (Bose and Satyanarayana 2017; Li et al. 2015; Sharma et al. 2009; Srivastava et al. 2015; Sundaram and Thakur 2018). CO_2 sequestration using microbes offers a reduction of the major greenhouse gas CO_2 and, hence, ameliorates global warming.

1.2.2 Plastic Pollution on Earth

Synthetic plastics represent the main anthropogenic waste entering and accumulating into the environment. Indeed, plastic pollution is now considered a global environmental threat, together with ozone depletion, ocean acidification, and climate changes (Barboza et al. 2018). Plastic is used in everyday life such as packaging material, clothes, water bottles, and carpets. Plastics such as polyethylene terephthalate (PET), polypropylene, and polyethylene present a serious risk to plant and animals growing in the marine ecosystem. The chemical bonds between the plastic monomer are stronger, so they are resistant to natural degradation.

While microplastics are plastic particles less than five millimeter, are of special concern due to their small size, high surface/volume ratio, long environmental persistence, and their ability to enter into the cells and cause adverse effects (Peixoto et al. 2019). The abiotic degradation of man-made plastic by temperature, oxygen, UV radiation, and physical stress (Gewert et al. 2015) slowly degrades plastic and generates microplastic which can spread into the environment by wind (Urbanek et al. 2018). Due to the agglomerations of plastics in the environment, microorganisms are evolving catabolic pathways and enzymes to partly degrade plastic (Yang et al. 2015).

In the literature biodegradation of polyethylene by different microbial strains such as *B. subtilis, Acinetobacter baumanni, Arthrobacter* sp., *Staphylococcus epidermidis and Flavobacterium* sp. was reported (Restrepo-Flórez et al. 2014; Vimala and Mathew 2016). A newly discovered bacteria *I. sakaiensis* enzyme PETase was reported that uses polyethylene terephthalate (PET) as a major energy and carbon source for growth and converts into nontoxic form (Yoshida et al. 2016). This enzyme thus offers a platform for further modification using directed evolution and protein engineering strategy to boost the efficiency of the enzyme, toward the persistent challenge of highly crystalline polymer degradation (Austin et al. 2018). Thermoset plastics such as polyester polyurethane and aliphatic polyester are simply attacked by microbes because of easily digestible ester and urethane bonds in their structures. Other enzymes secreted by microbes that show biodegradable activity include the esterases, lipases, dehydratases, depolymerases, cutinases, ureases, and proteinases (Dang et al. 2018; Masaki et al. 2005; Sood et al. 2016; Zheng et al. 2005). Nowadays bioplastic made from renewable natural resources has received a lot of attention and can be used to replace the plastic (Mostafa et al. 2018). However, bioplastic has not completely replaced the petroleum-based plastic due to various economic and manufacturing challenges.

1.2.3 Chemical Pesticides as a Pollutant on Earth

Currently, pesticides are applied in agricultural production to halt the growth of pests and associated diseases. The most commonly used pesticides include atrazine, lindane, chlordane, DDT, aldrin, cypermethrin, and heptachlor (Pereira et al. 2015). The rise in the use of pesticides/chemical fertilizers in agriculture practices has led to contamination of water, air, and land and adverse effect on human health

(Craig 2019; Li 2018). Although pesticides play a vital role in agriculture, these are recalcitrant to biodegradation and persist in the ecosystem for many years (Kumar et al. 2018; Nicolopoulou-Stamati et al. 2016).

Various microbial species such as *Arthobacter, Aspergillus, Chorella, Penicillium, Pseudomonas, and Flavobacterium* have shown a capability to degrade pesticides into a less toxic product by using enzymes (Kumar et al. 2018). Various enzymes have been isolated from microorganism such as diisopropyl fluorophosphatase, parathion hydrolase, phosphotriesterase, esterase, and paraoxonase, to study the pathways involved in the biotransformation of these xenobiotic compounds (Cycoń and Piotrowska-Seget 2016; Lu et al. 2013; Singh and Walker 2006; Zuo et al. 2015). These indigenous microbes have limited degradation efficiency, so at present several bacteria containing pesticide-degrading gene can be used for constructing genetically engineered bacteria (Hong et al. 2010). For environmental sustainability, the development of biological pesticides has become the key to safeguard the health of human and agricultural development.

1.2.4 Pharmaceutical Pollution and Increased Antimicrobial Resistance

The excessive use of antibiotics in animal and human medicine, as well as in agriculture, has not only led to their accumulation in the environment but also developed the broad range of highly antibiotic-resistant microorganisms (Almakki et al. 2019). Most bacteria develop resistance against commonly used antibiotics like methicillin-resistant *S. aureus, erythromycin-resistant Streptococcus, penicillin-resistant Pneumococcus, and tetracycline-resistant Shigella* (*Sengupta* et al. 2013; Ventola 2015). Wastewater effluents from pharmaceutical industry enter into the rivers, rich in antibiotics, steroids, hormones, and analgesic components, which have adversely affected the microbial ecology (Ding and He 2010). The continuous exposure of terrestrial and aquatic microorganisms to pharmaceuticals products has affected the genetic composition of microbial genera and developed the antimicrobial resistant genes.

However, biological methods for converting pharmaceutical pollutant into non-toxic forms are attractive because they are inexpensive and environment-friendly (Zur et al. 2018). Many bacterial and fungal strains such as *Klebsiella, Penicillium, Pseudomonas, Aspergillus, Sphingomonas* sp., *Bacillus, Enterobacter, Aeromonas, and Streptomyces* have been reported for biotransformation of pharmaceutical pollutant (Rana et al. 2017). Biotransformation results in the formation of the end product that is less toxic and more stable than the initial compound.

1.2.5 *Heavy Metal Pollution on Earth*

Heavy metals are metalloids that have density more than 5 g/cm^3 such as mercury, arsenic, and lead (Tchounwou et al. 2012). Heavy metals occur naturally in the earth's crust and anthropogenic activities such as smelting, mining, petroleum combustion, and burning of coal in power plant, and use of fertilizer increases its existence in the environment (He et al. 2005). In natural systems, heavy metals affect cellular organelles like endoplasmic reticulum, lysosome, cell membrane, nuclei, mitochondria, and various enzymes involved in detoxification, metabolism, and damage repair. Metal ions interact with nuclear proteins and DNA, causing conformational changes and DNA damage that leads to apoptosis or carcinogenesis (Wang and Shi 2001). Due to the persistence of metal in the terrestrial environment, heavy metal pollution poses a risk to animal, plant, and human health (Mishra 2017).

Heavy metal bioremediation by microorganism is emerging as an efficient technique. Different mechanisms used by microorganisms to tolerate the metal toxicity are extrusion, biotransformation, use of enzymes, and synthesis of metallothioneins and biosurfactants (Igiri et al. 2018; Ramasamy et al. 2007). *P. putida* is cadmium-tolerant strain and has the intracellular ability to sequester zinc, copper, and cadmium, by using cysteine-rich low molecular weight proteins (Higham et al. 1986). *Bacillus pumilus*, *Alcaligenes faecalis*, *Brevibacterium iodonium*, and *Pseudomonas aeruginosa* were reported for the removal of cadmium and lead (De et al. 2008). Microbial bioremediation is a cost-effective and eco-friendly technology for the clean-up of heavy metals.

1.3 Restoration of Soil Health Using Microbes

Soil respiration, microbial biomass, enzyme activities, and microbial diversity are the major biological indicators of soil health. Healthy soils are necessary for the integrity of terrestrial ecosystem or to recover from trouble, such as climate change, pest infestation, drought, pollution, and human exploitation including agriculture (Ellert et al. 1997). Moreover, with the continuous increase in the world's population, the demand for food production has increased (Fageria et al. 2008). But nowadays agricultural practices include the use of potentially dangerous chemical fertilizers that affects the soil and human health (Glick 2018). Protection of soil is therefore of high priority, and a thorough understanding of ecosystem processes is a critical factor in assuring that soil remains healthy. These enzymes catalyze many vital reactions necessary for the life processes of soil microorganisms and also help in the stabilization of soil structure. Although microorganisms are the primary source of soil enzymes, plants and animals also contribute to the soil enzyme pool. Soil enzymes respond rapidly to any changes in soil management practices and environmental conditions. Their activities are closely related to the physiochemical and biological properties of the soil. Hence, soil enzymes are used as sensors for soil microbial status, for soil physiochemical conditions, and for the influence of soil treatments or climatic factors on soil fertility. Overall the microbe and enzyme profile in the soil must be stabilized in order to maintain the health of the soil, its fertility and to sustain agricultural growth. At present, the biogeochemical cycles in the agricultural ecosystem have been disturbed as a result of increased pollution and toxic level in the environment. The rate of organic content decomposition becomes lesser due to the extinction of various soil microflora. Therefore, it becomes necessary to find the suitable troubleshoot methods to prevent the increase in soil pollution to restore the decreased fertility of soil using biological approaches.

1.4 Conclusion

The soil microflora and enzyme quantity vary with external factors, physical and chemical conditions. Whichever may be the source of pollution either industries, fertilizers, pesticides, or urbanization and automobiles, each source has seriously damaged the microbial population and enzyme activity of the soil. This is also very complex to determine the exact level of microbial existence and enzyme activity in various geographical areas. Undoubtedly, molecular biology techniques, such as directed evolution and recombinant DNA technology, have revolutionized the speed of enzyme and microbe engineering which could be a milestone to restore the microbial population and enzyme activity in the soil. In this book chapter, we have given an overview of various types of pollutants which are affecting the soil or water bodies on the earth and how various microbes and enzymes are correlated with the existence or increasing level of this pollution.

Acknowledgments The financial support from the Jaypee University of Information Technology, Waknaghat, to undertake this study is thankfully acknowledged. Further, the authors have no conflict of interest either among themselves or with the parent institution.

References

Almakki A, Jumas-Bilak E, Marchandin H, Licznar-Fajardo P (2019) Antibiotic resistance in urban runoff. Sci Total Environ 667:64–76

Austin HP, Allen MD, Donohoe BS, Rorrer NA, Kearns FL, Silveira RL, Pollard BC, Dominick G, Duman R, El Omari K, Mykhaylyk V, Wagner A, Michener WE, Amore A, Skaf MS, Crowley MF, Thorne AW, Johnson CW, Woodcock HL, McGeehan JE, Beckham GT (2018) Characterization and engineering of a plastic-degrading aromatic polyesterase. Proceedings of the National Academy of Sciences of the United States of America, Medium: ED; Size: Article No. 201718804

Barboza LGA, Dick Vethaak A, Lavorante BRBO, Lundebye A-K, Guilhermino L (2018) Marine microplastic debris: an emerging issue for food security, food safety and human health. Mar Pollut Bull 133:336–348

Bilal M, Adeel M, Rasheed T, Zhao Y, Iqbal HMN (2019a) Emerging contaminants of high concern and their enzyme-assisted biodegradation – a review. Environ Int 124:336–353

Bilal M, Rasheed T, Nabeel F, Iqbal HMN, Zhao Y (2019b) Hazardous contaminants in the environment and their laccase-assisted degradation – a review. J Environ Manag 234:253–264

Bose H, Satyanarayana T (2017) Utility of thermo-alkali-stable gamma-CA from polyextremophilic bacterium Aeribacillus pallidus TSHB1 in biomimetic sequestration of CO2 and as a virtual peroxidase. Environ Sci Pollut Res Int 24(11):10869–10884

Craig K (2019) A review of the chemistry, pesticide use, and environmental fate of sulfur dioxide, as used in california. Rev Environ Contam Toxicol 246:33–64

Cycoń M, Piotrowska-Seget Z (2016) Pyrethroid-degrading microorganisms and their potential for the bioremediation of contaminated soils: a review. Front Microbiol 7:1463

Dang TCH, Nguyen DT, Thai H, Nguyen TC, Hien Tran TT, Le VH, Nguyen VH, Tran XB, Thao Pham TP, Nguyen TG, Nguyen QT (2018) Plastic degradation by thermophilic Bacillus sp. BCBT21 isolated from composting agricultural residual in Vietnam. Adv Nat Sci Nanosci Nanotechnol 9(1):015014

De J, Ramaiah N, Vardanyan L (2008) Detoxification of toxic heavy metals by marine bacteria highly resistant to mercury. Mar Biotechnol (NY) 10(4):471–477

Ding C, He J (2010) Effect of antibiotics in the environment on microbial populations. Appl Microbiol Biotechnol 87(3):925–941

Ellert BH, Clapperton MJ, Anderson DW (1997) An ecosystem perspective of soil quality. In: Gregorich G, Carter MR (ed) Soil quality for crop production and ecosystem health. Elsevier, Amsterdam

Eloka-Eboka AC, Inambao FL (2017) Effects of CO2 sequestration on lipid and biomass productivity in microalgal biomass production. Appl Energy 195:1100–1111

Fageria NK, Baligar V, Li Y (2008) The role of nutrient efficient plants in improving crop yields in the twenty first century. J Plant Nutr 31:1121–1157

Garcia-Garcia JD, Sanchez-Thomas R, Moreno-Sanchez R (2016) Bio-recovery of non-essential heavy metals by intra- and extracellular mechanisms in free-living microorganisms. Biotechnol Adv 34(5):859–873

Gewert B, Plassmann MM, MacLeod M (2015) Pathways for degradation of plastic polymers floating in the marine environment. Environ Sci: Processes Impacts 17(9):1513–1521

Glick BR (2018) Soil microbes and sustainable agriculture. Pedosphere 28(2):167–169

He ZL, Yang XE, Stoffella PJ (2005) Trace elements in agroecosystems and impacts on the environment. J Trace Elem Med Biol 19(2–3):125–140

He Q, Ji L, Yu B, Yan S, Zhang Y, Zhao S (2018) Renewable aqueous ammonia from biogas slurry for carbon capture: chemical composition and CO2 absorption rate. Int J Greenhouse Gas Control 77:46–54

Higham DP, Sadler PJ, Scawen MD (1986) Cadmium-binding proteins in Pseudomonas putida: pseudothioneins. Environ Health Perspect 65:5–11

Hong Y, Zhou J, Hong Q, Wang Q, Jiang J, Li S (2010) Characterization of a fenpropathrin-degrading strain and construction of a genetically engineered microorganism for simultaneous degradation of methyl parathion and fenpropathrin. J Environ Manag 91(11):2295–2300

Igiri BE, Okoduwa SIR, Idoko GO, Akabuogu EP, Adeyi AO, Ejiogu IK (2018) Toxicity and bioremediation of heavy metals contaminated ecosystem from tannery wastewater: a review. J Toxicol 2018:1–16

Jacob JM, Karthik C, Saratale RG, Kumar SS, Prabakar D, Kadirvelu K, Pugazhendhi A (2018) Biological approaches to tackle heavy metal pollution: a survey of literature. J Environ Manag 217:56–70

Jang J, Jeon BW, Kim YH (2018) Bioelectrochemical conversion of CO2 to value added product formate using engineered Methylobacterium extorquens. Sci Rep 8(1):7211

Kumar S, Kaushik G, Dar MA, Nimesh S, LÓPez-Chuken UJ, Villarreal-Chiu JF (2018) Microbial degradation of organophosphate pesticides: a review. Pedosphere 28(2):190–208

Kuppusamy S, Thavamani P, Venkateswarlu K, Lee YB, Naidu R, Megharaj M (2017) Remediation approaches for polycyclic aromatic hydrocarbons (PAHs) contaminated soils: Technological constraints, emerging trends and future directions. Chemosphere 168:944–968

Li Z (2018) Health risk characterization of maximum legal exposures for persistent organic pollutant (POP) pesticides in residential soil: an analysis. J Environ Manag 205:163–173

Li C-X, Jiang X-C, Qiu Y-J, Xu J-H (2015) Identification of a new thermostable and alkali-tolerant α-carbonic anhydrase from Lactobacillus delbrueckii as a biocatalyst for CO2 biomineralization. Bioresour Bioprocess 2(1):44

Liu L, Bilal M, Duan X, Iqbal HMN (2019) Mitigation of environmental pollution by genetically engineered bacteria — current challenges and future perspectives. Sci Total Environ 667:444–454

Lloyd JR, Lovley DR, Macaskie LE (2003) Biotechnological application of metal-reducing microorganisms. Adv Appl Microbiol 53:85–128

Lu P, Li Q, Liu H, Feng Z, Yan X, Hong Q, Li S (2013) Biodegradation of chlorpyrifos and 3,5,6-trichloro-2-pyridinol by Cupriavidus sp. DT-1. Bioresour Technol 127:337–342

Masaki K, Kamini NR, Ikeda H, Iefuji H (2005) Cutinase-like enzyme from the yeast Cryptococcus sp. strain S-2 hydrolyzes polylactic acid and other biodegradable plastics. Appl Environ Microbiol 71(11):7548–7550

Mishra GK (2017) Microbes in heavy metal remediation: a review on current trends and patents. Recent Pat Biotechnol 11(3):188–196

Mondal M, Khanra S, Tiwari ON, Gayen K, Halder GN (2016) Role of carbonic anhydrase on the way to biological carbon capture through microalgae—a mini review. Environ Prog Sustain Energy 35(6):1605–1615

Mostafa NA, Farag AA, Abo-dief HM, Tayeb AM (2018) Production of biodegradable plastic from agricultural wastes. Arab J Chem 11(4):546–553

Nicolopoulou-Stamati P, Maipas S, Kotampasi C, Stamatis P, Hens L (2016) Chemical pesticides and human health: the urgent need for a new concept in agriculture. Front Public Health 4:148–148

Pavlik D, Zhong Y, Daiek C, Liao W, Morgan R, Clary W, Liu Y (2017) Microalgae cultivation for carbon dioxide sequestration and protein production using a high-efficiency photobioreactor system. Algal Res 25:413–420

Peixoto D, Pinheiro C, Amorim J, Oliva-Teles L, Guilhermino L, Vieira MN (2019) Microplastic pollution in commercial salt for human consumption: a review. Estuar Coast Shelf Sci 219:161–168

Pereira L, Franco-Bernardes M, Pazin M, Dorta D (2015) Impact of pesticides on environmental and human health, Cells, drugs and environment, Ana Cristina Andreazza and Gustavo Scola, Intechopen 195–234

Ramasamy K, Kamaludeen, Banu SP (2007) Bioremediation of metals: microbial processes and techniques In: Singh SN, Tripathi RD (eds) Environmental Bioremediation Technologies. Springer, Berlin, Heidelberg.

Rana RS, Singh P, Kandari V, Singh R, Dobhal R, Gupta S (2017) A review on characterization and bioremediation of pharmaceutical industries' wastewater: an Indian perspective. Appl Water Sci 7(1):1–12

Rasheed T, Bilal M, Nabeel F, Adeel M, Iqbal HMN (2019) Environmentally-related contaminants of high concern: potential sources and analytical modalities for detection, quantification, and treatment. Environ Int 122:52–66

Restrepo-Flórez J-M, Bassi A, Thompson MR (2014) Microbial degradation and deterioration of polyethylene – a review. Int Biodeterior Biodegradation 88:83–90

Sengupta S, Chattopadhyay MK, Grossart HP (2013) The multifaceted roles of antibiotics and antibiotic resistance in nature. Front Microbiol 4:47

Sharma A, Bhattacharya A, Singh S (2009) Purification and characterization of an extracellular carbonic anhydrase from Pseudomonas fragi. Process Biochem 44(11):1293–1297

Sharma A, Sharma T, Meena KR, Kumar A, Kanwar SS (2018) High throughput synthesis of ethyl pyruvate by employing superparamagnetic iron nanoparticles-bound esterase. Process Biochem 71:109–117

Singh BK, Walker A (2006) Microbial degradation of organophosphorus compounds. FEMS Microbiol Rev 30(3):428–471

Sood S, Sharma A, Sharma N, Kanwar S (2016) Carboxylesterases: sources, characterization and broader applications. Insight Enzymme Res 1(1): 1–11

Srivastava S, Bharti RK, Verma PK, Thakur IS (2015) Cloning and expression of gamma carbonic anhydrase from Serratia sp. ISTD04 for sequestration of carbon dioxide and formation of calcite. Bioresour Technol 188:209–213

Sundaram S, Thakur IS (2018) Induction of calcite precipitation through heightened production of extracellular carbonic anhydrase by CO2 sequestering bacteria. Bioresour Technol 253:368–371

Tchounwou PB, Yedjou CG, Patlolla AK, Sutton DJ (2012) 2012. Heavy metal toxicity and the environment. Experientia Suppl 101:133–164

Urbanek AK, Rymowicz W, Mironczuk AM (2018) Degradation of plastics and plastic-degrading bacteria in cold marine habitats. Appl Microbiol Biotechnol 102(18):7669–7678

Ventola CL (2015) The antibiotic resistance crisis: part 1: causes and threats. Pharm Ther 40(4):277–283

Vimala PP, Mathew L (2016) Biodegradation of polyethylene using bacillus subtilis. Proced Technol 24:232–239

Wang S, Shi X (2001) Molecular mechanisms of metal toxicity and carcinogenesis. Mol Cell Biochem 222(1–2):3–9

Yang Y, Yang J, Wu W-M, Zhao J, Song Y, Gao L, Yang R, Jiang L (2015) Biodegradation and mineralization of polystyrene by plastic-eating mealworms: part 1. Chemical and physical characterization and isotopic tests. Environ Sci Technol 49(20):12080–12086

Yoshida S, Hiraga K, Takehana T, Taniguchi I, Yamaji H, Maeda Y, Toyohara K, Miyamoto K, Kimura Y, Oda K (2016) A bacterium that degrades and assimilates poly(ethylene terephthalate). Science 351(6278):1196–1199

Zhao Q, Bilal M, Yue S, Hu H, Wang W, Zhang X (2017) Identification of biphenyl 2, 3-dioxygenase and its catabolic role for phenazine degradation in Sphingobium yanoikuyae B1. J Environ Manag 204(Pt 1):494–501

Zheng Y, Yanful EK, Bassi AS (2005) A review of plastic waste biodegradation. Crit Rev Biotechnol 25(4):243–250

Zuo Z, Gong T, Che Y, Liu R, Xu P, Jiang H, Qiao C, Song C, Yang C (2015) Engineering Pseudomonas putida KT2440 for simultaneous degradation of organophosphates and pyrethroids and its application in bioremediation of soil. Biodegradation 26(3):223–233

Zur J, Pinski A, Marchlewicz A, Hupert-Kocurek K, Wojcieszynska D, Guzik U (2018) Organic micropollutants paracetamol and ibuprofen-toxicity, biodegradation, and genetic background of their utilization by bacteria. Environ Sci Pollut Res 25(22):21498–21524

Chapter 2
Microbes and Processes in Bioremediation of Soil

Tarun Gangar, Kamal Kumar Bhardwaj, and Reena Gupta

Abstract Environmental pollution has been increasing at an alarming rate since the beginning of the twenty-first century. There is an enormous increase in the production and use of xenobiotic compounds that have created new sites of environmental contamination and problem worsens as many of such xenobiotic compounds are either persistent or recalcitrant to microbial breakdown. The presence of anthropogenic organic compounds/chemicals in the environment is a matter of significant concern because of their potential toxicity, mutagenicity, and bioconcentration (biomagnification) in higher organisms. This is of immense concern and hence provides impetus to the development of certain remediation techniques. Various microorganisms play a key role in the bioremediation of soil and may range from bacteria majorly to a few actinomycetes and fungi. Bioremediation can be carried out *via* two main approaches, ex situ and in situ, and choice of method depends largely on site characteristics, concentration, and type of pollutants present. To enhance the remediation process, a more recent approach called bioaugmentation is also practiced. Bioaugmentation trials have met varying degrees of success. This chapter will largely focus on various microorganisms which are potent biomediators and also the processes involved in the same.

Keywords Bioremediation · Xenobiotic compounds · Bioconcentration · Bioaugmentation

2.1 Introduction

Increasing the standard of living and urbanization has posed a great degradative threat to the environment and ecosystem. A typical example is the monstrous sized heaps of waste dumped daily into the dumping yards of cities. Also, the

T. Gangar · K. K. Bhardwaj · R. Gupta (✉)
Department of Biotechnology, Himachal Pradesh University,
Shimla, Himachal Pradesh, India

A. Kumar, S. Sharma (eds.), *Microbes and Enzymes in Soil Health and Bioremediation*, Microorganisms for Sustainability 16,
https://doi.org/10.1007/978-981-13-9117-0_2

advancements in science, technology, and industrial sector have led to the production of waste ranging from municipal sewage to nuclear waste, and also this has rendered our ecosystem unfit for the survival of life forms on earth (Lin et al. 2018; Ontanon et al. 2018; Parewa et al. 2018). Previously, the use of conventional techniques was practiced such as waste disposal by landfilling, dumping in open grounds, etc. With rapid and ever-growing waste disposal problems, conventional methods failed to cope up with this issue. New methods like incineration and chemical decomposition are being developed, but the use of such methods is either uneconomical or not environment-friendly. Such problems lead to the development of newer and better technologies which may better solve the purpose. Modern-day bioremediation is one such method (Karigar and Rao 2011).

The literal meaning of "bioremediation" is "biological treatment". So, bioremediation by definition means the use of biological agents such as microorganisms (majorly bacterial and fungi) and/or plants (in case of phytoremediation) for the treatment of contaminated soil and water, so as to make it fit for reuse by all the biological entities. Some of the remediation processes used for the treatment of contaminated area include natural attenuation, composting, biopiling, bioventing, landfarming, thermal desorption, landfilling, soil washing, and incineration (USEPA 2014). Till now, majorly the success to bioremediation and biodegradation has been provided by the indigenous microbes thriving in that very environment, and this is highly dependent upon the growth characteristics and nutritional requirements of the microbes used for the purpose (Verma and Jaiswal 2016). There are several factors that define the choice of bioremediation techniques, e.g., nature of pollutant, degree of pollution, geographical location, the cost involved in the process, etc. (Frutos et al. 2012; Smith et al. 2015). Biological treatment of soil using various biological agents primarily plants and microorganisms is considered as one of the cheapest and safest methods to remove the hazardous contaminants from the soil. Plants have the capability to neutralize various types of harmful chemicals in the soil by direct utilization, followed by the biotransformation of such compounds into nontoxic products which are harmful neither to the environment nor to any other form of life (Macek et al. 2008).

The major focus of this chapter is the microbes which play a vital role in effective bioremediation of soil and also on processes involved in the biodegradation. Microorganisms have this inherent capability to catalyze the degradation and mineralization of various contaminating xenobiotic compounds, thus converting them into nontoxic by-products (Seshadri and Heidelberg 2005; Head et al. 2006; Gomez et al. 2017). Such a conversion process is often a result of consortia of microorganisms. Recently, biodegradation of total petroleum hydrocarbons was investigated in slurry phase bioreactor using aged refuse (Fu Chen et al. 2019). Bioremediation can only be effective when the environmental conditions permit the microbial growth and activity, or conversely, there is a need to manipulate certain environmental parameters to allow the growth of microbes so that degradation could proceed at a faster pace (Vidali 2001). Most of the bioremediation procedures run under complete aerobic environment, but to treat certain recalcitrant molecules, the system may run under anaerobic environment (Colberg and Young 1995).

2.2 The Basic Approach to Bioremediation

Mostly bioremediation proceeds through a process of oxidation-reduction reactions (redox), whereby a chemical species donates an electron to a different species that accepts an electron. Bioremediation procedures can be broadly classified as aerobic and anaerobic bioremediation.

2.2.1 *Aerobic Bioremediation*

Aerobic bioremediation is the most practiced and most prevalent form of oxidative bioremediation. As the name suggests here oxygen acts as the terminal electron acceptor for the oxidation of various contaminants such as polyaromatic hydrocarbons (PAHs), phenols, petroleum, etc. Preference of oxygen points toward higher oxidative potential of oxygen and its requirement by some enzyme systems to initiate the degradation process.

Under ideal conditions, the biodegradation rates of aliphatic, alicyclic, and aromatic compounds (low to moderate molecular weight) can be very high. With an increase in the molecular weight of the compound, its resistance toward biodegradation increases (Norris 1993).

There are several physical methods for aerating the soil above water table, e.g., landfarming, composting, and bioventing (Frutos and Fernandez 2010). Approaches for aeration of soil below water table include flushing aerated water through treatment zone, air sparging, and the addition of molecular oxygen or peroxide.

2.2.2 *Anaerobic Bioremediation*

This technique can be employed to remediate a broad range of already oxidized contaminating pollutants including ethenes (PCE, TCE, DCE, VC), chlorinated ethanes (TCA, DCA), chloromethanes (CT, CF), chlorinated cyclic hydrocarbons, various energetics (e.g., perchlorate, RDX, TNT), and nitrate.

Anaerobic bioremediation occurs in two steps:

1. Depletion of background electron acceptors like oxygen (O), nitrate (NO_3^-), ferric ion (Fe^{3+}), etc.
2. Stimulation of biochemical reduction of oxidized contaminating pollutants.

2.2.3 The Fate of Organic Contaminating Pollutants

The knowledge of various catabolic pathways involved in the degradation of contaminating pollutants of both aerobic and anaerobic microorganisms has a major beneficial impact in the development of *in situ* and *ex situ* bioremediation protocols. The intrinsic chemical consideration that limits the biodegradability of aromatic pollutants in both aerobic and anaerobic environments was reviewed by Field et al. 1995.

The aerobic microorganisms make use of oxygenase enzyme to initiate the electrophilic attack on aromatic molecules. The process is greatly suppressed by the presence of electron withdrawing groups such as chloro, azo, and nitro (Dorb and Knackmuss 1978; Knackmuss 1981). The microorganisms involved in the aerobic degradation are *Candida*, *Anabaena*, *Nostoc*, *Chlamydomonas*, *Microcoleus*, *Oscillatoria*, *Saccharomyces*, *Chlorella*, and *Phormidium* (DorotaWolicka et al. 2009). On the other hand, anaerobic microorganisms proceed with the degradation of aromatic pollutants in a completely reciprocal manner, i.e., under anaerobic conditions, the microorganisms make use of enzymes to initiate an electrophilic attack on the aromatic molecules. So here, the presence of electron withdrawing group will enhance the initial reductive attack on aromatic contaminants (Knackmuss 1992, Dolfing and Harrison 1993, Ann-Kathrin Ghattas et al. 2017). Conversely, electron donating groups will hinder the anaerobic transformation of aromatic compounds but will favor the aerobic biotransformation process (Field et al. 1995). However, the complete absence of electron withdrawing as well as electron donating group will enhance the recalcitrance of hydrocarbon in the anaerobic environment (Schink 1985; Schink 1988). It has also been noted that the resulting products of anaerobic biodegradation of complex molecules such as polychlorinated and polynitroaromatic compounds are appropriate products for aerobic mineralization but they resist further anaerobic biodegradation (Zitomer and Speece 1993).

2.3 Microbes Involved in Bioremediation of Soil

The contamination of soil, sediment, and water from industrial and other human inputs is widespread and poses a threat to human and ecological health. Bioremediation is the use of microbes for the beneficial removal of contaminants of concern. The microbial processes involved in bioremediation are normally natural components of respiration or adaptation, often a component of carbon cycling or metal redox cycling. Thus, bioremediation often occurs without direct intervention; however, biostimulation (the addition of nutrients or adjustment of conditions) and bioaugmentation (the addition of microbes capable of bioremediation) are however important for the complete removal of contaminants within an economical timeframe. Various microorganisms involved in bioremediation of various contaminating pollutants are listed in Table 2.1.

Table 2.1 Microbes involved in the bioremediation of contaminants

Contaminants	Microorganisms	References
Monocyclic aromatic hydrocarbons	*Penicillium chrysogenum*	Abdulsalam et al. (2013) and Pedro et al. (2014)
BTEX	*P. chrysogenum*	
Phenols	*Bacillus subtilis*, *Penicillium chrysogenum*, *Corynebacterium propinquum*, *Alcaligenes odorans*, *Pseudomonas aeruginosa*	Singh et al. (2013)
Petrol and diesel oil	*Penicillium alcaligenes*, *P. putida*, *P. veronii*, *P. mendocina*, *Achromobacter*, *Flavobacterium*, *Acinetobacter*	Safiyanu et al. (2015) and Sani et al. (2015)
PAHs	*Pseudomonas putida*, *Pseudomonas* sp., *Coprinellus radians*, *Ralstonia* sp., *Microbacterium* sp.	Sarang et al. (2013), AI-Jawhari (2014) and Safiyanu et al. (2015)
Biphenyls and triphenylmethanes	*Phanerochaete chrysosporium*	Erika et al. (2013)
Hydrocarbons	*Aspergillus niger*, *A. fumigatus*, *F. solani*, *P. funiculosum*, *Tyromyces palustris*, *Gloeophyllum trabeum*, *Trametes*, *Versicolor*	Karigar and Rao (2011), AI-Jawhari (2014) and Xu et al. (2017)
Methylnaphthalene and dibenzofurans	*Coprinellus radians*	Aranda et al. (2010)
Phenanthrene and benzopyrene	*Candida viswanathii*	Hesham et al. (2012)
Oil	*Alcaligenes odorans*, *Bacillus subtilis*, *Fusarium* sp., *Corynebacterium propinquum*, *Penicillium chrysogenum*, *Pseudomonas aeruginosa*	Hidayat and Tachibana (2012) and Singh et al. (2013)
Crude oil	*Aspergillus niger*, *Saccharomyces cerevisiae*, *Candida glabrata*, *Candida krusei*, *B. brevis*, *P. aeruginosa* KH6, *B. licheniformis*, *B. sphaericus*	Aliaa et al. (2016) and Burghal et al. (2016)
Paints (oil based)	*B. subtilis* strain NAP2, NAP1, NAP4	Phulpoto et al. (2016)
Industrial dyes	*Myrothecium roridum* IM 6482, *Pycnoporus sanguineus*, *Phanerochaete chrysosporium*, *Trametes trogii*, *Penicillium ochrochloron*	Shedbalkar and Jadhav (2011) and Hassan et al. (2013)
Textile dyes	*Micrococcus luteus*, *Nocardia atlantica*, *Bacillus* spp. ETL-2012, *Pseudomonas aeruginosa*, *Bacillus pumilus* HKG212, *Listeria denitrificans*	Hassan et al. (2013), Maulin et al. (2013), Das et al. (2015) and Yogesh and Akshaya (2016)
Black liquor	*Bacillus firmus*, *Staphylococcus aureus*, *Bacillus macerans*, *Klebsiella oxytoca*	Adebajo et al. (2017)
Lead, nickel, and mercury	*Saccharomyces cerevisiae*	Chen and Wang (2015), Infante et al. (2014)

(continued)

Table 2.1 (continued)

Contaminants	Microorganisms	References
Fe^{2+}, Zn^{2+}, Pb^{2+}, Mn^{2+}, and Cu^{2+}	*Pseudomonas fluorescence*, *P. aeruginosa*	Paranthaman and Karthikeyan (2015)
Co, Cu, Cr, and Pb	*Lysinibacillus sphaericus* CBAM-5	Peña-Montenegro et al. (2015)
Cadmium	*Aspergillus versicolor*, *Trichoderma* sp., *A. fumigatus*, *Paecilomyces* sp., *Microsporum* sp., *Cladosporium* sp.	Priyalaxmi et al. (2014) and Soleimani et al. (2015)
Endosulfan	*Bacillus*, *Staphylococcus*	Mohamed et al. (2011)
Chlorpyrifos	*Enterobacter*	Niti et al. (2013)
Ridomil MZ 68MG, Fitoraz WP76, Decis 2.5EC, Malation	*Pseudomonas putida*, *Arthrobacter* sp., *Acinetobacter* sp.	Hussaini et al. (2013) and Mónica et al. (2016)
Chlorpyrifos and methyl parathion	*Acinetobacter* sp., *Photobacterium* sp., *Pseudomonas* sp., *Enterobacter* sp.	Ravi et al. (2015)

2.3.1 Bioaugmentation

Bioaugmentation is the process of enhancing/stimulating the rate of bioremediation by addition of single strain or consortia of microorganisms as to mimic the competitiveness among the indigenous microflora and also to remove/decrease adaptation/acclimatization time (Bourier and Zeahnder 1993; Liu and Suflita 1993; Singleton 1994). This technique may involve single strain or consortia of microorganisms but also involve genetically engineered microorganisms (GEMs) within certain strict international rules and regulations. Although GEMs are very efficient in such processes, their accidental release into the environment may pose a serious threat to mankind. Keeping in mind such negative impacts, the use of GEMs has been limited to laboratory-based bioreactor applications.

Bioaugmentation strategy used as per model proposed by Forsyth et al. (1995) for soil is:

1. Where the number of degrading microorganisms is low or sub-detectable.
2. Contaminating pollutants which require a multitude of processes to degrade contaminants.
3. Small-scale contaminated site where non-biological treatment processes are not economical.

2.3.1.1 Factors Affecting Bioaugmentation

Although bioaugmentation has solved a number of issues pertaining to bioremediation of contaminants which are aromatic in nature primarily, still there are a number of ecological constraints which hamper its effectiveness and have kept it to a minimal level. One of the major difficulties that arises during the process is the survival of non-native microbial species which are introduced to the contaminated site. Studies have revealed that the number of exogenous microorganisms has reduced shortly after the inoculation of soil. Hence both abiotic and biotic factors are shown to cause such decrease (Cho et al. 2000; Bento et al. 2005; Wolski et al. 2006). Various abiotic factors include temperature, moisture, pH, and organic content of the soil, and biotic factors include aeration, amount of nutrients, and type of soil.

There are various studies and examples which may prove above mentioned points.

The effect of moisture content in the soil on the survival of *Achromobacter piechaudii* TBPZ and degradation of tribromophenol (TBP) indicating minimum 25% water content was required for rapid degradation, whereas soils with 10% moisture content show limited activity (Ronen et al. 2000). Low moisture content in the soil decreases the efficiency with which microorganisms perform the degradation of contaminants, such effect can be attributed to the fact that the decreased bacterial activity is due to the diffusional limitation of substrate supply and adverse physiological effects associated with cell dehydration (Mashregi and Prosser 2006).

Other most crucial factors influencing the efficiency of bioaugmentation is the organic content of the soil. It plays an important role in the bioavailability of contaminants and hence impairs the survival of inoculated strains and ultimately their availability to degrade contaminants, e.g., the rate of 2,4-D degradation was lower in the soil with high organic content but was considerably higher in soils with lower organic content (Greer and Shelton 1992). Conversely, when the soil was combusted to remove the organic content, microbes completely lost their degradative activity. This indicated that there is presence of some components of insoluble organic matter that is nutritionally beneficial for microorganisms involved in BTEX degradation (Kim et al. 2008).

Other factors including competition primarily between indigenous and exogenous microorganisms for limited C-sources and also antagonistic interaction and predation by protozoa and bacteriophages also play an essential role in the final results of bioaugmentation. All these interactions greatly decrease the number of inoculated cells (England et al. 1993; Sorensen et al. 1999).

2.3.1.2 Microbes in Bioaugmentation

Before performing augmentation in the soil for the purpose of enhanced biodegradation, one should fully know the type and level of contaminants and about the strains of microorganisms and their consortia which play active role in the process. The following features should be kept in mind before augmenting soil:

1. The organism should be easily cultivable.
2. The organism used for the purpose should be able to grow fast under given environmental and nutritional conditions.
3. The organism should be able to withstand a high concentration of contaminants and also should be able to survive in varying environmental conditions.

In case of contaminants such as PAHs, it is especially necessary to use organisms which are capable of producing surfactants, so that these contaminants are more accessible and the process becomes more feasible (Forsyth et al. 1995; van der Gast et al. 2003; Gentry et al. 2004).

Several approaches can be followed to select for the microorganisms useful in bioaugmentation. First being, isolating microorganisms from a contaminated site in question and then growing it under laboratory conditions. Finally, this pre-adapted pure culture is returned to the contaminated site. The process is called reinoculation and involves the use of indigenous microflora. The second approach involves the use of microorganisms from the contaminated site having similar kind of contamination. Various studies revealed that microbial consortia for degradation of aromatic contaminants are effective as compared to selected single strains (Goux et al. 2003; Ghazali et al. 2004).

Both Gram-positive and Gram-negative bacteria play a major role in the bioaugmentation. Experiments pertaining to bioaugmentation were done using both the organisms belonging to genera *Pseudomonas*, *Alcaligenes*, *Flavobacterium*, *Achromobacter*, and *Sphingobium* (Gram-negative bacteria) and *Mycobacterium*, *Bacillus*, and *Rhodococcus* (Gram-positive). Potentially useful fungi in bioaugmentation are represented by genera *Aspergillus*, *Penicillium*, *Absidia*, *Mucor*, *Acremonium*, and *Verticillium*.

2.3.1.3 Delivery of Inoculum

The efficiency of bioaugmentation entirely depends upon the number of microorganisms and total biomass introduced in the soil. The delivery of microbes is also another important factor responsible for efficient bioaugmentation. The conventional delivery mechanisms make use of liquid culture for the introduction of microorganisms into the contaminated site. But nowadays various modifications to such systems have been made. The basic idea of such modifications aimed at maintaining optimum activity of inoculum over an extended period of release which was significantly hampered in case of liquid culture introduction methods. Various modifications include the use of certain carrier material which enhances the activity of microbes and also provides nutrition to the growing microbial population (van Veen et al. 1997). Example of carrier materials includes charcoal-amended soil (Beck 1991), chitin or chitosan (Gentili et al. 2006; Chen et al. 2007), nylon (Heitkamp and Steward 1995), zeolite (Liang et al. 2009), and clay (Omar et al. 1990). A study on activated carbon and zeolite in the treatment of site contaminated with crude oil showed that these materials increased microbial growth and enhanced hydrocarbon degradation (Liang et al. 2009). This revealed that dehydrogenase activity was three times higher in activated carbon than in zeolite. Such an increase in overall activity can be attributed to biocarriers as they improve the diffusion of oxygen, nutrient uptake, and water retention capacity.

Other entirely different approaches primarily used for biodegradation of aromatic compound make use of immobilized cells. This method offers a protective environment to the inoculated microorganisms and provide protection from environmental conditions not suited for their growth (improper pH and presence of toxic contaminants) and also eliminates competition with indigenous microflora (Lin and Wang 1991). Moreover, immobilization is known to increase the stability of cells (DNA, plasmids) (Cassidy et al. 1992). Immobilization is usually performed using both synthetic (poly(carbamoyl) sulfate, polyacrylamide, and polyvinyl alcohol) and natural materials (Dextran, agar, agarose, k-carrageenan, alginate) (Cassidy et al. 1992; Jen et al. 1996; Gardin and Pauss 2001).

2.4 Processes Involved in Bioremediation

The basic ideology of bioremediation is to treat or inactivate the toxic-contaminating pollutants to less toxic or completely nontoxic products which will not cause any deteriorating effect on the environment but in turn may have a positive effect. Bioremediation is generally done by consortia of microorganisms which generally reside in that very habitat which requires treatment. This is an example of in situ bioremediation. This enables the microorganisms to work efficiently because the local environment in which the microorganisms are already growing proves beneficial to their growth and also no adaptory phase is required. Ex situ bioremediation techniques, on the other hand, are primarily based upon physical manipulation of the contaminants, and in this case, there is no direct involvement of microorganisms for remediation procedures.

2.4.1 *In Situ Bioremediation Procedures*

Over the last few decades, the major thrust area in the field of bioremediation is to understand various nutritional requirements of microorganisms, their biosynthetic and degradative pathways, and their enzymatic machinery. This has led to an increased interest of many scientists, and now many people are working on how to enhance the rate of bioremediation. This is because the waste is generated on a daily basis throughout the world at an alarming rate. So now various strategies are being opted to enhance the biodegradation. Such enhanced *in situ* bioremediation methods have proved beneficial, and various processes have been designed as explained in the following sections.

As the conventional methods of *in situ* bioremediation cannot treat such a heavy load of contaminants of today's world, so now enhanced bioremediation techniques are being developed so that the biodegradation can be carried out at an enhanced rate. Enhanced *in situ* bioremediation of organic contaminants requires the stimulation of biodegradative activities of microbial population thriving in that particular environment by the involvement of certain nutrients or external electron acceptors. For this purpose, the microorganisms are provided with some combination of oxygen, nutrients, and moisture and controlling temperature and pH. There are various procedures through which this can be achieved and are as follows:

2.4.1.1 Bioventing

This is the cheapest mode of *in situ* bioremediation. As the name suggests, bioventing involves supplying the oxygen-rich air into the soil so as to increase the rate of degradation of contaminating organic pollutants in the soil. As already mentioned in the previous section, oxygen acts as the terminal electron acceptor for the oxidation of various organic contaminants. This technology is a choice for the treatment of petroleum waste and another similar kind of recalcitrant toxins (McCauly 1999). There are several kinds of bioventing remediation technologies, one of which includes air sparging, which includes forcing the compressed air into saturated soil, whereas venting technology uses low-pressure air and is more focused toward the deeper unsaturated zone of the soil. The simple bioventing setup consists of a blower or air compressor which is connected to air supply wells and soil-gas monitoring wells which are connected in series (Sellers 1999). Bioventing is the gentlest and stripped up a form of bioremediation because it occurs without intervention into the natural environment of microorganisms. But at the same time, the process of degradation is enhanced by the addition of oxygen (Leahy and Erickson 1995; US EPA 1995b).

2.4.1.1.1 Methods Involved in the Process

a. Injection system: These systems are generally economically feasible to install and simple to operate primarily because this system requires limited/no treatment of off-gas. Injection systems are set up at those locations which are away from various installments such as buildings/property boundary because they, upon injection of air, may push the contaminants deep or away from the actual site. This actually is the beneficial property of the injection system that when the contaminants are moved away, their concentration decreases per unit area, also increasing the contact with layer area of soil with microorganisms. This will result in enhanced biodegradation (US EPA 1995b; Sellers 1999).

b. Extraction system: This system works in a completely opposite way. Extraction system actually sucks out the contaminants. This system is installed in densely populated areas, but there are various side effects, few of which include that it causes the water table to rise and may cause contamination and also it requires treatment of off-gas (Fig. 2.1).

2.4.1.1.2 Applications

The principal compound of crude petroleum is the hydrocarbons; because of this reason, they have become a significant substrate for microbial oxidation (Rosenberg and Ron 1996). Hence, bioventing is preferably used for the treatment of oil spills and has proved to be an excellent option for petrochemical contaminants.

Bioventing is mostly preferred with hydrocarbons whose volatility is very low. Because of effective bioremediation of these petrochemicals, the rate of volatilization

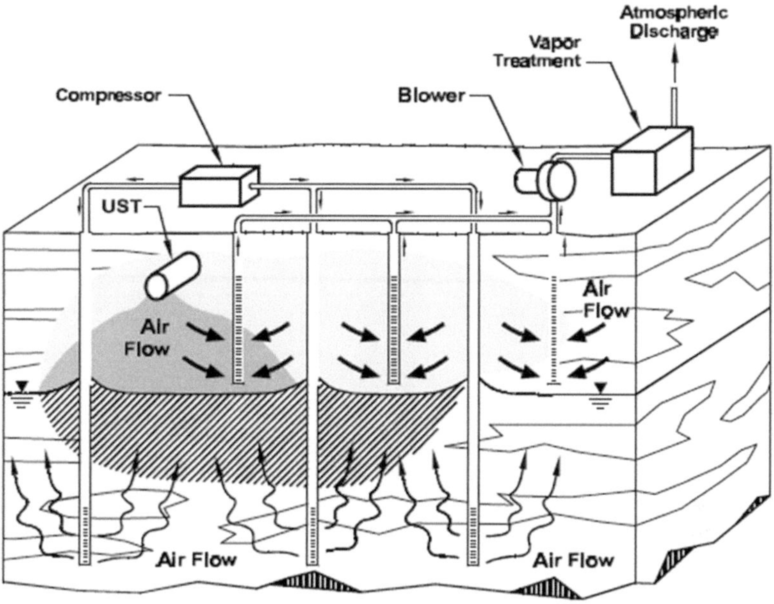

Fig. 2.1 Bioventing showing injection and extraction system (NMED 2010)

should be maintained at optimal level which should be lower than the rate of biodegradation. Low volatility also reduces the chances of degradation because air injection by the process of bioventing will push contaminants into the surrounding environment.

Majorly, gasoline, fuel oil, and bitumen are efficiently reduced by bioventing. Also, bioventing has shown to effectively reduce toluene, benzene, ethylbenzene, and xylenes to the levels below the detection within 1 year (US EPA 1995a). A laboratory test showed that bioventing is quite superior in remediating toluene and decane than other methods of bioremediation (Malina et al. 1998).

2.4.1.2 Biosparging

Biosparging/air sparging is the process of blowing compressed air (composed primarily of oxygen) directly into the saturated subsurface. As a result of this, the bubbles thus formed result in the physical separation of contaminants from groundwater (i.e., stripping) and are thus carried up into the unsaturated zone, where the contaminants are biodegraded by the process of *in situ* bioremediation. This process is further stimulated because of the addition of oxygen-rich air.

A conventional biosparging unit consists of air injection well, an air compressor, monitoring points and wells, and a vapor extraction system which is optional (Fig. 2.2).

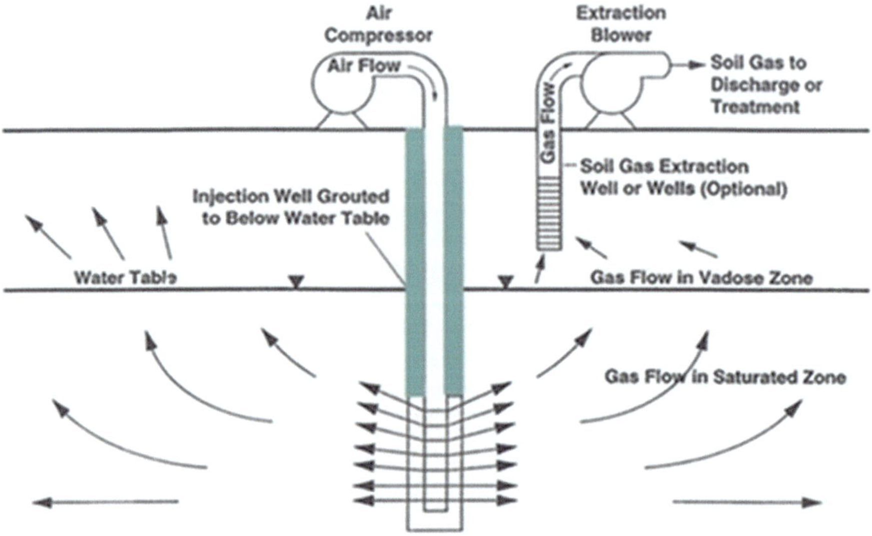

Fig. 2.2 A biosparging unit showing air injection well, monitoring points, deeper extraction system, and air compressor

The air injection wells are generally vertical and are dug to the depths below groundwater table to prevent further mixing of groundwater and contaminants. When compressed air is sparged below the groundwater level, this results in the rising up of gas bubbles thus formed. If the medium is homogeneous (i.e., soil particles are sandy), this may result in the homogeneous flow of air, this is rarely seen as there exists some kind of heterogeneity as non-uniform airflow is quite common.

The compressed system is used to supply compressed air into the injection well, and the choice of the compressor system depends entirely on the nature of the bed below the groundwater (e.g., clay, sand, etc.) and also the pressure required. Biosparging is most effective against contaminants which have higher Henry's law constant, such as benzene, toluene, ethylbenzene and xylenes, and TCE and PCE. However, it can be used to target less volatile compounds by enhancing the biodegradation of compounds like diesel fuel and waste oils (Anderson and Ward 1995; Miller 1995).

2.4.1.3 Anaerobic Bioremediation

Maximum bioremediation technologies focused on the addition of oxygen which acts as a terminal electron acceptor and hence enhances the process of bioremediation. But as this process of delivery of oxygen to subsurface contaminated sites is difficult and also the solubility of oxygen in water is also very low, alternate terminal electron acceptors are required to solve this purpose. A number of oxy-anions substituting oxygen can act as terminal electron acceptors and solve the purpose of microbial degradation of organic compounds. These include salts of iron III, sulfate, and nitrate. Also, there exist wide consortia of anaerobic bacteria which can use these electron acceptors to degrade the organic contaminants (Anderson and Ward 1995; Qencrantz et al. 1995).

Nitrates are highly soluble in water, are less reactive, and are much more mobile than oxygen. Such properties of nitrates make it suitable electron acceptor for anaerobic bioremediation. Sulfate is also highly soluble in water, and in comparison to its mass, it is having higher electron accepting capacity. Its inexpensiveness and nontoxicity to microorganisms make it highly suitable for use in anaerobic bioremediation (Freedman et al. 1995; Sherwood et al. 1995).

2.4.1.4 Phytoremediation

The basic concept of phytoremediation is that the plants are grown in the contaminated site and, in turn, when they grow to extract various contaminating pollutants from the site and concentrate them in the biomass (bioextraction) (Fig. 2.3). Such plants can further be burnt to produce energy. This way it is also possible to extract some metals from plant biomass (phytomining) (Meager 2000). Not only plants are able to extract some of the toxic minerals but also are able to accumulate a variety of organic contaminants, e.g., PCB (polychlorinated biphenyls), ammunition wastes (TNT, GTN), halogenated hydrocarbons, etc. These organic toxins then further

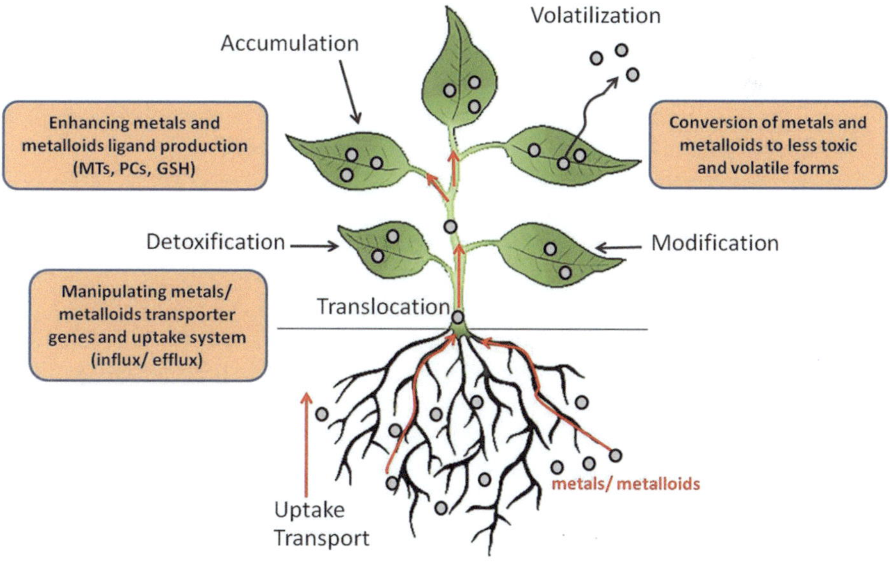

Fig. 2.3 Outline of phytoremediation showing transport, accumulation, volatilization, and detoxification of contaminants. (Adapted from Dhankher et al. (2011))

undergo metabolism in plant body and are converted into less toxic or nontoxic by-products (Salt et al. 1998; Meager 2000; Dietz and Schnoor 2001).

Rarely, there are some plants in kingdom Plantae which are known to accumulate large concentration of metals (Pajak et al. 2017; Pajak et al. 2018). Some of the hyperaccumulating or metal-resistant species are *Silene vulgaris*, *Arabidopsis halleri*, *Alyssum lesbiacum*, and *Brassica* spp. (Clemens et al. 2002; Kramer 2003). These species are known to accumulate high concentrations of essential as well as non-essential metals such as copper (Cu), iron (Fe), zinc (Zn), selenium (Se), cadmium (Cd), mercury (Hg), lead (Pb), aluminum (Al), and arsenic (As) (Salt et al. 1998, Meagher 2000, Clemens 2001,Guerrinot and Salt 2001, Clemens et al. 2002, Hall 2002, McGrath 2003).

2.4.1.4.1 Mechanism of Phytoremediation

The first step in the process is the uptake of metal ions from the root to the root cells. This process is primarily performed by organic acids in the plant system such as citrate, oxalate, formate, etc. (Michael and Christopher 2007). But at the same time there are certain examples of organic acids only which are known to cause a strong inhibition to this process (Guerra et al. 2011), e.g., avenic acids and mugineic acids are released by certain species of plants to increase the bioavailability of heavy metals for root uptake as is reported in some species of family Gramineae (Jakagi et al. 1984). On the other hand, zinc, copper, and aluminum uptake is inhibited by the formate, oxalate, and malate collectively (Dehaize et al. 1993; Kochian et al. 2007; Qin et al. 2007). The next step after root uptake is the vacuolar sequestration inside the plant cell. The first step includes the transport of heavy metals inside the cytosol, and this is mediated by ZIP (zinc−/iron-regulated transporter) proteins. This further stimulates phytochelatin synthetase for the production of phytochelatin from glutathione. This results in the formation of the heavy metal-phytochelatin complex which actually is transported inside the vacuole of a plant cell.

2.4.1.4.2 Translocation of Heavy Metals from Roots to Shoots and Shoot Metabolism

Heavy metals follow the path from roots to epidermal tissue, to pericycle, and finally to xylem parenchyma; from xylem parenchyma it is transported by transmembrane channels to xylems (Palmer and Guerinot 2009). In *Arabidopsis* species, ATPases HMA2 and HMA4 proteins are responsible for transportation and accumulation of zinc from roots to shoots (Hanikenne et al. 2008; Wong and Cobbett 2009). Some amino acids such as histidine (hyperaccumulator) and some organic acids such as citrate and malate play an active role in the translocation of metals such as zinc, copper, and cadmium (Pilon et al. 2009).

The excessive accumulation of heavy metals (redox active and non-redox active) inside the plant cells cause a huge amount of oxidative damage and stress by replacement of metal ions in pigments and another essential molecule such as chlorophyll. To counteract this, plants have an inbuilt anti-oxidative defense mechanism based upon the enzymes and reducing metabolites (GSH) which regulate redox status. GSH binds to metals and metalloids and eliminates reactive oxygen species (ROS) whose production is stimulated by heavy metals and thus maintains homeostasis for metabolism (Foyer and Noctor 2005).

2.4.2 *Ex Situ Bioremediation*

Ex situ bioremediation technology makes use of aerobic treatment of soil to make it free from contaminants. The major difference between *in situ* and *ex situ* bioremediation technique is that *in situ* remediation involves the treatment of contaminated soil on site, whereas *ex situ* methods involve physical extraction of media/soil from a contaminated site and move it to another location for efficient and controlled treatment. One of the major advantages of this method over the *in situ* method is its efficiency and certainty of control treatment due to the ability to uniformly screen and homogenize the soil mixture. However, the *ex situ* method involves the complete removal of contaminated soil and its transport to a different new location for treatment which makes this treatment method less cost-effective.

2.4.2.1 Nonbiological Methods of *Ex Situ* Bioremediation

2.4.2.1.1 Dig and Dump (Landfilling)

This is the most conventional method for ex situ bioremediation for the purpose of the following: first, a target land is engineered and prepared to receive the dumped waste. The site is so engineered so that it is able to receive inert, solid, and hazardous waste with a degradation rate between 5000 and 230,000 metric ton per year and also no leachate should leach into the dumping ground and contaminate groundwater table. To prevent the problem of leaching, covering and capping are usually preferred, but now newer technology makes use of leachate system which is an underground pipe network to collect leachate. One upgraded form of landfilling is called landfill bioreactor. This type of landfill makes use of microorganisms for a quick breakdown of all the contaminants present in the waste disposed into it. They may be aerobic or anaerobic and can also be hybrid (aerobic-anaerobic). Advantages of this type of landfilling include:

(a) Reduce greenhouse gas emission.
(b) The end product produced during the process does not require further landfilling.
(c) Leachate treatment cost is decreased exponentially.
(d) The decrease in overall landfilling cost over long periods of treatment.

Several disadvantages include:

(a) Open landfills emit certain kind of greenhouse gases, which may indirectly pose a serious threat to the environment (Gong et al. 2018).
(b) Excavation of landfills is very dangerous as the waste dumped into it is not pretreated.
(c) The cost of excavated material transfer to the final destination site is very huge and hence makes this process less economical.
(d) Finally, the landfill gas although have the advantage of being used as a biofuel but, if released into the open environment, proves harmful for local flora and fauna.

2.4.2.1.2 Incineration

Incineration is the method of treatment of solid waste by thermal energy. Actually, it involves the combustion of organic matter in waste material (Silva-Castro et al. 2012). Incineration is also called as thermal treatment, as direct heat is involved in the process. The result of incineration is the production of three major components, i.e., ash, flue gas, and heat.

(a) Ash mainly constitute inorganic matter from the waste which is left after burning. It forms solid clumps and is carried away by flue gas.
(b) Flue gas actually is the gas which is emitted after combustion of waste, and its composition depends entirely upon the constituents of the solid waste.
(c) There is a considerable amount of heat produced when the waste is combusted, and the calorific value here also depends upon the constituents of the solid waste. This heat is nowadays channelized into various fields with newer advanced technologies.

The method utilizes very high temperature (750–1200 °C) for disposal of solid waste. There are various incinerator systems which are infrared combustors using silicon carbide rods; fluidize bed combustors (high-velocity air with infrared heat source is used to attain temperature of 850 °C); circulating bed combustors (utilizes the kinetic energy of high-velocity air to create high turbulent combustion zone to attain temperature of 850 °C); and rotary kilns (consist of inclined rotating cylinders with an afterburner that burns at 980 °C) (FRTR 2012).

2.4.2.1.3 Soil Washing

Mechanical scrubbing, physical separation, soil scrubbing, and attrition scrubbing are some of the other names used to refer to soil washing. As the name suggests, soil washing involves physical separation of contaminants from the soil particles via various methods. This technology uses aqueous-based separation and physical separation units which are used to minimize the toxic levels of contaminated site (age-prone) to site-specific objectives. It is worth noting that there is no detoxification of pollutants, but by various means, it involves a concentration of hazardous material into smaller soil fraction or transfers the contaminants from the soil into washing fluids for further treatments (Dermont et al. 2008).

Soil washing involves the following procedures:

1. Mechanical screening (sorting, crushing, physical processing which includes soaking, spraying and tumbling attrition scrubbing).
2. Treatment of coarse- and fine-grained soil portions (includes leaching and physical separations).
3. Further treatment of generated toxins.

Hence, this is considered a stand-alone approach (Fig. 2.4).

Soil washing can be used for a wide variety of contaminants including heavy metals, PCBs, SVOCs, PAHs, petroleum, as well as fuel residues and pesticides. Soil washing is considered one of the cheapest methods of *ex situ* bioremediation as it limits the final fraction of soil which requires further treatment which in turn minimizes the post-remedial expenditures (Urum et al. 2003).

One of the major limitations is that this technique is highly unsuitable for soils containing more than 40% silt and clay. This is because homogenization is not feasible for such soils. Also, multicomponent mixture hampers the effectiveness of the method.

2.4.2.2 Biological Methods of *Ex Situ* Bioremediation

2.4.2.2.1 Solid Phase Bioremediation Technologies

Land farming

It is one of the most widely used conventional remediation technologies because of two main reasons: one being low technological input and the other being its

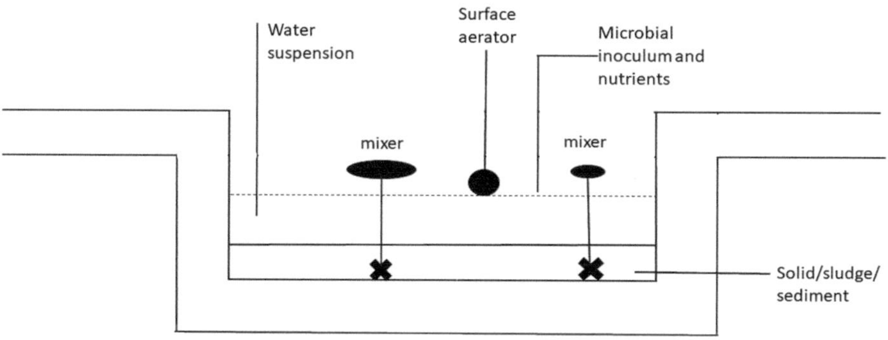

Fig. 2.4 Typical soil washing procedure

cost-effectiveness. It involves the excavation of a contaminated site and places it in the form of a thin bed of about 0.46 m which are lined by high-density polyethylene (HDPE) sheets or other such impermeable material. These are used to prevent leachate from coming in contact with groundwater through infiltration. Landfarming is based upon the principle that the indigenous microflora is used to aerobically degrade the contaminating pollutants present in the soil under the treatment process. For this purpose, the beds are provided with sufficient aeration by periodically tilling the bed and turning over until the bioremediation process is complete. The chemistry behind the chemical breakdown remains the same as discussed under aerobic bioremediation.

The optimized parameters for landfarming include moisture content (40–85%), aeration (periodic tilling), pH (6–8), temperature (20–40 °C), and C:N ratio (9:1) (Khan et al. 2004). There exist a plethora of contaminants which can be treated via landfarming and includes aliphatic and aromatic hydrocarbons, oily sludge from refineries, pesticides, etc. To enhance the rate of biodegradation, various strategies can be applied which include the regular addition of soils contaminated with hydrocarbons primarily to replenish the supply of hydrocarbons, co-substrates, and bulky agents can also be added to stimulate microbial metabolism (Straube et al. 2003; Maila et al. 2005). Silva-Castro et al. (2012) reported that a consortium of four bacteria (*Bacillus pumilus*, *Alcaligenes faecalis*, *Micrococcus luteus*, and *Enterobacter* sp.) can remediate 100% of PAH in 7 months from PAH-contaminated soil with organic fertilizers. The use of additive (kitchen waste compost), activated sludge, bully agent (rusk-husk), and petroleum-degrading bacteria removes 92.4% total petroleum hydrocarbons (TPH) in 25 days (Kuo et al. 2012).

2.4.2.2.2 Composting

It is the most primitive technology for the treatment of agricultural, municipal solid waste, and sewage sludge. The principle of operation consists of mixing of contaminated soil with nontoxic organic waste, another agricultural waste, manure, etc. This mix encourages the growth of aerobic microorganisms and hence biodegrades the toxic contaminants into nontoxic end products. The biodegradation occurs via co-metabolic pathways. The process is purely aerobic in nature and makes use of heat generated during the oxidative exothermic reaction to speed up the process. So this can be considered as an autonomously driven process. The positive characteristic of composting is that the product, i.e., mature compost, can be used as fertilizers in the field and also may be used for land restoration purposes (Antizar-Ladislao et al. 2006).

For the purpose of composting, there are various approaches, but the cheapest one makes use of windrows which are actually long mounds in which the entire composting mixture is kept for bioremediation. The optimal size for windrows is (3 l × 4w × 1.5 h)m^3. Other approaches include vessels and engineered windrows which are also used for biopiling. Engineered vessels can also be called as solid phase bioreactors in which all the physical and chemical parameters can be controlled. But such installations require high capital input.

Thermophilic composting is capable of reducing levels of monoaromatics (BTEX), phenols, PAHs, petroleum hydrocarbons, PCB, PCP, etc. (Nadeef et al. 2012). Metal-contaminated soil can also be treated by composting methods, e.g., van Herwizen et al. (2007) remediated 80% metal polluted soil with mineral-amended composts.

Limiting factors include the requirement of space and post-excavation treatment of contaminated soil. Management of order and the problem of leachate pose a major problem during composting.

2.4.2.2.3 Biopiling

Biopiling is a combination of two techniques (landfarming and composting) that provides a favorable environment for indigenous aerobic and anaerobic microorganisms and also controls physical losses of contaminants by leaching and volatilization (Kumar et al. 2011; Mani and Kumar 2014). There are various other names for biopile such as bioheaps, biomounds, biocells, compost cells, etc. A wide range of petrochemical contaminants in soils and sediments have been remediated by extensive use of biopiles (Germaine et al. 2014). In this method, contaminated soils are piled up or heaped, and then the microbiological activity is stimulated by aeration along with the addition of water and nutrients.

Its similarity with landfarming can be stated as it also involves the remediation of soil above the ground, and moreover, this system utilizes the aerobic environment to stimulate microbial activity. It differs from landfarming with respect to the control it provides over different physical as well as chemical parameters so that rapid biodegradation may occur (McCarthy et al. 2004). In comparison to both landfarming and composting, the mass transfer efficiency of nutrients, water, and air in biopiles offers a better potential for treatment of contaminating pollutants.

Biopiles are generally operated up to a height of 0.9–3.1 m, and also various strategies can be used to prevent volatilization, runoff, and evaporation, which includes covering the biopile with an impervious lining, which also promotes thermal heating up of biopile and enhances the microbial activity.

Biopiles are capable of degrading various contaminants such as pesticides, halogenated VOCs/SVOCs, and non-halogenated VOCs. Lighter petroleum products like gasoline are removed at the time of air injection or pile turning. Heavier petroleum products generally take more time to biodegrade than lighter petroleum products. Soil characteristics, climatic conditions, and contaminant characteristics are deciding factors for the efficiency of biopile (Giasi and Morelli 2003).

2.4.2.2.4 Slurry-Based Bioremediation (Bioreactors)

This is currently the most advanced system for the treatment of soil. The main advantage of this approach is the fine control of various physical and chemical parameters. The slurry bioreactors are so designed so as to provide a very efficient

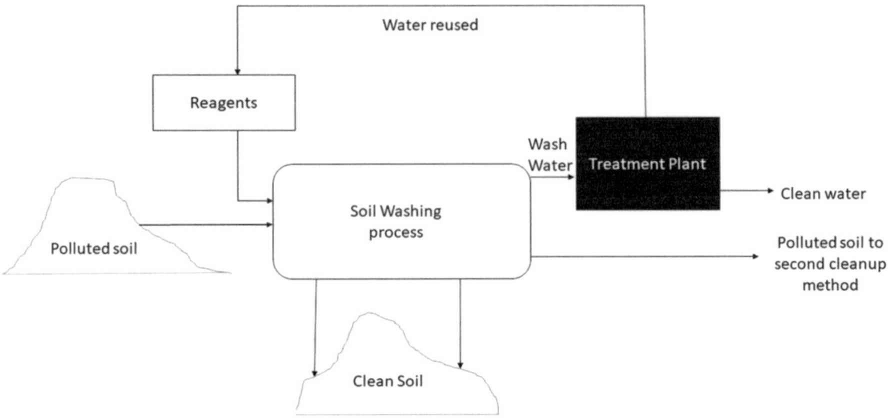

Fig. 2.5 Slurry-phase bioreactor (a typical open system)

system for optimizing and controlling critical operating parameters. The major disadvantage being the high cost of bioreactor as well as the operating cost which needs justification for each particular application.

The slurry-based bioreactor system is a stand-alone approach (Fig. 2.5) which is able to remediate the soil completely without the intervention or use of other approaches. The working principle includes first the mixing of contaminated soil with water/wastewater and other additives in the bioreactor vessel. Next, various critical operating parameters are set (Latha and Reddy 2013; Bhardwaj and Kapley 2015). Slurry-based soil treatment requires mechanical mixing, grinding, and volumetric classification before initializing biological treatment process for degradation of pollutants (Volf 2007). The formation of slurry depends upon how much soil is mixed with a specific amount of water which in turn depends upon the concentration of pollutants, soil type, and rate of biodegradation (Pavel and Gavrileseu 2008). Normally, a slurry is 10–30% solids by weight. The slurry is maintained under optimum conditions by providing oxygen via the aeration facility of bioreactor. Nutrients are also added depending upon the total amount of pollutants present in the soil and thus to maintain the optimal ratio between carbon, phosphorus, and nitrogen. After the bioremediation process is over, the slurry is drained via various downstream processing approaches (filtration (vacuum/pressure), centrifugation, etc.).

As bioreactors are closed systems in which all the environmentally critical parameters are maintained at an optimal level, hence such bioreactors provide accelerated and enhanced treatment rates (Bhardwaj and Kapley 2015; Kuppusamy et al. 2016). Slurry phase bioreactors are normally operated under batch, continuous, or fed-batch modes, and they may be operated under aerobic, anaerobic, or anoxic conditions depending upon the type of contaminants being treated.

Slurry bioreactors are efficient in the treatment of aerobically degradable compounds such as SVOCs, recalcitrant pesticides, explosive substances, aromatic hydrocarbons, chlorinated organic compounds, and PAHs. This technique is able to treat soils which are otherwise difficult to treat via other processes for, e.g., soils with high clay content (>40%).

2.5 Conclusion

The degradative potential of microorganisms has been used to a much greater extent, and this fundamental principle forms the basis of bioremediation of organic toxic-contaminated pollutants. For a successful bioremediation procedure, one must be equipped with the knowledge of physiological characteristics, biochemical capabilities, ecology, and genetic plasticity of the microorganism or consortia being utilized in the process. Also, the knowledge of contaminants with which the site is contaminated also plays a vital role in the efficient bioremediation of the soil. However, a complete and efficient process development requires a multi-disciplinary approach involving from chemical sciences through physical sciences to biological sciences. The increase in interest of various scientific communities towards this field has promised a very bright and promising future.

Acknowledgments The Fellowship granted to Mr. Kamal Kumar Bhardwaj in the form of SRF from Council of Scientific and Industrial Research under Ministry of Human Resource Development, Government of India, is thankfully acknowledged.

Conflict of Interest The authors declare that there is no conflict of interests regarding the publication of this article.

References

Abdulsalam S, la SS A, Bugaje IM et al (2013) Bioremediation of soil contaminated with used motor oil in a closed system. J Bioremed Biodegr 3:100–172

Adebajo SO, Akintokun AK, Balogun SA (2017) Screening and characterisation of biosurfactant producing bacteria from soil samples in Ogun-state, Nigeria. Int J Microbial Res 18:1–12

Al-Jawhari IFH (2014) Ability of some soil fungi in biodegradation of petroleum hydrocarbon. J Appl Environ Microbiol 2:46–52

Aliaa ME, Eltayeb KM, Mostafa AR et al (2016) Biodegradation of industrial oil-polluted wastewater in Egypt by bacterial consortium immobilized in different types of carriers. Pol J Environ Stud 25:1901–1909

Anderson WC, Ward CH (1995) Innovative site remediation technologies In: Vol I Bioremediation. american academy of environmental engineers, WASTECH

Antizar-Ladislao B, Lopez-Real J, Beck AJ (2006) Investigation of organic matter dynamics during in-vessel composting of an aged coal-tar emission spectroscopy. Chemosphere 64(5):839–847. https://doi.org/10.1016/j.chemosphere.2005.10.036

Aranda E, Ullrich R, Hofrichter M (2010) Conversion of polycyclic aromatic hydrocarbons, methyl naphthalenes and dibenzofuran by two fungal peroxygenases. Biodegradation 21:267–281

Beck DP (1991) Sustainability of charcoal amended mineral soil as carrier for Rhizobium inoculants. Soil Biol Biochem 23:41–44

Bento FM, Camargo FAO, Okeke BC et al (2005) Comparative bioremediation of soils contaminated with diesel oil by natural attenuation, biostimulation and bioaugmentation. Bioresour Technol 96:1049–1055

Bhardwaj P, Kapley A (2015) Remediation of pesticide-contaminated soil: emerging options vol 1 In: Kalia VC (ed) Microbial factories Biofuel waste treatment. pp 293–313

Bourier EJ, Zeahnder AJB (1993) Bioremediation of organic compounds putting microbial metabolism toward. Trends Biotechnol 11:360–367

Burghal AA, Najwa MJA, Al-Tamimi WH (2016) Mycodegradation of crude oil by fungal species isolated from petroleum contaminated soil. Int J Innov Res Sci Eng Technol 5:1517–1524

Cassidy MB, Mullineers H, Lee H et al (1992) Mineralization of pentachlorophenol in a contaminated soil by *Pseudomonas* sp. UG30cells encapsulated in K-carrageenan. J Ind Microbiol Biotechnol 3:232–243

Chen C, Wang JL (2015) Characteristics of Zn^{2+} biosorption by *Saccharomyces cerevisiae*. Biomed Enviro Sci 20:478–482

Chen YM, Lin TF, Huang C et al (2007) Degradation of phenol and TCE using suspended and chitosan bed immobilized *Pseudomonas putida*. J Hazard Mater 148:660–670

Chen F, Li X, Zhu Q, Ma J, Hou H, Zhang S (2019) Bioremediation of petroleum-contaminated soil enhanced by aged refuse. Chemo 222:98–105

Cho YG, Rhee SK, Lee ST (2000) Effect of soil moisture on bioremediation of chlorophenol contaminated soil. Biotechnol Lett 22:915–919

Clemens S (2001) Molecular mechanism of plant metal tolerance and homeostasis. Planta 212:475–486

Clemens S, Palmgren MG, Kramer U (2002) A long way ahead: understanding and engineering plant metal accumulation. Trends Plant Sci 7:309–315

Colberg PJS, Young LL (1995) Anaerobic degradation of nonhalogenated homocyclic aromatic compounds coupled with nitrate, iron, or sulfate reduction. In: Young LL, Cerniglia CE (eds) Microbial transformation and degradation of toxic organic chemicals. Wiley-Liss, New York, pp 307–330

Das A, Mishra S, Verma VK (2015) Enhanced biodecolorization of textile dye remazol navy blue using an isolated bacterial strain *Bacillus pumilus* HKG212 under improved culture conditions. J Biochem Technol 6:962–969

Dehaize E, Ryan PR, Randall PJ (1993) Aluminum tolerance in wheat (*Triticum aestivum* L.) (II. Aluminum stimulated excretion of malic acid from root apices). P Physiol 103:695–702

Dermont G, Bergeron M, Mercier G, Richer-Laflèche M (2008) Soil washing for metal removal: a review of physical/chemical technologies and field applications. J Hazard Mater 152(1):1–31

Dhankher OP, Pilon-Smits EAH, Meagher RB, Doty S (2011) Biotechnological approaches for phytoremediation. In: Altman A, Hasegawa PM (eds) Plant Biotechnology and Agriculture. Academic Press, Oxford, pp 309–328

Dietz AC, Schnoor JL (2001) Advances in phytoremediation. Environ Health Perspect 109:163–168

Dolfing J, Harrison BK (1993) Redox and reduction potentials as parameters to predict the degradation pathway of chlorinated benzenes in anaerobic environments. FEMS Microbiol Ecol 13:23–30

Dorb E, Knackmuss HJ (1978) Chemical structure and biodegradability of halogenated aromatic compounds: substituent effects of 1, 2-dioxygenation of catechol. Biochem J 174:85–94

DorotaWolicka AS, Bockowski A, Bieleka A (2009) Application of aerobic microorganisms in bioremediation *in-situ* of soil contaminated by petroleum products. Bioresour Technol 100(13):3321–3227

England LS, Lee H, Trevor JT (1993) Bacterial survival in soil: effect of clays and protozoa. Soil Biol Biochem 25:523–531

Erika AW, Vivian B, Claudia C et al (2013) Biodegradation of phenol in static cultures by *Penicillium chrysogenum* erk1: catalytic abilities and residual phytotoxicity. Rev Argent Microbiol 44:113–121

Field JA, Stams AJM, Kato M et al (1995) Enhanced biodegradation of aromatic pollutants in coculture of aerobic and anaerobic bacterial consortia. Anton Laeuw 67:47–77
Forsyth JV, Tsao YM, Bleam RD (1995) Bioremediation: why is augmentation needed. In: Fredrkso J, Hinchee RE (eds) Bioaugmentation for site remediation. Battelle Press, Columbus
Foyer CH, Noctor G (2005) Redox homeostasis and antioxidant signaling: a metabolic interface between stress perception and physiological responses. Plant Cell 17:1866–1875
Freedman DL, Lasecki M, Hashsham S et al (1995) Accelerated biotransformation of carbon tetrachloride and chloroform by sulfate reducing enrichment cultures. In: Herson DS, Baker HK (eds) Bioremediation. McGraw Hill, New York, pp 123–128
Frutos FJG, Fernandez MD (2010) Bioventing remediation and ecotoxicity evaluation of phenanthrene-contaminated soil. J Hazard Mater 183(1–3):806–813. https://doi.org/10.1016/j.jhazmat.2010.07.098
Frutos FJG, Pe'rez R, Escolano O et al (2012) Remediation trials for hydrocarbon-contaminated sludge from a soil washing process: evaluation of bioremediation technologies. J Hazard Mater 199:262–271. https://doi.org/10.1016/j.jhazmat.2011.11.017
Gardin H, Pauss A (2001) K-carrageenan/gelatin gel beads for the co-immobilization of aerobic and anaerobic microbial communities degrading 2, 4, 6-trichlorophenol under air limited conditions. Appl Microbiol Biotechnol 56:517–523
Gentili AR, Cubitto MA, Ferrero M et al (2006) Bioremediation of crude oil polluted sea water by a hydrocarbon degrading bacterial strain immobilized on chitin and chitosan flakes. Int Biodeterior Biodegrad 57:222–228
Gentry TJ, Rensing C, Pepper IL (2004) New approaches for bioaugmentation as a remediation technology. Crit Rev Environ Sci Technol 34:447–494
Germaine KJ, Byrne J, Liu X et al (2014) Ecopiling: a combined phytoremediation and passive biopiling system for remediating hydrocarbon impacted soil at field scale. Front Plant Sci 5:756
Ghattas A-K, Fischer F, Wick A, Ternes TA (2017) Anaerobic biodegradation of (emerging) organic contaminants in aquatic environment. Water Res 116(2017):268–295
Ghazali FM, Rahman RNZA, Salleh AB et al (2004) Biodegradation of hydrocarbons in soil by microbial consortium. Int Biodeterio Biodegrad 54:61–67
Giasi CI, Morelli A (2003) A landfarming application technique used as environmental remediation for coal oil pollution. J Environ Sci Health A Tox Hazard Subst Environ Eng 38:1557–1568
Gomez MJ, Pazos F, Guijarro FJ et al (2017) The environmental fate of organic pollutants through the global microbial metabolism. Mol Syst Biol 3:1–11
Gong XQ, Cai LL, Li SY et al (2018) Bamboo biochar amendment improves the growth and reproduction of *Eisenia fetida* and the quality of green waste vermicompost. Ecotox Environ Safe 156:197–204
Goux S, Shapir N, EL Frantrocessi S et al (2003) Long term maintenance of rapid atrazine degradation in soils inoculated with atrazine degraders. Water Air Soil Pollut Focus 3:131–142
Greer LE, Shelton DR (1992) Effect or inoculant strain and organic matter content on kinetics of 2, 4-dichlorophenoxyacetic acid degradation in soil. Appl Environ Microbiol 58:1459–1465
Guerra L, Guidi R, Slot I et al (2011) Bacterial genotoxin triggers FEN1-dependent RhoA activation cytoskeletal remodelling and cell survival. J Cell Sci 124:2735–2742
Guerrinot ML, Salt DE (2001) Fortified foods and phytoremediation: two sides of same coin. Plant Physiol 125:164–167
Hall JL (2002) Cellular mechanism for heavy metal detoxification and tolerance. J Exp Bot 53:1–11
Hanikenne M, Talke IN, Haydon MJ et al (2008) Evolution of metal hyperaccumulation required Cu-regulatory changes and triplication of HMA4. Nature 453:391–395
Hassan MM, Alam MZ, Anwar MN (2013) Biodegradation of textile azo dyes by bacteria isolated from dyeing industry effluent. Int Res J Biological Sci 2:27–31
Head IM, Martin Jones D, Röling WFM (2006) Marine microorganisms make a meal of oil. Nat. Rev Microbiol 4:173–182
Heitkamp MA, Steward WP (1995) A novel porus nylon biocarrier for immobilized bacteria. Appl Microbiol Biotechnol 62:4662–4669

Hesham A, Khan S, Tao Y et al (2012) Biodegradation of high molecular weight PAHs using isolated yeast mixtures: application of metagenomic methods for community structure analyses. Environ Sci Pollut Res Int 19:3568–3578

Hidayat A, Tachibana S (2012) Biodegradation of aliphatic hydrocarbon in three types of crude oil by Fusarium sp. F092 under stress with artificial sea water. J Environ Sci Technol 5:64–73

Hussaini S, Shaker M, Asef M (2013) Isolation of bacterial for degradation of selected pesticides. Bull Environ Pharmacol Life Sci 2:50–53

Infante JC, De Arco RD, Angulo ME (2014) Removal of lead, mercury and nickel using the yeast *Saccharomyces cerevisiae*. Rev MVZ Córdoba 19:4141–4149

Jakagi SI, Nomoto K, Takemoto T (1984) Physiological aspect of mugineic acid: a possible phytosiderophore of graminaceous plants. J Plants Nutri 7:1–5

Jen AC, Wake MC, Mikos AG (1996) Review: hydrogels for cell immobilization. Biotechnol Bioeng 50:254–272

Karigar CS, Rao SS (2011) Role of microbial enzymes in the bioremediation of pollutants: a review. Enzyme res 2011:1–11. https://doi.org/10.4061/2011/805187

Kim JM, Le NT, Cheng BS et al (2008) Influence of soil components on biodegradation of benzene, toluene, ethylbenzene and o-, m- and p-xylenes by newly isolated bacterium *Pseudoxanthomonas spadex* RD-a59. Appl Environ Microbiol 74:7313–7320

Knackmuss HJ (1981) Degradation of halogenated and sulfonated hydrocarbons. In: Hutter R, Cook AM, Neusch J, Leisinger T (eds) Microbial degradation of xenobiotic and recalcitrant compounds. Academic, London, pp 189–212

Knackmuss HJ (1992) Potentials and limitations of microbes to degrade xenobiotics. In: Soil decontamination using biological processes. Karlsruhe, Germany, pp 3–9

Kochian LV, Hoekenga OA, Pineros MA (2007) How do crop plants tolerate acid soils? Mechanism of aluminum tolerance and phosphorus efficiency. Ann Rev Plant Biol 55:459–493

Kramer U (2003) Phytoremediation to phytochelatin- plant trace metal homeostasis. New Physiol 158:4–6

Kumar A, Bisht BS, Joshi VD et al (2011) Review on bioremediation of polluted environment: a management tool. Int J Environ Sci 1(6):1079–1093

Kuo YC, Wang SY, Kao CM et al (2012) Stabilization of As, Cr, Cu, Pb and Zn in soil using amendments-a review. Waste Manag 28:215–225

Kuppusamy S, Palanisami T, Magharaj M et al (2016) *Ex-situ* remediation technologies for environmental pollutants: a critical perspective. Rev Environ Contam Toxicol 236:117–192

Latha AP, Reddy SS (2013) Reviews on bioremediation potential tool for removing environmental pollution. Int J Basic Appl Chem Sci 3(3):21–33

Leahy MC, Erickson GP (1995) Bioventing reduces soil cleanup costs. Hydrocarb Process 8:63–64

Liang Y, Zhang X, Dai D et al (2009) Porous biocarrier enhanced biodegradation of crude oil contaminated site. Int Biodeterior Biodegrad 63:80–87

Lin JE, Wang HY (1991) Use of co-immobilized biological systems to degrade toxic organic compounds. Biotechnol Bioeng 38:273–279

Lin Y, Watts DB, Kloepper JW et al (2018) Influence of plant growth-promoting rhizobacteria on corn growth under different fertility sources. Commun Soil Sci Plant Anal 49(10):1239–1255. https://doi.org/10.1080/00103624.2018.1457155

Liu S, Suflita JM (1993) Ecology and evolution of microbial populations for bioremediation. Trends Biotechnol 11:344–352

Macek T, Kotrba P, Svatos A et al (2008) Novel roles for genetically modified plants in environmental protection. Trends Biotechnol 26:146–152

Maila MP, Randima P, Surridge K et al (2005) Evaluation of microbial diversity of different soil layer at a contaminated diesel site. Int Biodeter Biodegrad 55:39–44

Malina GJ, Grotenhus TC, Rulkins WH et al (1998) Soil vapour extraction versus bioventing of toluene and decane in bench scale soil columns. Environ Technol 19:977–991

Mani D, Kumar C (2014) Biotechnological advances in remediation of heavy metals contaminated ecosystems: an overview with special reference to phyto remediation. Int J Environ Sci Technol 11(3):843–872

Mashregi M, Prosser JI (2006) Survival and activity of lux-marked phenanthrene degrading *Pseudomonas stutzeri* P16 under different conditions. Iran J Sci Technol Trans A 30:71–80

Maulin PS, Patel KA, Nair SS et al (2013) Microbial degradation of Textile Dye (Remazol Black B) by *Bacillus* spp. ETL-2012. J Bioremed Biodegr 4:1–5

McCarthy K, Walker L, Vigoren et al (2004) Remediation of spilled hydrocarbons by *in-situ* land-farming at an arctic site. Cold Region Sci Technol 40:31–39

McCauly, P (1999) Bioventing for enhanced biodegradation of PAHs. https://clu-in.org/products/newsltrs/ttrend/view.cfm?issue=tt0899.htm. Accessed Aug 1999

McGrath Z (2003) Phytoextraction of metals and metalloids from contaminated soil. Curr Opin Biotechnol 14:277–282

Meager RB (2000) Phytoremediation of toxic elemental and organic pollutants. Curr Opin Plant Biol 3:153–162

Michael JH, Christopher SC (2007) Transporters of ligand for essential metal ions in plants. New Phytol 174:499–506

Miller R (1995) Technology overview report: air sparging, ground water remediation technologies Analysis under GWRTAC. Pilsburgh. pp 10

Mohamed AT, El Hussein AA, El Siddig MA et al (2011) Degradation of oxyfluorfen herbicide by soil microorganisms: biodegradation of herbicides. Biotechnol 10:274–279

Mónica P, Darwin RO, Manjunatha B et al (2016) Evaluation of various pesticides-degrading pure bacterial cultures isolated from pesticide-contaminated soils in Ecuador. Afr J Biotechnol 15:2224–2233

Nadeef Y, Chitimus AD, Mosnegutu E et al (2012) Technology for soil remediation methods and techniques for environmental protection. J Bioremed Biodegr 79:13–20

Niti C, Sunita S, Kamlesh K (2013) Bioremediation: An emerging technology for remediation of pesticides. Res J Chem Environ 17:88–105

NMED (2010) New Mexico Environment Department. Clean up technologies. Petroleum Storage Tank Bureau

Norris R (1993) Handbook of bioremediation. CRC Press, Boca Raton

Omar SH, Buderker U, Rehm HJ (1990) Degradation of oily sludge from a floatation unit by free and immobilized microorganisms. Appl Microbiol Biotechnol 34:259–273

Ontanon OM, Fernandez M, Agostini E et al (2018) Identification of the main mechanisms involved in the tolerance and bioremediation of Cr(VI) by *Bacillus* sp. SFC 500-1E. Environ Sci Pollut Res 25(16):16111–16120. https://doi.org/10.1007/s11356-018-1764-1

Pajak M, Halecki W, Gąsiorek M (2017) Accumulative response of Scots pine (*Pinussylvestris* L.) and silver birch (*Betula pendula* Roth) to heavy metals enhanced by Pb-Zn ore mining and processing plants: explicitly spatial considerations of ordinary kriging based on a GIS approach. Chemosphere 168:851–859. https://doi.org/10.1016/j.chemosphere.2016.10.125

Pajak M, Blonska E, Szostak M et al (2018) Restoration of vegetation in relation to soil properties of spoil heap heavily contaminated with heavy metals. Water Air Soil Poll 229(12):392. https://doi.org/10.1007/s11270-018-4040-6

Palmer CM, Guerinot ML (2009) Facing challenges of Cu, Fe and Zn homeostasis in plants. Nat Chem Biol 5:333–340

Paranthaman SR, Karthikeyan B (2015) Bioremediation of heavy metal in paper mill effluent using *Pseudomonas* spp. Int J Microbiol 1:1–5

Parewa HP, Meena VS, Jain LK et al (2018) Sustainable crop production and soil health management through plant growth-promoting Rhizobacteria. Role Rhizospheric Microbes Soil 2018:299–329. https://doi.org/10.1007/978-981-10-8402-7_12

Pavel LV, Gavrileseu M (2008) Overview of *ex-situ* decontamination techniques for soil cleanup. Environ Eng Manag J 7(6):815–834

Pedro P, Francisco JE, Joao F et al (2014) DNA damage induced by hydroquinone can be prevented by fungal detoxification. Toxicol Rep 1:1096–1105

Peña-Montenegro TD, Lozano L, Dussán J (2015) Genome sequence and description of the mosquitocidal and heavy metal tolerant strain *Lysinibacillus sphaericus* CBAM5. Stand Genomic Sci 10:1–10

Phulpoto H, Qazi MA, Mangi S et al (2016) Biodegradation of oil-based paint by *Bacillus* species monocultures isolated from the paint warehouses. Int J Environ Sci Technol 13:125–134

Pilon M, Cohu CM, Ravet K et al (2009) Essential transition metal homeostasis in plants. Curr Opin Plant Biol 12:347–357

Priyalaxmi R, Murugan A, Raja P et al (2014) Bioremediation of cadmium by *Bacillus safensis* (JX126862), a marine bacterium isolated from mangrove sediments. Int J Curr Microbiol Appl Sci 3:326–335

Qencrantz JE, Johnson JG, Koenringberg SS (1995) Enhanced intrinsic bioremediation of dissolved phase hydrocarbons using an oxygen releasing compound. Ground Water Monit R 6(4):99–114

Qin R, Hirano Y, Brunner I (2007) Exudation of organic acid anions from poplar roots after exposure to Al, Cu and Zn. Tree Physiol 27:313–320

Ravi RK, Pathak B, Fulekar MH (2015) Bioremediation of Persistent Pesticides in Rice field Soil Environment Using Surface Soil Treatment Reactor. Int J Curr Microbiol App Sci 4:359–369

Rosenberg E, Ron EZ (1996) Bioremediation of petroleum contamination. In: Crawgord DL, Crawford RL (eds) Bioremediation principles and applications. Cambridge University Press, Cambridge

Safiyanu I, Isah AA, Abubakar US et al (2015) Review on comparative study on bioremediation for oil spills using microbes. Res J Pharm Biol Chem Sci 6:783–790

Salt DE, Smith RD, Raskin I (1998) Phytoremediation. Annu Rev Plant Physiol Plant Mol Biol 49:643–668

Sani I, Safiyanu I, Rita SM (2015) Review on bioremediation of oil spills using microbial approach. Int Sci Eng Res 3:41–45

Sarang B, Richa K, Ram C (2013) Comparative study of bioremediation of hydrocarbon fuel. Int J Biotechnol Bioeng Res 4:677–686

Schink B (1985) Degradation of unsaturated hydrocarbons by methanogenic enrichment cultures. FEMS Microbiol Ecol 31:9–77

Schink B (1988) Principles and limits of anaerobic degradation:environmental and technological aspects. In: Zehnder AJB (ed) Biology of anaerobic microorganisms. Wiley, New York, pp 771–846

Sellers K (1999) Fundamentals of hazardous waste site remediation. Lewis Publisher, New York

Seshadri R, Heidelberg J (2005) Bacteria to the rescue. Nat Biotechnol 23:1236–1237

Shedbalkar U, Jadhav J (2011) Detoxification of malachite green and textile industrial effluent by *Penicillium ochrochloron*. Biotechnol Bioprocess Eng 16:196–204

Sherwood JC, Peterson JN, Skee RS et al (1995) Effect of nitrate availability on chloroform production during CT destruction. In: Leeson A, Semprint L, Hinchee RE (eds) Bioremediation of chlorinated solvents. Batelle, Columbus, pp 85–89

Silva-Castro G, SantaCruz-Calvo L, Uad I et al (2012) Treatment of diesel-polluted clay soil employing combined biostimulation in microcosms. Int J Environ Sci Technol 9:535–542

Singh A, Kumar V, Srivastava JN (2013) Assessment of bioremediation of oil and phenol contents in refinery waste water via bacterial consortium. J Pet Environ Biotechnol 4:1–4

Singleton I (1994) Microbial metabolism of xenobiotics fundamental and applied research. J Chem Tech Biotechnol 59:9–23

Smith E, Thavamani P, Ramadass K et al (2015) Remediation trials for hydrocarbon-contaminated soils in arid environments: evaluation of bioslurry and biopiling techniques. Int Biodeterior Biodegrad 101:56–55. https://doi.org/10.1016/j.ibiod.2015.03.029

Soleimani N, Fazli MM, Mehrasbi M et al (2015) Highly cadmium tolerant fungi: their tolerance and removal potential. J Environ Health Sci Eng 13:1–9

Sorensen SJ, Schyberg T, Ronn R (1999) Predation by protozoa on Escherichia coli K12 in soil and transfer of resistant plasmid RP4 to indigenous bacteria in soil. App Soil Ecol 11:79–90
Straube W, Nestler C, Hansen et al (2003) Remediation of polyaromatic hydrocarbons (PAHs) through landfarming with biostimulation and bioaugmentation. Acta Biotechnol 23:179–196
United States Environmental Protection Agency (1995a) Bioventing principles and practice Vol I: Bioventing Principles. Washington DC
United States Environmental Protection Agency (1995b) Bioventing principles and practice Vol II: Bioventing design. Washington DC
Urum K, Pekdemir T, Gopur M (2003) Optimum conditions for soil washing of crude-oil contaminated with soil with biosurfactant solution. Process Saf Environ Protect 81:203–209
van der Gast CJ, EHiteley AS, Starkey M et al (2003) Bioaugmentation strategies for remediating mixed chemical effluents. Biotechnol Porg 19:1156–1161
van Herwizen R, Hutchings TR, Al-Tabbaa A et al (2007) Remediation of metal contaminated soil with mineral-amended composts. Environ Pollut 150:347–354
van Veen JA, van Overbeek LS, van Elas JD (1997) Fate and activity of microorganisms introduced into the soil. Microbiol Mol Bio Rev 61:121–135
Verma JP, Jaiswal DK (2016) Book review: advances in biodegradation and bioremediation of industrial waste. Front Microbiol 6:1–2. https://doi.org/10.3389/fmicb.2015.01555
Vidali M (2001) Bioremediation. An Overview Pure Appl Chem 73(7):1163–1172
Volf I (2007) Pin Elasi (ed) Elemental biotechnology and bioremediation
Wolski EA, Murialdo SE, Gonzales JF (2006) Effect of pH and inoculum size on pentachloro phenol degradation by *pseudomonas* sp. Water SA 32:1–5
Wong CKE, Cobbett CS (2009) HMA type ATPases are the major mechanisms for root to shoot translocation in *Arabidopsis thaliana.* New Phytol 181:71–78
Xu X, Zhai Z, Li H et al (2017) Synergetic effect of bio-photocatalytic hybrid system: g-C3N4, and Acinetobacter, sp. JLS1 for enhanced degradation of C16 alkane. Chem Eng J 323:520–529. https://doi.org/10.1016/j.cej.2017.04.138
Yogesh P, Akshaya G (2016) Evaluation of Bioremediation Potential of Isolated Bacterial Culture YPAG-9 (*Pseudomonas aeruginosa*) for Decolorization of Sulfonated di-azodye Reactive Red HE8B under Optimized Culture Conditions. Int J Curr Microbiol App Sci 5:258–272
Zitomer DH, Speece RE (1993) Sequential environments for enhanced biotransformation of aqueous contaminants. Environ Sci Technol 27:227–224

Chapter 3
Unique Microorganisms Inhabit Extreme Soils

Maria-Guadalupe Sánchez-Otero, Rodolfo Quintana-Castro, Jorge Guillermo Domínguez-Chávez, Carolina Peña-Montes, and Rosa María Oliart-Ros

Abstract Natural extreme soils are widely distributed on the Earth in all types of ecosystems; the permafrost, the cold soils of the poles, the dry sands of the deserts, the saline soils of marshes, the hot areas associated to volcanoes and hot springs, acidic solfatara fields, soda lakes, hydrothermal marine vents, or mud pots are the habitat of extraordinary organisms, capable to withstand the harsh physicochemical conditions that prevail in those extreme environments, namely, high or low temperatures, acidic or alkaline pH, high salt concentration, and the presence of heavy metals, among others. Those organisms are named extremophiles and constitute a potential and promising source of biomolecules such as polymers, antibiotics, and enzymes; the latter are called extremozymes and are able to perform their natural activity and other interesting reactions under industrial process conditions. Due to their endurance, biomolecules produced by extremophiles are also potential candidates to be used in soil bioremediation.

Keywords Extremophiles · Extremozymes · Oil degradation · Waste treatment · Extreme soils

3.1 Introduction

The soil is a heterogeneous and discontinuous system with an infinite number of discrete microhabitats. The same kind of soil from distinct geographic spots may have a different microbial composition, and little is known about how microbial diversity affect soil function (Nannipieri et al. 2017).

M.-G. Sánchez-Otero · R. Quintana-Castro · J. G. Domínguez-Chávez
Facultad de Bioanálisis, Universidad Veracruzana, Veracruz, México

C. Peña-Montes · R. M. Oliart-Ros (✉)
Unidad de Investigación y Desarrollo en Alimentos, Tecnológico Nacional de México/Instituto Tecnológico de Veracruz, Veracruz, México
e-mail: roliart@itver.edu.mx

A. Kumar, S. Sharma (eds.), *Microbes and Enzymes in Soil Health and Bioremediation*, Microorganisms for Sustainability 16,
https://doi.org/10.1007/978-981-13-9117-0_3

The global microbial diversity has been estimated to be over a trillion (10^{12}) species; however, less than 10^5 species have been identified and represented by classified sequences (Locey and Lennon 2016). In soil, it has been estimated that 1 gram of soil or sediment possesses 10^9 individual prokaryotic microorganisms belonging to 10,000 different species (Torsvik et al. 2002), including the Archaea and Bacteria domains. However, only 1% of those bacteria are cultivable by traditional techniques due to numerous limitations which arise not only from the complexity of their metabolisms and adaptive mechanisms but also from the heterogeneity of the communities, as well as their spatial distribution that is difficult to emulate in the laboratory. These limitations have been resolved to some extent by the development of molecular culture-independent techniques known as "omic" technologies (metagenomics, single-cell genomics, transcriptomics, proteomics, and metabolomics), which have facilitated the description of microbial communities and their interactions within, and with, the environment (Cowan et al. 2015). Particularly challenging is the study of microorganisms that inhabit in environments considered as extreme due to the physicochemical conditions that prevail in those environments, such as extremes of temperature, pH, pressure, salinity, desiccation, and radiation, which hinders the isolation and identification of microorganisms by traditional methods. The microorganisms that thrive in environments that are considered by man as extreme are called extremophiles (Rothschild and Mancinnelli 2001; Satyanarayana et al. 2005; Oliart-Ros et al. 2016; Nannipieri et al. 2017).

3.2 Extremophilic Microorganisms

The physicochemical conditions of extreme environments are far from the values in which the life of many organisms is possible (Rothschild and Mancinnelli 2001; Gomes and Steiner 2004). Extreme conditions can be defined as those near the limits, for enzyme activities or for cell functioning, without damage to biomolecules. Liquid water, energy supply and control, environmental oxide-reduction conditions, and organic chemistry are essential for life, and extremophiles can live and thrive within extreme parameters or might be able to keep them regulated intracellularly (Rothschild and Mancinnelli 2001; Torsvik and Øvreås 2008). In fact, the discovery of extremophiles has favored the search for the most extreme conditions that can be compatible with some form of life and has made life outside of our planet more plausible. The study of life and environmental conditions that existed on the early Earth, and the physical and chemical limits of life nowadays, is necessary in order to develop missions for planetary explorations and the establishment of possible permanent planet stations (Javaux 2006; Ferrer et al. 2007; Champdoré et al. 2007; Harrison et al. 2013; Rampelotto 2013; Madigan et al. 2015).

Microbial species that have been identified and isolated from environments with multiple extremes are called polyextremophiles (Rothschild and Mancinnelli 2001; Harrison et al. 2013; Seckbach and Rampelotto 2015). Soil, as other ecological reservoirs, might have two or more extreme conditions, and the studies that deal with their influence in microbial life are scarce (Harrison et al. 2013).

Extremophiles can be found in the three taxonomic domains, although most of the known ones belong to Archaea and Bacteria. However, there are some multicellular extremophiles such as the Himalayan fly, which survives at −18 °C, and the nematode *Panagrolaimus davidi* that can withstand the freezing of its body's water (Rothschild and Mancinnelli 2001; Harrison et al. 2013).

Extremophiles are classified according to the extreme physical or chemical condition of the environment where they inhabit (Table 3.1).

An interesting aspect of extremophiles is the cellular components and biomolecules they possess, which are stable and functional under the extreme conditions of their environment. Examples are the lipid composition and type of chemical bonds of membranes and cell walls, the production of compatible solutes, and the highly efficient protein and DNA repair systems (Otohinoyi and Omodele 2015; Elleuche et al. 2014). From an industrial perspective, the most considerable interest in extremophiles lies in the enzymes they produce, called extremozymes (Gomes and Steiner 2004).

Table 3.1 Extremophile's classification

Parameter(s)	Extremophile	Definition
Temperature		Optimum growth
	Hyperthermophile Extreme thermopile Thermophile Psychrophile	80–110 °C 60–80 °C 50–60 °C <5 °C
pH		Optimum growth
	Alkaliphilic Acidophilic	pH > 8 pH < 5
Radiation	Radiophile	Resistant to high levels of ionizing and ultraviolet radiation (up to 5 Mrad)
Pressure	Piezophile (formerly known as barophiles)	Grow in high-pressure conditions levels (38 MPa)
Metal concentration	Metallophile	Resistance to mM concentrations of Co, Pb, Cu, Hg, Ni, among other metals
Low organic carbon concentration	Oligotroph/oligophile	Growth in 0.2–16.8mc of organic carbon/L
Desiccation	Xerophile	Prokaryotes tolerate up to a_w 0.75 Some fungi and yeast up to a_w 0.61
Salinity	Halophile	Require high salt concentration (up to NaCl 2–5 M)
Temperature and pH		Optimum growth
	Alkalitermophile Acidothermophile	pH > 8 and 50–85 °C Low pH and T > 50 °C
pH and salinity	Haloalkaliphile	Optimum growth pH > 8 and up to NaCl 33% m/V

Nies (2000), Rothschild and Mancinnelli (2001), Gomes and Steiner (2004), Grant (2004), Satyanarayana et al. (2005), and Seckbach and Rampelotto (2015)

Extremozymes are catalytically active under extreme conditions, which are similar to those of several industrial processes. This has fostered the research on the development of extremophile's culture techniques, which, in parallel with the improvement of cloning and heterologous expression systems, has increased the number of biotransformations in the chemical, food, and pharmaceutical industries where extremozymes are used. Extremozymes also have been used as a model to design and construct proteins with new properties of interest for industrial applications (Mirete et al. 2016). The future in biotechnology is now seen as the "next-generation industrial biotechnology" or "NGIB," where strategies must be developed to overcome the disadvantages of the current industrial biotechnology, including the reduction in energy and water consumption and the lowering of the installation of equipment and facilities cost, among others. Extremophilic microorganisms and their enzymes have capabilities that fit this new approach, such as growth in the presence of toxic compounds as aldehydes or heavy metals or the utilization of unusual substrates as methane or H_2 (Chen and Jiang 2018).

3.3 Extremophilic Microorganisms from Soils

Soils are physically constituted by an ordered collection of sizable and interacting structures and possess physicochemical characteristics that define their microbial diversity; soils from diverse ecosystems might have one, two, or more extreme conditions and so be inhabited by extremophilic microorganisms (Fig. 3.1). Examples are soils, from polar or cold environments, such as permafrost or the Antarctic soil; soils from hot environments, such as hot springs, geysers, or mud volcanoes; soils from dry environments, namely, deserts; soils from saline and hypersaline environments such as salt marshes; and acidic and alkaline soils, such as those found in solfatara fields and soda lakes, commonly associated with high temperatures or the presence of heavy metals (Torsvik and Øvreås 2008). Extreme environments are distributed worldwide; some of the most representatives are depicted in Fig. 3.2.

3.4 Extremophiles from Cold Soils

Cold soils (polar soil-permafrost, glaciers, and high alpine soils) are widely distributed in the biosphere. Low temperature and permanent ice that remains year-round are characteristics of high alpine regions, of permafrost soils that make up to 20% of Earth's surface and are permanently below 0 °C, and of polar regions (Arctic and Antarctic) (Satyanarayana et al. 2005; Yadav et al. 2017). Cold soils host a wide spectrum of psychrophilic microorganisms (optimal growth temperature below 5 °C) including cyanobacteria, bacteria, archaea, and fungi (Zhang et al. 2013; Yadav et al. 2017).

Fig. 3.1 Examples of extreme environments (**a**) desert region in Baja California Sur, México; (**b**) Guerrero Negro Salt Marches in Baja California Sur, México; (**c**) thermal pools "Los Baños," Veracruz, México; (**d**) and (**e**) acid/rich in heavy metal soil; mud spring, respectively, from Yellowstone National Park, Montana, USA; (**f**) geothermal field "Los Azufres," Michoacán, México. (Credits (**a**), (**c**), (**d**), (**e**), and (**f**) from personal archives; (**b**) Francisco Romero Ríos)

The presence of oligotrophic and non-spore-forming viable psychrophile bacteria from different genus including *Subtercola*, *Arthrobacter*, *and Glaciimonas* has been demonstrated in permafrost that has been demonstrated in 3.5–5 million years old permafrost, such as the permafrost located in some regions of Siberia (Vishnivetskaya et al. 2000; Zhang et al. 2013; Margesin et al. 2016). Psychrophile bacteria that belong to the genera *Sphingobacterium*, *Hymenobacter*, *Cryobacterium*, *Pedobacter*, and *Psychrobacter* have been isolated from Antarctic soils (Shivaji

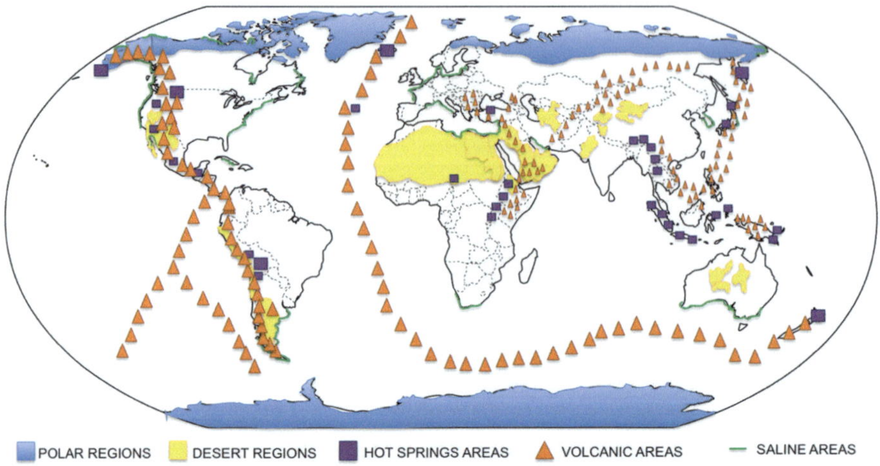

Fig. 3.2 The geographical location of main areas with extreme environments in the world. (Prepared by Manuel Juárez-Arias based on various public sources)

et al. 1992; Hirsch et al. 1998; Reddy et al. 2010; Zhou et al. 2012; Kim et al. 2012). Bacterial genera commonly found in mesophilic environments can also inhabit in cold soils, for example, the presence of *Pseudomonas*, *Acinetobacter*, *Paenibacillus*, and *Planococcus* has been detected in soil samples of the Arctic region from Ny-Ålesund (Chattopadhyay et al. 2014).

Several studies in psychrophilic microorganisms from the polar regions have been undertaken to decipher their cellular and molecular adaptations to cold habitats, which include the production of pigments and solutes associated with osmotic stress response, the composition of membrane lipids which facilitates the maintenance of membrane fluidity, and structural properties of enzymes that enables them to perform their catalytic activity at temperatures below 0 °C (Casanueva et al. 2010; De Pascale et al. 2012). All these features and capabilities make psychrophilic organisms candidate for several biotechnological applications. For instance, Álvarez-Guzmán and coworkers (2016) reported the ability of the psychrophilic strain *Polaromonas rhizosphaerae* G088, isolated from Collins glacier soil in Antarctica (García-Echauri et al. 2011), to produce biohydrogen using glucose as substrate; the high yields obtained made the authors propose this psychrophilic strain as a potential candidate to be used in the production of this environmental energy carrier.

Another potential application of soil and sediment psychrophiles is power generation in microbial fuel cells (MFCs), participating in the oxidation of carbon substrates and the reduction of iron or nitrate as electron acceptors (Dunaj et al. 2012). Microbes from the soils of Hanover, NH, USA, with a mean annual temperature of 7.8 °C, were used as inoculum for a terrestrial MCF that functioned in a temperature range of 8–35 °C. Authors found in electrodes a mixed microbial community of bacterial and archaeal strains, with a relative abundance of known electrogenic bacteria such as members of *Geobacteraceae* (Barbato et al. 2017).

Enzymes from psychrophiles are interesting for industrial applications particularly in biocatalytic procedures that can be carried out at low temperatures, eliminating the expensive heating steps and facilitating the control of the reaction by temperature changes (Gerday et al. 2000; Van der Burg 2003; Chattopadhyay et al. 2014; Siddiqui 2015). Some research groups have focused efforts in cold-active hydrolases such as proteases, lipases, amylases, pectinases, xylanases, cellulases, β-glucosidases, and chitinases from psychrophilic genera, with application in laundry detergents for low-temperature processes (Van der Burg Van den Burg 2003; Yadav et al. 2017). Among hydrolases, one of the most important groups of enzymes from psychrophilic microorganisms is cold-active proteases that have been applied for the synthesis of fine chemicals, the detergent industry, for silk degumming, and for pharmaceutical industries waste management, in textile industry, in peptide synthesis, and in bioremediation (Kuddus and Ramteke 2012; Joshi and Satyanarayana 2013). Interesting examples of these cold-active proteases are those produced by a *Bacillus* strain isolated from the Antarctic soil (Park and Cho 2011) or produced by *Pseudomonas* and *Flavobacterium* strains from the Uruguayan Antarctic Base (Rosales and Sowinski 2011). Many of these hydrolytic enzymes may be active in a broad range of temperatures, for example, the extracellular proteases produced by three psychro-tolerant *Stenotrophomonas maltophilia* strains isolated from soil-surface samples from an Argentinean research station near the Antarctic Peninsula, which showed mesophilic-like properties (Vázquez et al. 2005).

Cold-active lipases are able to catalyze chemical transformations that differ from their natural physiological reactions either by the types of bonds on which they act or by the catalytic mechanisms by which these bonds are formed or cleaved. These types of reactions include the formation of carbon-carbon, carbon-heteroatom, heteroatom-heteroatom, and oxidative processes such as prehydrolysis and direct epoxidation of alkenes (Joseph et al. 2007; Kapoor and Gupta 2012). The main reasons why lipases are considered one of the most important groups of biocatalysts for organic chemistry besides the characteristics mentioned above are the high enantio- and regioselectivity that these enzymes exhibit for specific substrates (Lasón and Ogonowski 2010). Cold-active lipases are being applied in detergent industry and food and beverage industry, in the resolution of racemic mixtures in order to produce pure enantiomers for pharmaceutical applications, and are widely used in biodiesel production (Daiha et al. 2015; Krüger et al. 2018). Of particular interest are the cold-active lipases CAL-A and CAL-B, produced by the yeast *Candida antarctica* that was isolated from a perennially ice-covered Vanda Lake in Antarctica. These lipases are commercially available and have been used for industrial and research purposes during decades because of their unique catalytic properties (Domínguez De María et al. 2005). In particular, CAL-B is a robust lipase able to catalyze synthetic reactions with regio- and enantio-selectivity, so it is used to produce fragrance and flavor esters, surfactants, biodiesel, bioplastics and modified triacylglycerides; CAL-B has also proven to be useful in the kinetic resolution of chiral compounds such as some drugs (Larios et al. 2004; Montanier et al. 2017).

3.5 Extremophiles from Hot Soils

High-temperature soils are not as widely distributed in our planet as cold or polar environments; they include geothermal areas near volcanoes or thermal springs and sun-heated litter or sediments. Besides temperature, other environmental parameters such as ionic strength, nutrient concentration, and pH influence the microbial diversity in these spots (Satyanarayana et al. 2005; Torsvik and Øvreås 2008; Urbieta et al. 2015). Since the isolation of *Thermus aquaticus* by Brock and Freeze in Yellowstone National Park, USA (Brock and Freeze 1969), the interest in discovering and characterizing new species of thermophiles has been increasing, and several species have been found in the thermal waters, submarine hot springs, mud springs, geothermal geysers, hydrothermal vents and fumaroles, or even in man-made thermal sites such as water heaters and sludges. Hot environments can also have other extreme physicochemical conditions such as acidic or alkaline pHs and high hydrostatic pressures, as found in submarine hydrothermal vents and mud springs, for example (Madigan and Marrs 1997; Haki and Rakshit 2003; Satyanarayana et al. 2005).

High extreme temperatures translate into a challenge for life, due to the denaturation of biomolecules, alterations in membrane fluidity, enzyme inactivation, or changes in oxygen solubility (Rothschild and Mancinnelli 2001). Thermophiles thrive in high-temperature environments, thanks to several morphological and molecular features, like the presence of chaperonins, which are proteins that help other proteins to acquire its active form, cell membranes with a high amount of saturated fatty acids, and reverse gyrases that produce super-curls in DNA increasing its denaturation temperature, among other adaptations (Rothschild and Mancinnelli 2001; Haki and Rakshit 2003).

Proteins from thermophiles and hyperthermophiles have been studied for decades in order to understand the mechanisms behind protein thermostability. Comparative studies in sequence and structure of proteins from mesophilic and thermophilic organisms have generated vast information showing that thermophilic proteins have larger hydrophobic core than their mesophilic counterparts. They also present changes in their quaternary structure or an increase in the number of disulfide bonds, salt bridges, and charged amino acids on their surface in order to increase its stability and prevent aggregation. Even though many of these modifications are considered entropically unfavorable for mesophilic proteins, in environments with high temperatures, entropic costs are virtually non-existent in exchange for the structural stability they can provide (Otohinoyi and Omodele 2015). Similarly, enzymes from thermophilic and hyperthermophilic organisms are usually more resistant to the action of denaturing agents, detergents, and organic solvents, as well as exposure to extreme pH; all these properties make them the ideal candidates for its use in biotechnological applications and as model proteins in studies aimed at understanding the mechanisms involved in protein stability (Champdoré et al. 2007; Sarmiento et al. 2015; Krüger et al. 2018).

Enzymes from thermophilic microorganisms have been the most widely used, since processes carried out at high temperatures present many advantages, such as the significant increase in the bioavailability and solubility of organic compounds, especially polymeric substrates, and the decrease of viscosity and the increase of the diffusion coefficients of organic compounds that result in higher reaction rates. Additionally, some pollutants present in the reaction medium, which are hardly biodegraded, may be more soluble, allowing efficient bioremediations. Another advantage is a reduction in the risk of contamination in the reaction media (Van den Burg 2003; Gomes and Steiner 2004; Raddadi et al. 2015; Krüger et al. 2018).

The most widely used thermophilic enzymes are those that degrade polymers, such as amylases, xylanases, proteases, and lipases. In the food industry, thermostable and thermophilic lipases have been used in the synthesis of flavors and the production of structured lipids; in the oil industry, they have been applied in desulphurization and the biodegradation of other toxic compounds. Lipases have also been successfully implemented in the synthesis of biopolymers and biodiesel and used for the production of pharmaceuticals, agrochemicals, cosmetics, detergents, and other organic compounds (Hasan et al. 2006; Sarmiento et al. 2015; Oliart-Ros et al. 2016). Among the thermophilic enzymes whose application has had more impact is the *Taq polymerase* from *Thermus aquaticus*, used for the polymerase chain reaction (PCR). This enzyme is considered the key element of this molecular technique, resisting temperatures above 90 °C without undergoing denaturation. Thermostable amylases and glucosidases have been applied in the processing of polysaccharides, which reduces the risk of contamination and media viscosity and lowers the costs and process times (Schiraldi and De Rosa 2002; Haki and Rakshit 2003; Turner et al. 2007; Sarmiento et al. 2015).

Hot springs are produced by the emergence of geothermally heated groundwater rising from the Earth's crust; they might be related to volcanic activity or not. Hot springs represent a unique environment for unusual forms of life and a source of thousands of genes and metabolites. Several reports describe strains of bacteria and archaea isolated from the waters of hydrothermal vents or detected by metagenomics or other molecular approaches; in contrast, fewer studies refer to microorganisms from soils in the vicinity of hydrothermal vents (Mahajan and Balachandran 2017). Classical reports in that respect are the isolation of *Clostridium* strains from mud of an alkaline hot spring in Yellowstone National Park, Wyoming, USA, by Wiegel and coworkers in 1979 (Wiegel et al. 1979), and the isolation of the fungus *Elaphocordyceps ophioglossoides* HF272 and its bioactive product (the tetramic antibiotic ophiosetin) from the Tsuchiyu Hot Spring soil in Fukushima, Japan (Putri et al. 2010). It is important to note that most of the molecules with antibiotic, anticancer, and anti-inflammatory activity have been obtained from soil *Actinomycetes* that inhabit most of the terrestrial area in our planet, including thermophilic species of *Actinomycetes* from hot sulfur springs (Pednekar et al. 2011; Mahajan and Balachandran 2017). Another report is the isolation of *Limisphaera ngatamarikiensis*, a thermophilic bacterial strain of the phylum *Verrucomicrobia*, from the geothermally heated sediment at the Ngatamariki Geothermal Field in New Zealand (Anders et al. 2015).

Interesting enzymes have also been obtained from soil thermophiles, such as a protease with a remarkable activity produced by a thermotolerant *Bacillus* sp. strain isolated from muddy thermoalkaline soil samples of the hot spring Tarabalo in Odisha, India (Panda et al. 2013), the thermophilic lipases produced by *Bacillus* and *Brevibacillus* strains isolated from soil samples of a hot spring in Malaysia (Norashirene et al. 2013), the thermostable lipases produced by *Bacillus* strains isolated from a hot spring soil at the south Persian Gulf that are able to withstand 37 °C for over 14 h (Rabbani et al. 2014), and the thermophilic lipase produced by a *Geobacillus* strain isolated from a hot spring vicinity soil in India, able to catalyze the synthesis of methyl salicylate at 55 °C with 87% yield (Bhardwaj et al. 2017).

Examples of other important hot sites are semi-arid soils, like the Cariris Velhos region in Paraiba, Brazil, from where 34 thermophilic bacteria with proteolytic and amylolytic activity were isolated, showing potential for biotechnological and bioremediation use (Gorlach-Lira and Coutinho Gorlach-Lira and Coutinho 2007), as well as mudflows, such as the Lapindo mudflow in Java, Indonesia, where sediments have a temperature of about 48 °C and a pH of 7.5 and are rich in heavy metals such as Pb, Hg, Fe, and Ni and from where two *Marinobacter* genus strains were isolated, *M. lutaoensis* and *M. hydrocarbonoclasticus*, considered as a source of thermophilic and metallotolerant enzymes (Dagdag and Asthervina 2015).

The areas nearby active volcanoes are termed "high-temperature fields"; in these soils, little liquid water comes out to the surface, and usually soil pH is in the acidic side (Satyanarayana et al. 2005). Members of genera that are commonly isolated from water and soil of different ecosystems have also been isolated from soils near volcanoes. For example, a strain of *Geobacillus thermoleovorans* was isolated from soil samples of a volcanic geothermal environment at the Pizzo Sopra la Fossa site on Stromboli Island, Italy (Romano et al. 2005), as well as a couple of thermotolerant and halotolerant strains of *Pseudomonas* and *Ochrobactrum* from soil near volcanoes in Andaman and Nicobar Islands, India (Mishra et al. 2017). Some radiotolerant members of genus *Rubrobacter* have also been found in volcanic soils; a novel strain of the moderate thermophilic oligotrophic bacteria *Rubrobacter spartanus* was isolated from volcanic soil samples collected from Kilauea in Hawaii Volcanoes National Park, USA (Norman et al. 2017).

Compost piles are human-made hot spots that can harbor a thermophilic microbiota. For example, members of *Geobacillus* and *Bacillus* genera isolated from compost piles have been considered candidates to be used in biological control of certain vegetable pathogens, since they can produce siderophores, salicylic acid, and hydrolases such as chitinases (Bosma et al. 2015; Sánchez San Fulgencio et al. 2018).

3.6 Extremophiles from Hypersaline Soils

Hypersaline environments include saltern evaporation ponds, the Dead Sea, the Great Salt Lake, African soda lakes, and deep-sea brines (Ventosa et al. 2008; Ma et al. 2010). Hypersaline soils are widely distributed in our planet, and most of them contain more than 0.2% (w/v) of soluble salt. Microorganisms that survive and grow optimally in these environments are referred to as halophiles (Ventosa et al. 2008). According to the salt concentration in which they live, halophiles can be divided in slight halophiles (0.3–0.8 M), moderate halophiles (0.8–3.4 M), and extreme halophiles (3.4–5.1 M) (Otohinoyi and Omodele 2015).

Halophiles include members of the three domains of life, for example, the *Halobacteriaceae* family or *Methanothermea* class, both from the phylum *Euryarchaeota* from Archaea; the *phyla Proteobacteria*, *Firmicutes*, *Actinobacteria*, *Cyanobacteria*, *Spirochaetes*, *and Bacteroidetes* from Bacteria; and even Eukarya such as the unicellular microalgae *Dunaliella salina*, several fungi species such as *Trimmatostroma salinum* or *Debaryomyces hansenii*, and members of the *Artemia* genus (Oren 2008; Gunde-Cimerman et al. 2009; Ma et al. 2010; Siglioccolo et al. 2011).

Many species of halophilic bacteria have been isolated and cultivated successfully in laboratory conditions. For example, Delgado-García et al. (2013) isolated 35 halophilic strains from saline soils at Coahuila State in México, including *Halobacillus trueperi*, *Staphylococcus succinus*, and some members of the *Bacillus* genus, such as *B. atrophaeus* and *B. licheniformis*. They also identified two strains of moderate halophilic organisms: *Halobacillus* sp. and *Oceanobacillus* sp. From the soils of Rambla Salada, Spain, a novel strain of *Roseovarius ramblicola* was isolated, cultivated, and characterized (Castro et al. 2018). In salt-affected soils of the East Anatolia Region of Turkey, 18 members of genera *Bacillus*, *Halobacillus*, *Halomonas*, *and Thalassobacillus*, among other salt-tolerant bacteria, were isolated and identified (Orhan and Gulluce 2015).

Several studies have isolated members of the phylum *Actinobacteria* from the soil. In the first attempt to study the extremophiles that live in the "Cave of Crystals" (aka "Naica") in Chihuahua state, México, rich in calcium sulfate crystals with a 50 °C temperature over the year, two groups of actinobacteria were isolated, including members of the family *Pseudonocardiaceae* (Quintana et al. 2013). Similarly, a novel halophilic actinomycete (*Nocardiopsis salina* YIM 90010 T) was isolated from a hypersaline habitat in Xinjiang Province, China (Li et al. 2004).

One of the strategies halophiles use to survive and thrive in saline and hypersaline environments is the regulation of the salt concentration in their cytoplasm. For example, they possess the ability to accumulate inorganic ions, such as K^+, and to reduce their osmotic pressure by concentrating low-molecular-weight neutral organic species, such as sugars and amino acids, to increase osmolarity (Otohinoyi and Omodele 2015; Babu et al. 2015). Concentrations of salt above 0.1 M make water less available, provoking that hydrophobic amino acids in a protein to lose water and tend to aggregate; therefore, some proteins of the halophilic microorganisms present either a reduced number of hydrophobic amino acids or a great abundance of negative charges on their surface that, when bound to solvated ions, cause a decrease in their hydrophobicity, preventing their precipitation at high concentrations of salt (Kumar et al. 2011; Oren 2011; Reed et al. 2013).

Halophiles are a potential source of bioactive compounds such as antioxidants and antibiotics and compounds that can be used as stress-protective agents (Chen and Jiang 2018). Some halophiles, such as some members of the genera *Haloferax*,

Haloarcula, *Halococcus*, and *Haloquadratum*, produce and accumulate polyhydroxyalkanoate (PHA) inside the cell as lipid inclusions. The physicochemical characteristics of PHA make it a potential substitute for some chemical synthesis-based plastics (Otohinoyi and Omodele 2015; Oliart-Ros et al. 2016; Chen and Jiang 2018). Another important application of halophiles is in bioremediation area.

3.7 Extremophiles from Desertic Soils

Region-denominated deserts are those that receive less amount of rain than the necessary to support life. Deserts receive an annual average of rain (AAR) minor to 400 mm per year and can be classified according to their AAR (Table 3.2) (Azúa-Bustos et al. 2012). All kinds of deserts are inhabited by microorganisms adapted to live with limited water availability and hot or cold temperatures, called xerophiles (Phoenix et al. 2006).

Of particular interest is the microbiota of the Atacama Desert because of its character of hyperarid desert. Phoenix et al. (2006) isolated *Chroococcidiopsis* spp., *Phormidium* spp., *Lyngbya* spp., *Calothrix* spp., and a cyanobacteria, growing endolithically in salt rocks of silice, at the Atacama Desert. Lester et al. (2007) described the isolation of *Actinobacteria*, *Proteobacteria*, and *Firmicutes* in the same Atacama Desert. More recently, members of *Planomicrobium*, *Bacillus*, *Streptomyces*, and *Agrococcus* have been isolated from the Sonora Desert in México (Paulino-Lima et al. 2016).

Enzymes produced by xerophiles are very attractive because they are catalytically active at high temperatures and/or low water concentrations. Some cellulases

Table 3.2 Desert classification according to the AAR (Azúa-Bustos et al. 2012)

Region	AAR (mm)	Desert type
Kalahari Desert	250	Subtropical desert
Mojave Desert	330	
Chihuahuan Desert	250	
Sonoran Desert	250	
[a]the Sahara Desert	20–100	
[b]Atacama Desert	1	Coastal desert
[a]the Namib Desert	10	
[a]Gobi Desert	194	Cold desert
[a]the Patagonian Desert	31–105	
Colorado Plateau Desert	260	
Arctic Desert	500	Polar desert
Antarctic Desert	200	

[a]A desert that receives less than 250 mm of AAR is denominated as "true desert"

[b]Deserts with an ARR less than 2 mm are denominated "hyperarid deserts"

and lipases producers that have been recovered from Saudi Arabian soils are members of the genera *Bacillus*, *Nocardia*, *Cupriavidus*, *Rhodococcus*, and *Streptomyces* (Röttig et al. 2016). From the Tengger Desert, Zhang et al. (2012) identified potential producers of several enzymes, namely, proteases, catalases, ureases, and polyphenol oxidases, in members of *Cyanobacteria*, *Acidobacteria*, and *Proteobacteria* phyla. Essoussi et al. (2010) reported the isolation of 25 strains from the Sahara Desert, categorized into three groups: *Geodermatophilus*, *Blastococcus*, and *Modestobacter*. The esterases produced by *Geodermatophilaceae* showed high resistance to thermal inactivation and alkaline pH, retaining 30 and 20% of activity after heating at 120 °C, pH 12 for 20 minutes.

Various fungi have also been found in desert's soils. In the Indian Desert, strains of *Aspergillus terreus* and *A. niger* were isolated, with the ability to secrete acid phosphatases, as well as strains of *A. niger*, *A. nidulans*, and *A. rugulosus* able to produce alkaline phospatases (Tarafdar et al. 1988). Lu and collaborators (Lu et al. 2013) reported the cloning of a β-mannanase in *Pichia pastoris* from *Thielavia arenaria* recovered from sand samples of the Yinchuan desert. The recombinant enzyme showed high activity (>50%) at a broad range of temperature 50–85 °C and good adaptability to acid and basic media, with properties for the application in the food industry, in coffee extraction, in oil drilling, in the textile industry, and in the bleaching of kraft pulp.

On the other hand, microalgae and cyanobacteria are good colonizers of deserts because of their ability to grow as autotrophs, heterotrophs, or mixotrophs (Subashchandrabose et al. 2013). The most frequent algal species in the deserts of Baja California, México, are *Nostoc commune* and *Schizothrix calcicola* (cyanobacteria), *Myrmecia astigmatica* (chlorophyte), and diatoms such as *Hantzschia amphioxys*, *H. amphioxys f. capitata*, *Luticola cohnii*, *L. mutica*, and *Pinnularia borealis* var. *scalaris* (Flechtner et al. 1998). Consortia of cyanobacteria, microalgae, and bacteria have been observed in desert soils, living in a mutualism or parasitism interaction. The use of these consortia for mitigation of climatic changes, for stabilizing desert soil from wind and water erosion, for fertilizing arid lands, and for bioremediation of oil-polluted areas, is currently being explored (Perera et al. 2018).

3.8 Extremophiles from Acid/Alkaline Soils

Natural acidic soils with pH values below 5.0 are widely distributed in the planet, covering 30% of ice-free surface; however only 4.5% is used for agriculture, since a pH < 5.5 might be detrimental to the plant health due to nutrient or cation (Ca^{2+} or K^{+}) deficiencies or ion toxicities (Läuchli and Grattan 2012; Quatrini and Johnson 2018). Extremely acidic areas with pH lower than 3.0 are characterized by elevated concentrations of sulfuric acid, pyrite, metals (e.g., iron, nickel, aluminum, manganese, copper), and metalloids (e.g., arsenic), conditions that are found in caves, volcanic and geothermal areas, and in acid mine drainages with pH values as low as 3.6 (Sharma et al. 2016; Mirete et al. 2016). The microorganisms that inhabit acidic environments are called acidophiles; they have been isolated from natural areas, such as solfatara fields, sulfuric pools, hot springs and geysers, hydrothermal marine vents, as well as from artificial man-made areas for mining of coal and metal ores, coal dumps, acidic soil of sulfur stockpiles, copper deposits on pyrite, and uranium mines, among others (Sharma et al. 2012).

Acidophiles can be found in the three domains of life, including archaea (Euryarchaeota and Cranearchaeota), bacteria (α, β, and γ proteobacteria, Acidithiobacillia, Nitrospirae, Aquificae, Verrucomucrobia, Actinobacteria and Firmicutes), and acidophilic and acidotolerant eukaryotic microorganisms such as green and red algae, diatoms, amoebas, ciliates, heliozoan, flagellates, fungi, yeasts and rotifers (Quatrini and Johnson 2018). The most studied acidophile genera are *Acidianus*, *Acidilobus*, *Acidiphilium*, *Acidithiobacillus*, *Alicyclobacillus*, *Desulphurolobus*, *Ferroplasma*, *Hydrogenobacter*, *Leptospirilum*, *Metallosphaera*, *Stygiolobus*, *Sulfobacillus*, *Sulfolobus*, *Sufurococcus*, *Sulphurisphaera*, *Thermoplasma*, *Thermogymomonas*, *Thiobacillus*, and *Picrophilus*, the most acidophilic with an optimum pH for growth of 0.7 (Bertoldo et al. 2004; Parashar and Satyanarayana 2018).

In addition to acidic pH, soil acidophiles usually encounter other extreme conditions like high or low temperatures, high osmotic potentials, elevated concentration of metals and metalloids, and variable oxidation-reduction potentials (Johnson and Quatrini 2011). Examples of these polyextremophilic environments are solfatara fields with sulfur acidic soils, acidic hot springs and boiling mud pots, and acidic and metal-enriched mine regions (Satyanarayana et al. 2005), which are inhabited by indigenous life forms adapted to live optimally under these conditions. Examples of them are the unicellular red algae of the Cyanidiophyceae class isolated from cryptoendolithic layers on rock, in environments with pH below 4.0 and the presence of heavy metals (Seckbach and Rampelotto 2015); the thermoacidophile archaea *Thermoplasma volcanium* isolated from solfataric fields that grows optimally at pH 2 and 55 °C; *Picrophilus oshimae* isolated from two hot solfataric locations in Northern Japan that has an optimum pH for growth of 0.7 at 60 °C (Satyanarayana et al. 2005); *Sulfolobus solfataricus*, which thrives in acidic volcanic hot springs growing optimally at approx. 80 °C and pH 2–4 (Huang et al. 2005); and the most thermoacidophilic archaeon reported, *Acidianus infernus* that grows at an optimum temperature of 90 °C in an optimum pH around 2. The wide range of metabolic capabilities that enable acidophiles to endure these extreme conditions make them an interesting group for biotechnological applications. For example, acidophiles have been exploited commercially for biomining and bioleaching, using single strains or in consortia (Quatrini and Johnson 2018).

Acidophilic organisms possess the ability to maintain an internal pH in neutral values (near 6.5), which is accomplished through active mechanisms as proton exclusion, exchange, pumping, consumption and neutralization, and passive mechanisms as cytoplasmic buffering using strategic changes in proton permeability of the membrane and cell surface. Examples of these possible passive systems have been reported for *Acidithiobacillus ferrooxidans*, namely, the presence of surface proteins with positive charges that could act as a transient proton repellent at the cell surface and adjustments of membrane lipids and porins to minimize inward proton leakage during acid stress (Navarro et al. 2013); they also possess small genomes for ease in maintenance, damage mitigation strategies for DNA repair, and the synthesis of acid-stable proteins (Gomes and Steiner 2004; Quatrini and Johnson 2018).

While cytoplasmic pH maintenance keeps metabolism enzymes on neutral conditions, extracellular enzymes are able to function optimally at low pHs due to biochemical modifications like the acidic composition of the amino acids on their surface, the presence of salt bridges, and their hydrophobicity (Ito 2011; Otohinoyi and Omodele 2015). Several extracellular enzymes from acidophiles have been studied due to the enormous potential they have for biotechnological and industrial applications, especially for polymer degradation in biofuel and ethanol production, textile industry, and fruit juice processes, although very few have been exploited for commercial purposes. For example, cellulolytic and xylanolytic thermoacidophilic enzymes have been used at high temperature and acidic conditions to help hydrolyze cellulolytic material and make them more manageable. Such is the case of the endo-β-glucanase SSO1949 and the endoxylanase produced by the thermoacidophilic archaeon *Sulfolobus solfataricus*, which thrives in acidic volcanic hot springs growing optimally at approx. 80 °C and pH 2–4. The endo-β-glucanase SSO1949 has a pH optimum of 1.8 and a temperature optimum of 80 °C and a half-life of 8 h at its optimal conditions (Huang et al. 2005). The xylanase exhibits an optimal temperature and pH of 90 °C and 7.0, respectively, displaying activity at pHs between 4 and 9. It is also highly thermostable, with a half-life of 47 min at 100 °C (Cannio et al. 2004). Other studied enzymes are α-amylases with application in industrial processes such as starch saccharification and hydrolysis of polysaccharides in plant biomass for bioethanol and sugar syrup production, as well as maltooligosaccharide production from raw starches useful as antistaling agents in baking industry, in fruit juice processing, or in the pharmaceutical industry in the elaboration of digestive aids (Parashar and Satyanarayana 2018). Examples are the ones produced by various *Alicyclobacillus* and *Bacillus* strains, with optimum temperatures and pHs of 75 °C and 3.0–4.2, respectively, and the amylase produced by *Pyrococcus furiosus* which is optimally active at 100 °C and pH 5.5–6 (Matzke et al. 1997; Bai et al. 2012; Asoodeh et al. 2014; Laderman et al. 1993; Sharma and Satyanarayana 2010). Glucoamylases from archaeal origin are interesting enzymes due to their extreme optimal conditions (90 °C and pH 2.0) and thermostability (half-life of 20–24 h at 90 °C). Examples are those produced by *Picrophilus torridus*, *P. oshimae*, and *Thermoplasma acidophilum* (Serour and Antranikian 2002). This kind of enzymes is used in the production of dextrose and fructose syrups, in the baking industry, in the brewing of low-calorie beer, and in whole grain hydrolysis in the alcohol industry (Sharma et al. 2012). Interesting acidic proteases have been found in *Sulfolobus acidocaldarius* (thermopsin) with maximum activity at pH 2.0 and 90 °C (Fusek et al. 1990) and the pepstatin-insensitive protease from *T. volcanium* with temperature and pH optima at 55 °C and 3.0. (Kocabayak and Ozel 2007). *Sulfolobus solfataricus* possess an extracellular thermopsin-like proteolytic system with optimal activity at 70 °C and pH 2.0, which is insensitive to common protease inhibitors and is probably implicated in signal transduction pathways (Gogliettino et al. 2014).

Acidic proteases such as rennin are used in the cheese industry (Sharma et al. 2012). Biotechnologically important enzymes are lipases and esterases, which have been reported in various extremophilic species. For example, two esterases from the thermoacidophilic euryarchaeon *Picrophilus torridus* display remarkable thermostability and chemostability, showing the highest activity at temperatures between 55 °C and 70 °C and pH 4–9 and a high thermostability for 24 h at 90 °C. The high resistance to the presence of organic solvents, detergents, and urea of these esterases and its ability to hydrolyze non-steroidal anti-inflammatory drugs make these enzymes important candidates for its application in the pharmaceutical industry (Hess et al. 2008).

Alkaline soils are those with a pH above 7.0. The increase in pH is due to microbial ammonification and sulfate reduction and by water derived from leached silicate minerals (Satyanarayana et al. 2005). Some calcareous and sodic soils are alkaline, but its alkalinity depends on the overall mineral composition. Sodic soils can be sulfate dominated (pH > 7.5), chloride dominated (pH 7.5–8.5), or bicarbonate dominated (pH > 8.2), and nutrient availability varies across the pH spectrum (Läuchli and Grattan 2012). The best studied alkaline environments are soda lakes and soda deserts (e.g., the East African Rift Valley, and the Indian Sambhar Lake). These are characterized by the presence of large amounts of Na_2CO_3 but are significantly depleted in Mg^{2+} and Ca^{2+} due to their precipitation as carbonates. There are also alkaline environments derived from industrial processes including cement manufacture, mining, disposal of blast furnace slag, electroplating, food processing, and paper and pulp manufacture (Satyanarayana et al. 2005). The microorganisms that inhabit in those environments are called alkaliphiles.

Alkaliphilic microorganisms typically grow well at pH 9, with the most extreme strains growing at pH 12–13. They include Bacteria, Archaea, and Eukarya (Horikoshi 1999). Several alkaliphilic bacteria and archaea have been isolated and studied from the genera *Bacillus*, *Pseudomonas*, *Paracoccus*, *Micrococcus*, *Aeromonas*, *Corynebacterium*, *Actinopolyspora*, *Exiguobacterium*, *Ancyclobacterium*, *Vibrio*, *Flavobacterium*, *Cyanospira*, *Chlorococcum*, *Pleurocapsa*, *Spirulina*, *Natronobacterium*, *Natronococcus*, *Methylobacter*, and *Methylomicrobium* (Satyanarayana et al. 2005).

Alkaline environments are usually accompanied by high concentrations of salts, as in saline soda soils in Tibet, Pakistan, India, Kenya, and Russia, or high temperatures as in alkaline hot springs and geysers, where haloalkaliphilic and thermoalkaliphilic organisms can be found. Cyanobacteria is one of the best adapted bacterial groups to polyextremes that lives successfully in hypersaline and alkaline lakes (Rampelotto 2013). *Clostridium*, *Halomonas*, and the *Alkalilimnicola/Alkalispirillum* group are also well-known haloalkaliphilic bacteria. Halophilic alkalithermophiles is a unique group of polyextremophiles that grow optimally under the combined conditions of extreme salinity, alkaline pH, and elevated temperatures, such as those found in the alkaline lakes of Wadi El Natrun, Egypt, and the Lake Magadi in Kenya. Representatives of halophilic alkalithermophiles are the bacterial genera *Natranaerobius*, *Natronovirga*, *Halonatronum*, and *Dichotomicrobium* and the archaea *Natronolimnobius* and *Natrialba* (Mesbah and Wiegel 2010).

To thrive in those environments, alkaliphilic organisms maintain their interior pH in neutral or slightly alkaline values, by means of the presence of Na^+/H^+ antiporter systems in plasma membranes that transport H^+ into the cell at the expense of Na^+ export from it. They also contain negatively charged cell wall polymers that stabilize the cell membrane by reducing the charge density at the cell surface and peptidoglycan layer that provides a shielding effect by tightening the cell wall (Horikoshi 1999; Wiegel and Kevbrin 2004; Satyanarayana et al. 2005).

Alkalophiles have been extensively studied since they are an important source of useful, stable enzymes and novel chemicals, including antimicrobials. In addition, cells can be used in bioremediation processes, such as the removal of H_2S from sour gas streams produced in the petrochemical industry; enzymes from alkalophiles not only display optimal activity at elevated pH but often show activities in a broad pH range, and at other extreme conditions as high or low temperatures or the presence of chaotropic compounds, which increase the range of applications in which they are catalytically competent (Preiss et al. 2015).

Several alkaline enzymes have been successfully applied in biotechnological and industrial processes, such as in laundry detergents, for efficient food processing, in the finishing of fabrics, and in pulp and paper industries, in particular, those produced by alkaliphilic *Bacillus* species because of their ubiquitous presence, nonpathogenicity, and high production capacity of extracellular enzymes which is advantageous in the industry due to the low production cost and the ease in enzyme recovery (Fujinami and Fujisawa 2010).

Alkaline enzymes have a dominant position in the global enzyme market as constituents of detergents. In particular, alkaline proteases, lipases, cellulases, and amylases are used as detergent additives to improve the washing power. The first high-alkaline protease was found by Horikoshi from the alkaliphilic *Bacillus* sp. strain 221 in 1971. This enzyme has an optimal pH of 11–12 and showed thermostability at 60 °C and tolerance to surfactants (Horikoshi 1971a, b; Kobayashi et al. 1995). From then, other alkaline proteases from alkaliphilic bacteria and fungi have been reported, showing their versatility in optimal activity conditions and applications. Examples are the alkaline protease from *Stenotrophomonas maltophilia* MTCC 7528 optimally active at pH 9.0 and 20 °C (Kuddus and Ramteke 2009), the salt and organic solvent-tolerant protease functional at low temperatures from *Pseudomonas aeruginosa* strain K (Rahman et al. 2010), the cold-active alkaline metalloprotease from *Pseudomonas lundensis* HW08 (Yang et al. 2010), and the laundry-detergent-stable alkaline protease from *Paenibacillus tezpurensis* AS-S24-II active at temperatures of 45–50 °C (Rai et al. 2010).

To date, several alkaline proteases have been commercialized, some of them in its native form (e.g., Savinase® from Novozymes, Purafect®from Gegegcor) and others genetically modified to meet special requirements, such as activity at low temperature (Kannase® from Novozymes, Properase® from Genencor) or in the presence of oxidant compounds (Durazyme® from Novozymes). Commercially produced alkaline proteases have been successfully used in detergent formulations, silk degumming, food and feed industry, photographic gelatin hydrolysis, leather dehairing, cosmetics, and pharmaceuticals (Furhan and Sharma 2014).

Alkaline amylases retain activity at the pH at which detergents function (8.0–11.0) and, therefore, are of practical use in the laundry industry (Ito et al. 1998). The first alkaline amylase was reported by Horikoshi obtained from the soil *Bacillus* strain A-40-2 (Horikoshi Horikoshi 1971a, b). From then, several other alkaline amylases have been characterized and applied in detergent formulation, in combination with alkaline debranching enzymes such as pullulanases, amylopullulanases, neopullulanase, and isoamylases (Sarethy et al. 2011). Examples of commercial alkaline amylases added to detergents to remove starch stains are Termamyl® and Duramyl® (Novozymes), derived from *Bacillus licheniformis* (Olsen and Falholt 1998).

Microbial lipases functionally stable at alkaline pH are being used for organic synthesis (Bornscheuer and Kazlauskas 2006), in food processing, and in the pharmaceutical and laundry industries (Gupta et al. 2004). Lipolase® and Lipolase Ultra®, obtained from the fungus *Thermomyces lanuginosus*, are examples of alkaline lipases used in detergents. Various alkaline and thermophilic/thermostable lipases have been isolated, characterized, and modified to improve their performance. Several members of *Pseudomonas* produce alkaline lipases compatible for use with detergents, pharmaceutical products, and processing of fat, for example, *Pseudomonas monteilii* TKU009 (Wang et al. 2009). One interesting example is the thermo-active alkaline lipase produced by the fungus *Thermomyces dupontii* (formerly classified as *Talaromyces thermophilus*), which is stable in the presence of different surfactants and has been overproduced heterologous expression in *Pichia pastoris* (Wang et al. 2019). This enzyme has been applied for biodiesel production (Romdhane et al. Romdhane et al. 2011) and the synthesis of a chiral intermediate of Pregabalin through the kinetic resolution of 2-carboxyethyl-3-cyano-5-methylhexanoic acid ethyl ester (CNDE) (Ding et al. 2018). Activity in organic solvents is a desirable property for lipolytic enzymes for their use in non-aqueous catalytic processes. Examples of alkaline lipases able to perform organic synthesis have been reported in *Acinetobacter* sp. EH28 (Ahmed et al. 2010), *Amycolatopsis mediterranei* DSM 43304 (Dheeman et al. 2010), *Acinetobacter baylyi* (Uttatree et al. 2010), *Pseudomonas mandelii* (Kim et al. 2013), *Streptomyces* sp. CS273 (Mander et al. 2013), and *Bacillus subtilis* strain Kakrayal-1 (Saraswat et al. 2018).

Horikoshi and coworkers described for the first time an extracellular carboxymethylcellulase functional at pH 5.0–10.0, from *Bacillus* sp. KSM-635 (Ito et al. 1989). This alkaline cellulase (Eg1K) was the first used as a detergent additive; to date, several alkaline cellulases are commercially available (e.g., Carezyme® from Novozymes produced by *Thermomyces lanuginosus*). They are applied as components of laundry detergents, in the finishing of fabrics and clothes, and in the paper recycling process, among other applications (Raddadi et al. 2013).

Other alkaline enzymes with important biotechnological applications are alkaline xylanases for the bleaching of lignin from kraft pulp as an eco-friendly substitute for chlorine (Damiano et al. 2006), for the conversion of lignocellulosic biomass to serve as source of biofuel, for the improvement of cereal food products and animal feedstocks, and for degumming of plant fibers (Li et al. 2005); alkaline keratinases used to recycle waste of feathers (Kojima et al. 2006); alkaline cyclomaltodextrin glucanotransferases (CGTases) to enhance the production of cyclodextrins, which are used in pharmaceuticals and foodstuffs and for chemical interactions (Komiyama and Terao 2008); and alkaline pectinases for plant fiber processing for usage in textiles and during the paper-making process (Furhan and Sharma 2014).

Biotechnological applications of acidophiles and alkaliphiles include the ability to accomplish redox transformation of iron and sulfur that has been already commercially used in bioprocessing for over 50 years and constitute new opportunities for new "biomining" technologies; in the food industry, cellulases and acidophilic amylases of *Fructobacillus* sp. have been used to improve digestion of ruminants (Otohinoyi and Omodele 2015; Quatrini and Johnson 2018). Alkaliphilic enzymes have been widely used as detergent additives and dehairing agents, in the food industry in juice clarification, and as biocatalyst, as in the synthesis of cyclodextrins with an alkaliphile cyclodextrin-glucanotransferase of *Bacillus macerans* (Horikoshi 1999).

3.9 Metagenomic Approaches

Metagenomics has become one of the most sophisticated tools for the identification of the phylogenetic composition and metabolic pathways of microbial communities in diverse ecosystems by means of the analysis of the total DNA present in an environmental sample (Foster et al. 2012).

Databases such as GOLD (Genomes Online Database) host the information on projects based on metagenomic approximations (Ansorge 2016). By October 2018, this database included 281 metagenomic projects and a total of 5112 studies of biological samples from soil (http://genomesonline.org/cgi-bin/GOLD/index.cgi).

Metagenomics, defined as the functional and sequence analysis of the genomes of a group of organisms in an environmental sample, has been used to increase the understanding of the remarkable complexity and versatility of extremophilic microbial communities. Several metagenomic studies describing microbial diversity of natural and human-made extreme soils have been published to date, unraveling the vast number of extremophilic microorganisms that remain resistant to traditional cultivation approaches and that represent more than 80% of the phylogenetic diversity of archaea and bacteria (Hedlund et al. 2014). Some examples are: the description of the thermophilic microbial community of marine sediments associated with hydrothermal oceanic sources from the Jan Mayen vent fields in Norwegian-Greenland Sea (Urich et al., 2014); the characterization of many uncultured bacteria genes from phyla Proteobacteria, Actinobacteria and Cyanobacteria recovered from the metagenomic project conducted in Manikaran hot spring of Himachal Pradesh, India (Kaur et al., 2018); the analysis of psychrophilic communities at the Antarctic maritime soil from the Mars Oasis at the Antarctic Peninsula (Pearce et al., 2012); the presence of archaeas from Euryarchaeota and Crenarchaeota phyla in the Antarctic continental shelf surficial sediment (Bowman and McCuaig, 2003; Satyanarayana et al., 2005); the acidophilic microorganisms that inhabit at the Rio

Tinto acid sediments in Spain (García-Moyano et al., 2012); the distribution of microorganisms in alkaline lake sediments from the Tibetan Plateau (Xiong et al., 2012)); the description of the bacterial communities that inhabit in heavy metals polluted soils near lead, zinc and chromium producing facilities in Poland (Gołębiewski et al., 2014); the description of the bacterial community structure of the Tapovan hot spring soil in India (Rawat and Joshi, 2018); the investigation of bacterial diversity of hot spring soil from Manikaran, Himachal Pradesh, India, where 61% of the recovered clones were uncultured bacteria genes from the phyla Proteobacteria, Actinobacteria and Cyanaobacteria (Kaur et al., 2018); and the microbial community structures in geothermal springs from the Chilas and Hunza areas from the Himalayan geothermal zone, Pakistán, that have temperatures up to 95 ºC and pHs up to 9.4 (Amin et al., 2017).

Additionally, metagenomic studies focused on extreme environments have helped to illustrate the existence of several yet uncultivated candidate phyla, encompassing putative acidophiles (Parvarchaeota), halophiles (Nanohaloarchaeota), thermophiles (Acetothermia, Aigarchaeota, Atribacteria, Calescamantes, Korarchaeota, and Fervidibacteria), and piezophiles (Gracilibacteria) (Baker and Dick 2013).

In addition to providing information about the prevailing taxonomic groups in specific locations, metagenomic studies give invaluable information about their metabolic diversity, the molecular adaptations to life in extreme environments, their various unique genomic features, and the presence of genes codifying for proteins with potential biotechnological applications (Hedlund et al. 2014; de Castro et al. 2016). By means of function-based metagenomics, which incorporates an activity-based screening after environmental DNA analysis, interesting extremoenzymes have been characterized after metagenomic approaches. An example is the thermoacidophilic, thermostable, and ionic liquid-tolerant cellulase obtained from an environmental sample collected from a saline alkaline lake in Inner Mongolia, China. The enzyme's high stability at pH 3.0–9.0, with maximum activity at pH 4.5, thermostability and maximal activity at 60–70 °C, and stability in the presence of 4 M NaCl 10% and 20% 1-butyl-3-methylimidazolium chloride, makes this cellulase applicable in the pre-treatment of lignocellulosic biomass in biofuel production processes (Zhao et al. 2018).

Lipolytic enzymes have also been obtained from metagenomic libraries, such as the thermophilic, alkali-, and salt-tolerant esterase from an oil-fed microcosm grown from a permafrost soil sample collected at the Kolyma Lowland region of northeastern Siberia (Petrovskaya et al. 2016). Also, a new alkaline protease was obtained after a metagenomic analysis of an oil-polluted mud flat sample collected from the Black Oil Mountain of Karamay, China. The proteases displayed good washing performance at low temperature (30 °C) and alkaline pH range (pH 8.0–11.0) suggesting a great potential for use as an additive in detergent formulation (Gong et al. 2017). The analysis of three small metagenomic libraries constructed from High Arctic marine sediment samples allowed the identification of the cold-active and salt-tolerant esterase Lip 3 with potential application in the food industry (de Santi et al. 2016). Similarly, a novel cold-active esterase (derived from a non-described Actinobacteria genome) was cloned from a metagenomic library constructed from cold soil samples of the National Park "Los Nevados," Colombia (Jiménez et al. 2012).

3.10 Extremophiles and Extremozymes as a Tool for Bioremediation

Traditional methods for the treatment of toxic and organic wastes are often expensive and sometimes generate new environmental difficulties. Microorganisms and their enzymes are now used to degrade and metabolize toxic residuals and contaminated soils; this action, called bioremediation, is more feasible with the use of extremophiles and extremozymes since both of them tolerate and work efficiently in harsh conditions.

One of the leading causes of pollution in soil worldwide is oil spills and the activities derived from its extraction and refinement, due to factors such as the generation of drilling mud, faults in old systems and pipes in lines of transport, the existence of old refineries with obsolete water, and waste treatment systems (Plaza et al. 2001). In this sense, thermophilic microorganisms have been studied for their application in oil degradation. Examples are some members of the *Bacillus* genus isolated from heavy oil fields, capable of degrading oil to lighter components (Al-Sayegh et al. 2015), and members of Archaea domain such as the thermophilic *Archaeoglobus fulgidus* and *Candidatus Syntrophoarchaeum butanivorans*, which are capable to degrade alkanes, which constitute the major component of oil (Park and Park 2018).

Extremozymes have also been applied for bioremediation. Examples are extremophilic and extremoalkaliphilic lipolytic enzymes that can be used to degrade lipids wastes from the food industry, i.e., esterases hydrolyzing short-chain acyl esters or lipases catalyzing the hydrolysis of long-chain acyl esters (Daiha et al. 2015; Krüger et al. 2018). The enzymes from thermophilic bacteria, such as members of the *Bacillus* genus or related species, constitute an important source of novel enzymes for bioremediation applications. A thermophilic strain of *Anoxybacillus rupiens*is 19S isolated from a hot spring in South Africa showed the ability to produce enzymes for the degradation of starch, proteins, and phenols which are potentially useful for bioremediation of wastewater from food industry (amylases, proteases), solubilization of phosphates (inorganic pyrophosphatases), and polyaromatic hydrocarbons (catalase, peroxidase, quinone oxidoreductase) (Jardine et al. 2018). Oil food waste production has steadily increased along with population growth and represents a problem that deserves proper attention. In this regard, 21 thermotolerant oil-degrading bacteria strains isolated from oil-spilled soil were tested in order to evaluate its oil-degrading capacity as pure strains or in consortia. The oil-degrading ability of the latter was higher compared with individual strains; the identification based on 16S rRNA gene sequencing of the members of the consortium showed the presence of *Brevibacillus borstelensis*, *Bacillus cereus*, *Bacillus licheniformis*, and *Brevibacillus agri* (Awasthi et al. 2018).

Among hydrolases, another important group of enzymes are cold-active proteases, since they are used to manage wastes from food industries, and thermophilic cellulases for agricultural residues and municipal wastes treatment (Kuddus and Ramteke 2012; Joshi and Satyanarayana 2013). An example is a thermostable cellulase from *Paenibacillus tarmensis* isolated from the Sahara Desert. The enzyme is stable at low water activity conditions, in the presence of organic solvents, high salt concentrations, heavy metals, and ionic liquids and is active at an optimum pH of 9.0 at 50 °C (Raddadi et al. 2013).

Halophiles are becoming a key biotechnological tool for many bioremediation processes since there is an increasing interest in the decontamination of high salt waters and soils, which are mostly influenced by the discharge of industrial effluents (Aracil-Gisbert et al. 2018). Halophiles also tolerate elevated levels of metals in the environment. In that respect, *Halobacterium* sp. NRC-1 and *Nesterenkonia* sp. MF2 were reported to remediate arsenic and chromate from wastewater, and *Salinicoccus iranensis* and *Halomonas* were applied to remediate phenol and toxic ions in polluted water from textile industries (Otohinoyi and Omodele 2015). Haloarchaea are characterized by their capacity to grow in media with a high salt concentration in a range of 12% to 30% salt (2–5 M NaCl). Their cellular machinery works in a high concentration of salt due to the accumulation of potassium ion to counteract the high concentration of sodium ion. The harsh environment in which archaea survive makes these a useful agent for bioremediation in water treatment processes and in saline and hypersaline environments contaminated with toxic compounds such as nitrate, nitrite and ammonia, chlorine compounds, hydrocarbons, or heavy metals (Bonete et al. 2015). An interesting example is *Haloferax mediterranei*, a denitrifying haloarchaea from the family Haloferacaceae that can grow in a broad range of NaCl concentrations (1.0–5.2 M). It was isolated from seawater evaporation ponds near Alicante, Spain (Rodríguez-Valera et al. 1980; Nájera-Fernández et al. 2012). It has been demonstrated that *Hfx. mediterranei* can remove most of the nitrogen compounds present in a treated medium, specifically in anoxic conditions after the induction of the denitrification pathway. It can resist up to 2 M and 50 mM of nitrate and nitrite concentrations, respectively. Furthermore, this organism can efficiently reduce bromate and (per)chlorate (Martínez-Espinosa et al. 2015).

Mercury detoxification is another critical task of bioremediation as many industrial wastes contain millions of tons of mercury compounds. In this sense, it has been reported the use of archaeal strains isolated from a hypersaline soil for mercury removal. Different haloarchaeal strains, which were initially isolated from a hypersaline (4 M NaCl) coastal area of the Arabian Gulf, were used (*Haloferax* sp. *HA1*, *Haloferax* sp. *HA2*, *Halobacterium* sp. *HA3*, and *Halococcus* sp. *HA4*). It was proved that these extremophiles could be useful biological systems for removing toxic forms of mercury effectively from mercury-contaminated, non-oily, hypersaline areas (Al-Mailem et al. 2011).

Recently, it has been developed as an expression system from a halophilic bacterium (*Salinicoccus salsiraiae*) which was isolated from a saline soil of Dagang Oilfield in Tianjin, China. This strain can tolerate high concentrations of sodium chloride and sodium acetate, showing great potentials in biological treatment of hypersaline wastewaters with high chemical oxygen demand (COD) (Zhao et al. 2017).

Another group of soil pollutant is hydrocarbons. Halophilic microorganisms used for this purpose comprise bacteria (e.g., *Marinobacter sedimentalis*, *Halomonas salina*, *Pseudomonas* sp.), archaea (e.g., Ha*lobacterium salinarum*, *Haloferax larsenii*), and fungi. All of these halophiles can biodegrade oil and pure aliphatic and aromatic hydrocarbons at high salinities (Al-Mailem et al. 2017).

Since many of the flocculants used in wastewater treatment are not readily biodegradable or produce toxic derivatives such as acrylamide, novel bioflocculants have been produced from halophilic bacteria from diverse genera, such as *Halomonas*, *Bacillus*, *Alcaligenes*, and *Enterobacter* (Nontembiso et al. 2011; Sam et al. 2011).

Concerning the use in bioremediation of xerophiles isolated from desertic soils, in the last few decades, several consortia of microorganisms have been identified. For instance, bacteria from *Aquabacterium*, *Bacillus*, *Massilia*, *Alkalispirillum*, *Salinimicrobium*, *Sphingomonas*, *Alkanindiges*, *Pseudomonas*, *Nocardiopsis*, and *Actinoalloteichus* genera were isolated from desert soil contaminated with oil during the Second Gulf War. The material was rich in hydrocarbonoclastic bacteria, and it was used for the bioaugmentation of seawater from the Arabian/Persian Gulf and desert soil from Kuwait (Dashti et al. 2018).

Another interesting study for the bioremediation of oily sludge and other oil wastes is the use of 40 bacterial isolates from Libyan Desert, from which 14 could grow up to 10% on crude oil. Some of the bacterial strains with increased ability to degrade hydrocarbons were identified as *Cellulosimicrobium cellulans*, *Brevibacterium liquefaciens*, *Brevibacterium mcbrellneri*, and *Enterococcus saccharolyticus* (Shaieb et al. 2015).

Metallophilic microorganisms can be used in heavy metal-polluted soil bioremediation, since they tolerate up to mM concentrations of iron, copper, cobalt, and nickel and possess the ability to reduce them to a lower redox state, producing metal species that have a lower bioactivity. Heavy metal pollution constitute a major threat to most life forms; therefore, there is an increased interest in generating strategies to remove these elements in soils, sediments, and wastewaters (Gomes and Steiner 2004). In this regard, a strain of *Citrobacter freundii* JPG1, isolated from a gold tailing pile in Jiapiguo, China, and able to grow under the pressure of multiple heavy metals such as Ag, Cd, Co, Cr, Cu and Ni, was recently proposed as a good candidate to be used in the treatment of copper-laden industrial wastes, due to its ability to bioaccumulate copper under aerobic and anaerobic conditions (Wang et al. 2018).

One of the most toxic heavy metals is mercury due to its high toxicity and bioaccumulation; remediation of Hg consists in the transformation of the toxic ionic and inorganic forms into elemental or vaporforms, or into mercury sulfides, which are less toxic and can be readily absorbed. In this regard, it was reported that the mercury-resistant bacteria *Sphingobium* sp. SA2 was able to produce a mercuric reductase enzyme that rapidly volatilized mercury, proposing this bacterium as suitable for mercury remediation in soils (Mahbub et al. 2016).

Extremophiles have also been used to degrade toxic synthetic compounds, such as pesticides, chemical catalysts, and antifouling preparations. That is the case of tributyltin (TBT), an organic tin compound that has been banned since 2008 due to its high toxicity, but, unfortunately, TBT is still present in many places. Extremophilic bacteria strains resistant to TBT from the Atacama Desert of Northern Chile have been isolated, and their capacity to degrade TBT using a debutylation pathway was assessed. An example is *Moraxella osloensis* UC-94, able to degrade TBT into the less toxic DBT (Yáñez et al. 2015).

Acidophilic microorganisms may be an alternative to treat mine wastes and recover heavy metals from them; microbial communities that can survive in soils that have been polluted with mine wastes include acidophilic iron- and sulfur-oxidizing autotrophs. The use of consortia of sulfate-reducing bacteria is a strategy that can be used in soil bioremediation since they catalyze the conversion of sulfate to sulfide, neutralize acidity, and immobilize heavy metals (Ayangbenro et al. 2018).

Acknowledgments We thank Manuel Juárez-Arias for his help schematizing the geographical location of the main extreme environments in the world; we also thank Francisco Romero Ríos for kindly providing the picture of Guerrero Negro Salt Marches in Baja California Sur, México.

References

Ahmed EH, Raghavendra T, Madamwar D (2010) An alkaline lipase from organic solvent tolerant Acinetobacter sp. EH28: application for ethyl caprylate synthesis. Bioresour Technol 101:3628–3634. https://doi.org/10.1016/j.biortech.2009.12.107

Al-Mailem DM, Al-Awadhi H, Sorkhoh NA, Eliyas M, Radwan SS (2011) Mercury resistance and volatilization by oil utilizing haloarchaea under hypersaline conditions. Extremophiles 15:39–44. https://doi.org/10.1007/s00792-010-0335-2

Al-Mailem DM, Al-Deieg M, Eliyas M, Radwan SS (2017) Biostimulation of indigenous microorganisms for bioremediation of oily hypersaline microcosms from the Arabian Gulf Kuwaiti coasts. J Environ Manag 193:576–583. https://doi.org/10.1016/j.jenvman.2017.02.054

Al-Sayegh A, Al-Wahaibi Y, Al-Bahry S, Elshafie A, Al-Bemani A, Joshi S (2015) Microbial enhanced heavy crude oil recovery through biodegradation using bacterial isolates from an Omani oil field. Microb Cell Factories 14:141–152. https://doi.org/10.1186/s12934-015-0330-5

Álvarez-Guzmán CL, Oceguera-Contreras E, Ornelas-Salas JT, Balderas-Hernández VE, de León-Rodríguez A (2016) Biohydrogen production by the psychrophilic G088 strain using single carbohydrates as substrate. Int J Hydrog Energy 41:8092–8100. https://doi.org/10.1016/j.ijhydene.2015.11.189

Amin A, Ahmed I, Salam N, Kim BY, Singh D, Zhi XY, Xiao M, Li WJ (2017) Diversity and distribution of thermophilic bacteria in Hot Springs of Pakistan. Microb Ecol 74:116–127. https://doi.org/10.1007/s00248-017-0930-1

Anders H, Power J, MacKenzie A, Lagutin K, Vyssotski M, Hanssen E, Moreau J, Stott M (2015) *Limisphaera ngatamarikiensis gen. nov., sp. nov.*, a thermophilic, pink-pigmented coccus isolated from subaqueous mud of a geothermal hot spring. Int J Syst Evol Microbiol 65:1114–1121. https://doi.org/10.1099/ijs.0.000063

Ansorge WJ (2016) Next generation DNA sequencing (II): techniques, applications. J Next Generat Sequenc Appl S1:0005. https://doi.org/10.4172/2469-9853.S1-005

Aracil-Gisbert S, Torregrosa-Crespo J, Martínez-Espinosa RM (2018) Recent trend on bioremediation of polluted salty soils and waters using Haloarchaea. Adv Bioremed Phytoremed:74–77. https://doi.org/10.5772/intechopen.70802

Asoodeh A, Emtenani S, Emtenani S, Jalal R, Housaindokht MR (2014) Molecular cloning and biochemical characterization of a thermoacidophilic, organic-solvent tolerant α-amylase from a Bacillus strain in *Escherichia coli*. J Mol Cat B Enzym 99:114–120. https://doi.org/10.1016/j.molcatb.2013.10.025

Awasthi MK, Selvam A, Chan MT, Wong JWC (2018) Bio-degradation of oily food waste employing thermophilic bacterial strains. Bioresour Technol 248:141–147. https://doi.org/10.1016/j.biortech.2017.06.115

Ayangbenro AS, Olanrewaju OS, Babalola OO (2018) Sulfate-reducing Bacteria as an effective tool for sustainable acid mine bioremediation. Front Microbiol 9:1–15. https://doi.org/10.3389/fmicb.2018.01986

Azúa-Bustos A, Urrejola C, Vicuña R (2012) Life at the dry edge: microorganisms of the Atacama Desert. FEBS Lett 586:2939–2945. https://doi.org/10.1016/j.febslet.2012.07.025

Babu P, Chandel AK, Singh OV (2015) Extremophiles and their applications in medical processes. Springer, Cham/Heidelberg

Bai Y, Huang H, Meng K, Shi P, Yang P, Luo H, Luo Y, Zhang W (2012) Identification of an acidic α-amylase from *Alicyclobacillus sp.* A4 and assessment of its application in the starch industry. Food Chem 131:1473–1478. https://doi.org/10.1016/j.foodchem.2011.10.036

Baker BJ, Dick GJ (2013) Omic approaches in microbial ecology: charting the unknown. Microbe 8:353–360. https://doi.org/10.1128/microbe.8.353.1

Barbato RA, Foley KL, Toro-Zapata JA, Jones RM, Reynolds CM (2017) The power of soil microbes: sustained power production in terrestrial microbial fuel cells under various temperature regimes. Appl Soil Ecol 109:14–22. https://doi.org/10.1016/j.apsoil.2016.10.001

Bertoldo C, Dock C, Antranikian G (2004) Thermoacidophilic microorganisms and their novel biocatalysts. Eng Life Sci 4:521–531. https://doi.org/10.1002/elsc.200402155

Bhardwaj KK, Saun NK, Gupta R (2017) Immobilization of lipase from *Geobacillus sp.* and its application in synthesis of methyl salicylate. J Oleo Sci 66:391–398. https://doi.org/10.5650/jos.ess16153

Bonete MJ, Bautista V, Esclapez-Espliego JM, García-Bonete MJ, Pire C, Camacho M, Martínez-Espinosa RM (2015) New uses of haloarchaeal species in bioremediation processes. Adv Bioremed Phytoremed:23–49. https://doi.org/10.5772/60667

Bornscheuer UT, Kazlauskas RJ (2006) Hydrolases in organic synthesis: regio- and stereoselective biotransformations, 2nd edn. Wiley-VCH, Weinheim

Bosma EF, van de Weijer AHP, Daas MJA, van der Oost J, de Vos WM, van Kranenburg R (2015) Isolation and screening of thermophilic bacilli from compost for electrotransformation and fermentation: characterization of *Bacillus smithii* ET138 as a new biocatalyst. Appl Environ Microbiol 81:1874–1883. https://doi.org/10.1128/AEM.03640-14

Bowman JP, McCuaig RD (2003) Biodiversity, community structural shifts, and biogeography of prokaryotes within Antarctic continental shelf sediment. Appl Environ Microbiol 69:2463–2483. https://doi.org/10.1128/AEM.69.5.2463-2483.2003

Brock TD, Freeze H (1969) *Thermus aquaticus* gen. n. and sp. n., a nonsporulating extreme thermophile. J Bacteriol 98(1):289–297

Cannio R, Di Prizito N, Rossi M, Morana A (2004) A xylan-degrading strain of *Sulfolobus solfataricus* : isolation and characterization of the xylanase activity. Extremophiles 8:117–124. doi: https://doi.org/10.1007/s00792-003-0370-3

Casanueva A, Tuffin M, Cary C, Cowan DA (2010) Molecular adaptations to psychrophily: the impact of 'omic' technologies. Trends Microbiol 18:374–381. https://doi.org/10.1016/j.tim.2010.05.002

Castro DJ, Cerezo I, Sampedro I, Martínez-Checa F (2018) *Roseovarius ramblicola sp. nov.*, a moderately halophilic bacterium isolated from saline soil in Spain. Int J Syst Evol Microbiol 68:1851–1856. https://doi.org/10.1099/ijsem.0.002744

Champdoré MD, Staiano M, Rossi R, D'Auria S (2007) Proteins from extremophiles as stable tools for advanced biotechnological applications of high social interest. J R Soc Interface 4:183–191. https://doi.org/10.1098/rsif.2006.0174

Chattopadhyay AN, Dey A, Roy P, Chatterjee S, Saha P, Mukhopadhyay SK (2014) Cold active extracellular hydrolytic enzyme producing culturable heterotrophic Bacteria from NY-ÅLESUND, Arctic. Int J Adv Biotechnol Res 5:271–278

Chen GQ, Jiang XR (2018) Next generation industrial biotechnology based on extremophilic bacteria. Curr Opin Biotechnol 50:94–100. https://doi.org/10.1016/j.copbio.2017.11.016

Cowan DA, Ramond J-B, Makhalanyane TP, De Maayer P (2015) Metagenomics of extreme environments. Curr Opin Microbiol 25:97–102. https://doi.org/10.1016/j.mib.2015.05.005

Dagdag EEA, Asthervina WP (2015) Isolation and characterization of isolate thermophilic bacteria from water and solid sediment of Lapindo Mud. Resour Environ 5:66–71. https://doi.org/10.5923/j.re.20150502.03

Daiha KG, Angeli R, de Oliveira SD, Almeida RV (2015) Are lipases still important biocatalysts? A study of scientific publications and patents for technological forecasting. PLoS One 10:e0131624. https://doi.org/10.1371/journal.pone.0131624

Damiano VB, Ward R, Gomes E, Alves-Prado HF, Da Silva R (2006) Purification and characterization of two xylanases from alkalophilic and thermophilic *Bacillus licheniformis* 77-2. Appl Biochem Biotechnol 129:289–302

Dashti N, Ali N, Salamah S, Khanafer M, Al-Shamy G, Al-Awadhi H, Radwan SS (2018) Culture-independent analysis of hydrocarbonoclastic bacterial communities in environmental samples during oil-bioremediation. Microbiol Open. https://doi.org/10.1002/mbo3.630

de Castro ME, Rodríguez-Belmonte E, González-Siso MI (2016) Metagenomics of thermophiles with a focus on discovery of novel thermozymes. Front Microbiol 7:1–21. https://doi.org/10.3389/fmicb.2016.01521

De Pascale D, De Santi C, Fu J, Landfald B (2012) The microbial diversity of polar environments is a fertile ground for bioprospecting. Mar Genomics 8:15–22. https://doi.org/10.1016/j.margen.2012.04.004

de Santi C, Altermark B, Pierechod MM, Ambrosino L, de Pascale D, Willassen NP (2016) Characterization of a cold-active and salt tolerant esterase identified by functional screening of Arctic metagenomic libraries. BMC Biochem 17:1–13. https://doi.org/10.1186/s12858-016-0057-x

Delgado-García M, De La Garza-Rodríguez I, Cruz-Hernández MA, Balagurusamy N, Aguilar C, Rodríguez-Herrera R (2013) Characterization and selection of halophilic microorganisms isolated from mexican soils. ARPN J Agric Biol Sci 8:457–458

Dheeman DS, Frias JM, Henehan GTM (2010) Influence of cultivation conditions on the production of a thermostable extracellular lipase from *Amycolatopsis mediterranei* DSM 43304. J Ind Microbiol Biot 37:1–17. https://doi.org/10.1007/s10295-009-0643-7

Ding X, Tang XL, Zheng RC, Zheng YG (2018) Efficient biocatalytic synthesis of chiral intermediate of Pregabalin using immobilized *Talaromyces thermophilus* lipase. Biomed Res Int 2018:1–8. https://doi.org/10.1155/2018/6192059

Domínguez de María P, Carboni-Oerlemans C, Tuin B, Bargeman G, van der Meer A, van Gemert R (2005) Biotechnological applications of Candida Antarctica lipase: state-of-the-art. J Mol Catal B Enzym 37:36–46. https://doi.org/10.1016/j.molcatb.2005.09.001

Dunaj SJ, Vallino JJ, Hines ME, Gay M, Kobyljanec C, Rooney-Varga JN (2012) Relationships between soil organic matter, nutrients, bacterial community structure, and the performance of microbial fuel cells. Environ Sci Technol 46:1914–1922. https://doi.org/10.1021/es2032532

Elleuche S, Schröder C, Sahm K, Antranikian G (2014) Extremozymes—biocatalysts with unique properties from extremophilic microorganisms. Curr Opin Biotechnol 29:116–123. https://doi.org/10.1016/j.copbio.2014.04.003

Essoussi I, Ghodhbane-Gtari F, Amairi H, Sghaier H, Jaouani A, Brusetti L, Boadabous A, Gtari M (2010) Esterase as an enzymatic signature of *Geodermatophilaceae* adaptability to Sahara Desert stones and monuments. J Appl Microbiol 108:1723–1732. https://doi.org/10.1111/j.1365-2672.2009.04580.x

Ferrer M, Golyshina O, Beloqui A, Golyshin PN (2007) Mining enzymes from extreme environments. Curr Opin Microbiol 10:207–214. https://doi.org/10.1016/j.mib.2007.05.004

Flechtner VR, Johansen JR, Clark WH (1998) Algal composition of microbiotic crusts from the central desert of Baja California, Mexico. Great Basin Nat 58:295–311

Foster JA, Bunge J, Gilbert JA, Moore JH (2012) Measuring the microbiome: perspectives on advances in DNA-based techniques for exploring microbial life. Brief Bioinform 13(4):420–429. https://doi.org/10.1093/bib/bbr080

Fujinami S, Fujisawa M (2010) Industrial applications of alkaliphiles and their enzymes--past, present and future. Environ Technol 31:845–856. https://doi.org/10.1080/09593331003762807

Furhan J, Sharma S (2014) Microbial alkaline proteases: findings and applications. Int J Inv Pharm Sci 2:823–834

Fusek M, Lin XL, Tang J (1990) Enzymatic properties of thermopsin. J Biol Chem 265:1496–1501

García-Echauri SA, Gidekel M, Gutiérrez-Moraga A, Santos L, De León-Rodríguez A (2011) Isolation and phylogenetic classification of culturable psychrophilic prokaryotes from the Collins glacier in the Antarctica. Folia Microbiol 56:209–214. https://doi.org/10.1007/s12223-011-0038-9

García-Moyano A, González-Toril E, Aguilera Á, Amils R, Aguilera A (2012) Comparative microbial ecology study of the sediments and the water column of Río Tinto, an extreme acidic environment. FEMS Microbiol Ecol 81:303–314. https://doi.org/10.1111/j.1574-6941.2012.01346.x

Gerday C, Aittaleb M, Bentahir M, Chessa JP, Claverie P, Collins T, D'Amico S, Dumont J, Garsoux G, Georlette D, Hoyoux A, Lonhienne T, Meuwis MA, Feller G (2000) Cold-adapted enzymes: from fundamentals to biotechnology. Trend Biotechnol 18:103–107

Gogliettino M, Riccio A, Cocca E, Rossi M, Palmieri G, Balestrieri M (2014) A new pepstatin-insensitive thermopsin-like protease overproduced in peptide-rich cultures of *Sulfolobus solfataricus*. Int J Mol Sci 15:3204–3219. https://doi.org/10.3390/ijms15023204

Gołębiewski M, Deja-Sikora E, Cichosz M, Tretyn A, Wróbel B (2014) 16S rDNA pyrosequencing analysis of bacterial community in heavy metals polluted soils. Microb Ecol 67:635–647. https://doi.org/10.1007/s00248-013-0344-7

Gomes J, Steiner W (2004) The biocatalytic potential of extremophiles and extremozymes. Food Technol Biotech 42:223–225

Gong BL, Mao RQ, Xiao Y, Jia ML, Zhong XL, Liu Y, Xu PL, Li G (2017) Improvement of enzyme activity and soluble expression of an alkaline protease isolated from oil-polluted mud flat metagenome by random mutagenesis. Enzym Microb Technol 106:97–105. https://doi.org/10.1016/j.enzmictec.2017.06.015

Gorlach-Lira K, Coutinho HDM (2007) Population dynamics and extracellular enzymes activity of mesophilic and thermophilic bacteria isolated from semi-arid soil of Northeastern Brazil. Braz J Microbiol 38:135–141. https://doi.org/10.1590/S1517-83822007000100028

Grant WD (2004) Life at low water activity. Philos Trans R Soc Lond Ser B Biol Sci 359:1249–1267. https://doi.org/10.1098/rstb.2004.1502

Gunde-Cimerman N, Ramos J, Plemenitaš A (2009) Halotolerant and halophilic fungi. Mycol Res 113:1231–1241. https://doi.org/10.1016/j.mycres.2009.09.002

Gupta R, Gupta N, Rathi P (2004) Bacterial lipases: an overview of production, purification and biochemical properties. Appl Microbiol Biotechnol 64:763–781. https://doi.org/10.1007/s00253-004-1568-8

Haki GD, Rakshit SK (2003) Developments in industrially important thermostable enzymes: a review. Bioresour Technol 89:17–34. https://doi.org/10.1016/S0960-8524(03)00033-6

Harrison JP, Gheeraert N, Tsigelnitskiy D, Cockell CS (2013) The limits for life under multiple extremes. Trends Microbio 21:204–212. https://doi.org/10.1016/j.tim.2013.01.006

Hasan F, Shah AA, Hameed A (2006) Industrial applications of microbial lipases. Enzyme and Microb Technol 39:235–251. https://doi.org/10.1016/j.enzmictec.2005.10.016

Hedlund BP, Dodsworth JA, Murugapiran SK, Rinke C, Woyke T (2014) Impact of single-cell genomics and metagenomics on the emerging view of extremophile "microbial dark matter". Extremophiles 18:865–875. https://doi.org/10.1007/s00792-014-0664-7

Hess M, Katzer M, Antranikian G (2008) Extremely thermostable esterases from the thermoacidophilic euryarchaeon *Picrophilus torridus*. Extremophiles 12:351–364. https://doi.org/10.1007/s00792-008-0139-9

Hirsch P, Ludwig W, Hethke C, Sittig M, Hoffmann B, Gallikowski CA (1998) *Hymenobacter roseosalivarius gen. nov., sp. nov*. from continental Antarctic soils and sandstone: bacteria of the *Cytophaga/Flavobacterium/Bacteroides* line of phylogenetic descent. Syst Appl Microbiol 21(3):374–383

Horikoshi K (1971a) Production of alkaline enzymes by alkalophilic microorganisms. II. Alkaline amylase produced by Bacillus No. A-40–2. Agric Biol Chem 35:1783–1791. https://doi.org/10.1271/bbb1961.35.1783

Horikoshi K (1971b) Production of alkaline enzymes by alkalophilic microorganisms. I. Alkaline protease produced by Bacillus no. 221. Agric Biol Chem 35:1407–1414. https://doi.org/10.1080/00021369.1971.10860094

Horikoshi K (1999) Alkaliphiles: some applications of their products for biotechnology. Microbiol Mol Biol Rev 63:735–750. doi:1092-2172/99/$04.0010

Huang Y, Krauss G, Cottaz H, Driguez H, Lipps G (2005) A highly acid-stable and thermostable endo-β-glucanase from the thermoacidophilic archaeon *Sulfolobus solfataricus*. Biochem J 385:581–588. https://doi.org/10.1042/BJ20041388

Ito S (2011) Alkaline Enzymes in Current Detergency. In: Horikoshi K, Antranikian G, Bull AT, Robb FT, Stetter KO (eds) Extremophiles handbook. Springer, Berlín, pp 230–251

Ito S, Shikata S, Ozaki K, Kawai S, Okamoto KI, Takei A, Ohta Y, Satoh T (1989) Alkaline cellulase for laundry detergents: production by *Bacillus* sp. KSM-635 and enzymatic properties. Agric Biol Chem 53:1275–1281. https://doi.org/10.1080/00021369.1989.10869489

Ito S, Kobayashi T, Ara K, Ozaki K, Kawai S, Hatada Y (1998) Alkaline detergent enzymes from alkaliphiles: enzymatic properties, genetics, and structures. Extremophiles 2(3):185–190

Jardine JL, Stoychev S, Mavumengwana V, Ubomba-Jaswa E (2018) Screening of potential bioremediation enzymes from hot spring bacteria using conventional plate assays and liquid chromatography – tandem mass spectrometry (Lc-Ms/Ms). J Environ Manag 223:787–796. https://doi.org/10.1016/j.jenvman.2018.06.089

Javaux EJ (2006) Extreme life on Earth—past, present and possibly beyond. Res Microbiol 157:37–48. https://doi.org/10.1016/j.resmic.2005.07.008

Jiménez DJ, Montaña JS, Álvarez D, Álvarez D, Baena S (2012) A novel cold active esterase derived from Colombian high Andean forest soil metagenome. World J Microbiol Biotechnol 28:361–370. https://doi.org/10.1007/s11274-011-0828-x

Johnson DB, Quatrini R (2011) Acidophile microbiology in space and time In: Quatrini R, Johnson DB (eds) Acidophiles: life in extremely acidic environments. Caister Academic Press, Norfolk, pp 3–16

Joseph B, Pramod W, Ramteke PW, Thomas G, Shrivastava N (2007) Standard review cold-active microbial lipases: a versatile tool for industrial applications. Biotechnol Mol Biol Rev 2:39–48

Joshi S, Satyanarayana T (2013) Biotechnology of cold-active proteases. Biology (Basel) 2:755–783. https://doi.org/10.3390/biology2020755

Kapoor M, Gupta MN (2012) Lipase promiscuity and its biochemical applications. Process Biochem 47:555–569. https://doi.org/10.1016/j.procbio.2012.01.011

Kaur R, Rajesh C, Sharma R, Boparai JK, Sharma PK (2018) Metagenomic investigation of bacterial diversity of hot spring soil from Manikaran, Himachal Pradesh. India Ecol Genet Genom 6:16–21. https://doi.org/10.1016/j.egg.2017.11.003

Kim SJ, Shin SC, Hong SG, Lee YM, Choi IG, Park H (2012) Genome sequence of a novel member of the genus *Psychrobacter* isolated from Antarctic soil. J Bacteriol 194:2403–2403. https://doi.org/10.1128/JB.00234-12

Kim J, Jang SH, Lee C (2013) An organic solvent-tolerant alkaline lipase from cold-adapted *Pseudomonas mandelii*: cloning, expression, and characterization. Biosci Biotechnol Biochem 77:320–323. https://doi.org/10.1271/bbb.120733

Kobayashi T, Hakamada Y, Adachi S, Hitomi J, Yoshimatsu T, Koike K, Kawai IS (1995) Purification and properties of an alkaline protease from alkalophilic *Bacillus* sp. KSM-K16. Appl Microbiol Biotechnol 43:473–481

Kocabayak S, Ozel H (2007) An extracellular-Pepstatin insensitive acid protease produced by *Thermoplasma volcanium*. Bioresour Technol 98:112–117. https://doi.org/10.1016/j.biortech.2005.11.016

Kojima M, Kanai M, Tominaga M, Kitazume S, Inoue A, Horikoshi K (2006) Isolation *and characterization of a feather degrading enzyme from Bacillus* pseudofirmus FA30-01. Extremophiles 10:229–235. https://doi.org/10.1007/s00792-005-0491-y

Komiyama M, Terao K (2008) Applied technology of cyclodextrin. CMC Publishing, Tokyo

Krüger A, Schäfers C, Schröder C, Antranikian G (2018) Towards a sustainable biobased industry – highlighting the impact of extremophiles. New Biotechnol 40:144–153. https://doi.org/10.1016/j.nbt.2017.05.002

Kuddus M, Ramteke PW (2009) Cold-active extracellular alkaline protease from an alkaliphilic *Stenotrophomonas maltophilia*: production of enzyme and its industrial applications. Can J Microbiol 55:1294–1301. https://doi.org/10.1139/w09-089

Kuddus M, Ramteke PW (2012) Recent developments in production and biotechnological applications of cold-active microbial proteases. Crit Rev Microbiol 38:330–338. https://doi.org/10.3109/1040841X.2012.678477

Kumar L, Awasthi G, Singh B (2011) Extremophiles: a novel source of industrially important enzymes. Biotechnology 10(2):121–135. https://doi.org/10.3923/biotech.2011.121.135

Laderman KA, Davis BR, Krutzsch HC, Lewis MS, Griko YV, Privalov PL, Anfinsen CB (1993) The purification and characterization of an extremely thermostable alpha-amylase from the hyperthermophilic archaebacterium *Pyrococcus furiosus*. J Biol Chem 268:24394–24301

Larios A, García HS, Oliart RM, Valerio-Alfaro G (2004) Synthesis of flavor and fragrance esters using *Candida antarctica* lipase. Appl Microbiol Biotechnol 65:373–376. https://doi.org/10.1007/s00253-004-1602-x

Lasón E, Ogonowski J (2010) Lipase - characterization, applications and methods of immobilization. CHEMIK 64:97–102

Läuchli A, Grattan S (2012) Soil pH Extremes. In: Shabala S (ed) Plant stress physiology chapter: soil pH extremes. CAB International, Boston, pp 194–209

Lester ED, Satomi M, Ponce A (2007) Microflora of extreme arid Atacama Desert soils. Soil Biol Biochem 39:704–708. https://doi.org/10.1016/j.soilbio.2006.09.020

Li WJ, Park DJ, Tang SK, Wang D, Lee JC, Xu LH, Kim CJ, Jiang CL (2004) *Nocardiopsis salina sp. nov.*, a novel halophilic actinomycete isolated from saline soil in China. Int J Syst Evol Microbiol 54:1805–1809. https://doi.org/10.1099/ijs.0.63127-0

Li XT, Jiang ZQ, Li LT, Yang QS, Feng WY, Fan JY, Kusakabe I (2005) Characterization of a cellulase-free, neutral xylanase from *Thermomyces lanuginosus* CBS 288.54 and its biobleaching effect on wheat straw pulp. Bioresour Technol 96:1370–1379. https://doi.org/10.1016/j.biortech.2004.11.006

Locey KJ, Lennon JT (2016) Scaling laws predict global microbial diversity. Proc Natl Acad Sci U S A 113:5970–5975. https://doi.org/10.1073/pnas.1521291113

Lu H, Zhang H, Shi P, Luo H, Wang Y, Yang P, Yao B (2013) A family 5 β-mannanase from the thermophilic fungus *Thielavia arenaria* XZ7 with typical thermophilic enzyme features. Appl Microbiol Biotechnol 97:8121–8128. https://doi.org/10.1007/s00253-012-4656-1

Ma Y, Galinski EA, Grant WD, Oren A, Ventosa A (2010) Halophiles 2010: life in saline environments. Appl Environ Microbiol 76:6971–6981. https://doi.org/10.1128/AEM.01868-10

Madigan MT, Marrs BL (1997) Extremophiles. Sci Am 276:82–87

Madigan MT, Martinko JM, Bender KS, Buckley DH, Stahl DA (2015) Brock biology of microorganisms. Pearson, Boston

Mahajan GB, Balachandran L (2017) Sources of antibiotics: hot springs. Biochem Pharmacol 134:35–41. https://doi.org/10.1016/j.bcp.2016.11.021

Mahbub KR, Krishnan K, Megharaj M, Naidu R (2016) Bioremediation potential of a highly mercury resistant bacterial strain *Sphingobium* SA2 isolated from contaminated soil. Chemosphere 144:330–337. https://doi.org/10.1016/j.chemosphere.2015.08.061

Mander P, Yoo HY, Kim SW, Choi YH, Cho SS, Yoo JC (2013) Transesterification of waste cooking oil by an organic solvent-tolerant alkaline lipase from *Streptomyces sp.* CS273. Appl Biochem Biotechnol 172:1377–1389. https://doi.org/10.1007/s12010-013-0610-7

Margesin R, Zhang DC, Frasson D, Brouchkov A (2016) *Glaciimonas frigoris sp. nov.*, a psychrophilic bacterium isolated from ancient Siberian permafrost sediment, and emended description of the genus *Glaciimonas*. Int J Syst Evol Microbiol 66:744–748. https://doi.org/10.1099/ijsem.0.000783

Martínez-Espinosa R, Richardson D, Bonete M (2015) Characterization of chlorate reduction in the haloarchaeon *Haloferax mediterranei*. Biochim Biophys Acta 1850:587–594. https://doi.org/10.1016/j.bbagen.2014.12.011

Matzke J, Schwermann B, Bakker EP (1997) Acidostable and acidophilic proteins: the example of the α-amylase from *Alicyclobacillus acidocaldarius*. Comp Biochem Physiol A Physiol 118:475–479

Mesbah NM, Wiegel J (2010) Alkaliphilic bacteria with impact on industrial applications, concepts of early life forms, and bioenergetics of ATP synthesis. Environ Technol 31:845–856. https://doi.org/10.1080/09593331003762807

Mirete S, Morgante V, González-Pastor JE (2016) Functional metagenomics of extreme environments. Curr Opin Biotechnol 38:143–149. https://doi.org/10.1016/j.copbio.2016.01.017

Mishra SK, Khan MH, Misra S, Dixit VK, Khare P, Srivastava S, Chauhan PS (2017) Characterization of *Pseudomonas spp.* and *Ochrobactrum* sp. isolated from volcanic soil. Antonie Van Leeuwenhoek 110:253–270. https://doi.org/10.1007/s10482-016-0796-0

Montanier CY, Chabot N, Emond S, Guieysse D, Remaud-Siméon M, Peruchb F, André I (2017) Engineering of *Candida antarctica* lipase B for poly(ε-caprolactone) synthesis. Eur Polym J 95:809–819. https://doi.org/10.1016/j.eurpolymj.2017.07.029

Nájera-Fernández C, Zafrilla B, Bonete M, Martínez-Espinosa R (2012) Role of the denitrifying haloarchaea in the treatment of nitrite-brines. Int Microbiol 15:111–119. https://doi.org/10.2436/20.1501.01.164

Nannipieri P, Ascher J, Ceccherini MT, Landi L, Pietramellara G, Renella G (2017) Microbial diversity and soil functions. Eur J Soil Sci 68:12–26. https://doi.org/10.1111/ejss.4_12398

Navarro CA, von Bernath D, Jerez CA (2013) Heavy metal resistance strategies of acidophilic bacteria and their acquisition: importance for biomining and bioremediation. Biol Res 46:363–371. https://doi.org/10.4067/S0716-97602013000400008

Nies D (2000) Heavy metal-resistant bacteria as extremophiles: molecular physiology and biotechnological use of *Ralstonia sp.* CH34. Extremophiles 4:77–82. https://doi.org/10.1007/s007920050140

Nontembiso P, Sekelwa C, Leonard MV, Anthony OI (2011) Assessment of bioflocculant production by *Bacillus sp. Gilbert*, a marine bacterium isolated from the bottom sediment of Algoa Bay. Mar Drugs 9:1232–1242. https://doi.org/10.3390/md9071232

Norashirene MJ, Sarah HU, Khairiyah MHS, Nurdiana S (2013) Biochemical characterization and 16S rDNA sequencing of lipolytic thermophiles from Selayang Hot Spring, Malaysia. IERI Procedia 5:258–264. https://doi.org/10.1016/j.ieri.2013.11.101

Norman JS, King GM, Friesen ML (2017) *Rubrobacter spartanus sp. nov.*, a moderately thermophilic oligotrophic bacterium isolated from volcanic soil. Int J Syst Evol Microbiol 67:3597–3602. https://doi.org/10.1099/ijsem.0.002175

Oliart-Ros RM, Manresa-Presas A, Sanchez-Otero MG (2016) Utilización de microorganismos de ambientes extremos y sus productos en el desarrollo biotecnológico. CienciaUAT 11(1):79–90
Olsen HS, Falholt P (1998) The role of enzymes in modern detergency. J Surfact Deterg 1:555–566. https://doi.org/10.1007/s11743-998-0058-7
Oren A (2008) Microbial life at high salt concentrations: phylogenetic and metabolic diversity. Saline Systems 15:1–13. https://doi.org/10.1186/1746-1448-4-2
Oren A (2011) Alkaline Enzymes in Current Detergency. In: Horikoshi K, Antranikian G, Bull AT, Robb FT, Stetter KO (eds) Extremophiles handbook. Springer, Berlín, pp 230–251
Orhan F, Gulluce M (2015) Isolation and characterization of salt-tolerant bacterial strains in salt-affected soils of Erzurum, Turkey. Geomicrobiol J 32:521–529. https://doi.org/10.1080/01490451.2014.962674
Otohinoyi DA, Omodele I (2015) Prospecting microbial extremophiles as valuable resources of biomolecules for biotechnological applications. Int J Sci Res 4:1042–1059
Panda MK, Sahu MK, Tayung K (2013) Isolation and characterization of a thermophilic *Bacillus sp.* with protease activity isolated from hot spring of Tarabalo, Odisha, India. Iran J Microbiol 5:159–165
Parashar D, Satyanarayana T (2018) An insight into ameliorating production, catalytic efficiency, thermostability and starch saccharification of acid-stable α-amylases from acidophiles. Front Bioeng Biotechnol 28:1–14. https://doi.org/10.3389/fbioe.2018.00125
Park I, Cho J (2011) Productions of an extracellular protease by an Antarctic bacterial isolate (*Bacillus sp.* JSP1) as a potential feed additive. Rev Colomb Cienc Pecu 24:3–10
Park C, Park W (2018) Survival and energy producing strategies of alkane degraders under extreme conditions and their biotechnological potential. Front Microbiol 9:1–15. https://doi.org/10.3389/fmicb.2018.01081
Paulino-Lima IG, Fujishima K, Navarrete JU, Galante D, Rodrigues F, Azúa-Bustos A, Rothschild LJ (2016) Extremely high UV-C radiation resistant microorganisms from desert environments with different manganese concentrations. J Photochem Photobiol B 163:327–336. https://doi.org/10.1016/j.jphotobiol.2016.08.017
Pearce DA, Newsham KK, Thorne MA, Calvo-Bado L, Krsek M, Laskaris P, Hodson A, Wellington EM (2012) Metagenomic analysis of a southern maritime Antarctic soil. Front Microbiol 3:403. https://doi.org/10.3389/fmicb.2012.00403
Pednekar P, Jain R, Mahajan G (2011) Anti-infective potential of hot-spring bacteria. J Glob Infect Dis 3:241–245. https://doi.org/10.4103/0974-777X.83529
Perera I, Subashchandrabose SR, Venkateswarlu K, Naidu R, Megharaj M (2018) Consortia of cyanobacteria/microalgae and bacteria in desert soils: an underexplored microbiota. Appl Microbiol Biotechnol 102:7351–7363. https://doi.org/10.1007/s00253-018-9192-1
Petrovskaya LE, Novototskaya-Vlasova KA, Spirina EV, Durdenko EV, Lomakina GY, Zavialova MG, Nikolaev EN, Rivkina EM (2016) Expression and characterization of a new esterase with GCSAG motif from a permafrost metagenomic library. FEMS Microbiol Ecol 92:fiw046. https://doi.org/10.1093/femsec/fiw046
Phoenix VR, Bennett PC, Engel AS, Tyler SW, Ferris FG (2006) Chilean high-altitude hot-spring sinters: a model system for UV screening mechanisms by early Precambrian cyanobacteria. Geobiology 4:15–28. https://doi.org/10.1111/j.1472-4669.2006.00063.x
Plaza G, Otero-Cabada M, Torres N, Velásquez M, Corbalan E, Rodríguez T (2001) Biorremediación en suelos contaminados con hidrocarburos. Av energ renov medio ambiente 5:163–167
Preiss L, Hicks DB, Suzuki S, Meier T, Krulwich TA (2015) Alkaliphilic Bacteria with impact on industrial applications, concepts of early life forms, and bioenergetics of ATP synthesis. Front Bioeng Biotechnol 3:1–16. https://doi.org/10.3389/fbioe.2015.00075
Putri SP, Kinoshita H, Ihara F, Igarashi Y, Nihira T (2010) Ophiosetin, a new tetramic acid derivative from the mycopathogenic fungus, *Elahocordyceps ophioglossoides*. J Antibiot 63:195–198. https://doi.org/10.1038/ja.2010.8

Quatrini R, Johnson DB (2018) Microbiomes in extremely acidic environments: functionalities and interactions that allow survival and growth of prokaryotes at low pH. Curr Opin Microbiol 43:139–147. https://doi.org/10.1016/j.mib.2018.01.011

Quintana ET, Flores-Badillo R, Maldonado LA (2013) Characterization of the first actinobacterial group isolated from a Mexican extremophile environment. Antonie Van Leeuwenhoek 104:63–70. https://doi.org/10.1007/s10482-013-9926-0

Rabbani M, Bagherinejad MR, Sadeghi HM, Shariat ZS, Etemadifar Z, Moazen F, Rahbari M, Mafakher L, Zaghian S (2014) Isolation and characterization of novel thermophilic lipase-secreting bacteria. Braz J Microbiol 44:1113–1119

Raddadi N, Cherif A, Daffonchio D, Fava F (2013) Halo-alkalitolerant and thermostable cellulases with improved tolerance to ionic liquids and organic solvents from *Paenibacillus tarimensis* isolated from the Chott El Fejej, Sahara Desert, Tunisia. Bioresour Technol 150:121–128. https://doi.org/10.1016/j.biortech.2013.09.089

Raddadi N, Cherif A, Daffonchio D, Neifar M, Fava F (2015) Biotechnological applications of extremophiles, extremozymes and extremolytes. Appl Microbiol Biotechnol 99:7907–7913. https://doi.org/10.1007/s00253-015-6874-9

Rahman RN, Geok LP, Wong CF, Basri M, Salleh AB (2010) Molecular investigation of a gene encoding organic solvent- tolerant alkaline protease from *Pseudomonas aeruginosa* strain K. J Basic Microbiol 50:143–149. https://doi.org/10.1002/jobm.200900133

Rai SK, Roy JK, Mukherjee AK (2010) Characterization of a detergent-stable alkaline protease from a novel thermophilic strain *Paenibacillus tezpurensis sp.* nov. AS-S24-II. Appl Microbiol Biotechnol 85:1437–1450. https://doi.org/10.1007/s00253-009-2145-y

Rampelotto PH (2013) Extremophiles and extreme environments. Life 3:482–485. https://doi.org/10.3390/life3030482

Rawat N, Joshi GK (2018) Bacterial community structure analysis of a hot spring soil by next generation sequencing of ribosomal RNA. Genomics. https://doi.org/10.1016/j.ygeno.2018.06.008

Reddy G, Pradhan S, Manorama R, Shivaji S (2010) *Cryobacterium roopkundense* sp. nov., a psychrophilic bacterium isolated from glacial soil. Int J Syst Evol Microbiol 60:866–870. https://doi.org/10.1099/ijs.0.011775-0

Reed CJ, Lewis H, Trejo E, Winston V, Evilia C (2013) Protein adaptations in archaeal extremophiles. Archaea 2013:1–14. https://doi.org/10.1155/2013/373275

Rodríguez-Valera F, Ruiz-Berraquero F, Ramos-Cormenzana A (1980) Isolation of extremely halophilic bacteria able to grow in defined inorganic media with single carbon sources. J Gen Microbiol 9:535–538

Romano I, Poli A, Lama L, Gambacorta A, Nicolaus B (2005) *Geobacillus thermoleovorans subsp. stromboliensis subsp. nov.*, isolated from the geothermal volcanic environment. J Gen Appl Microbiol 51:183–189

Romdhane BB, Romdhane ZB, Gargouri A (2011) Belghith H esterification activity and stability of *Talaromyces thermophilus* lipase immobilized onto chitosan. J Mol Catal B Enzym 68:230–239. https://doi.org/10.1016/j.molcatb.2010.11.010

Rosales CM, Sowinski SC (2011) Antarctic bacterial isolates that produce cold-active extracellular proteases at low temperature but are active and stable at high temperature. Polar Res 30:7123–7130. https://doi.org/10.3402/polar.v30i0.7123

Rothschild LJ, Mancinnelli RL (2001) Life in extreme environments. Nature 409:1092–1101. https://doi.org/10.1038/35059215

Röttig A, Hauschild P, Madkour MH, Al-Ansari AM, Almakishah NH, Steinbüchel A (2016) Analysis and optimization of triacylglycerol synthesis in novel oleaginous *Rhodococcus* and *Streptomyces* strains isolated from desert soil. J Biotechnol 225:48–56. https://doi.org/10.1016/j.jbiotec.2016.03.040

Sam S, Kucukasik F, Yenigun O, Nicolaus B, Oner ET, Yukselen MA (2011) Flocculating performances of exopolysaccharides produced by a halophilic bacterial strain cultivated on agro-industrial waste. Bioresour Technol 102:1788–1794

Sánchez San Fulgencio N, Suárez-Estrella F, López MJ, Jurado MM, López-González JA, Moreno J (2018) Biotic aspects involved in the control of damping-off producing agents: the role of the thermotolerant microbiota isolated from composting of plant waste. Biol Control 124:82–91. https://doi.org/10.1016/j.biocontrol.2018.04.015

Saraswat R, Bhushan I, Gupta P, Kumar V, Verma V (2018) Production and purification of an alkaline lipase from *Bacillus sp.* for enantioselective resolution of (±)-Ketoprofen butyl ester. 3Biotech 8:1–12. https://doi.org/10.1007/s13205-018-1506-6

Sarethy IP, Saxena Y, Kapoor A, Sharma M, Sharma SK, Gupta V, Gupta S (2011) Alkaliphilic bacteria: applications in industrial biotechnology. J Ind Microbiol Biotechnol 38:769–790. https://doi.org/10.1007/s10295-011-0968-x

Sarmiento F, Peralta R, Blamey JM (2015) Cold and hot extremozymes: industrial relevance and current trends. Front Bioeng Biotechnol 3:1–15. https://doi.org/10.3389/fbioe.2015.00148

Satyanarayana T, Raghukumar C, Shivaji S (2005) Extremophilic microbes: diversity and perspectives. Curr Sci 89:78–90. https://doi.org/10.1038/35059215}

Schiraldi C, De Rosa M (2002) The production of biocatalysts and biomolecules from extremophiles. Trends Biotechnol 20(12):515–521. https://doi.org/10.1016/S0167-7799(02)02073-5

Seckbach J, Rampelotto PR (2015) Polyextremophiles. In: Bakermans C (ed) Microbial evolution under extreme conditions, Life in extreme environments, vol 2. De Gruyter, Berlin, pp 154–170

Serour E, Antranikian G (2002) Novel thermoactive glucamylases from the thermoacidophilic Archaea *Thermoplasma acidophilum, Picrophilus torridus* and *Picrophilus oshimae*. Antonie Van Leewenhock 81:73–83

Shaieb FM, Elghazawani AH, Issa A (2015) Studies on crude oil-degrading bacteria isolated from Libyan desert. Int J Curr Microbiol Appl Sci 4:920–927

Sharma A, Satyanarayana T (2010) High maltose-forming, Ca2+−independent and acid-stable α-amylase from a novel acidophilic bacterium, *Bacillus acidicola*. Biotechnol Lett 32:1503–1507. https://doi.org/10.1007/s10529-010-0322-9

Sharma A, Kawarabayasi Y, Satyanarayana T (2012) Acidophilic bacteria and archaea: acid stable biocatalysts and their potential applications. Extremophiles 16:1–19. https://doi.org/10.1007/s00792-011-0402-3

Sharma A, Parashar D, Satyanarayana T (2016) Acidophilic microbes: biology and applications. In: Rampelotto PH (ed) Biotechnology of extremophiles, Grand challenges in biology and biotechnology. Springer Cham, Heidelberg, pp 215–241

Shivaji S, Ray MK, Rao NS, Saisree L, Jagannadham MV, Kumar GS, Reddy GSN, Bhargava PM (1992) *Sphingobacterium antarcticus sp. nov.*, a psychrotrophic bacterium from the soils of Schirmacher oasis, Antarctica. Int J Syst Evol Microbiol 42:102–106

Siddiqui KS (2015) Some like it hot, some like it cold: temperature dependent biotechnological applications and improvements in extremophilic enzymes. Biotechnol Adv 33:1912–1922. https://doi.org/10.1016/j.biotechadv.2015.11.001

Siglioccolo A, Paiardini A, Piscitelli M, Pascarella S (2011) Structural adaptation of extreme halophilic proteins through decrease of conserved hydrophobic contact surface. Struct Biol 11:1–12. https://doi.org/10.1186/1472-6807-11-50

Subashchandrabose SR, Ramakrishnan B, Megharaj M, Venkateswarlu K, Naidu R (2013) Mixotrophic cyanobacteria and microalgae as distinctive biological agents for organic pollutant degradation. Environ Internat 51:59–72. https://doi.org/10.1016/j.envint.2012.10.007

Tarafdar JC, Rao AV, Bala K (1988) Production of phosphatates by fungi isolated from desert soils. Folia Microbiol 33:453–457. https://doi.org/10.1007/BF02925770

Torsvik V, Øvreås L (2008) Microbial diversity, life strategies, and adaptation to life in extreme soils. In: Dion P, Nautiyal CS (eds) Microbiology of extreme soils, Soil biology, vol 13. Springer, Berlin, pp 15–43

Torsvik V, Øvreås L, Thingstad TF (2002) Prokaryotic diversity–magnitude, dynamics, and controlling factors. Science 296:1064–1066. https://doi.org/10.1126/science.1071698

Turner P, Mamo G, Karlsson EN (2007) Potential and utilization of thermophiles and thermostable enzymes in biorefining. Microb Cell Factories 6:1–23

Urbieta MS, Donati ER, Chan KG, Shahar S, Sin LL, Goh KM (2015) Thermophiles in the genomic era: biodiversity, science, and applications. Biotechnol Adv 33:633–647. https://doi.org/10.1016/j.biotechadv.2015.04.007

Urich T, Lanzén A, Stokke R, Pedersen RB, Bayer C, Thorseth IH, Schleper C, Steen IH, Ovreas L (2014) Microbial community structure and functioning in marine sediments associated with diffuse hydrothermal venting assessed by integrated meta-omics. Environ Microbiol 16:2699–2710. https://doi.org/10.1111/1462-2920.12283

Uttatree S, Winayanuwattikun P, Charoenpanich J (2010) Isolation and characterization of a novel thermophilic-organic solvent stable lipase from Acinetobacter baylyi. Appl Biochem Biotechnol 162:1362–1376. https://doi.org/10.1007/s12010-010-8928-x

Van den Burg B (2003) Extremophiles as a source for novel enzymes. Curr Opin Microbiol 6:213–218

Vázquez SC, Ruberto L, Mac Cormack WP (2005) Properties of extracellular proteases from three psychrotolerant *Stenotrophomonas maltophilia* isolated from Antarctic soil. Polar Biol 28:319–325. https://doi.org/10.1007/s00300-004-0673-6

Ventosa A, Mellado E, Sanchez-Porro C, Carmen-Marquez M (2008) Halophilic and Halotolerant micro-organisms from soil. In: Dion P, Nautiyal CS (eds) Microbiology of extreme soils. Soil biology, Springer, Berlin, pp 87–115

Vishnivetskaya T, Kathariou S, McGrath J, Gilichinsky D, Tiedje JM (2000) Low-temperature recovery strategies for the isolation of bacteria from ancient permafrost sediments. Extremophiles 4:165–173

Wang SL, Lin YT, Liang TW, Chio SH, Ming LJ, Wu PC (2009) Purification and characterization of extracellular lipases from *Pseudomonas monteilii* TKU009 by the use of soybeans as the substrate. J Ind Microbiol Biot 36:65–73. https://doi.org/10.1007/s10295-008-0473-z

Wang X, Huang N, Shao J, Hu M, Zhao Y, Huo M (2018) Coupling heavy metal resistance and oxygen flexibility for bioremoval of copper ions by newly isolated *Citrobacter freundii* JPG1. J Environ Manag 226:194–200. https://doi.org/10.1016/j.jenvman.2018.08.042

Wang J, Zhang T, Li Y, Li L, Wang Y, Yang B, Wang Y (2019) High-level expression of *Thermomyces dupontii* thermo-alkaline lipase in *Pichia pastoris* under the control of different promoters. 3. Biotech 9:1–8. https://doi.org/10.1007/s13205-018-1531-5

Wiegel J, Kevbrin VV (2004) Alkalithermophiles. Biochem Soc Trans 32:193–198. https://doi.org/10.1042/bst0320193

Wiegel J, Ljungdahl LG, Rawson JR (1979) Isolation from soil and properties of the extreme thermophile *Clostridium thermohydrosulfuricum*. J Bacteriol 139:800–810

Xiong J, Liu Y, Lin X, Zhang H, Zeng J, Hou J, Yang Y, Yao T, Knight R, Chu H (2012) Geographic distance and pH drive bacterial distribution in alkaline lake sediments across Tibetan Plateau. Environ Microbiol 14:2457–2466. https://doi.org/10.1111/j.1462-2920.2012.02799.x

Yadav AN, Verma P, Kumar V, Sachan SG, Saxena AK (2017) Extreme cold environments: a suitable niche for selection of novel Psychrotrophic microbes for biotechnological applications. Adv Biotech Micro 2:1–4. https://doi.org/10.19080/AIBM.2017.02.555584

Yáñez J, Riffo P, Santander P, Mansilla HD, Mondaca MA, Campos V, Amarasiriwardena D (2015) Biodegradation of tributyltin (TBT) by extremophile bacteria from Atacama Desert and speciation of tin by-products. Bull Environ Contam Toxicol 95:126–130. https://doi.org/10.1007/s00128-015-1561-1

Yang C, Wang F, Hao J, Zhang K, Yuan N, Sun M (2010) Identification of a proteolytic bacterium, HW08, and characterization of its extracellular cold-active alkaline metallopro- tease Ps5. Biosci Biotechnol Biochem 74(6):1220–1225. https://doi.org/10.1271/bbb.100011

Zhang W, Zhang G, Liu G, Dong Z, Chen T, Zhang M, Dyson PJ, An L (2012) Bacterial diversity and distribution in the southeast edge of the Tengger Desert and their correlation with soil enzyme activities. J Environ Sci 24:2004–2011

Zhang DC, Brouchkov A, Griva G, Schinner F, Margesin R (2013) Isolation and characterization of Bacteria from Ancient Siberian permafrost sediment. Biology 2:85–106. https://doi.org/10.3390/biology2010085

Zhao D, Yang H, Chen J, Cheng F, Kumar S, Han J, Li M, Zhou J, Xiang H (2017) Development of the first gene expression system for *Salinicoccus* strains with potential application in bioremediation of hypersaline wastewaters. Appl Microbiol Biotechnol 101:7249–7258. https://doi.org/10.1007/s00253-017-8428-9

Zhao J, Guo C, Zhang L, Tian C (2018) Biochemical and functional characterization of a novel thermoacidophilic, heat and halo-ionic liquids tolerant endo-β-1,4-glucanase from saline-alkaline lake soil microbial metagenomic DNA. Int J Biol Macromol 15:1035–1044. https://doi.org/10.1016/j.ijbiomac.2018.06.141

Zhou Z, Jiang F, Wang S, Peng F, Dai J, Li W, Fang C (2012) *Pedobacter arcticussp. nov.*, a facultative psychrophile isolated from Arctic soil, and emended descriptions of the genus *Pedobacter, Pedobacter heparinus, Pedobacter daechungensis, Pedobacter terricola, Pedobacter glucosidilyticus* and *Pedobacter lentus*. Int J Syst Evol Microbiol 62:1963–1969. https://doi.org/10.1099/ijs.0.031104-0

Chapter 4
Effect of Pollution on Physical and Chemical Properties of Soil

Prakriti Singh and Gayatri Dhumal

Abstract Soil acts as a base for sustaining life on Earth. Through various means, humans exploited every possible resource on the planet and emitted out the used wastes that became pollutants. These pollutants altered the normal functioning of ecological services, especially affecting soil health and productivity, and all kinds of pollution are interrelated with soil. The application of waste or polluted water into the soil alters its physical and chemical, thereby affecting the growth of agricultural crops and other living organisms. Even the extensive agricultural practices such as fertilizer and pesticide application deteriorate the soil quality. Dumping of garbage and municipal solid waste openly on the land results in leaching of unwanted elements into the soil, hence intoxicating it to an extent that is irreversible. Moreover, air pollution too, directly or indirectly, affects the soil properties.

Keywords Soil pollution · Pollutants · Sewage · Wastewater · Agricultural practices · Soil health · Soil properties · Nutrients · Industrial waste · Solid waste · Soil quality

4.1 Introduction

The science of soil developed with human evolution. With the rudimentary knowledge of soil utilization and its use for agricultural practices at around 11,000 B.C.E., global civilizations persuaded toward advanced soil utilization techniques by the fourth century C.E. which include irrigation, the use of terraces to control erosion, various ways of improving soil fertility, various complex structures for water harvesting, and ways to maintain productivity of soils.

P. Singh (✉) · G. Dhumal (✉)
Department of Environmental Science, Dr. Y.S. Parmar University of Horticulture and Forestry, Nauni, Himachal Pradesh, India

A. Kumar, S. Sharma (eds.), *Microbes and Enzymes in Soil Health and Bioremediation*, Microorganisms for Sustainability 16,
https://doi.org/10.1007/978-981-13-9117-0_4

A scientific approach to the soil was not developed during the early historical times. During those times, observations helped in most of the soil-based practices. This provided a solid base for the knowledge about soil handling and basic practices carried out on it. In modern society, the knowledge of soils has become an important source for nonagricultural land use evaluations, such as construction, community building planning, environmental works, and other such practices (Brevik 2005).

Soil is an important component of sustaining life. It provides a base to cultivate a thousand types of crops and lay down various industries and settlements. The quality and health of soil definitely play an important role for not only producing food product but also sustaining the quality of every aspect of the environment. Such kind of awareness about soil has stimulated the curiosity in assessing its health and quality (Glanz 1995). Soil quality stands for the capability of the soil to be able to sustain the life of plants and animals along with humans in a sustainable manner. The continuous agricultural activities, as well as conversion of land areas, ultimately reduce organic carbon that finally causes changes in mineralization (Pennock and Van Kessel 1997 and Janzen and Molina-Ayala 1997). The sustainable management of land and even the quality of environment (National Research Council 1993), food security, and availability (Lal 1999) can be estimated in accordance with the health and quality of the soil.

An assessment of the impact of changing soil management should therefore ideally include some measure of soil health or quality, as this is inseparable from issues of sustainability (Doran and Safely 1997). In this way, soil quality has become a tool for assessing the sustainability of soil management systems (Leininger et al. 2006). Soil health in one form can be referred to the quality that is a manifestation of various natural and anthropogenic activities like agricultural implementations or addition of various organic and inorganic substance in order to get or improve better productivity so as to set maximum output in a sustainable manner. While talking about the agricultural fields, the addition of FYM, manures, and fertilizers according to the soil requirements and conditions helps in maintaining a healthy soil state to yield a healthy produce. But most of the times, it is not the case with a forest or community land or unused soils. They are most of the times ignored and used in an unaware process. Many scientists believe that the health of soil depends upon soil nutrients (mostly carbon, nitrogen, phosphorus, potassium, and other macro- and micronutrients), proper cycling or circulation of these nutrients, adequate soil structures and composition, and lastly proper regulation of unwanted pests causing diseases and deformations. All of these factors are needed to be in balance with each other so as to harness maximum output.

Proper and adequate management of soil is mandatory in every type of agricultural systems, but at the same time, evidence of deterioration of cultivated soils because of erosion, loss of organic nutrients, contamination by pollutants, compaction, and other anthropogenic harms is necessary to be pointed out (European Commission 2002).

Soil quality refers to the continued capacity of soil to function as a vital living ecosystem that sustains plants, animals, and humans. The physical properties of soil include soil texture (sand, silt, clay), structure (blocky, prismatic, columnar, platy), bulk density, pore space, color, and permeability. The chemical properties include pH, electrical conductivity, and ion exchange properties. The changes in the land use influence the nutrient supply of an ecosystem; hence, the soil nutrient status is better in abandoned land than in orchards (Ye et al. 2009). Conversion of forest land to artificial surfaces such as urban settlements, agricultural use, and industrialization can have several environmental impacts on soil, water, and biodiversity resources.

Soil pollution or contamination is degradation of land, or soil, because of anthropogenic activities or any other alteration of the natural soil environment. In most of the cases, the soil polluters are fertilizers, pesticides, and insecticides as in the case of agricultural activities; industrial effluents from industrial plants; sewage and solid waste dumping by human settlements; and open waste dumping of plastics, metals, and other nonbiodegradable wastes. The contamination of land and soil is directly correlated with the increasing population, urbanization, industrialization, agricultural field expansions, and the use of nonbiodegradable resource in daily life.

4.2 Soil Pollution

Soil pollution in simple terms refers to the depletion in the productivity of soil because of the presence of pollutants. Soil pollutants such as fertilizers, chemicals, pesticides, radioactive wastes, plastics, and sewage wastes reduce its productivity and create an adverse impact on the physical, chemical, and biological properties of the soil. The industrial wastes comprise of chemicals like iron, lead, mercury, copper, zinc, cadmium, aluminum, cyanides, and acids which end up reaching the soil either directly with water or indirectly through acid rain.

4.3 Types of Soil Pollution

The different types of soil pollution can be distinguished according to the type of soil pollutants such as wastewater, sewage, agricultural and industrial wastes, and airborne pollutants.

I. *Sewage pollution*

- Inefficient municipal sewage waste disposal systems lead to accumulation of pollutants in undesired places.
- Improper management of dumping areas results in pollutant runoff.
- Excessive waste dumping in a single site increases the growth of harmful microorganisms.

II. *Agricultural pollution*

- Use of fertilizers without soil testing makes soil polluted.

- Pesticides have a harmful impact on soil microflora which creates an imbalance in the soil.
- Monoculture reduces soil productivity, and to balance the fertility of soil fertilizers, chemicals are used which further degrades the soil quality.

III. *Solid waste pollution*

- Improper dumping of wastes.
- Electronic goods, broken furniture, junk papers, polythene bags, plastic cans, bottles, wastewater, toxic waste from the hospital, etc. are examples of solid wastes which pollute the soil. Most of this litter is nonbiodegradable. These wastes affect the soil structure by being blocked in it for long periods.

IV. *Air pollution*

- Heavy metals in the atmosphere are mainly from gas and dust produced by energy, transport, metallurgy, and production of construction materials. Except for mercury, heavy metals basically go into the atmosphere in the form of aerosol and deposit to the soil through natural sedimentation and precipitation (Havugimana et al. 2015).

4.4 Effect of Different Types of Pollution on Soil Properties

Pollutants mix in soil and make it toxic, and the chemical changes in the natural form of the soil begin to take place. Due to chemical fertilizers and biochemicals, an imbalance in the entire ecosystem is created. All types of pollution have direct or indirect effects on ground soil properties.

4.4.1 Effect of Water Pollution on Physical and Chemical Properties of Soil

Fresh water is a limiting resource, and with the ever-increasing human population, the demand for water has increased too. In many parts of the world, life is threatened because of a shortage of water. With the increasing need of conserving fresh water, new techniques are being developed in order to recycle the water that has already been spent or polluted.

The major sources of water pollution are household sewage, industrial and commercial chemical plants, and agricultural runoff. Across the globe, these sources of water are to be treated with certain chemicals in accordance with the prescribed values, so as to make them less polluting and more ecofriendly. The wastewater is thoroughly screened before disposing into different water sinks. Generally, this practice is not followed up to the mark, resulting in mixing up of the unwanted polluting material into water resources, especially in the developing countries. The polluted water from these sources is then used for irrigation purposes, construction, and aquaculture and even for groundwater recharge, without any prior treatment, hence disturbing the natural properties of soil in most of the developing countries. The disposal of wastewater is one of the most key problems faced by municipalities. The main problem is not the wastewater, as it can be reused in many ways while it is enriched with nutrients, but its proper management. The key for successful utilization of the wastewater is awareness and proper adherence of rules and regulations for the purpose of disposal and recycling of wastewater.

The wastewater is the liquid waste coming out of residences, institutes, hospitals, commercial buildings, industries, and factories which gets mixed with fresh waters of rivers, streams, lakes, and even groundwater. The wastewater many-a-times get leached out by rainwater or through surface runoffs. Most of the rivers of India and the world are heavily polluted by the municipal and industrial wastes. As the population is increasing, the agricultural, industrial, and urban sectors are also expanding extensively. This not only exerts pressure on resources but at the same time creates an ever-increasing demand for clean and healthy resource. Approximately, 38,354 million liters of sewage is generated per day in major cities of India, but the state-owned sewage treatment plants and the basic industrial waste treatment plants of India have a capacity of treating only 11,786 million liters of sewage per day. Moreover, only 60% of industrial wastewater is treated (Giannakis et al. 2014). But what is important, which makes wastewater dangerous for soil environment, is its composition. Generally, the wastewater from all those sources contains organic and inorganic substances. The organic substances comprise of plant material, feces, paper, ceramics, and salts mixed and incorporated with many toxic chemicals such as cleaners, pesticides, detergents, and millions of bacteria, fungi, viruses, and other microorganisms. Most of the macro-particles get sieved onto the surface only, and the micro ones find their place in the subsurface soil layers. The nutrients leach from the surface layer and spread out in the lower ones, combining and reacting with the other nutrients that are already present there. The reactions may cause formation of hard pans and change in soil structure, hence resulting in reduced porosity and fertility, salinization, or acidification from the chemical reactions. The effects of wastewater on physical and chemical properties of soil are discussed below in detail.

4.4.1.1 Effect of Wastewater Pollution on Soil pH

The irrigation with wastewater alters the soil pH; this fact was presented by many research works (Osaigbovo et al. 2006; El-Hady 2007; Gupta and Mitra 2002). To know the effect of pollutants from irrigation of wastewater on the pH of the soil, long-term studies are considered to be the most influential. One such study for a term of 50 years brought out the fact that the soils in the state of Tamil Nadu, when continuously irrigated with wastewater from textile industries, resulted in 0.4 unit increase in pH as compared to the one irrigated with groundwater (Saravanamoorthy and Ranjitha-Kumari 2007). A study on the soils of Kolkata irrigated by sewage water showed similar results of an increase in pH by 0.5 units in 50–60 years (Gupta and Mitra 2002). Similarly, the pH of the soils of Nigeria showed an increase of 0.5 units when irrigated with wastewater from fertilizer-manufacturing plant (Ana and Sridhar 2002).

All these studies conclude that wastewater from sewage, textile industries, fertilizer plants, pharmaceutical, and other kinds of industries that throw out a considerable amount of $Ca^{2+}$and Mg^{2+} tend to alter and, thus, increase the pH of the soils to which they are irrigated.

On the other hand, oxidation of organic compounds and nitrification of ammonia cause a decrease in pH of soils when irrigated with wastewater from distilleries units and sewage. The production of organic acids because of anaerobic decomposition of organic matter is one of the major reasons for the decrease of pH because of wastewater irrigation (Wang et al. 1967). Similarly, another research observed a decrease in soil pH with the distillery, steel industry, and other industrial effluents (Nawaz et al. 2006).

4.4.1.2 Effect of Wastewater Pollution on Soil Electrical Conductivity

In a study, it was observed that the soils irrigated with polluted water had an electrical conductivity of 0.2 dSm^{-1}at a depth of about 61 cm, which was more than the soil irrigated with normal water (Hayes et al. 1990). The soluble salts from the application of municipal wastewater got accumulated in the lower surface layers up to 60 cm deep because of the leaching effect (Rusan et al. 2007). This resulted in a considerable high EC in a long run. In some other study conducted in a potato field in Jalandhar (Punjab, India), an increase in soil EC from 0.28 to 0.64 $dSm^{-1}$was observed. The soil was irrigated with municipal and leather manufacturing industrial wastewater, and because of this, there was a considerable decrease in soil EC (Brar et al. 2002). In contrast to his findings, many other researches (Aghabarati et al. 2008) suggested a significant increase in surface soil EC when polluted water is applied on soils. Similar results were observed in arid and calcareous soils (Hayes et al. 1990).

So, it can be concluded from the above research results that soils when irrigated with any kind of polluted water trap the nutrients in its surface and subsurface layers, hence increasing its electrical conductivity to a significant level which can alter the growth and yield of agricultural produce.

4.4.1.3 Effect of Wastewater Pollution on Soil Organic Carbon

Soil organic carbon represents the quality and quantity of nutrients utilized and stored by a plant. In soil, the source of organic carbon is numerous that can be easily discerned from the fact that every living organism and everything derived from it contains carbon. So, every kind of pollution be it water, air, or agricultural contributes to adding up more carbon in the soil. And the most important contribution of them all can be ascertained to the disposal of wastewater which is enriched with many kinds of organic materials. The result of using wastewater for irrigation purposes is most of the times the increase in soil OC to a considerable amount as compared to the soils that are irrigated with normal water (Osaigbovo et al. 2006; Saravanamoorthy and Ranjitha-Kumari 2007). Even the amount of increased SOC depends upon the kind of wastewater which is used for irrigation. For instance, because of the use of the sewage water for irrigation for around 20 years in subtropical soils of India, SOC had increased from 38% to 79% (Rattan et al. 2005). At another instance, sewage water when used for irrigation of vertisols in Mexico City for around 80 years resulted in an increase in SOC by 2.5 times as compared to soils that were irrigated with groundwater.

The soil organic carbon content is found to be significantly lower in textile mill effluent-affected soil as compared to normal rainfed soils. Also, soils are found to be enriched with organic carbon due to the addition of industrial effluents (Jadhav and Savant 1975). An increase in pH, EC, organic matter, and total nitrogen content of the soil is observed with increase in the concentration of paper mill wastewater (Mishra and Sahoo 1989). Similarly, increase in SOM from 1 to 1.08% in Nigeria was reported when the soil was irrigated with pharmaceutical waste for 8 weeks (Osaigbovo et al. 2006); 0.5 to 4.07% increase because of irrigation with distillery wastes in Nepal was reported (Ale et al. 2008); and 0.19 to 0.37% increase in Kolkata (Gupta and Mitra 2002) and 1.24 to 1.7% increase in Kurukshetra, Haryana, India (Yadav et al. 2002), by irrigating with sewage wastewater for 60 and 25 years, respectively.

So, all these instances suggest that wastewater, when used for irrigating fields for longer durations, result in considerable accumulation of organic carbon in soils, which is helpful for soil productivity in some way. But at the same time, the accumulation of organic matter from sewage wastewater irrigation creates an anaerobic condition that reduces the decomposition process, especially of organic carbon because of which organic carbon will get deposited more and more in soil layers (Dheri et al. 2007).

4.4.1.4 Effect of Wastewater Pollution on Calcium Carbonate

Calcium carbonate is one of the important constituents of soil that helps in defining the soil pH. More $CaCO_3$ will be the soil pH and vice versa. There are various researches to back the fact that the content of $CaCO_3$ increases in agricultural soil when it is irrigated with wastewater. The use of wastewater for irrigation in Egypt uplifted the concentration of $CaCO_3$ (El-Hady 2007). However, from another study area, completely opposite results were observed. The $CaCO_3$ content decreased by 1.42% because of the constant use of wastewater for irrigation. This is noted to be a result of anaerobic activities in soil-producing acids that decreases soil pH (Wang et al. 1967). The organic acids produced during such processes solubilize $CaCO_3$ which then leach deep inside the earth crust (McClean et al. 2003).

4.4.1.5 Effect of Wastewater Pollution on Soil Macronutrients

The continuous use of wastewater, whether sewage, industrial, or commercial, is proved to add up nutrients, such as OC and NPK, in the soil (Rusan et al. 2007; Osaigbovo et al. 2006; El-Hady 2007; Yadav et al. 2002). A considerable increase in NPK content of surface (0.20 cm) soil was observed after 10 years of use of sewage wastewater for irrigation (as shown in Table 4.1) (Rusan et al. 2007). Also from the same table, it can be ascertained that any kind of wastewater, from municipal, textile, pharmaceutical, and fertilizer industries to distillation plants, tends to increase the concentration of macronutrients, i.e., NPK in surface soils in the different parts of the world.

The content of total nitrogen, total phosphorus, and total potassium was observed to be increased in soils from 1141, 1120, and 146,000 mg kg^{-1} to 2200, 2000, and 25,940 mg kg^{-1}, when irrigated with sewage wastewater (Gupta and Mitra 2002). Similar results were documented in the soils of Egypt, in which the total nitrogen content increased by 4.40 times, available phosphorus by 5.3 times, and $NH_4 - N$ content by 2.55 times (Hayes et al. 1990).

4.4.1.6 Effect of Wastewater Pollution on Soil Micronutrients and Heavy Metals

It is evident from various researches that wastewater from any source has a considerable amount of macro- and micronutrients and heavy metals (Rusan et al. 2007; Osaigbovo et al. 2006). Heavy metals such as Fe, Cu, Cd, Mn, Zn, and Pb have registered their presence multiple times in the soils irrigated with wastewater from sewage, industries, fertilizer plants, and municipalities. As shown in Table 4.2, it is evident under various agricultural systems around the globe utilizing wastewater for irrigation that the concentration of micronutrients has elevated because of the addition of heavy metals. In one study, a comparison was made of two cities in the same region with their agricultural lands irrigated with different kinds of water (Khurana et al. 2003). One in Ludhiana (Punjab, India) was irrigated with sewage wastewater and the other in Sangrur (Punjab, India) with normal water. The results proved that sewage wastewater added extra nutrients such as Pb, Ni, Cd, Zn, and Fe in the magnifying concentration of heavy metals in the surface layers of agricultural soils. Similar results were confirmed in different regions for Pb and Cd from the surface soil layer (Aghabarati et al. 2008). There exist many pieces of evidence compiled from across the globe over a period of time to confirm that wastewater carries a considerable amount of heavy metals which when used into soil result in their accumulation which with time keep on increasing because of continuous addition and availability of fewer sinks. An increased concentration of Cd, Pb, Zn, Cr, and Cu in the volcanic soils of Turkey was found with enrichment factor of Cd (1.8), Cr (1.7), Cu (2.3), Zn (2.0), and Pb (1.9), because of continuous irrigation with sewage water (Liu et al. 2005) .

Table 4.1 Effect of wastewater irrigation on soil fertility parameters

Location	The time period of study	Wastewater source	Parameter	Heavy metal content		Δ (Times)
				Wastewater	Groundwater	
Tamil Nadu, India	50 days	Textile wastewater	pH	7.92	7.85	1.00
			N(kg ha^{-1})	118	117	1.00
			P(kg ha^{-1})	13.45	11.10	1.21
			K(kg ha^{-1})	56	55.0	1.01
			OM (%)	1.29	0.29	4.45
Benin City, Nigeria	8 weeks	Pharmaceutical wastewater	pH	7.15	5.98	1.20
			P(kg ha^{-1})	6.00	3.07	1.95
			K(cmol kg^{-1})	0.32	0.41	0.78
			OC (%)	1.28	1.00	1.28
			Ca(cmol kg^{-1})	10.06	0.07	143.7
			Na (cmol kg^{-1})	1.10	0.05	22.0
			CEC (cmol kg^{-1})	31.58	0.61	51.8
Onne, Rivers state, Nigeria	–	Fertilizer plant	pH	7.76	7.23	1.07
			N (g 100 g^{-1})	0.54	0.09	6.0
			P(g 100 g^{-1})	0.29	0.21	1.68
			K (g 100 g^{-1})	32.3	2.55	12.7
			Ca(g 100 g^{-1})	0.19	0.86	0.22
			Na(g 100 g^{-1})	1.50	1.20	1.25
			Mg(g 100 g^{-1})	0.98	0.68	1.44

(continued)

Table 4.1 (continued)

Location	The time period of study	Wastewater source	Parameter	Heavy metal content		Δ (Times)
				Wastewater	Groundwater	
El-Khashab, Egypt	–	Mixed domestic and industrial wastewater	pH	8.4	7.9	1.06
			TN(mg kg^{-1})	2200	500	4.40
			$NH_4 - N$(mg kg^{-1})	51	20	2.55
			$NO_3 - N$(mg kg^{-1})	155	67.5	2.30
			P(mg kg^{-1})	11.7	2.2	5.30
			K(mg kg^{-1})	10.2	13.2	0.77
Khajura, Nepal	–	Distillery effluent	pH	6.4	7.2	0.88
			Om(%)	4.07	0.52	7.83
			TN(%)	0.17	0.10	1.70
			$P_2O_5 - P$ (kg ha^{-1})	183	84	2.18
			$K_2O - K$(kg ha^{-1})	590	209	2.82
Calcutta, India	50–60 years	Sewage effluents	pH	7.7	7.2	1.07
			OC (%)	0.37	0.19	1.95
			TN (%)	0.10	0.06	1.67
			TP (%)	0.10	0.05	2.00
			TK (%)	0.13	0.07	1.86
Kurukshetra, India	25 years	Sewage effluents	pH	8.1	8.3	0.98
			OC (%)	1.73	1.24	1.40
			TN (%)	0.15	0.08	1.89
			TP (%)	0.88	0.56	1.57
			TK (%)	0.24	0.18	1.33

Table 4.2 Effect of different types of wastewater irrigation on heavy metal concentration (mg kg^{-1}) of soil

Location	The time period of study	Wastewater source	Heavy metal	Heavy metal content		Δ (times)
				Wastewater	Groundwater	
Esfahan, Iran	–	Steel industry wastewater	Fe	22.8	4.02	5.67
			Cu	2.67	1.07	2.50
			Mn	15.4	9.90	1.56
			Zn	5.10	1.83	2.79
			Cd	0.42	0.12	3.50
			Pb	4.48	1.76	2.55
El Sadat City, Egypt	5	Industrial wastewater	Fe	52.9	30.8	1.72
			Mn	26.3	10.6	2.48
			Zn	11.5	4.23	2.72
			Cu	6.96	1.55	4.49
			Co	1.91	0.45	4.24
			Ni	1.88	0.99	1.90
			Pb	6.82	2.89	2.36
Tehran, Iran	7	Domestic and industrial waste	Zn	187.3	94.7	1.98
			Pb	78.4	50.0	1.57
			Cr	82.8	34.7	2.39
			Ni	46.0	27.4	1.68
			Cu	4.2	0.99	4.24
Tehran, Iran	7	Municipal effluents	Pb	78.4	50.0	1.57
			Cr	82.8	34.7	2.39
			Ni	46.0	27.4	1.68
			Zn	30.8	3.6	85.6
			Cu	37.1	2.4	15.46

Even the concentration of Zn, Cu, Fe, Ni, and Pb increased to whooping 208%, 170%, 170%, 63%, and 29%, respectively, when sewage wastewater was used for irrigating soils for 20 years (Rattan et al. 2005). Also, the concentration of Mn dramatically decreased by 31%. Leafy vegetables grown in heavy metal-contaminated soils have been noticed to accumulate higher amounts of metals than those grown in uncontaminated (Al Jassir et al. 2005). Vegetables can absorb toxic metals from the soil as well as from deposition on parts of vegetables exposed to the air from polluted environments (Haiyan and Stuanes 2003). The toxic metals in soils were reported to inhibit root and shoot growth, nutrient uptake, and homeostasis and were noticed to be accumulated frequently by agriculturally important crops. So, the contamination of leafy vegetables (*Brassica* species) with Cu, Zn, Cd, Ni, Pb, and Cr and subsequent human exposure risks were determined at two sites in the city of Hazare, where wastewater is used for irrigating vegetables (Mapanda et al. 2005). And the same theory applies for the soils of Jordan, which showed no considerable difference in concentration of Pb and Cd when irrigated with municipal wastewater.

From almost every type of study, it is evident that sewage wastewater is capable of adding a considerable amount of heavy metals in the soil as compared to other sources, as visible in Tables 4.2 and 4.3. At this instance, time plays an important role as it is directly proportional to the amount of deposition. The more duration of application of wastewater, the more will be the accumulation of nutrients especially heavy metals. Most considerable study results are from the soils of Kolkata, India

Table 4.3 Total micronutrient and heavy metal concentration ($mg\ kg^{-1}$)in surface soils with wastewater irrigation in Indian cities

Location	Years of study	Wastewater type	Heavy metal	Heavy metal content		Δ (times)
				Wastewater	Groundwater	
Ludhiana, India	–	Leather and sewage effluents	Fe	39.7	13.8	2.88
			Mn	18.9	14.3	1.32
			Zn	14.7	2.9	5.07
			Al	6.5	3.63	1.79
			As	2.1	1.87	1.12
			Cr	1.7	0.81	2.10
			Ni	1.27	0.56	2.27
			Zn	187.3	94.7	1.98
Kurukshetra, India	25	Sewage effluents	Fe	22.7	17.8	1.28
			Mn	7.2	5.8	1.24
			Pb	1.65	0.99	1.67
			Ni	3.3	2.4	1.38
			Cr	6.8	2.2	3.09
			Cu	17.0	15.5	1.10
			Pb	16.5	11.0	1.50
			Cd	40.5	26.5	1.53
Kolkata, India	50–60	Sewage effluents	Fe	22,120	9090	2.43
			Zn	1210	26	46.5
			Cu	198	52	3.81
			Mn	382	446	0.86
			Cd	3.72	0.04	93.0
			Pb	385	24.2	15.9
			Co	46.6	12.0	3.88
			Ni	61.0	25	2.44
			Cr	164	24.8	6.61
Faridabad, India	20	Sewage water	Cd	0.12–0.20	0.15	0.80–1.33
			Fe	2207	966	2.28
			Zn	261	53	4.92
			Cu	60	23	2.61
			Mn	241	188	1.28
			Cd	4.2	1.1	3.82
			Ni	73	19	3.84
			Cr	79	23	3.43

(Gupta and Mishra). The study period lasted for 60 years for analyzing soils irrigated with sewage wastewater. And then it was revealed that the total heavy metal content, i.e., Cd, Zn, Pb, Cr, Co, Cu, Ni, and Fe, increases surprisingly by 93, 46.5, 15.9, 6.61, 3.88, 3.81, 2.44, and 2.43 times, respectively. The deposition of heavy metals on the soil surface is due to sorption reactions of negatively charged soil colloids for these cationic heavy metals.

4.4.2 Effect of Agricultural Pollution on Physical and Chemical Properties of Soil

As human civilization evolved in time, the agricultural practices are modified simultaneously. Many hybrid species are produced, many different types of techniques are experimented, and many working practices are carried on in order to get qualitative and quantitative yield. But in the competitive environment with ever-increasing demands, the struggle to have a quality product in a great quantity at the same time has resulted in incorporating different means that may harm the environment, especially soil health.

Everyone is aware of the negative impacts that the chemical agricultural products, such as pesticides and fertilizers, have on soil and environment. There exists a natural system of balancing growth and fertility. But because of human greed to have more in limited time, they altered and modified the natural phenomenons. And the biggest impact of the anthropogenic activities could be seen on the surroundings especially on the soil. Moreover, disturbing an ecological service gradually disturbs other functioning too. For instance, overusage of chemical fertilizers for getting higher yields leads to agricultural runoff. The nutrients from the fertilizers either get leached to groundwater or get mixed with nearby water sources through surface runoff. The land use change influences the nutrient supply of an ecosystem; hence, the soil nutrient status is better in abandoned land and least in orchards. The land conversion and agriculture activities have been reported to decrease soil organic carbon because of decrease in soil organic carbon inputs, soil redistribution, and changes in mineralization (Pennock and Van Kessel 1997 and Janzen and Molina-Ayala 1997).

Starting from initial soil working during agricultural implementation, since clearing and tilling the land, many important soil components, which define soil fertility and health, change their properties. With time and by tilling actions, the lower surface layers of soil got tossed to the surface exposing major nutrients to direct sunlight and air. Many of the organic substances and nutrients are lost in the process. As a continuous cycle of cultivation carries on through time, soil fertility declines as the cultivated plants consume most of them (Leigh and Johnston 1994). In order to obtain continuous quality productions, most of the agricultural practices incorporate addition of extra chemical nutrients that not only enhance the quality of their products but also increase the yield except for a few agricultural techniques like slash-and-burn system, in which a patch of forest land is cleared by burning and then desired crops are grown continuously until the soil gets devoid of nutrients. After which the land is left as such and a new patch is cleared, and the process is repeated again. But these chemical nutrients degrade the soil health in the long run. They make soil sick that it is unable to carry out its natural cycle of producing nutrients. As the soil organic matter content reduces, the ion exchange capacity decreases simultaneously resulting in less binding of nutrients and more leaching of them to the ground.

Because of the anthropogenic disturbances of soil, the micro- and macroscopic flora and fauna too get affected. For instance, the diverse presence of almost six endemic species of earthworms in the Amazon forest region of Brazil was replaced by a single exotic species because of producing a compact cast, which was not helpful for tunneling effect (which otherwise help in improving soil porosity), thus resulting in soil compaction. This resulted in the creation of anaerobic conditions during rains proceeding to denitrification and then to methane emissions. The agricultural practices that damage the ecological cycle the most are tillage, fertilizers, and pesticides. These three practices have become an integral part of agricultural practices. But they impede many direct or indirect negative impacts on soil and environment.

To cater the ever-increasing problem of soil erosion and loss of soil microflora and fauna because of tillage practices, the concept of zero tillage has evolved in the last two decades especially in American continents (Landers et al. 2001). The implementation of zero tillage has helped in improving soil health and water retention capacity. Moreover, natural litter manuring of residue crops reduces the costs of soil working and cultural practices in order to get quality produce (Van Doren and Allmaras 1978). With tillage the weed growth increased, enhancing the ground litter inviting different species of decomposers and scavengers which feed on decaying organic matter which, otherwise, are a pest to an agricultural field. In order to check the unwanted growth of these pests, insecticides, herbicides, and any kind of pesticides are used over them. The use of herbicide in a specified quantity in such conditions is not harmful, but this process helps in providing more organic matter to the soil (Wardle 1995). Moreover, there is the least effect of herbicides on the soil.

The overuse and intensive cultivation practices reduce the SOM content in the soil depleting the soil quality to a large extent. This further is linked with a reduction in soil microbial activity which is a major cause of distortion of soil fertility and productivity. Moreover, this leads to a drastic reduction of biodiversity and microbial activity in the soil (Krishna et al. 2016).

4.4.2.1 Effect of Tillage

As discussed earlier, tillage is an important factor that modifies soil properties. By tilling and tossing out soil strata, the lower nutrients and macro-fauna-rich soil get exposed to open air. The nutrients get lost during rainwater runoff or through surface evaporation, and the micro- and macrofauna become exposed to predators like birds. So, substantially tillage does not prove beneficial to the soil. High SOC content in forest soils was reported because of its undisturbed nature as compared to agricultural systems wherein higher decomposition rates due to heavy tilling operations result in less SOC build up (Zegeye 1999; Kater et al. 1992; Rhodes 1995; Sanneh 2007; and Minj 2008). Tilling was advocated as the cause of depleting SOC and diminishing its stock (Burke et al. 1995; Grandy and Robertson 2006; Mann 1986; Schlesinger and Andrews 2000; Schnitzer et al. 2006). The minimum amount of SOC stock was observed under urban and agricultural land use system due to less return of organic material to such systems and its further loss through runoff as prevailing loose surfaces.

4.4.2.2 Effect of Pesticides

Pesticides refer to a large variety of chemical solutions that are designed specifically to eradicate any kind of pest. Pesticide includes insecticides, herbicides, weedicides, and fungicides.

Herbicides are considered safe for soil health by many scientists as they are considered as target chemicals that effectively check the growth of specific weeds, even though they have indirectly affected the growth of few insects such as beetles and collembola. Even earthworms are known to get affected by herbicides/weedicides because they thrive on soils with weedy conditions filled with more ground litter. Relatively higher contents of heavy metals under orchard and agriculture land use than forest land were observed (Delbari and Kulkarni 2011; Shiva Kumar and Srikantaswamy 2012), which was attributed to being due to the addition of chemicals in the form of fertilizers and pesticides.

Even during some researches, it was found that because of the use of some pesticides, the soil biota gets disturbed. The growth and interactions of plants and microorganisms such as phosphorus-solubilizing and nitrogen-fixing rhizobacteria are inhibited by the effect of pesticides at molecular levels. Many other studies revealed that it has an adverse effect on the normal cycling of nutrients and organic matter. The ecological processes like nitrification, denitrification, phosphorus solubilization, ammonification, and methanogenesis get modified or altered because of the addition of certain chemicals that are present in some pesticides.

The excessive use of strong pesticides and other chemicals to eradicate certain pests also has negative impacts on other life forms in the surrounding areas. They alter their normal growths by killing their preys, and because of this, the ecological balance gets disturbed. With the suppression of microbiota, the soil health deteriorates, which gradually in time lead to depletion of soil quality and its ability to produce a good yield of agricultural crops.

The impact of pesticides on soil microbial biomass also depends upon certain factors such as temperature, pH, moisture content and other environmental factors, amount of nutrients, and intensity of soil working practices and helps in governing the extent of chemical damage. Also, the type of chemical or ingredients of the pesticides also determines the magnitude of damage done because of chemical reactions. In a study, it was found that the usage of Herbogil as a herbicide on soil resulted in the initiation of nitrogen and carbon mineralization and inhibition of all other biological activities (Engelen et al. 1998).

4.4.2.3 Effect of Grazing

Grazing has negative as well as positive impacts on soil which are most of the times indirectly observable. The pasture lands over which cattle and livestock graze result in surface compaction which reduces the soil porosity. The surface compaction in the grasslands or pasture lands is an important factor that governs the water holding capacity of the soil. During rains, these kinds of soils are most prone to erosions and landslides.

4.4.2.4 Effect of Fertilizers

The use of chemical fertilizers no doubt enhances the growth of agricultural crops, but their usefulness is short lived and their aftereffects are long term. The addition of sulfur and nitrogenous fertilizers in soil tends to initiate chemical reactions in the soil which reduce the pH of the soil. This result in a great loss of soil biota, especially the microorganisms, shows the immediate impacts because of soil acidification. Even long-term acidification of soil reduces its capacity to decompose organic litter, and hence a thick mat of undecomposed matter forms over the surface of the soil, under which very limited microbiota survives.

To cater to the problem of reduced pH or acidification, liming has been used and recommended under various agricultural practices. But it has been observed that long-term usage of lime for soil reduces the C/N ratio (Belkacem and Nys 1995; Marschner and Wilczynski 1991). It also affects the organic carbon content and carbon storage (Derome et al. 1986; Persson et al. 1995). Even tree growths are restricted due to the stimulated growth of weeds and other vegetations (Rodenkirchen 1998). Even with soil microbial activity (Rodenkirchen 1998), nitrification increases when lime is applied to the soil (Neale et al. 1997; De Boer et al. 1993).

Higher content of available N was observed in soils under agricultural and orchard land use, which was found to be due to high input of nitrogenous fertilizers and organic manures (Jaipaul et al. 2011; Gopinath and Mina 2011; and Datt et al. 2003). The long-term use of strong fertilizers such as urea and phosphorus has severe impacts on soil health. The presence of a variety of heavy metals, such as Cd, Hg, Pb, and Cr, results in toxic formations in the soil.

4.4.3 Effect of Solid Waste Pollution on Physical and Chemical Properties of Soil

Solid waste comprises of garbage, rubbish, industrial waste, and pathological waste. It is considered as the third most important type of pollution that seeks immediate action. Disposal of solid waste on open land is a common practice in the areas where proper dumping methodologies are not followed especially in developing nations. What makes open dumping worse is rainfall, which mixes the harmful toxins with water, helping them to leach through soil surface to the deeper layers, contaminating the soil. The solid waste generation in an Indian town is 0.34 $kg\ day^{-1}$per capita. According to a report, the urban waste constituted 46.5% organic matter, 43% moisture, 6% paper, 0.7% glass, 3.2% rags, and 1.1% plastic (Maudgal 1995). It is proved by various researches that the open waste dumping directly or indirectly affects the living organisms, especially plants that leads to deterioration of the environment which is not reversible (Phil-Eze 2010).

In a study conducted at a landfill site in Bengaluru, India, it was observed that the moisture content of the soil was considerably more than the normal soil, even the amount of chloride in the soil was found to be 108.46 mg/l as compared to the normal soil that had 40 mg/l of chloride.

The open dumping of garbage results in leaching of certain chemical substances which affects physical as well as chemical properties of soil. In certain areas, such as metropolitan cities, the solid wastes or garbage dumps are thrown into large ditches and then compacted with soil. There the chances exist that the chemical substances from the dump would leach to the below surface soil layers, finally mixing up with groundwater. Those chemical leachates react with soil components in a manner that it changes the soil pH (gets reduced in calcareous soils) and affect the content of organic matter (Krishna et al. 2016). The garbage dump comes from many different sources and is the end product of many different activities. They include biodegradable articles such as paper, cardboard, wood, kitchen waste, and agricultural waste, which are not harmful to the soil over which they reside because they get decomposed and add organic matter to the soil, thereby improving soil structure and porosity. On the other hand, the garbage dump also contains a significant amount of nonbiodegradable wastes such as metals, plastics, and electronic articles, which not only stay intact on the same place for hundreds of years but also leach out a lot of elements such as Fe, Zn, Cr, Cu, Mn, Hg, Pb, and Ar which are known to toxicate the surroundings by percolating into the water sources.

The open dumping of solid waste on land alters the ground compositions in many ways. It increases the soil permeability by the process of flocculation that alters the composition from clay to silt, which increases the pore space, and hence the chemical leachates easily percolate to the lower ground layers and then to the groundwater sources.

The municipal solid waste when used for preparing compost results in a product that had lower levels of nutrients when compared to EPA limits (Stillwell and David 1993). Solid waste pollutants serve as an external force affecting the physiochemical characteristics of soil ultimately contributing toward the poor production of vegetation (Papageorgiou 2006). The pollutants from the open dumping disturb the normal metabolism and functioning of plants that result in its distorted growth. At open garbage dumping sites, relatively high concentration of organic matter with an average mean value of 1.54 was found which lead to an increase in the pH. This is mainly due to the process of mineralization during which there is a release of exchangeable cations (Woomer et al. 1994; Anikwe and Nwobodo 2002).

During a study of soil from a garbage dumping site in Islamabad City, Pakistan, a considerable variation in the mean values of EC and TDS was observed (Sayeda et al. 2014). It was calculated to be lower in normal sites as compared to waste disposal sites where it was found at higher levels. Also, the concentration of Na, Pb, and K was lower at control sites but higher at dumping sites. It is known that the higher concentration of NPK, micronutrients, and heavy metals in soil is harmful to the growth and development of plants and other living organisms. Similarly, the concentrations of Cu, Ni, Cr, and Zn were found to be significantly higher in the soils of waste disposal sites as compared to the soils of normal control sites.

Many environmental institutes and agencies advise and promote the use of municipal and industrial solid waste for making composts to be used as fertilizer in the agricultural fields. This help to resolve the ever-increasing problem of waste, by providing strong environmental and economic advantages in comparison with the traditional management practices, such as combustion and landfilling (Hargreaves et al. 2008). At the same time, they also help in restoring SOM and soil structure, enhancing the microbial activity, and providing essential nutrients (Garcia Gil et al. 2000). There exist many risks to the health and ecology of plants and other living organisms because of nutrient leaching and accumulation of heavy metals in the soil (Pierzynski and Gehl 2005; Smith 2009). So, in order to use MSW compost, careful methods and procedures are to be utilized, which are followed by many countries (Barral and Paradelo 2011). Interestingly, when applying MSW compost, the availability of N for the plants has been found to be very low in the first application period resulting in considerably lower yield (Eriksen et al. 1999). The reduced yields are due to the low release of N from the compost into the soil (Iglesias-Jimenez and Alvarez 1993; Mkhabela and Warman 2005). And in order to compensate with the low N availability due to reduced mineralization and immobilization by microbes, more amount of MSW compost is added in agricultural soils (Garcia Gil et al. 2000).

4.4.4 Effect of Air Pollution on Physical and Chemical Properties of Soil

The Earth is made up of various components that balance each other in order to sustain life on it. One such important component is air, which resides almost everywhere, and from the atmosphere, hydrosphere, and to the lithosphere, only the amount and composition vary. The composition of the atmosphere is fixed which is known to all. The presence of oxygen and other gases in a balanced proportion helps to assist life. However, the anthropogenic activities add unwanted polluting substances into the atmosphere which adversely affect the health and development of living beings on Earth. The major sources of anthropogenic pollutants into the atmosphere are:

- Vehicular fuel combustion
- Smelting of ores
- Burning agricultural residues
- Petroleum refining
- Industrial plants

The industrial plants have become one of the most important sources of pollutants in the atmosphere. Various industries such as asphalt and brick plants, boiling and heating installations, cement, fertilizer manufacturing, mineral acid, paper and pulp, thermal, nuclear power plants, sewage treatment plants, engineering workshops, and another such kind of industries are known as the stationary point sources of air pollution. The vehicles such as cars, scooters, auto rickshaw, trucks, and buses which are moving are considered as the mobile non-point sources of air pollution.

According to a study conducted in order to observe the impact caused by brick plants on the soils of towns of Saran division in north Bihar, certain facts were noticed and presented. The pollution intensity on the upper soil layer was higher than the lower surface layers (Srivastava and Singh 2012). Moreover, a sudden drop in pH in the surface layer was observed that in the lower one. They also observed that the acidity was higher in the soils which were near to the brick plant as compared to the ones present far away from it. It may be concluded that effluent gases are absorbed more when these come in the contact with the soil at the nearest distance from the plant site. The reduction of pH makes the soils more acidic, which reduces their ability to retain essential nutrients, like Ca, Mg, and K. This results in leaching out of these nutrients into the nearest water sources making them less available for land organisms. In the same way, the mobilization of heavy metals in the soil increases with the increase in soil acidity. They too get leached out into lakes, rivers, and streams. Most of the heavy metals, like Al, Hg, Pb, Cr, and Ar, are toxic to the aquatic species. Through the food chain, they enter into the living world, and by the process of bioaccumulation and biomagnification, these toxic substances get transferred from one trophic level to the other and expand in amount while passing through it.

4.5 Conclusion

Soil is a basic necessity for sustaining life on Earth. And humans have utilized it in every way possible. At the same time, the explicit use of various resources that causes pollution has altered the normal functioning of ecological services, especially affecting soil health and productivity. The use of wastewater for irrigating agricultural soil alters its physical and chemical properties affecting the growth of agricultural crops and other living organisms. Even the extensive agricultural practices such as fertilizer and pesticide application deteriorate the soil quality. The open dumping of garbage and municipal solid waste results in leaching of unwanted elements into the soil, hence intoxicating it to an extent that is irreversible. Moreover, air pollution too, directly or indirectly, affects the soil properties.

References

Aghabarati A, Hosseini SM, Maralian H (2008) Heavy metal concentration of soil and olive trees (*Olea europaea* L.) in suburban areas of Tehran, Iran. Res J Environ Sci 2:323–329

Al Jassir MS, Shaker A, Khaliq MA (2005) Deposition of heavy metals on green leafy vegetables sold on roadsides of Riyadh City Saudi Arabia. Bull Environ Contam Toxicol 75:1020–1027

Ale R, Jha PK, Belbase N (2008) Effects of distillery effluents on some agricultural crops: a case of environmental injustice to local farmers in Khajura VDC Banke. Scientific World, 6:68–75

Ana GREE, Sridhar MKC (2002) Soil quality near a chemical fertilizer industry at Port Harcourt Nigeria. Afr J Environ Assess Monitor 4:50–57

Anikwe MAN, Nwobodo KCA (2002) Long term effect of municipal waste disposal on soil properties and productivity of sites used for urban agriculture in Abakaliki, Nigeria. Bioresour Technol 83:241–250

Barral MT, Paradelo R (2011) Trace elements in compost regulation: the case of Spain. Waste Manag 31:407–410

Belkacem S, Nys C (1995) Consequences of liming and gypsum topdressing on nitrogen and carbon dynamics in acid forest soils with different humus forms. Plant Soil 173:79–88

Brar MS, Khurana MPS, Kansal BD (2002) Effect of irrigation by untreated sewage effluents on the micro and potentially toxic elements in soils and plants paper, p 198

Brevik EC (2005) A brief history of soil science. Global sustainable development, Theme, 1. Land use, land cover and soil sciences-VI, pp 1–10

Burke I, Lauenoth W, Coffin D (1995) Soil organic matter recovery in semiarid grasslands: implications for the Conservation Reserve Program. Ecol Appl 5:793–801

Datt N, Sharma RP, Sharma GD (2003) Effect of supplementary use of farmyard manure along with chemical fertilizers on productivity and nutrient uptake by vegetable pea (*Pisum sativum var. arvense*) and buildup of soil fertility in Lahaul valley of HP. Indian J Agricult Sci 73(5):266–268

De Boer W, Hundscheid MPJ, Schotman JMT, Troelstra SR, Laanbroek HJ (1993) In situ net N transformations in pine, fir, and oak stands of different ages on acid sandy soil, 3 years after liming. Biol Fertil Soils 15:120–126

Delbari AS, Kulkarni DK (2011) Seasonal variations in heavy metal concentrations in agriculture soils in Tehran-Iran. Biosci Discov 2(3):333–340

Derome J, Kukkola M, Malkonen E (1986) Forest living on mineral soils. Results of finish experiments. Natl Swedish Environ Protect Board Rep 3084:101–107

Dheri GS, Brar MS, Malhi SS (2007) Heavy-metal concentration of sewage-contaminated water and its impact on underground water, soil and crop plants in alluvial soils of Northwestern India. Commun Soil Sci Plant Anal 38:1353–1370

Doran JW, Safely M (1997) Defining and assessing soil health and sustainable productivity. In: Biological indicators of soil health. CAB International, New York, pp 1–28

El-Hady (2007) Compare the effect of polluted and river Nile irrigation water on contents of heavy-metals of some soils and plants. Res J Agricult Biol (3):287–294

Engelen B, Meinken K, Wintzingerode FV, Heuer H, Malkomes HP, Backhaus H (1998) Monitoring impact of a pesticide treatment on bacterial soil communities by metabolic and genetic fingerprinting in addition to conventional testing procedures. Appl Environ Microbiol 64(8):2814–2821

Eriksen GN, Coale FJ, Bollero GA (1999) Soil nitrogen dynamics and maize production in municipal solid waste amended. Soil Agron J 91:1009–1016

Garcia Gil JC, Plaza C, Soler Rovira P, Polo A (2000) Long-term effects of municipal solid waste compost application on soil enzyme activities and microbial biomass. Soil Biol Biochem 32:1907–1913

Giannakis GV, Kourgialas NN, Paranychianakis NV, Nikolaidis NP, Kalogerakis N (2014) Effects of municipal solid waste compost on soil properties and vegetables growth. Compost Sci Utiliz 22:116–131

Glanz JT (1995) Saving our soil: solutions for sustaining Earth's vital resource. Johnson Books, Boulder

Gopinath KA, Mina BL (2011) Effect of organic manure on agronomic and economic performance of garden pea (*Pisum sativum L.*) and on soil properties. Indian J Agricult Sci 81(3):236–239

Grandy S, Robertson P (2006) Initial cultivation of a temperate-region soil immediately accelerates aggregate turnover and CO_2 and N_2O fluxes. Global Change Biol 12:1507–1520

Gupta SK, Mitra A (2002) Advances in land resource management for 21st century soil conservation society of India New Delhi India, pp 446–460

Haiyan W, Stuanes A (2003) Heavy metal pollution in air-water-soil-plant system of Zhuzhou City, Hunan Province, China. Water Air Soil Pollut 147:79–107

Hargreaves JC, Adl MS, Warman PR (2008) A review of the use of composted municipal solid waste in agriculture. Agric Ecosyst Environ 123:1–14

Havugimana E, Bhople BS, Kumar A, Byiringiro E, Mugabo JP, Kumar A (2015) Soil pollution – major sources and types of soil pollutants. In: Environmental science and engineering (vol. 11). Soil pollution and phytoremediation, pp 53–86

Hayes AR, Mancino CF, Pepper LL (1990) Irrigation of turf grass with secondary sewage effluent. Soil Leachate Water Qual Agron J 82:939–943

Iglesias-Jimenez E, Alvarez CE (1993) Apparent availability of nitrogen in composted municipal refuse. Biol Fertil Soils 16:313–318

Jadhav HD, Savant NK (1975) Analysis of total nitrogen in organic wastes. Indian J Agricult Chem 8:73–84

Jaipaul SS, Dixit AK, Sharma AK (2011) Growth and yield of capsicum (*Capsicum annum*) and garden pea (*Pisum sativum L.*) as influenced by organic manures and biofertilizers. Indian J Agricult Sci 81(7):637–642

Janzen HH, Molina-Ayala M (1997) Carbon storage in soils under continuous grain cropping: N fertilizer and straw. CRC Press, Boca Raton, pp 235–254.

Kater LJM, Kante S, Budelman A (1992) Karite (*Vitellaria paradoxa*) and nere (*Parkia biglobosa*) associated with crops in South Mali. Agroforest Syst 18(2):89–105

Khurana MPS, Nayyar VK, Bansal RL, Singh MV (2003) Heavy metal pollution in soils and plants through untreated sewage water. In: Ground Water Pollution (ed) Proceedings of the international conference on Water and Environment Management of carbon sequestration in soil. Lewis Publishers/CRC Press, Boca Raton

Krishna M, Chaitra BR, Kumari J (2016) Effect of MSW leachate on the quality of soil. Intl J Eng Sci Invent 5(6):69–72

Lal R (1999) Soil quality and food security: the global perspective. In: R Lal (ed) Soil quality and erosion. Soil Water Conservation Society Inc., Ankeny, Iowa, pp 3–16

Landers JN, Dec Barros GSA, Manfrinato WA, Weiss JS, Rocha MT (2001) Environmental benefits of zero-tillage in Brazilian agriculture a first approximation. In: Conservation agriculture a worldwide challenge, Cordoba, Italy, pp 317–326

Leigh RA, Johnston AE (1994) Long-term experiments in agricultural and ecological sciences. CAB International Wallingford

Leininger S, Urich T, Schloter M, Schwark L, Qi J, Nicol GW, Prosser JI, Schuster SC, Schleper C (2006) Archaea predominate among ammonia-oxidizing prokaryotes in soils. Nature 442:806–809

Liu W, Zhao J, Ouyang Z, Soderlund L, Liu G (2005) Impacts of sewage irrigation on heavy metal distribution and contamination in Beijing China. Environ Intl 31:805–812

Mann L (1986) Changes in soil carbon storage after cultivation. Soil Sci 142:279–288

Mapanda F, Mangwayana EN, Nyamangara J, Giller KE (2005) The effect of long term irrigation using wastewater on heavy metal contents of soils under vegetables in Harare Zimbabwe. Agricult Ecosyst Environ 107:151–165

Marschner B, Wilczynski A (1991) The effect of liming on quantity and chemical composition of soil organic matter in a pine forest in Berlin, Germany. Plant Soil 137:229–236

Maudgal S (1995) Waste management in India. J Indian Assoc Environ Manage 22(3):203–208

McClean CJ, Cresser MS, Smart RP, Aydinalp C, Karkat AV (2003) Unsuitable irrigation practices in the bursa plain turkey in diffuse pollution and basin management In: Proceedings of the 7th international specialized IWA conference. International Water Association, Dublin, pp 60–65

Minj A V(2008) Carbon sequestration potential of agroforestry systems-an evaluation in low & mid hills of western Himalayas. PhD thesis Dr. YS Parmar University of Horticulture and Forestry Nauni, Solan (HP) India, p 124

Mishra K, Sahoo S (1989) Agro-potentiality in paper mill wastewater in soil pollution and soil organism. Ashish Publication House, New Delhi, India

Mkhabela MS, Warman PR (2005) The influence of municipal solid waste compost on yield soil phosphorus availability and uptake by two vegetable crops grown in a pugwash sandy loam soil in nova scotia. Agricult Ecosyst Environ 106:57–67

Nawaz A, Khurshid K, Arif MS, Ranjha AM (2006) Accumulation of heavy metals in soil and rice plant (*Oryza sativa* L.) irrigated with industrial effluents. Intl J Agricult Biol (3):391–393

Neale SP, Shah Z, Adams WA (1997) Changes in microbial biomass and nitrogen turnover in acidic organic soils following liming. Soil Biol Biochem 29:1463–1474

Osaigbovo A, Ulamen O, Ehirobert (2006) Influence of pharmaceutical effluent on some soil chemical properties and early growth of maize (*Zea mays L)* Afr J Biotech 5:612–617

Papageorgiou M (2006) Public community partnerships for waste collection in three Indian cities, an exercise in world making best student essays of 2005–06. Inst Soc Stud 24:104–117

Pennock DJ, Van Kessel C (1997) Effect of agriculture and of clear-cut forest harvest on landscape scale soil organic carbon storage in Saskatchewan. J Soil Sci 77:211–218

Persson T, Rudebeck A, Wiren A (1995) Pools and fluxes of carbon and nitrogen in 40 year old forest liming experiments in Southern Sweden. Water Air Soil Pollut 85:901–906

Phil-Eze PO (2010) Variability of soil properties related to vegetation cover in a tropical rainforest landscape. J Geogr Reg Plan 3(7):177–184

Pierzynski GM, Gehl KA (2005) Plant nutrient issues for sustainable land application. J Environ Qual 34:18–28

Rattan RK, Datta SP, Chhonkar PK, Suribabua K, Singh AK (2005) Long-term impact of irrigation with sewage effluents on heavy metal content in soils, crops and groundwater. Agricult Ecosyst Environ 109:310–322

Rhodes C (1995) Seasonal pattern of nitrogen mineralization and soil moisture beneath *Faidherbiaalbida* (snyn. *Acacia albida*) in Central Malawi. Agrofor Syst 29:133–145

Rodenkirchen H (1998) Evidence for a nutritional disorder of *Oxalis acetosella* L. on acid forest soils. Plant Soil 199:153–166

Rusan MJM, Hinnawi S, Rousan L (2007) Long term effect of waste water irrigation of forage crops on soil and plant quality parameters. Desalinization 215:143–152

Sanneh A (2007) Status of carbon stock under different land use systems in wet temperate North Western Himalaya. MSc thesis Dr. YS Parmar University of Horticulture and Forestry Nauni, Solan (HP) India, p 81

Saravanamoorthy MD, Ranjitha-Kumari BD (2007) Effect of textile water on morphology and yield on two varieties of peanut (*Arachis hypogea L.)*. J Agricult Technol (3):335–343

Schlesinger WH, Andrews JA (2000) Soil respiration and the global carbon cycle. Biogeochemistry 48:7–20

Schnitzer M, McArthur DFE, Schulten H, Kozak LM, Huang PM (2006) Long term cultivation effects on the quantity and quality of organic matter in selected Canadian prairie soils. Geoderma 130:141–156

Shiva Kumar D, Srikantaswamy S (2012) Heavy metals pollution assessment in industrial area soil of Mysore city Karnataka India. Intl J Appl Sci Eng Res 1(4):604–611

Smith SR (2009) A critical review of the bioavailability and impacts of heavy metals in municipal solid MSW compost and vegetables growth and quality waste composts compared to sewage sludge. Environ Intl 35:142–156

Srivastava KP, Singh VK (2012) Impact of air pollution on pH of soil of Saran, Bihar, India. Res J Recent Sci 1(4):9–13

Stillwell, David E (1993) Evaluation of a low grade rock phosphate composted with some agricultural wastes for use in a crop rotation. Agril Waste 18:73–79

Syeda MA, Aroma P, Beenish A, Naima H, Azra Y (2014) Open dumping of municipal solid waste and its hazardous impacts on soil and vegetation diversity at waste dumping sites of Islamabad city. J King Saud University Sci 26:59–65

Van Doren DM, Allmaras MR (1978) Effect of residue management practices on the soil physical environment, microclimate, and plant growth. In: Oschwald WR (ed) Crop residue management system. Society of Agronomy, Madison, pp 49–83

Wang TSC, Cheng S-Y, Tung H (1967) Dynamics of soil organic acids. Soil Sci 104:138–144

Wardle DA (1995) Impacts of disturbance on detritus food webs in agro-ecosystems of contrasting tillage and weed management practices. Adv Ecol Res 26:105–185

Woomer PL, Martin A, Albrecht A, Resck DVS, Scharpenseel HW (1994) The importance and management of soil organic matter in tropics. In: The biological management of tropical soil fertility. Wiley, Chichester, pp 47–80

Yadav RK, Goyal B, Sharma RK, Dubey SK, Minhas PS (2002) Post-irrigation impact of domestic sewage effluent on composition of soils, crops and ground water – a case study. Environ Intl 28:481–486

Ye R, Wright AL, Inglett K, Wang Y, Ogram AV, Reddy KR (2009) Land-use effects on soil nutrient cycling and microbial community dynamics in the Everglades Agricultural Area, Florida. Commun Soil Sci Plant Anal 40(17–18):2725–2742

Zegeye MW (1999) Tree crop interaction studies in agri-horti-silviculture system. MSc thesis Dr. YS Parmar University of Horticulture and Forestry, Nauni, Solan (H.P.) India

Chapter 5
Role of Soil Microbiome and Enzyme Activities in Plant Growth Nutrition and Ecological Restoration of Soil Health

Esther Jamir, Rajiv Das Kangabam, Kasmika Borah, Abhijit Tamuly, H. P. Deka Boruah, and Yumnam Silla

Abstract The soil consists of a plethora of both biotic and abiotic matter and provides the medium for plant growth, habitat for microorganism, which plays an important role in maintaining and contributing to the ecosystem service. The soil microbial community is an important component as they are involved in organic matter decomposition, nutrient cycles, and soil health. Likewise, enzymes in soil are also known for substantial role in energy transfer, catalyzing reactions necessary for all life processes, and also used as indicator of soil health. Soil microbes and enzyme are co-dependent with one another enhancing soil fertility by increasing the nutrient availability for plant growth. For instance, microbes like bacteria, fungi, and algae form association with soil rhizosphere like plant growth-promoting rhizobacteria which regulate plant growth by producing hormones and protect plant from disease-causing pathogen. In addition, microbes and enzymes play key role in the removal of harmful chemical or pollutants present in the soil through bioremediation. Herein, the details of various role of soil microbes and enzyme activities in improving the soil health and maintaining the soil structure were discussed and further, the application of new technologies like metagenomics and metaproteomics in indentification of noval microbes and their applications with huge agroeconomical potential were highlighted.

Keywords Soil ecosystem · Enzymes · Microbial community · Bioremediation · Metagenomics · Metaproteomics.

E. Jamir · K. Borah · A. Tamuly · H. P. Deka Boruah · Y. Silla (✉)
Biological Sciences & Technology Division, CSIR-North-East Institute of Science and Technology, Jorhat, Assam, India
e-mail: ysilla@neist.res.in

R. D. Kangabam
Department of Agricultural Biotechnology, Assam Agricultural University, Jorhat, Assam, India

A. Kumar, S. Sharma (eds.), *Microbes and Enzymes in Soil Health and Bioremediation*, Microorganisms for Sustainability 16,
https://doi.org/10.1007/978-981-13-9117-0_5

5.1 Introduction

Soil is the naturally prevailing substructure of the ecosystem and a habitat for numerous microorganisms such as bacteria, fungi, algae, annelids, invertebrates, and plants and higher animals. These inhabitants and the enzymatic activities play an important role in sustaining the quality of the soil for all the living organisms (Aislabie and Deslippe 2013). Different applications of soil can be assessed as (1) it promotes growth of plants; (2) prevents air and water pollution by degrading harmful chemicals or pollutants from industrial, organic wastes, agricultural runoff wastes, etc.; (3) and prevents watershed by controlling separation and partitioning of precipitation (Cunningham and Sumner 1997). In addition, soil matrix influences the structure and functions of microbial community as well as enzyme activity in soil, and this interaction between the soil biosphere and abiotic factors implements the benefits for environment as well as human (Schulz et al. 2013).

Soil microorganisms are the integral part of the soil ecosystem. A large amount of diverse soil microbial community plays important role in nutrient transformation including decomposition, mineralization, and retention of available nutrients. Soil also possess different activities such as parasitic activity, act as phytopathogens (Compant et al. 2005), and secrete several key microbial enzymes; thus they are thought to be a good model for ecological strengthening of living system on earth (Jackson et al. 2012). There are numerous microorganisms in the soil which form an association or interaction with the soil bio-system and produce soil enzymes. Like microbes, these soil enzymes also play a key role in most of the soil biochemical processes such as detection of soil quality, removal of soil pollutants, and decomposition of organic matters and also involved in nutrient recycling (Dong et al. 2015; Han et al. 2007; Zhang et al. 2017). For instance, intracellular and extracellular soil enzymes such as β-glucosidase and hydrolase initiate the breakdown of organic matter, while enzymes like amylase, urease, phosphatase, and sulfatase are involved in nutrient mineralization (Yuan and Yue 2012; Vasconcellos et al. 2013; Sherene 2017). In addition catalase enzymes like dehydrogenase and phosphatase are also involved in degradation of heavy metals in soil which are utilized in bioremediation (Khan et al. 2007).

Further microorganisms and soil enzymes promote availability and uptake of mineral nutrient for plant growth. In the current scenario, with the increasing demand of crop production, soil management strategy is mainly based on inorganic chemical-based fertilizer which is the main cause of serious health and environmental issues (Bhardwaj et al. 2013).This addresses an immediate attention for the exploitation of beneficial soil microbes as bio-fertilizer in sustainable agricultural industry by improving soil fertility. A variety of bacteria, fungi, and algae are associated in soil rhizosphere and act as bio-fertilizer for stimulating plant growth. Plant growth-promoting rhizobacteria (PGPR) can act as growth regulator, protect plants against pathogen, enhance nutrient availability, detoxify heavy metals, and degrade xenobiotic compounds (Ahemad and Khan 2012a; Ahemad and Malik 2011; Hayat 2010; Rajkumar et al. 2012; Braud et al. 2009a).

Furthermore, microbes and enzyme play key role in eco-restoration of soil health. They help in cleaning polluted soil with petroleum hydrocarbons, halogenated organic chemicals, and toxic metalloid(s) (Rohrbacher and St-Arnaud 2016). Various living microorganisms mainly bacteria and fungi are used in biotransformation of environmental contaminants into less toxic substances through bioremediation process. This book chapter highlights the key role of soil microbes and enzymes in nutrient availability for plant growth and discusses microbial-based eco-friendly methods for restoration of soil health. In addition we also address high-throughput "omics" approaches to explore diverse soil microbial community profile and their future aspects.

5.2 Soil Microbial Community

Soil microbial community mainly composes of bacteria, fungi, algae, protozoa, actinomycetes, etc., and each microorganism has different characteristic roles in decomposition of organic matter, nutrient recycling, and maintenance of soil health (Yang et al. 2018). Figure 5.1 depicted the major key microorganisms present in soil. A brief description of the most common microorganisms found in the soil is given below:

Bacteria Soil bacteria can be divided into four groups on the basis of their source of carbon and energy which are as follows:

1. Photoautotrophs like cyanobacteria acquire energy from the sunlight and fix carbon as nutrient source, and they take part in nitrogen fixation (Aislabie and Deslippe 2013; Dijkhuizen and Harder 1984).
2. Photoheterotrophs in presence of an electron donor assimilate carbon dioxide during photosynthesis (Dijkhuizen and Harder 1984).

Fig. 5.1 The microbial community in soil consisting of bacteria, actinomycetes, fungi, protozoa, archaea, and algae; each microorganism plays an important role in maintaining soil health. Microbial pictures were adapted from multiple study (please refer reference list; (Barker 2010; Ingham 1999; Naorungrote et al. 2011)

3. Chemoautotrophs obtain carbon and energy by using inorganic compounds and are involved in nitrification, e.g., nitrosomonas and nitrobacteria (Boschker et al. 2014).
4. Chemoheterotrophs can grow by utilizing the pre-formed organic matter as a source of carbon and energy (Aislabie and Deslippe 2013; Boschker et al. 2014).

Fungi The different types of fungi are associated with the soil environment like saprotrophic fungi that produce enzymes from the hyphae and increase the availability of substrates for the other soil organisms, thus enhancing the biomass and diversity of the soil, and along with it these fungi also play a role in carbon cycle (Thomas 2012). Mycorrhizal fungi form symbiotic association with the roots of the plant and exchange their source of nutrient and carbon. For example, ectomycorrhizal fungi are important for the plant to get access to the inorganic soil phosphorus (Smith et al. 2003).

Lichens are symbiotic mutualistic association between the fungus and the pigmented algae, e.g., cyanobacteria (blue-green algae) which are involved in nitrogen fixation and carbon source (Aislabie and Deslippe 2013; Abdel-Raouf et al. 2012).

5.2.1 General Role of Soil Microbiota

The microbial community in the soil plays key role in almost all the soil types. Microorganisms are required for the function of soil, especially in nutrient cycling decomposition and enhancing soil fertility in all the aspects (Johns 2017; Harris 2009).

Different functions of microbial community in the soil and their relationship with the soil environment are described as follows:

1. Microorganisms take part in nutrient cycle and decomposition of soil which increase organic matters and make the soil suitable for all the biotic and abiotic components (Arrigo 2005; Conley et al. 2009).
2. Soil microbes produce enzymes which are useful for determining pollutants and also produce some antimicrobial agents use in pharmaceutical industry (Karigar and Rao 2011; Mohammadian et al. 2017).
3. Promote plant growth by enhancing availability and uptake of soil nutrient (Aislabie and Deslippe 2013; Supramaniam et al. 2016).
4. Microbes take part in mineralization, immobilization, and detoxification (Arrigo 2005; Francis et al. 2007; Lugtenberg and Kamilova 2009).
5. Soil microbes enhance soil moisture which increases the soil ability to filter the contaminants (Karigar and Rao 2011).

6. Soil microbial community is the foundation of soil food web, and thus they are major structural component of the soil biodiversity (Herndl and Weinbauer 2003).
7. Some microbes are widely associated in surpassing pathogens (David and Jos 2002).
8. Bacteria and fungi are involved in denitrification, and they also produce or consume methane which regulates the emission of nitrous oxide and methane in soil (Aislabie and Deslippe 2013; Yang et al. 2018).

5.2.2 Microbiome in Enhancing Soil Fertility or Soil Health

Microbes facilitate various activities for maintaining soil health. In soil environment, they provide physical support, buffering water flow, nutrient cycling, recycling of waste and detoxification, filtering of pollutants, regulation of greenhouse gas with storage of carbon, etc. *(*Johns 2017*;* Kibblewhite et al. 2008*)*. Activities of nutrient cycling result in the decomposition of the organic matter releasing carbon phosphorous sulfur from the pollutants to the soil which is then absorbed by the plants for their growth leading to bioremediation (Yuan and Yue 2012; Arrigo 2005). Key roles of microorganisms involved in maintaining soil fertility are follows:

1. *Fixing atmospheric nitrogen*: In symbiosis, rhizobia or bradyrhizobia fix atmospheric nitrogen gas and make it accessible to legume (Conley et al. 2009; Shridhar 2012).
2. *Increasing phosphorous availability*: Plant root forms a symbiotic relationship with fungi like arbuscular mycorrhizal fungi which increase phosphorus uptake by the plant (Richardson and Simpson 2011).
3. *Controlling pathogen*: Some protozoa in soil consume disease-causing pathogenic fungi, thereby helping in controlling certain plant disease (David and Jos 2002; Saha et al. 2016).
4. *Improving soil structure*: During organic matter disintegration, some bacteria and fungi produce substances that chemically and physically integrate with soil particles and improve soil structure (Torsvik and Ovreas 2002a).
5. *Degrading pesticide*: Degradation of agricultural pesticides in soil is mainly performed by microorganisms. Microorganisms in soil produced enzymes that can break down agricultural pesticides or other toxic substances in soil (Iqbal and Bartakke 2014).
6. *Releasing nutrient from organic matter*: Soil microorganisms are responsible for a large amount of nutrient discharge from organic matter (Arrigo 2005). They decompose and degrade a large amount of plant and animal waste and residue into functional organic particles in soil (Shraddha et al. 2011).

5.3 Soil Enzyme

Soil enzymes play an important biochemical function in catalyzing several important reactions for microbes and maintenance of soil health, structure, and ecology (McLaren 1975; Dick et al. 1996). They are continuously produced and being associated in soil, thus playing an important role in overall process of decomposition of organic matters, nutrient recycling, and plant growth (Sinsabaugh et al. 1991). Generally two major types of enzymes occur in nature, and both the types are found in the soil.

1. *Constitutive enzymes* are eternally present in the organism with a constant amount for its metabolic activity. These types of enzymes are not affected by addition or presence of any particular substrate (Das and Varma 2011a). For example, enzymes involve in glycolysis pathway like hexokinase, phosphofructokinase, phosphoglucose isomerase, pyruvate kinase, etc. are constitutive enzymes (Maitra and Lobo 1971). In soil, phosphatase and urease enzymes are found as constitutive enzymes (Das and Varma 2011a; Nannipieri et al. 2011; Margalef et al. 2017).
2. *Inductive or inducible enzymes* are present in a very low amount or sometimes not contained at all in a cell; however their concentration may vary and increase rapidly when the substrate is present. For example, amidase (Das and Varma 2011a) and cellulase enzyme (Kandeler 2015) are some of the inducible enzymes found in soil.

5.3.1 *Importance of Soil Enzyme, Its Activity, and Application*

Soil enzymes catalyze several reactions necessary for maintaining soil ecosystem including the decomposition of organic matters and recycling of nutrients (Doran and Safley 1997; Waldrop et al. 2000; Kourtev et al. 2002). They are the major driving force in the circulation of nutrient material and the flow of energy in the agroecosystems. Some of the enzymes such as hydrolase (Bautista-Cruz and Ortiz-Hernandez 2015) and glucosidase (Almeida et al. 2015) enhance breakdown of organic matters, while enzymes like amidase (Das and Varma 2011a), urease (Corstanje et al. 2007), phosphatase (Nannipieri et al. 2011), and arylsulfatase (Karaca et al. 2010) are involved in nutrient mineralization.Major functions and biological application of soil enzymes are mentioned as follows:

1. It catalyzes the degradation of organic matters or decomposition of plant or animal residues.
2. Mineralization and biochemical cycling of important elements in soil such as carbon, nitrogen, phosphorus, and sulphur.
3. Determination of soil health and microbial activity.
4. Soil enzyme as sensitive indicator or response to change of environmental condition and measures of pollution level.

Among the different soil enzymes, certain enzymes such as acid and alkaline phosphatase, urease, dehydrogenase, and β-glucosidase have been thoroughly studied for their specific importance in organic matter transformation process in various agricultural practices (Das and Varma 2011a; Wick et al. 1998; Jin et al. 2009; Fierer 2017; Saha et al. 2008; Piotrowska-Dlugosz and Wilczewski 2014; Adetunji et al. 2017). Details of various soil enzymes, environmental factors affecting their activity, and potential industrial applications were shown in Table 5.1.

5.3.2 *Determination of Soil Enzyme and Factors Affecting Enzyme Activity*

Measurements of enzyme activity in soil have been carried out by using various methods and techniques such as microcalorimetry (Cenciani et al. 2011), fluorimetry (Darrah and Harris 1986), spectrophotometry (Upson et al. 1996; Kheyrodin 2014), fluorescence, and radiolabeling (Pritsch et al. 2004; Bell et al. 2013). Soil enzymatic assay can be used in which the substrate is added to the soil system and the amount of product form is measured or the enzymatic assay of the enzyme extracted from the soil can be carried out indirectly (Fierer 2017; Saha et al. 2008). This method mainly detects the amount of degradation of the target substrate and formation of the product. Substrate-induced respiration assay have been developed to measure enzyme activities during production of CO_2 or consumption of O_2 (Piotrowska-Dlugosz and Wilczewski 2014; Adetunji et al. 2017; Cenciani et al. 2011; Darrah and Harris 1986). The advancement in high-throughput molecular biology techniques such as genomics, proteomics, and transcriptomics approaches provide various platforms to estimate the presence of specific enzymes and expression of enzyme-coding genes in the soil environment (Upson et al. 1996).

Several physicochemical factors affect the activities of soil enzymes including temperature, pH, chemicals and pesticides, nature of soil such as soil organic matter or composition, soil texture, soil fertility, synthesis or secretion of soil enzyme, diversity of microbes and plant community, etc. (Kheyrodin 2014; Acosta-Martinez et al. 2007; Utobo and Tewari 2015) which are described in Table 5.2. Poor enzyme activity of some of the enzymes particularly the pesticide-degrading enzymes can lead to the accumulation of harmful chemicals or hazardous materials to the environment that further inhibit soil enzyme activity (Bell et al. 2013; Chibuike and Obiora 2014; Rani and Dhania 2014; Javaid et al. 2016).

Soil enzymes are an enhancer or catalyst for several biochemical reactions, so suppression or lack of soil enzyme largely affects the soil fertility and agronomic productivity. In order to improve the soil enzyme activity, it is suggested to have the following applications (James Cook 2006; Rana and Rana 2011; Guilpart et al. 2017):

1. Use of animal manure and less soil disturbance.
2. Modification of agronomic methods such as crop rotation, plantation timing, cropping system (e.g., legume-based system), and plant/animal residue removal.

Table 5.1 The enzymes in the soil are produced via various activities of plants and microorganisms. These enzymes have important application in numerous fields

Soil enzymes	Source	Indicator of microbial activity	Factors influencing enzyme activity	Applications	References
α-Amylase	Plants, animals, microorganism	C-cycling	Management practices, type of vegetation, environment, soil types	Food and pharmaceutical industry, used in manufacturing detergents	Zeman and McCrea (1985), Mitidieri et al. (2006), Mobini-Dehkordi and Javan (2012), and Das and Varma (2010))
Dehydrogenase	Microorganisms	C-cycling, microbial oxidative activity	Soil water content, temperature, pesticides, trace elements, management practices, pollution, etc.	Food industry, paper and pulp industry, textile industry, pharmaceutical industry, and bioremediation	Singh et al. (2016), Das and Varma (2010), and Rao et al. (1998)
β-glucosidase	Fungi, bacteria	C-cycling	Temperature, pH, water, O_2 contents, quality and location of organic matter, mineral elements, and fungicides	Cellulose biodegradation, protein engineering, biofuel production, flavor enhancement, isoflavones, and hydrolysis	Ahmed et al. (2017) and Das and Varma (2010))
Phenol oxidase	Plants and microorganisms	C-cycling	Soil pH, mean annual precipitation and temperature, SOM content, management practices, N enrichment, etc.	Biocatalysts in several food processing, plant biotechnology in pharmaceutical sciences, food industry, etc.	Burton (1994) and Das and Varma (2010))
Urease	Microorganisms, plants	N-cycling	Cropping history, organic matter content, soil depth, management practices, heavy metals, temperature, pH, etc.	Blood urea analyses, urea content analysis in blood, urine, alcoholic beverages, environmental wastewater	Das and Varma (2010)

Phosphatase	Microorganisms, plants	P-cycling	Organic matter content, pH, management practices, pollution, crop species, and varieties	ELISA applications, precipitation of heavy metals and enzyme immunoassays	Das and Varma (2010) and Rani et al. (2012)
Alkaline phosphatase	Mainly bacteria	P-cycling	Organic matter content, pH, management practices, pollution, crop species, and varieties	Baking and processed cheese, ELISA applications, precipitation of heavy metals, enzyme immunoassays, etc.	(Rani et al. (2012) and Das and Varma (2010)
Acid phosphatase	Plants, fungi, and bacteria	P-cycling	Organic matter content, pH, management practices, pollution, crop species, and varieties	In vitro diagnostic use, biochemical marker, etc.	Bull et al. (2000)
Aryl sulfatase	Microorganisms, plants, and animals	S-cycling	Heavy metal pollution, pH, organic matter content and composition, and availability of organic sulfate esters	Analytical endocrinology	Zsolt et al. (2015)
Protease	Microorganisms and plants	N-cycling	Humic acid concentration, availability of C and N, etc.	Leather, laundry, biocatalyst, and bioremediation	Das and Varma (2010) andRao et al. (1998)
Chitinase	Plants and microorganisms	C- and N-cycling	Availability of N, soil depth, atmospheric CO_2 levels, etc.	Protein engineering, fertilizer, the production of non-allergenic, nontoxic, biocompatible, and biodegradable materials	Das and Varma (2010)and Shapira et al. (1989)

(continued)

Table 5.1 (continued)

Soil enzymes	Source	Indicator of microbial activity	Factors influencing enzyme activity	Applications	References
Cellulase	Bacteria and fungi, termites, Ruminant	C-cycling	Temperature, pH, enzyme concentration, substrate concentration, and the presence of any inhibitors or activators	Textile industry, in detergents, pulp and paper industry, improving digestibility of animal feeds, and food industry	Das and Varma (2010)and Bennet et al. (2002)
Amidase	Both Prokaryotes and eukaryotes	N-cycling	Humic acid concentration, availability of C and N	Therapeutic applications, food industry, analytical applications, manufacture of fine chemicals	Das and Varma (2010) and El-Ghonemy (2015)
Monooxygenase	Both eukaryotic and prokaryrotic	C-cycling	Temperature, pH, enzyme concentration, substrate concentration, and the presence of any inhibitors or activators	Application in biodegradation and biotransformation of aromatic compounds, act as biocatalysts in bioremediation process, protein engineering	Das and Varma (2010) and Arora et al. (2010)
Dioxygenase	Soil bacteria	C-cycling	Temperature, pH, enzyme concentration, substrate concentration, and the presence of any inhibitors or activators	Synthetic chemistry, pharmaceutical industry, bioremediation, etc.	Fetzner and Lingens (1994)
Laccase	plants, fungi, insects, and bacteria	C- and N-cycling	Availability of N, soil depth, atmospheric CO_2 levels, etc.	Food industry, paper and pulp industry, textile industry, nanotechnology, synthetic chemistry, bioremediation, and cosmetics	Das and Varma (2010) and Mai et al. (2000)
Peroxidase	Plants, fungi, and prokaryotes	C- and N-cycling	Humic acid concentration, availability of C and N	Food industry, paper and pulp industry, textile industry, pharmaceutical industry and bioremediation.	Das and Varma (2010) and Wariishi et al. (1992)

Lignin peroxidase	White rot fungus	C- and N-cycling	Humic acid concentration, availability of C and N, etc.	Food industry, paper and pulp industry, textile industry, pharmaceutical industry, bioremediation, etc.	Das and Varma (2010) and Piontek et al. (2001)
Manganese peroxidase	Lignin-degrading basidiomycetes fungus	C- and N-cycling	Humic acid concentration, availability of C and N, etc.	Food industry, paper and pulp industry, textile industry, pharmaceutical industry, bioremediation, etc.	Das and Varma (2010)and Yoshida (1998)
Versatile peroxidase	*Nicotiana tabacum*, *Brassica oleracea*, *Capitata* var. Alba L.	C- and N-cycling	Humic acid concentration, availability of C and N, etc.	Industrial biocatalyst and bioremediation	Das and Varma (2010) and Tsukihara et al. (2006)
Lipase	Bacteria, plant, actinomycetes, and animal cell	C- and N-cycling	Humic acid concentration, availability of C and N, etc.	Control of oil spills, detergent production, baking industry, paper and pulp industry, personal care products, etc.	Sharma et al. (2011)

C Carbon, *N* nitrogen, *P* phosphorous, *S* sulfur, *SOM* soil organic matter, *ELISA* enzyme-linked immunosorbent assay

Table 5.2 List of bacterial and fungal strains used for stimulation of plant growth

Bacteria	PGPR/ bio-fertilizer	Mode of action	Crops	References
Bacillus pumilus	Used as PGPR	IAA, siderophores, HCN, ammonia	Tobacco	Wani and Khan (2010)
Bacillus licheniformis	Used as PGPR	Systemic resistance	Tomato, Pepper	Lucas et al. (2004) and García et al. (2004)
Bacillus subtilisRJ46	Used as PGPR	IAA, phosphate solubilization	Black gram and garden pea	Saikia et al. (2018) and Zaidi et al. (2006)
Bacillus subtilus	Used as PGPR	Antifungal activity, IAA, phosphate solubilization	Tomato	Zaidi et al. (2006), Murphy et al. (2000), and Cazorla et al. (2007)
Bacillus cereus	Used as PGPR	Lowers the toxicity of Chromium (Cr) to seedlings by reducing Cr(VI) to Cr(III)	Red pepper	Joo et al. (2005)
Pseudomonas sp. *(A3R3, RJ15*	Used as both PGPR and Bio-fertilizer	Phosphate solubilization, IAA, siderophore, HCN, biocontrol potentials, ACC deaminase, antifungal activity, N2- fixation, heavy metal solubilization	Wheat, rice, maize, black gram, and garden pea	Saikia et al. (2018), Tank and Saraf (2009), Poonguzhali et al. (2008), Indira Gandhi et al. (2008), Rajkumar (2008), and Ma and Rajkumar (2009)
Pseudomonas putida	Used as both PGPR and Bio-fertilizer	IAA, siderophores, HCN, ammonia, phosphate solubilization, antifungal activity, Pband Cd resistance	Wheat, rice, maize, black gram, and garden pea	Ahemad and Khan (2012a, b), Pandey et al. (2006), and Tripathi et al. (2005)
Pseudomonas aeruginosa	Used as both PGPR and bio-fertilizer	IAA, siderophores, HCN, ammonia, phosphate solubilization, ACC deaminase	Mung bean	Siddiqui et al. (2001), and Ahemad and Kibretb (2014)
Pseudomonas fluorescens	Used as both PGPR and bio-fertilizer	Siderophores, ACC deaminase, phosphate solubilization, antifungal activity	Soybean, tobacco	Ahemad and Kibretb (2014) and Garcia de Salamone et al. (2001)

(continued)

Table 5.2 (continued)

Bacteria	PGPR/ bio-fertilizer	Mode of action	Crops	References
Burlkholderia cepacia	Used as PGPR	ACC deaminase, IAA, siderophore, heavy metal solubilization, phosphate solubilization	Alfalfa, barley, beans, clover, cotton, maize, peas, sorghum, wheat	Ahemad and Khan (2009) and Bhattacharyya and Jha (2012)
Azospirillum amazonense	Used as bio-fertilizer	IAA, nitrogenase activity	Cereals, wheat, oat, barley, sorghum	Rodrigues et al. (2008)
Aeromonas veronii	Used as PGPR	IAA	Rice	Mehnaz et al. (2010) and Barazani and Friedman (1999)
Agrobacterium sp	Used as both PGPR and bio-fertilizer	IAA	Lettuce, fruit, nut, ornamental nursery stock, and trees	Barazani and Friedman (1999) and Bhattacharyya and Jha (2012)
Alcali genes piechaudii	Used as both PGPR and bio-fertilizer	IAA	Lettuce	Barazani and Friedman (1999) and Bhattacharyya and Jha (2012)
Azospirillum brasilense	Used as both PGPR and bio-fertilizer	IAA, phosphorous solubilization, nitrogenase activity, antibiotic resistance	Wheat	Kaushik et al. (2000)
Bradyrhizobium sp.	Used as PGPR	IAA, siderophores, HCN, ammonia, exo-polysaccharides	Radish	Antoun et al. (1998), Wani and Khan (2010), and Bhattacharyya and Jha (2012)
Bradyrhizobium sp. 750	Used as PGPR	Heavy metal mobilization, IAA	Radish	Ahemad and Kibretb (2014), and Dary et al. (2010)
Enterobacter cloacae	Used as PGPR	ACC deaminase, IAA, siderophore, phosphate solubilization	Rice, chickpea	Mehnaz et al. (2010)

(continued)

Table 5.2 (continued)

Bacteria	PGPR/ bio-fertilizer	Mode of action	Crops	References
Enterobacter asburiae	Used as PGPR	IAA, siderophores, HCN, ammonia, exo-polysaccharides, phosphate solubilization	Rice, chickpea	Ahemad and Kibretb (2014) and Hynes et al. (2008)
Rhizobium sp.	Used as both PGPR and bio-fertilizer	IAA, siderophores, HCN, ammonia, exo-polysaccharides	All the types of pea, lentil	Ahemad and Kibretb (2014)
Rhizobium leguminosarum	Used as both PGPR and bio-fertilizer	Cytokinin	Radish, rape and lettuce	Antoun et al. (1998) and Noel et al. (1996)
Paenibacillus polymyxa	Used as PGPR	IAA, siderophores	Wheat	Timmusk et al. (1999), and Phi et al. (2010)
Azotobacter sp.	Used as both PGPR and bio-fertilizer	IAA, siderophore, antifungal activity, ammonia production, HCN	Cereal, wheat, oat, barley	Ahemad and Kibretb (2014)
Azotobacter chroococcum	Used as both PGPR and bio-fertilizer	IAA, siderophores, gibberellin, kinetin, IAA, P-solubilization	Beans, pea	Wani and Khan (2010) and Verma et al. (2001)
Burkholderia sp.	Used as PGPR	ACC deaminase, IAA, siderophore, heavy metal solubilization, phosphate solubilization	Rice	Bhattacharyya and Jha (2012) and Dou et al. (2008)
Gluconacetobacter sp.	Used as PGPR	Zinc solubilization	Sorghum, sugarcane	Isopi et al. (1995) and Boddey et al. (2001)
Pseudomonas syringae	Used as PGPR	IAA production	Citrus and pome fruit	Bhattacharyya and Jha (2012)
Agrobacterium spp.	Used as PGPR	IAA, biosynthesis	Fruit, nut, ornamental nursery stock and trees	Bhattacharyya and Jha (2012)

(continued)

Table 5.2 (continued)

Bacteria	PGPR/ bio-fertilizer	Mode of action	Crops	References
Azospirillum brasilense	Used as PGPR	IAA, P solubilization, nitrogenase activity, antibiotic resistance	Turf and forage crops	Bhattacharyya and Jha (2012) and Thakuria et al. (2004)
Kluyvera ascorbata	Used as PGPR	Siderophore, ACC deaminase, metal resistance	Indian mustard, tomato	Burd et al. (2000) and Genrich et al. (1998)
Ochrobactrum pseudogrignonenseRJ12	Used as PGPR	IAA, siderophores, HCN, ammonia	Black gram and garden pea	Saikia et al. (2018)
Klebsiella oxytoca	Used as PGPR	IAA, phosphate solubilization, nitrogenase activity	Triticum aestivum	Ahemad and Khan (2010) and Jha and Kumar (2007)
Mesorhizobium sp.	Used as both PGPR and bio-fertilizer	IAA, siderophores, HCN, exo-polysaccharides, IAA, antifungal activity, ammonia production	Chickpea	Ahemad and Kibretb (2014)
Mesorhizobium ciceri	Used as both PGPR and bio-fertilizer	IAA, siderophores	Chickpea	Wani et al. (2007)
Acinetobacter spp.	Used as PGPR	IAA, phosphate solubilization, siderophores	Pennisetum glaucum	Rokhbakhsh-Zamin et al. (2011)
Psychrobacter sp. *SRS8*	Used as PGPR	Heavy metal mobilization	Ricinus communis,	Ma and Rajkumar (2009)
Ralstonia metallidurans	Used as PGPR	Siderophores	Maize	Braud et al. (2009b)
Serratia marcescens	Used as PGPR	IAA, siderophore, HCN	Summer quash	Ahemad and Kibretb (2014) and Selvakumar et al. (2008)
Rhodococcus sp., *Flavobacterium*	Used as bio-fertilizer	IAA and siderophores	Brassica juncea	Belimov et al. (2005)
Pseudomonas aeruginosa strains AS03 and NA 108	Used as PGPR	IAA and HCN	Tea	Roy et al. (2013)

PGPR Plant growth-promoting rhizobacteria, *IAA* indole-3-acetic acid, *HCN* hydrogen cyanide, *ACC* aminocyclopropane-1-carboxylate

3. Changes in soil organic matter content and microbial biomass.
4. Modification of soil physical properties like soil liming changes in pH, soil moisture, temperature, etc.

5.4 Role of Soil Microorganism and Enzyme in Plant Growth

Nutrient availability is one of the important factors for controlling growth of plant and net productivity. In global soil environment, there is a direct proportion of soil biomass (microbes and enzymes) and plant biomass which are interlinked to maintain the soil health. A schematic representation of key functional interaction of soil, plant, and microbes was depicted in Fig. 5.2. Soil represents the habitats in which a diverse microbial community exists which produced extracellular soil enzymes that act as biological catalyst to degrade plant and animal waste residue to nutrient component, thereby promoting growth of the plant. For instance, hydrolytic enzymes in soil play key role in degradation of many biological macromolecules such as cellulose, hemicelluloses, chitin, and several soil-associated proteins that are found in soil and plant litter (Allison et al. 2007).

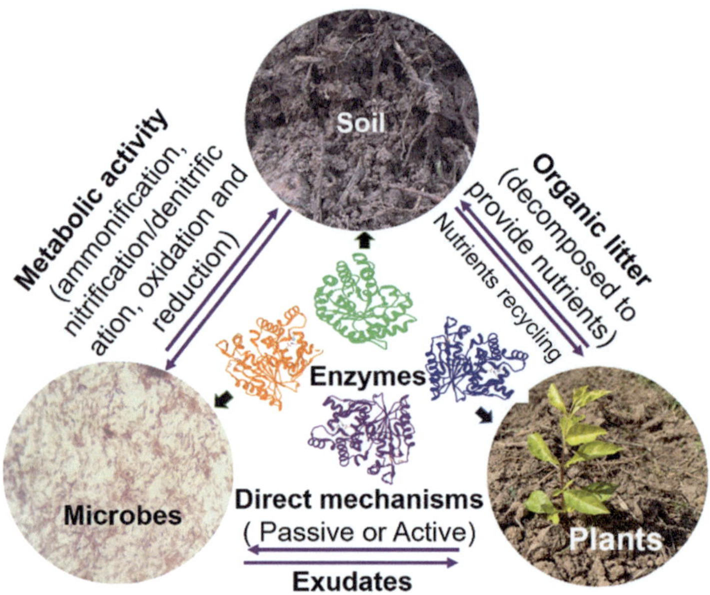

Fig. 5.2 Soil, microbe, enzyme, and plant interaction. Metabolic activity with the influence of enzymes takes place between the microbial community and the soil which leads to production of organic matters, essential for plant growth

5.4.1 Bio-fertilizers

Bio-fertilizers or so-called "microbial inoculants" are formulated by different types of living microorganism and major component of integrated soil nutrient management (Mohammadi and Sohrabi 2012). Bio-fertilizers play a key role in productivity and sustainability of soil fertility which is directly applied to the soil surface and different parts of plant including seed and enhances yield production. Microorganisms in bio-fertilizers promote plant growth by providing nutrition through biological process such as nitrogen fixation and solubilization of rock phosphate (Rokhzadi et al. 2008) and protect plants from pest, disease, and stress (El-Yazeid et al. 2007). Microbes that are commonly used as component of bio-fertilizers are nitrogen-fixing bacteria (N-fixer), potassium solubilizer (K-solubilizer), and phosphorus solubilizer (P-solubilizer) (Mohammadi and Sohrabi 2012). Some of important soil bacteria that are used as a component of bio-fertilizer are listed in Table 5.2.

5.4.2 Plant Growth-Promoting Rhizobacteria

Several soil bacteria are grown in rhizosphere of plants and attached with plant tissues, resulting in stimulation of plant growth by a plethora of mechanism (Vessey 2003). These beneficial bacteria are collectively called plant growth-promoting rhizobacteria (PGPR) (Kloepper and Schroth 1978; Suslow et al. 1979; Schroth and Hancock 1982). PGPRs are believed to promote plant growth either by providing nutrient to the host plant or by stimulating root growth as well as morphology and by aiding beneficial symbiotic relationships (Vessey 2003). However, the application of PGPR as bio-fertilizer in crop production is still a matter of debate, since not all PGPR are bio-fertilizer as many of them stimulate growth of plant by controlling the effect of pathogenic organism (Vessey 2003; Whipps 2001; Zehnder et al. 2001). Some of the key PGPRs which stimulate plant growth are listed in Table 5.2.Based on the mode of action, Somer et al. (Somers et al. 2004) classified PGPR as follows:

1. Bio-fertilizer enhances the nutrient availability and promotes growth of the plant.
2. Phytostimulators act as phytohormones and stimulate plant growth.
3. Rhizoremediators help in degradation of organic pollutants and provide more efficiency of phytoremediation or bioaugmentation.
4. Biopesticides mainly produced antibiotics and antifungal metabolites to control pest and plant disease (Antoun and Prévost 2005).

5.5 Role of Microbes and Enzymes in Eco-restoration of Soil

5.5.1 Microbes and Enzymes as Indicator of Soil Health

Soil health is the overall property of soil which represents the capacity of the soil to function as a system so that it continues to support and sustain the productivity of both biotic and abiotic ecosystem (Kibblewhite et al. 2008; Das and Varma 2011a; Alkorta et al. 2017). Microorganisms in the soil and enzymes produced by them act as a complete factor in determining soil health. Microbes respond rapidly to any change in the physical and chemical property of the soil; thus the microbial activity and microbiome population are used as a good soil indicator which cannot be obtained from other superior organisms (Shonkor and Das 2011).

In addition to microorganisms, soil enzymes play a different role in maintaining soil health which indicates soil microbial level, physicochemical status and exhibit useful sensor to environmental changes of soil fertility (Wick et al. 1998; Chen 2003) as well as nutrient availability (Asmar et al. 1994). Enzymes can easily detect the change in soil environment and are widely used to study soil condition and sustainability (Shonkor and Das 2011). One of the first applications of enzyme as a soil indicator is analyzing the soil condition and types of crops with response to the effect of pesticides, herbicides, and waste present in soil (Riah et al. 2014). As an example, dehydrogenase is used as an indicator of microbial activity in soil reflecting the presence of organic matter in the soil (Kumar et al. 2013; Kaczynska et al. 2015). Similarly, few enzymes like cellulase, amidase, dehydrogenase, and urease are documented to be used as soil indicator in various diverse soil samples (Dick et al. 1996; Tabatabai 1994; Das and Varma 2011b). Growth of microorganisms in soil depends greatly on the availability of cellulose and the enzyme cellulase play an important role in global recycling of cellulose (Abdel-Raouf et al. 2012). One of the most important applications of soil enzymes such as oxidoreductases, oxygenases, laccases, hydrolytic enzymes, etc. are used in the detection of toxicity or pollutants in soil (Karigar and Rao 2011; Okolo et al. 2007). These enzymes have the ability to detect, convert, and detoxify toxic substances in the environment and restore its suitable condition, thus playing an important role in determining the soil health (Whiteley and Lee 2006).

5.5.2 Microbes and Enzymes in Bioremediation of Soil

Bacteria, fungi and plants produce huge number of microbial enzymes which have been associated with biodegradation of deadly untreated pollutants. Generally, Bioremediation is an eco-friendly and cost-effective technology that relies on microorganisms, which enzymatically assault the pollutants and change them to less harmful products (Karigar and Rao 2011; Megharaj et al. 2017).

5.5.2.1 Bioremediation Techniques and Applications

Bioremediation is appropriate for the treatment of organic chemicals such as volatile organic compounds, benzene, toluene, ethyl benzene, xylene (BTEX) compounds, polycyclic aromatic hydrocarbons (PAHs) (particularly the simpler aromatic compounds), petroleum hydrocarbons, and nitro aromatic compounds (Soil bioremediation-EPA guideline 2005). The technique is divided into two main types based on application as in situ and ex situ bioremediation (Fig. 5.3).

5.5.2.2 In Situ and Ex Situ Bioremediation Treatment

In situ type of treatment is applied directly to the site of contaminated soil or water through a natural spontaneous process in which the naturally occurring soil microorganisms decompose toxic pollutants and make it less harmful. The technique does not disturb soil structure (Margesin and Schinner 2001), and for the smooth functioning of natural bioremediation process, it requires some of the environmental factors such as optimal temperature, pH, oxygen, nutrient, etc. (Iturbe and López

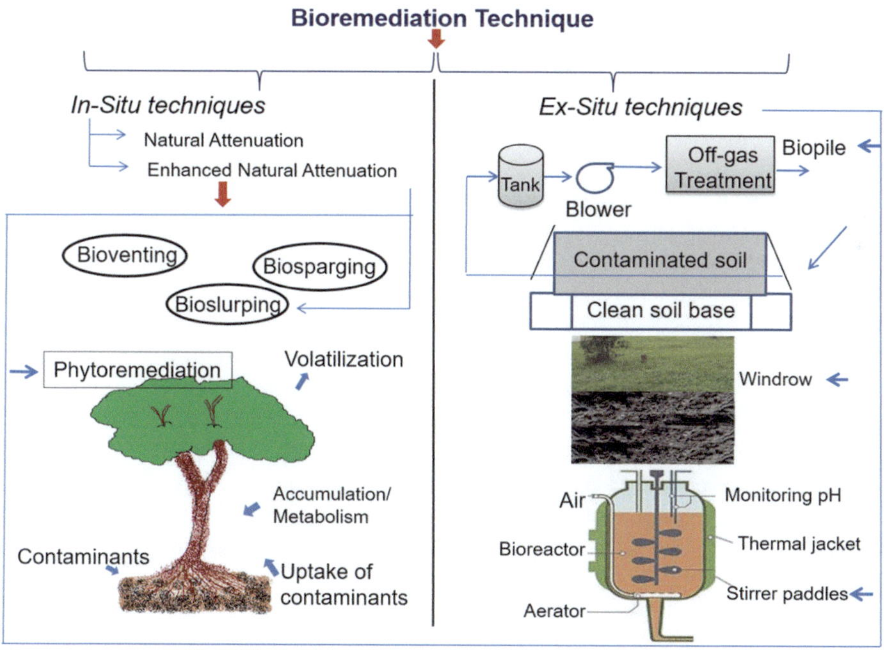

Fig. 5.3 Representation of the techniques involved in bioremediation. In situ techniques involve the process of natural attenuation of contaminants at the place of contamination mainly by phytoremediation. Ex situ technique involves the process of removing contaminants at the site of treatment via biopile, window, and bioreactor techniques. (The flowchart of the bioremediation technique mentioned in the above figure was adapted from Torsvik and Ovreas (2002b))

2015; Mulligana and Yongb 2004). Whereas, the ex situ bioremediation technique requires excavation or pumping of pollutants from polluted soil or groundwater to the site of the bioremediation treatment. Ex situ technique is mainly based on the cost of the treatment, types and degree of pollutant, and geographical location (Philp and Atlas 2005). Different methods of in situ and ex situ bioremediation treatments are listed in Table 5.3.

5.5.2.3 Role of Microbial Enzymes in Soil Bioremediation

Soil enzymes have an immense ability to convert and detoxify toxic substances because they have been known to be able to transform pollutants at an obvious rate and are naturally appropriate to restore polluted environments. Some of the important enzymes that are commonly used for bioremediation of polluted environments are described in Table 5.4.

5.6 High-Throughput Functional "Omics" Approaches to Study Soil Microbiome

Soil microbial community is the most complex, heterogeneous with highest state of prokaryotic diversity in any environmental ecology (Azubuike et al. 2016; Mary Kensa 2011; Park et al. 2006). It is reported that forest soil was found to contain an estimated amount of 4×10^7 prokaryotic cells per one gram of soil (Richter and Markewitz 1995), whereas one gram cultivated soil was found to contain approximately 2×10^9 prokaryotic cells (Paul and Clark 1989), altogether representing a source of gene pool for different applications of soil. However less than 1% of this microbial diversity was considered to be utilized by traditional methods (Delmont et al. 2011; Kuiper et al. 2004). With the emergence of advances in sequences technique, the high-throughput approaches like "meta-omics" have immersed widespread application in environmental microbiome study as they allow to characterize insight into microbial diverse component and ecosystem function (Maron et al. 2007; Becher et al. 2013; Armengaud et al. 2013). Over the last decades, metagenomics (Zhang et al. 2017; Vasconcellos et al. 2013; Torsvik and Ovreas 2002b; Delmont et al. 2011; Alexander and Svenning 2014; Chaparro et al. 2012) and metaproteomics (Maron et al. 2007; Becher et al. 2013) approaches have been applied to explore a wide range of soil environmental microbiology.

5.6.1 Metagenomics

The term metagenomics was first introduced by Handelsman et al. with the aim to explore collective genome and the biosynthetic pathway of soil microflora (Handelsman et al. 1998). Soil metagenomics represents the collective DNA

Table 5.3 Different methods of in situ and ex situ bioremediation techniques

In situ treatment			
Types of process	**Mode of action**	**Application site/ field**	**References**
Bioventing	Involved stimulation of airflow to unsaturated zone. Addition of nutrients and moisture to the contaminated site and transformed to harmless form	Used in polluted site with light spilled petroleum products	Mulligana and Yongb (2004), Philp and Atlas (2005), Azubuike et al. (2016), and Mary Kensa (2011)
Bioslurping	Involved soil vapor extraction, vacuum-enhanced pumping, etc. for soil and groundwater treatment	Used in polluted site with volatile, semi-volatile organic compounds	Philp and Atlas (2005), Azubuike et al. (2016), and Mary Kensa (2011)
Biosparging	Air spray is inserted into soil subsurface to manage microbial actions	Petroleum hydrocarbon spill site	Johns (2017), Philp and Atlas (2005), Azubuike et al. (2016), and Mary Kensa (2011)
Phyto-remediation	Plants are used to reduce the contaminated effects of the toxic pollutants	Soil and water remediation, heavy metal polluted sites	Azubuike et al. (2016), Kuiper et al. (2004), and Meagher (2000)
Ex situ treatment			
Method	**Mode of action**	**Application site**	
Biopile	Aboveground piling of excavated polluted soil and improved nutrient supply by increasing microbial action	Polluted extreme environments, e.g., cold region	Philp and Atlas (2005), and Whelan et al. (2015)
Windrow	Spinning of piled contaminated soil to bioremediate by indigenous hydrocarbonoclastic bacteria	Applied in contaminated oil and coal mine site	Whelan et al. (2015), and Barr (2002)
Bioreactor	Carried out in a vessel under specific control to mimic natural environment for the growth of microbes. It involved series of reaction to transform polluted soil to less harmful product	Crude oil-polluted soil	Johns (2017) and Barr (2002)

information of indigenous soil microorganisms which allow assessing largely untapped genetic diversity of uncultivated microbial species. This method comprises isolation of soil DNA, enrichment, construction and screening of clone library, deep sequencing, and data analysis which lead to discover abundance of microbial taxonomic units like novel phyla, classes, genera, species, and diverse gene pool (Delmont et al. 2011; Daniel 2005). Soil metagenomics study provided opportunity to capture novel pharmaceutical bioactive molecules such as new antibiotic (Ling et al. 2015) mining of bioproducts, biofuels, and enzymes. Although metagenomics have been used to decipher soil microbial community and diversity, it still remains a significant challenges to overcome functional gene annotation and identification of key

Table 5.4 List of important enzymes used for bioremediation

Types of enzymes	Mode of action	Example	References
Oxidoreductases	Detoxification and humification of toxic organic compounds through oxidative coupling reaction	*Flavobacterium* sp., *Phanerochaete chrysosporium*	Fierer 2017, Philp et al. (2005), Park et al. (2006), Rubilar et al. (2008), and Diez and Gianfreda (2008)
Oxygenases	Degradation/ detoxification of organic substrates	*Methylosinus trichosporium*, *Bacillus subtilis*	Fierer 2017, Vidali (2001), Arora et al. (2009), Bhagavan and Ha (2015), van Hellemond et al. (2007), and Muthukamalam et al. (2017)
Laccases	Catalyze the oxidation of a broad range of reduced phenolic and aromatic substrates	*Theiophora terrestris*, *Marinomonas mediterranea*	Shraddha et al. (2011), Karigar and Rao (2011), Fierer 2017, and Viswanath et al. (2014)
Peroxidases	Catalyze oxidation of organic/inorganic compounds	*Bacillus sphaericus*	Bansal and Kanwar (2013)
Hydrolytic enzymes	Hydrolysis of organic pollutants by breaking the bonds and reduce the toxicity of the compound	*Bacillus thuringiensis*, *Klebsiella* sp.	Karigar and Rao (2011) and Margesin et al. (1999)

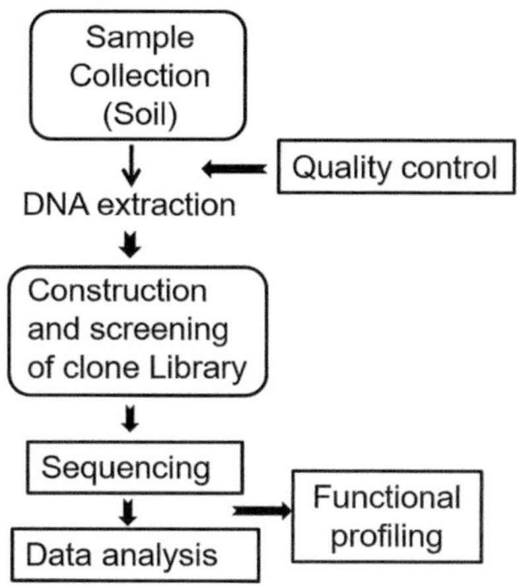

Fig. 5.4 A schematic work flow for experimental design of metagenomics

metabolites and proteins/enzymes. Recent advances in mass spectrometry techniques facilitate higher sensitive approach for identification of key metabolites and proteins from environmental samples which provide an emerging platform for soil proteomics study (Becher et al. 2013; Jansson and Baker 2016). A schematic work flow for experimental design of metagenomics was depicted in Fig. 5.4.

5.6.2 *Metaproteomics*

Metaproteomics or community proteogenomics has appeared as a complementary approach to metagenomics due to its large-scale profiling and characterization of proteins of environmental microbiota. It provides more information on linking genetics and diversity to gain insight into the functional characteristic of the microbial component (Maron et al. 2007; Becher et al. 2013). This approach utilizes the mass spectrometry (MS)-based highly sensitive technique and allows the quantitative and qualitative assessment of the thousands of protein component from the microbial community (Nannipieri and Smalla 2006; Simon and Daniel 2011; Lee et al. 1999). An early gel-based sodium dodecyl sulfate polyacrylamide gel electrophoresis (SDS-PAGE) soil microbiome protein study was conducted by Singleton et al. (2003) to investigate soil microbial protein and biomass as indicator of metal contamination (Singleton et al. 2003). In 2004, Wilmes and Bond studied first shotgun proteomics using 2D-PAGE and Q-ToF-MS (quadrupole time-of-flight mass spectrometry) with de novo peptide sequencing of a mixed bacterial community and identified depth coverage of proteins (over 2000 proteins) (Wilmes and Bond 2009); later several metaproteomics studies were carried out to reveal functional metabolic diversity, key metabolites, and enzymes (Maron et al. 2007; Becher et al. 2013). Metaproteomics work package typically comprises three basic steps: (1) protein extraction, purification, and enrichment of the concentration of protein, (2) protein/peptide separation and acquisition of both MS and MS/MS (tandem MS) level peptide, and (3) data analysis and microbiome community and functional characterization. A schematic work flow for experimental design of metaproteomics is depicted in Fig. 5.5.

Besides its immense potential, metaproteomics study of a highly complex environmental samples, like soil and water, is facing huge challenges and becomes restricted to certain extent. The possible reason for these limitations are as follows: (1) large complex heterogeneity of the sample (Whiteley and Lee 2006; Wilmes and Bond 2009) (e.g., soil and water); (2) low yield of protein, requiring well-established soil/water protein extraction protocol (Bastida et al. 2009; Chourey et al. 2010); (3) wide range of protein abundant level; and (4) requirement of high-end computer power and advanced bioinformatics algorithms for data search and interpretation.

5.7 Conclusion

Enhancing soil productivity through application of beneficial microbe sand enzymes without disturbing soil ecological structure is one of the key challenges in the current scenario as various anthropogenic activities led to increase environmental issues. In this chapter we discussed how microorganisms and soil system are correlated and co-dependent in almost all the biological regulations and maintenance of the ecosystem. To function as a sustainable environment for all the living and

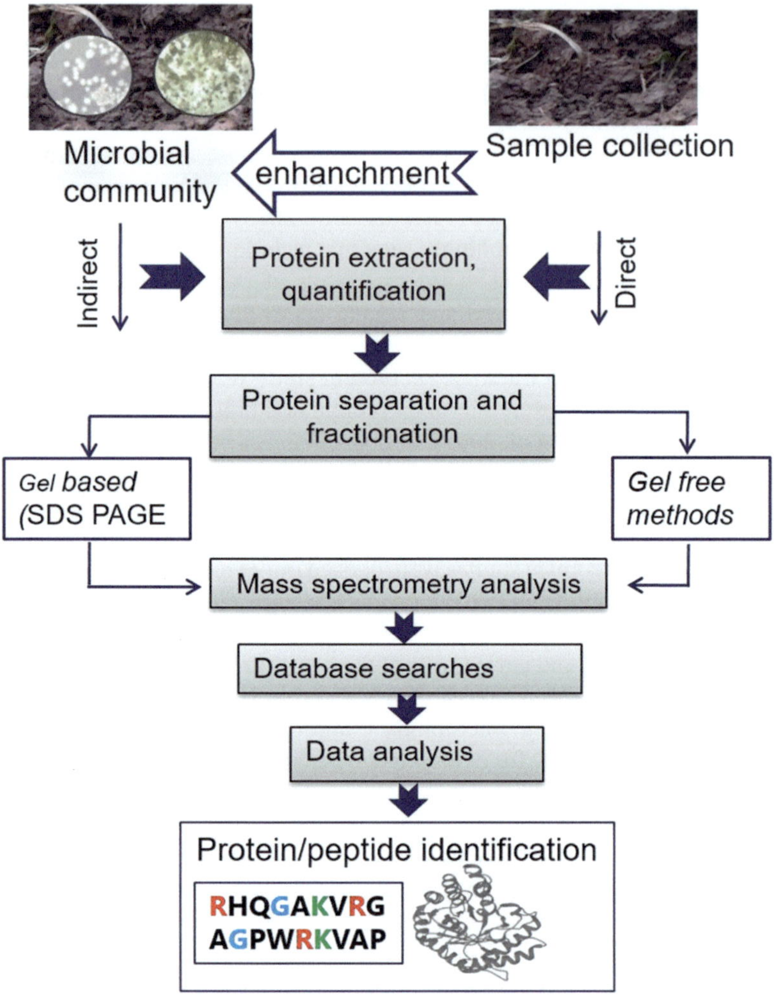

Fig. 5.5 A schematic work flow for experimental design of metaproteomics study

non-living system, soil greatly depend on the microbial community and its activities. In addition microbes play an important role in nutrient availability of plant growth as various types of rhizobacteria are living in or around soil rhizosphere, associated with plants tissues, and play an amazing role in crop production. Further microbes and enzymes also possess potential activity to detoxify harmful chemicals or pollutants present in soil through bioremediation process, thus playing as one of the most important factors in regulating soil health. The emergence of new sequencing technology through different "omics approaches" have provided a massive amount of soil microbial community profiling data to extract novel microbial pathways, metabolites, and antibiotics which would have a huge agroeconomical, industrial, and pharmaceutical potential to mankind.

Acknowledgments We thank the Director, CSIR-North East Institute of Science and Technology (CSIR-NEIST) for the support. We acknowledge Bioinformatics Infrastructure Facility (BIF) Centre of CSIR- NEIST, Jorhat, for providing computational research facilities. EJ, R. Das K acknowledge DST Govt. of India and KB, AT acknowledges DBT, Govt. of India, for providing their fellowship.

References

Abdel-Raouf N, Al-Homaidan AA, Ibraheem IBM (2012) Agricultural importance of algae. Afr J Biotechnol 11(54):11648–11658

Acosta-Martinez V, Mikha MM, Vigil MF (2007) Microbial communities and enzyme activities in soils under alternative crop rotations compared to wheat – fallow for the Central Great Plains. Appl Soil Ecol 37(1–2):41–52

Adetunji AT, Lewu FB, Mulidzi R, Ncube B (2017) The biological activities of β-glucosidase, phosphatase and urease as soil quality indicators: a review. J Soil Sci Plant Nutr 17:794–807

Ahemad M, Khan MS (2009) Effect of insecticide-tolerant and plant growth promoting Mesorhizobium on the performance of chickpea grown in insecticide stressed alluvial soils. J Crop Sci Biotechnol 12:213–222

Ahemad M, Khan MS (2010) Ameliorative effects of Mesorhizobium sp. MRC4 on chickpea yield and yield components under different doses of herbicide stress. Pestic Biochem Physiol 98:183–190

Ahemad M, Khan MS (2012a) Effect of fungicides on plant growth promoting activities of phosphate solubilizing Pseudomonas putida isolated from mustard (Brassica compestris) rhizosphere. Chemosphere 86:945–950

Ahemad M, Khan MS (2012b) Evaluation of plant growth promoting activities of rhizobacterium *Pseudomonas putida* under herbicide-stress. Ann Microbiol 62:1531–1540

Ahemad M, Kibretb M (2014) Mechanisms and applications of plant growth promoting rhizobacteria: current perspective. J King Saud Univ 26(1):1–20

Ahemad M, Malik A (2011) Bioaccumulation of heavy metals by zinc resistant bacteria isolated from agricultural soils irrigated with waste water. Bacteriology:12–21

Ahmed A, Nasim FUH, Batool K, Bibi A (2017) Microbial β-glucosidase: sources, production and applications. J Appl Environ Microbiol 5(1):31–46

Aislabie J, Deslippe JR (2013) Soil microbes and their contribution to soil services. Soil Microb Diver:142–161

Alkorta I, Epelde L, Garbisu C (2017) Environmental parameters altered by climate change affect the activity of soil microorganisms involved in bioremediation. FEMS Microbiol Lett 364:19

Allison SD, Gartner T, Holland K, Weintraub M, Sinsabaugh R (2007) Soil enzymes: linking proteomics and ecological processes. In: Hurst C, Crawford R, Garland J, Lipson D, Mills A, Stetzenbach L (eds) Manual of environmental microbiology. ASM Press, Washington, DC: 704–711

Almeida RFD, Naves ER, Mota RPD (2015) Soil quality: enzymatic activity of soil β- glucosidase. Glob J Agricult Res Rev 3:146–150

Antoun H, Prévost D (2005) Ecology of plant growth promoting rhizobacteria. In: Siddiqui ZA (ed) PGPR: biocontrol and biofertilization. Springer, Dordrecht:1–38

Antoun H, Beauchamp CJ, Goussard N, Chabot R, Lalande R (1998) Potential of Rhizobium and Bradyrhizobium species as plant growth promoting rhizobacteria on non-legumes: effect on radishes (*Raphanus sativus* L.). Mol Microb Ecol Soil:57–67

Armengaud J, Marie HE, Bland C (2013) Proteogenomics for environmental microbiology. Proteomics 13:2731–2742

Arora PK, Kumar M, Chauhan A, Raghava GP, Jain RK (2009) OxDBase: a database of oxygenases involved in biodegradation. BMC Res Notes 2:67

Arora PK, Srivastava A, Singh VP (2010) Application of Monooxygenases in dehalogenation, desulfurization, denitrification and hydroxylation of aromatic compounds. J Bioremed Biodegr 1:1–8

Arrigo KR (2005) Marine microorganisms and global nutrient cycles. Nature 437:349–355

Asmar F, Eiland F, Nielson NE (1994) Effect of extracellular-enzyme activities on solubilization rate of soil organic nitrogen. Biol Fertil Soils 17:32–38

Azubuike CC, Chikere CB, Okpokwasili GC (2016) Bioremediation techniques–classification based on site of application: principles, advantages, limitations and prospects. World J Microbiol Biotechnol 32(11):180

Bansal N, Kanwar SS (2013) Peroxidase (s) in environment protection.The Scientific World Journal 2013, 714639:1–9

Barazani O, Friedman J (1999) Is IAA the major root growth factor secreted from plant-growth-mediating bacteria? J Chem Ecol 25(10):2397–2406

Barker DM (2010) Archaea: salt-lovers, methane-makers, thermophiles, and other archaeans. Crabtree, St. Catharines:48p

Barr D (2002) Biological methods for assessment and remediation of contaminated land: case studies. Construction Industry Research and Information Association, London

Bastida F, Moreno JL, Nicolas C, Hernandez T et al (2009) Soil metaproteomics: a review of an emerging environmental science. Significance, methodology and perspectives. Eur J Soil Sci 60:845–859

Bautista-Cruz A, Ortiz-Hernandez YD (2015) Hydrolytic soil enzymes and their response to fertilization: a short review. Comunicata Scientiae 6:255

Becher D, Bernhardt J, Fuchs S, Riedel K (2013) Metaproteomics to unravel major microbial players in leaf litter and soil environments: challenges and perspectives. Proteomics 13(18–19):2895–2909

Belimov AA, Hontzeas N, Safronova VI, Demchinskaya SV, Piluzza G, Bullitta S, Glick BR (2005) Cadmium-tolerant plant growth-promoting bacteria associated with the roots of Indian mustard (*Brassica juncea* L. Czern.). Soil Biol Biochem 37(2):241–250

Bell CW, Fricks BE, Rocca JD, Steinweg JM, McMahon SK, Wallenstein MD (2013) High-throughput fluorometric measurement of potential soil extracellular enzyme activities. J Vis Exp 81:50961

Bennet JW, Wunch KG, Faison BD (2002) Use of fungi biodegradation. ASM Press, Washington, DC: 960–971

Bhagavan NV, Ha C-E (2015) Chapter 15-protein and amino acid metabolism. Essent Med Biochem:227–268

Bhardwaj D, Ansari MW, Sahoo RK, Tuteja N (2013) Bio-fertilizers function as key player in sustainable agriculture by improving soil fertility, plant tolerance and crop productivity. Microbial Cell Factor 1:66

Bhattacharyya PN, Jha DK (2012) Plant growth-promoting rhizobacteria (PGPR): emergence in agriculture. World J Microbiol Biotechnol 28(4):1327–1350

Boddey RM, Polidoro JC, Resende AS, Alves BJR, Urquiaga S (2001) Use of the 15N natural abundance technique for the quantification of the contribution of N2 fixation to sugar cane and other grasses. Funct Plant Biol 28:889–895

Boschker HTS, Vasquez-Cardenas D, Bolhuis H, Moerdijk-Poortvliet TWC, Moodley L (2014) Chemoautotrophic carbon fixation rates and active bacterial communities in intertidal marine sediments. Plos One 9:e101443

Braud A, Jézéquel K, Bazot S, Lebeau T (2009a) Enhanced phytoextraction of an agricultural Cr-, Hg- and Pb-contaminated soil by bioaugmentation with siderophore-producing bacteria. Chemosphere 74:280–286

Braud A, Jézéquel K, Bazot S, Lebeau T (2009b) Enhanced phytoextraction of an agricultural Cr-and Pb-contaminated soil by bioaugmentation with siderophore-producing bacteria. Chemosphere 74(2):280–286

Bull H, Murray PG, Thomas D, Fraser AM, Nelson PN (2000) Acid phosphatases. Mol Pathol 55(2):65

Burd GI, Dixon DG, Glick BR (2000) Plant growth promoting bacteria that decrease heavy metal toxicity in plants. Can J Microbiol 46:237–245

Burton SG (1994) Biocatalysis with polyphenol oxidase: a review. Catal Today 22(3):459–487

Cazorla FM, Romero D, Pérez-García A, Lugtenberg BJJ, de Vicente A, Bloemberg G (2007) Isolation and characterization of antagonistic Bacillus subtilis strains fromthe avocado rhizoplane displaying biocontrol activity. J Appl Microbiol 103(5):1950–1959

Cenciani K, Freitas SS, Critter SAM, Airoldi C (2011) Enzymatic activity measured by microcalorimetry in soil amended with organic residues. R Bras Ci Solo 35:1167–1175

Chaparro JM, Amy M, Danial K, Jorge M (2012) Manipulating the soil microbiome to increase soil health and plant fertility. Biol Fertil Soils 48:489–499

Chen HJ (2003) Phosphatase activity and P fractions in soils of an 18–year old Chinese fir (*Cunninghamia lanceolata*) plantation. For Ecol Manag 178:301–310

Chibuike GU, Obiora SC (2014) Heavy metal polluted soils: effect on plants and bioremediation methods. Appl Environ Soil Sci 752708:12

Chourey K, Jansson J, Ver Berkmoes N, Shah M et al (2010) Direct cellular lysis/protein extraction protocol for soil metaproteomics. J Proteome Res 9:6615–6622

Compant S, Duffy B, Nowak J, Clement C, Barka EA (2005) Use of plant growth-promoting bacteria for biocontrol of plant diseases: principles, mechanisms of action, and future prospects. Appl Environ Microbiol 71:4951–4959

Conley DJ, Paerl HW, Howarth RW, Boesch DF, Seitzinger SP et al (2009) Controlling eutrophication: nitrogen and phosphorus. Ecology 323(5917):1014–1015

Corstanje R, Schulin R, Lark R (2007) Scale. dependent relationships between soil organic carbon and urease activity. Eur J Soil Sci 58:1087–1095

Crowther TW, Boddy L, Jones TH (2012) Functional and ecological consequences of saprotrophic fungus–grazer interactions. Intl Soc Microbial Ecol 6:1992–2001

Cunningham SJT, Sumner ME (1997) Assessing soil quality for environmental purposes: roles and challenges for soil scientists. J Environ Qual 26:20–25

Daniel R (2005) The metagenomics of soil. J Nat Rev Microbiol 3(6):470–478

Darrah PR, Harris PJ (1986) Afluorometric method for measuring the activity of soil enzymes. Plant Soil 92:81–88

Dary M, Chamber-Pérez MA, Palomares AJ, Pajuelo E (2010) "In situ" phytostabilisation of heavy metal polluted soils using *Lupinus luteus* inoculated with metal resistant plant-growth promoting rhizobacteria. J Hazard Mater 177:323–330

Das SK, Varma A (2010) Roleofenzymesinmaintainingsoilhealth. Soilenzymology:25–42

Das SK, Varma A (2011a) Role of enzymes in maintaining soil health. In: Shukla G, Varma A (eds) Soil enzymology. Springer, Berlin, pp 43–60

Das SK, Varma A (2011b) Role of enzymes in maintaining soil health. Soil Enzymology, Soil biology. Springer, Berlin :25–42

David M, Jos M (2002) Raaijmakers, Brian B. McSpadden Gardener, Linda S. Thomashow, Microbial populations responsible for specific soil suppressiveness to plant pathogens. *Annu. Rev*. Phytopathology 40:309–348

Delmont TO, Robe P, Clark I, Simonet P, Vogel TM (2011) Metagenomic comparison of direct and indirect soil DNA extraction approaches. J Microbiol Methods 86:397–400

Dick RP, Breakwell DP, Turco RF (1996) Soil enzyme activities and biodiversity measurements as integrative microbiological indicators. In: Doran JW, Jones AJ (eds) Methods of assessing soil quality. Soil Science Society of America, Madison, pp 247–271

Diez MC, Gianfreda L (2008) Transformation of chlorinated phenolic compounds by White Rot Fungi Olga Rubilar. J Critic Rev Environ Sci Technol 38:4

Dijkhuizen L, Harder W (1984) Current views on the regulation of autotrophic carbon dioxide fixation via the Calvin cycle in bacteria. Antonie Van Leeuwenhoek 50:473–487

Dong WY, Zhang XY, Liu XY, Fu XL, Chen FS et al (2015) Responses of soil microbial communities and enzyme activities to nitrogen and phosphorus additions in Chinese fir plantations of subtropical China. Biogeosciences 12:5537–5546

Doran JW, Safley M (1997) Defining and assessing soil health and sustainable productivity. In: Pankhurst CE, Doube BM, Gupta VVSR (eds) Biological indicators of soil health. CAB International, Wallingford: 2–23

Dou D, Kale SD, Wang X, Jiang RHY, Bruce NA, Arredondo FD, Zhang X, Tyler BM (2008) RXLR-mediated entry of Phytophthora sojae effector Avr 1b into soybean cells does not require pathogen-encoded machinery. Plant Cell 20(7):1930–1947

El-Ghonemy DH (2015) Microbial amidases and their industrial applications: a review. J Med Microbiol Diag 4(1):1

El-Yazeid AA, Abou-Aly HA, Mady MA, Moussa SAM (2007) Enhancing growth, productivity and quality of squash plants using phosphate dissolving microorganisms (bio phosphor) combined with boron foliar spray. Res J Agric Biol Sci 3(4):274–286

Fetzner S, Lingens F (1994) Bacterial dehalogenases: biochemistry, genetics, and biotechnological applications. Microbiol Rev 58(4):641–685

Fierer N (2017) Embracing the unknown: disentangling the complexities of the soil microbiome. Nat Rev Biotechnol:1–12

Francis CA, Beman JM, Kuypers MMM (2007) New processes and players in the nitrogen cycle: the microbial ecology of anaerobic and archaeal ammonia oxidation. Intl Soc Microb Ecol 1:19–27

Garcia de Salamone IE, Hynes RK, Nelson LM (2001) Cytokinin production by plant growth promoting rhizobacteria and selected mutants. Can J Microbiol 47:404–411

García JAL, Probanza A, Ramos B, Palomino M, Manero FJG (2004) Effect of inoculation of Bacillus licheniformis on tomato and pepper. Agronomie 24:169–176

Genrich IB, Dixon DG, Glick BR (1998) A plant growth-promoting bacterium that decreases nickel toxicity in seedlings. Appl Environ Microbiol 64:3663–3668

Guilpart N, Grassini P, Sadras VO, Timsina J, Cassman KG (2017) Estimating yield gaps at the cropping system level. Field Crop Res 206:21–32

Han W, Kemmitt SJ, Brookes PC (2007) Soil microbial biomass and activity in Chinese tea gardens of varying stand age and productivity. Soil Biol Biochem 39:1468–1478

Handelsman J, Rondon MR, Brady SF, Clardy J, Goodman RM (1998) Molecular biological access to the chemistry of unknown soil microbes: a new frontier for natural products. Chem Biol:245–249

Harris J (2009) Soil microbial communities and restorationecology: facilitators or flowers? Science 325:573–574

Hayat R, Ali S, Amara U, Khalid R, Ahmed I (2010) Soil beneficial bacteria and their role in plant growth promotion: a review. Ann Microbiol 60:579–598

Herndl GJ, Weinbauer MG (2003) Marine microbial food web structure and function. Mar Sci Front Europe:265–277

Hynes RK, Leung GC, Hirkala DL, Nelson LM (2008) Isolation, selection, and characterization of beneficial rhizobacteria from pea, lentil and chickpea grown in western Canada. Can J Microbiol 54:248–258

Indira Gandhi P, Anandham R, Madhaiyan M, Sa TM (2008) Characterization of plant growth–promoting traits of bacteria isolated from larval guts of diamondback moth Plutella xylostella (Lepidoptera: Plutellidae). Curr Microbiol 56(4):327–333

Ingham ER (1999). Soil Protozoa, The Soil Biology Primer Chapter 4. Natural Resources Conservation Service Soils. United States Department of Agriculture

Iqbal MA, Bartakke KV (2014) Isolation of pesticide degrading microorganisms from soil. Adv Biores 4:164–168

Isopi R, Fabbri P, Del Gallo M, Puppi G (1995) Dual inoculation of *Sorghum bicolor* (L.) Moench ssp. bicolor with vesicular arbuscular mycorrhizas and Acetobacter diazotrophicus. Symbiosis Philadelphia 18:43–55

Iturbe R, López J (2015) Bioremediation for a Soil Contaminated with Hydrocarbons. J Pet Environ Biotechnol 6:1–6

Jackson LE, Bowles TM, Hodson AK, Lazcano C (2012) Soil microbial-root and microbial- rhizosphere processes to increase nitrogen availability and retention in agroecosystems. Curr Opin Environ Sustain 4:517–522

James Cook R (2006) Towardcroppingsystemsthatenhanceproductivityandsustainability. PNAS 103:18389–18394

Jansson JK, Baker ES (2016) A multi-omic future for microbiome studies. J Nat Microbiol 1:16049

Javaid MK, Ashiq M, Tahir M (2016) Potential of biological agents in decontamination of agricultural soil. Scientifica (Cairo) 2016:1598325

Jha PN, Kumar A (2007) Endophytic colonization of Typha australis by a plant growth- promoting bacterium Klebsiella oxytoca strain GR-3. J Appl Microbiol 103:1311–1320

Jin K, Sleutel S, Buchan D, De Neve S, Cai D, Gabriëls D et al (2009) Changes of soil enzyme activities under different tillage practices in the Chinese Loess Plateau. Soil Till Res 104:115–120

Johns C (2017) Living soils: the role of microorganisms in soil health. Fut Direct Intl:1–7

Joo GJ, Kin YM, Kim JT, Rhee IK, Kim JH (2005) Lee IJ Gibberellins-producing rhizobacteria increase endogenous gibberellins content and promote growth of red peppers. J Microbiol 43(6):510–515

Kaczynska G, Borowik A, Wyszkowska J (2015) Soil dehydrogenases as an indicator of contamination of the environment with petroleum products. Water Air Soil Pollut 226:372

Kandeler E (2015) Physiological and biochemical methods for studying soil biota and their functions. In: Soil microbiology, ecology and biochemistry. Elsevier:187–222

Karaca A, Cetin SC, Turgay OC, Kizilkaya R (2010) Soil enzymes as indication of soil quality. Soil Enzymol:237–261

Karigar CS, Rao SS (2011) Role of microbial enzymes in the bioremediation of pollutants: a review. Enzyme Res 805187:11

Kaushik R, Saxena AK, Tilak KVBR (2000) Selection of Tn5: lacZ mutants isogenic to wild type *Azospirillum brasilense* strains capable of growing at sub-optimal temperature. World J Microbiol Biotechnol 16:567–570

Khan S, Cao Q, Hesham AE-L, Xia Y, He J-Z (2007) Soil enzymatic activities and microbial community structure with different application rates of Cd and Pb. Environ Sci 19:834–840

Kheyrodin H (2014) Methodology for measurement of enzyme activity in soil. World J Biol Med Sci 1:18–25

Kibblewhite MG, Ritz K, Swift MJ (2008) Soil health in agricultural systems. Phil Trans R Soc B 363:685–701

Kloepper JW, Schroth MN (1978) Plant growth-promoting rhizobacteria on radishes. In: Proceedings of the 4th international conference on plant pathogenic bacteria, Station de Pathologie Vegetale et de Phytobacteriologie, INRA, Angers, France 2: 879–882

Kourtev PS, Ehrenfeld JG, Huang WZ (2002) Enzyme activities during litter decomposition of two exotic and two native plant species in hardwood forests of New Jersey. Soil Biol Biochem 34(9):1207–1218

Kuiper I, Lagendijk EL, Bloemberg GV, Lugtenberg BJJ (2004) Rhizoremediation: a beneficial plant-microbe interaction. Mol Plant-Microbe Interact 7:6–15

Kumar S, Chaudhuri S, Maiti SK (2013) Soil dehydrogenase enzyme activity in natural and mine soil – a review. Middle-East J Sci Res 13:898–906

Lee N, Nielsen PH, Andreasen KH, Juretschko S et al (1999) Combination of fluorescent in situ hybridization and microautoradiography—a new tool for structure-function analyses in microbial ecology. Appl Environ Microbiol 65:1289–1297

Ling LL, Schneider T, Peoples AJ, Spoering AL, Engels I, Conlon BP, Mueller A, Schaberle TF, Hughes DE, Epstein S et al (2015) A new antibiotic kills pathogens without detectable resistance. Nature 22:455

Lucas GJA, Probanza A, Ramos B, Palomino MR, Gutierrez Manero FJ (2004) Effect of inoculation of *Bacillus licheniformis* on tomato and pepper. Agronomie 24:169–176

Lugtenberg B, Kamilova F (2009) Plant-growth-promoting rhizobacteria. Annu Rev Microbiol 63:541–556

Ma Y, Rajkumar M (2009) H. Freitas Isolation and characterization of Ni mobilizing PGPB from serpentine soils and their potential in promoting plant growth and Ni accumulation by Brassica spp. Chemosphere 75:719–725

Mai C, Schormann W, Milstein O, Huttermann A (2000) Enhanced stability of laccase in the presence of phenolic compounds. Appl Microbiol Biotechnol 54(4):510–514

Maitra PK, Lobo Z (1971) A kinetic study of glycolytic enzyme synthesis in yeast. J Biol Chem 246:475–488

Margalef O, Sardans J, Fernandez-Martinez M, Molowny Horas R, Janssens IA, Ciais P, Goll D, Richter A, Obersteiner M, Asensio D, Penuelas J (2017) Global patterns of phosphatase activity in natural soils. Sci Rep 7:1337.26

Margesin R, Schinner F (2001) Bioremediation (Natural Attenuation and Biostimulation) of Diesel-Oil-Contaminated Soil in an Alpine Glacier Skiing Area. Appl Environ Microbiol 67:3127–3133

Margesin R, Zimmerbauer A, Schinner F (1999) Soil lipase activity – a useful indicator of oil biodegradation. Biotechnol Tech 13:859–863

Maron PA, Ranjard L, Mougel C, Lemanceau P (2007) Metaproteomics: a new approach for studying functional microbial ecology. Microb Ecol 53:486–493

Mary Kensa V (2011) Bioremediation – an overview. J Indust Pollut Contr 27:161–168

McLaren AD (1975) Soil as a system of humus and clay immobilised enzymes. Chem Scripta 8:97–99

Meagher RB (2000) Phytoremediation of toxic elemental organic pollutants. Curr Opin Plant Biol 3:153–162

Megharaj M et al (2017) Soil and brown field bioremediation. Microb Biotechnol 5:1244–1249

Mehnaz S, Baig DN, Lazarovits G (2010) Genetic and phenotypic diversity of plant growth promoting rhizobacteria isolated from sugarcane plants growing in Pakistan. J Microbiol Biotechnol 20:1614–1623

Mitidieri S, Martinelli AHS, Schrank A, Vainstein MH (2006) Enzymatic detergent formulation containing amylase from *Aspergillus niger*: a comparative study with commercial detergent formulations. Bioresour Technol 97:1217–1224

Mobini-Dehkordi M, Javan FA (2012) Application of alpha-amylase in biotechnology. J Biol 1:39–50

Mohammadi K, Sohrabi Y (2012) Bacterial bio-fertilizers for sustainable crop production: a review. ARPN J Agricult Biol Sci 7(5): 307–316

Mohammadian E, Ahari AB, Arzanlou M, Oustan S, Khazaei SH (2017) Tolerance to heavy metals in filamentous fungi isolated from contaminated mining soils in the Zanjan Province, Iran. Chemosphere 185:290–296

Mulligana CN, Yongb RN (2004) Natural attenuation of contaminatedsoils. Environ Int 30:587–601

Murphy JF, Zehnder GW, Schuster DJ, Sikora EJ, Polston JE (2000) Kloepper JW Plant growth-promoting rhizobacterial mediated protection in tomato against tomato mottle virus. Plant Dis 84:779–784

Muthukamalam S, Sivagangavathi S, Dhrishya D, Rani SS (2017) Characterization of dioxygenases and biosurfactants produced by crude oil degrading soil bacteria. J Microbiol 48(4):637–647

Nannipieri P, Smalla K (eds) (2006) Role of stabilised enzymes in microbial ecology and enzyme extraction from soil with potential applications in soil proteomics. Ser Soil Biol 8:75–94

Nannipieri P, Giagnoni L, Landi L, Renella G (2011) Role of phosphatase enzymes in soil. Soil Biol Ser:215–243

Naorungrote S, Chunglok W, Monthon L, Bangrak P (2011) Actinomycetes producing anti-methicillin resistant *Staphylococcus aureus* from soil samples in Nakhon Si Thammarat. J Walailak 8:2

Noel TC, Sheng C, Yost CK, Pharis RP, Hynes MF (1996) Rhizobiumleguminosarum as a plant growth-promoting rhizobacterium: direct growth promotion of canola and lettuce. Can J Microbiol 42:279–283

Okolo JC, Nweke CO, Nwabueze RN, Dike CU, Nwanyanwu CE (2007) Toxicity of phenolic compounds to oxidoreductases of Acinetobacter species isolated from a tropical soil. Sci Res Essay 2(7):244–250

Pandey P, Trivedi B, Kumar LMS (2006) Palni Characterization of a phosphate solubilizing and antagonistic strain of *Pseudomonas putida* (B0) isolated from a Sub-Alpine Location in the Indian Central Himalaya. Curr Microbiol 53:102–107

Park J-W, Park B-K, Kim J-E (2006) Remediation of soil contaminated with 2, 4- dichlorophenol by treatment of minced shepherd's purse roots. Arch Environ Contam Toxicol 50(2):191–195

Paul EA, Clark FE (1989) Soil microbiology and biochemistry. Academic, San Diego:272p

Phi Q-T, Park Y-M, Seul K-J, Ryu C-M, Park S-H, Kim J-G, Ghim S-Y (2010) Assessment of root-associated *Paenibacillus polymyxa* groups on growth promotion and induced systemic resistance in peppera. J Microbiol Biotechnol 20:1605–1613

Philp JC, Atlas RM (2005) Bioremediation of contaminated soils and aquifers. Bioremediation: applied microbial solutions for real-world environmental cleanup. American Society for Microbiology (ASM) Press, Washington, DC, pp 139–236

Philp JC, Bamforth SM et al (2005) Environmental pollution and restoration: a role for bioremediation. Bioremed Appl Microb Solut Real-World Environ Cleanup:1–48

Piontek K, Smith AT, Blodig W (2001) Ligninperoxidasestructureandfunction. Biochem Soc Trans 29(2):111–116

Piotrowska-Dlugosz A, Wilczewski E (2014) Changes in enzyme activities as affected by green-manure catch crops and mineral nitrogen fertilization. Zemdirbyste (Agriculture) 101:139–146

Poonguzhali S, Madhaiyan M, Sa T (2008) Isolation and identification of phosphate solubilizing bacteria from chinese cabbage and their effect on growth and phosphorus utilization of plants. J Microbiol Biotechnol 18(4):773–777

Pritsch K, Raidl S, Marksteiner E, Blaschke H, Agerer R, Schloter M, Hartmann A (2004) A rapid and highly sensitive method for measuring enzyme activities in single mycorrhizal tips using 4-methylumbelliferone-labelled fluorogenic substrates in a microplate system. J Microbiol Methods 58:233–241

Rajkumar M (2008) H. Freitas Effects of inoculation of plant growth promoting bacteria on Ni uptake by Indian mustard. Bioresour Technol 99:3491–3498

Rajkumar M, Ae N, Prasad MNV (2012) H. Freitas Potential of siderophore-producing bacteria for improving heavy metal phytoextraction. Trends Biotechnol 28:142–149

Rana SS, Rana MC (2011) Cropping system. Department of Agronomy, College of Agriculture, CSK Himachal Pradesh Krishi Vishvavidyalaya, Palampur: 80p

Rani K, Dhania G (2014) Bioremediation and biodegradation of pesticide from contaminated soil and water – a noval approach. Inter J Cur Micro App Sci 3:23–33

Rani K, Datt S, Rana R (2012) Brief review on alkaline phosphatases: an overview. Intl J Microbiol Bioinform 2(1):1–4

Rao MB, Tanksale AM, Ghatge MS, Deshpande VV (1998) Molecular and biotechnological aspects of microbial proteases. Microbiol Mol Biol Rev 62(3):597–635

Riah W, Laval K, Laroche-Ajzenberg E (2014) Effects of pesticides on soil enzymes: a review. Environ Chem Lett 12:257–273

Richardson AE, Simpson RJ (2011) Soil microorganisms mediating phosphorus availability. Plant Physiol 156:989–996

Richter DD, Markewitz D (1995) How deep is soil? Bioscience 45:600–609

Rodrigues EP, Rodrigues LS, de Oliveira ALM, Baldani VLD, Teixeira KRS, Urquiaga S, Reis VM (2008) Azospirillum amazonense inoculation: effects on growth, yield and N2 fixation of rice (*Oryza sativa* L.). Plant Soil 302:249–261

Rohrbacher F, St-Arnaud M (2016) Root exudation: the ecological driver of hydrocarbon rhizoremediation. Agronomy 6:19

Rokhbakhsh-Zamin F, Sachdev D, Kazemi-Pour N, Engineer A, Pardesi KR, Zinjarde S, Dhakephalkar PK, Chopade BA (2011) Characterization of plant-growth- promoting traits of Acinetobacter species isolated from rhizosphere of *Pennisetum glaucum*. J Microbiol Biotechnol 21:556–566

Rokhzadi A, Asgharzadeh A, Darvish F, Nourmohammadi G, Majidi E (2008) Influence of plant growth-promoting rhizobacteria on dry matter accumulation and yield of chickpea (*Cicer arietinum* L.) under field condition. Am-Euras J Agric Environ Sci 3(2):253–257

Roy AS, Yenn R, Singh AK, Deka Boruah HP, Saikia N, Deka M (2013) Bioremediation of crude oil contaminated tea plantation soil using two *Pseudomonas aeruginosa* strains AS 03 and NA 108 Afr J Biotechnol 12(19):2600–2610

Rubilar O, Diez MC, Gianfreda L (2008) Transformation of chlorinated phenolic compounds by white rot fungi. Crit Rev Environ Sci Technol 38(4):227–268

Saha S, Prakash V, Kundu S, Kumar N, Mina BL (2008) Soil enzymatic activity as affected by long term application of farm yard manure and mineral fertilizer under a rainfed soybean – wheat system in N-W Himalaya. Eur J Soil Biol 44:309–315

Saha S, Loganathan M, Rai AB, Singh A, Garg R (2016) Role of microbes in soil health improvement. SATSA Mukhapatra Ann Tech Iss 20:53–62

Saikia J, Sarma RK, Dhandia R, Yadav A, Gupta VK, Bharali R, Saikia R (2018) Alleviation of drought stress in pulse crops with ACC deaminase producing rhizobacteria isolated from acidic soil of Northeast India. Scientific Rep 8:3560

Schroth MN, Hancock JG (1982) Disease-suppressive soil and root-colonizing bacteria. Science 216:1376–1381

Schulz S, Brankatschk R, Dumig A, Kogel-Knabner I, Schloter M et al (2013) The role of microorganisms at different stages of ecosystem development for soil formation. Biogeosciences 10:3983–3996

Selvakumar G, Mohan M, Kundu S, Gupta AD, Joshi P, Nazim S, Gupta HS (2008) Cold tolerance and plant growth promotion potential of Serratia marcescens strain SRM (MTCC 8708) isolated from flowers of summer squash (*Cucurbita pepo*). Lett Appl Microbiol 46(2):171–175

Shapira R, Ordentlich A, Chet I, Oppenheim AB (1989) Control of plant diseases by chitinase expressed from cloned DNA in *Escherichia coli*. Phytopathology 79:1246–1249

Sharma D, Sharma B, Shukla AK (2011) Biotechnological approach of microbial lipase: a review. Biotechnology 10(1):23–40

Sherene T (2017) Role of soil enzymes in nutrient transformation: a review. Bio Bull 3:109–131

Shonkor K, Das AV (2011) Role of enzymes in maintaining soil health. In: Soil enzymology, Soil biology. Springer, Berlin

Shraddha RS, Sehgal S, Kamthania M, Kumar A (2011) Laccase: microbial sources, production, purification, and potential biotechnological applications. Enzyme Res 2011:217861

Shridhar BS (2012) Review: nitrogen fixing microorganisms. Intl J Microbiol Res 3:46–52

Siddiqui IA, Ehteshamul-Haque S, Shaukat SS (2001) Use of rhizobacteria in the control of root rot-root knot disease complex of mungbean. J Phytopathol 149:337–346

Simon C, Daniel R (2011) Metagenomic analyses: past and future trends. Appl Environ Microbiol 77:1153–1161

Singh R, Kumar M, Mittal A (2016) Amylases: a note on current applications. Intl Res J Biol Sci 5(11):27–32

Singleton I, Merrington G, Colvan S, Delahunty JS (2003) The potential of soil protein- based methods to indicate metal contamination. Appl Soil Ecol 23:25–32

Sinsabaugh RL, Antibus RK, Linkins AE (1991) An enzymic approach to the analysis of microbial activity during plant litter decomposition. Agric Ecosyst Environ 34:42–54

Smith SE, Smith SE, Smith FA, Jakobsen I (2003) Mycorrhizal fungi can dominate phosphate supply to plants irrespective of growth responses. Plant Physiol 133:16–20

Soil bioremediation-EPA guideline (2005) November 2005, pp 1–9

Somers E, Vanderleyden J, Srinivasan M (2004) Rhizosphere bacterial signalling, a love parade beneath our feet. Crit Rev Microbiol 30:205–235

Supramaniam Y, Chong C-W, Silvaraj S (2016) Irene Kit-Ping Tana, Effect of short term variation in temperature and water content on the bacterial community in a tropical soil. Appl Soil Ecol 107:279–289

Suslow TV, Kloepper JW, Schroth MN, Burr TJ (1979) Beneficial bacteria enhance plant growth. Calif Agric 33:15–17

Tabatabai MA (1994) Soil enzymes. In: Weaver RW, Angle JS, Bottomley PS (eds) Methods of soil analysis, part 2. Microbiological and biochemical properties. Soil Science Society of America, Madison, pp 775–833

Tank N, Saraf M (2009) Enhancement of plant growth and decontamination of nickel-spiked soil using PGPR. J Basic Microbiol 49(2):195–204
Thakuria D, Talukdar NC, Goswami C, Hazarika S, Boro RC, Khan MR (2004) Characterization and screening of bacteria from rhizosphere of rice grown in acidic soils of Assam. Curr Sci 86:978–985
Timmusk S, Nicander B, Granhall U, Tillberg E (1999) Cytokinin production by *Paenibacillus polymyxa*. Soil Biol Biochem 31:1847–1852
Torsvik V, Ovreas L (2002a) Microbial diversity and function in soil: from genes to ecosystems. Curr Opin Microbiol 5:240–245
Torsvik V, Ovreas L (2002b) Microbialdiversityandfunctioninsoil: fromgenestoecosystems. Curr Opin Microbiol 5:240–245
Tripathi M, Munot HP, Shouch Y, Meyer JM, Goel R (2005) Isolation and functional characterization of siderophore-producing lead- and cadmium-resistant *Pseudomonas putida* KNP9. Curr Microbiol 5:233–237
Tsukihara T, Honda Y, Sakai R, Watanabe T, Watanabe T (2006) Exclusive overproduction of recombinant versatile peroxidase MnP2 by genetically modified white rot fungus, *Pleurotus ostreatus*. J Biotechnol 126(4):431–439
Tveit AT, Urich T, Svenning MM (2014) Metatranscriptomic analysis of arctic peat soil microbiota. Environ Microbiol 80:18
Upson RH, Haugland RP, Malekzadeh MN, Haugland RP (1996) Anal Biochem 243:41–45
Utobo E, Tewari L (2015) Soil enzymes as bioindicators of soil ecosystem status. Appl Ecol Environ Res 2015(13):147–169
van Hellemond EW, Janssen DB, Fraaije MW (2007) Discovery of a novel styrene monooxygenase originating from the metagenome. Appl Environ Microbiol 73:5832–5839
Vasconcellos RLF, Bonfim JA, Andreote FD, Mendes LW, Baretta D, Cardoso EJBN (2013) Microbiological indicators of soil quality in a riparian forest recovery gradient. Ecol Eng 53:313–320
Verma A, Kukreja K, Pathak DV, Suneja S, Narula N (2001) *In vitro* production of plant growth regulators (PGRs) by Azorobacter chroococcum. Indian J Microbiol 41:305–307
Vessey JK (2003) Vessey Plant growth promoting rhizobacteria as bio-fertilizers. Plant Soil 255:571–586
Vidali M (2001) Bioremediation. An overview. Pure Appl Chem 73:1163–1172
Viswanath B, Rajesh B, Janardhan A et al (2014) Fungal laccases and their applications in bioremediation. Enzyme Res 2014(163242):21
Waldrop MP, Balser TC, Firestone MK (2000) Linking microbial community composition to function in a tropical soil. Soil Biol Biochem 32:1837–1846
Wani PA, Khan MS (2010) Bacillus species enhance growth parameters of chickpea (*Cicer arietinum* L.) in chromium stressed soils. Food Chem Toxicol 48:3262–3267
Wani PA, Khan MS, Zaidi A (2007) Synergistic effects of the inoculation with nitrogen fixing and phosphate solubilizing rhizo- bacteria on the performance of field grown chickpea. J Plant Nutr Soil Sci 170:283–287
Wariishi H, Valli K, Gold MH (1992) Manganese (II) oxidation by manganese peroxidase from the basidiomycete Phanerochaete chrysosporium. Kinetic mechanism and role of chelators. J Biol Chem 267(33):23688–23695
Whelan MJ, Coulon F, Hince G et al (2015) Fate and transport of petroleum hydrocarbons in engineered biopiles in polar regions. Chemosphere 131:232–240
Whipps JM (2001) Microbial interactions and biocontrol in the rhizosphere. J Exp Bot 52:487–511
Whiteley CG, Lee DJ (2006) Enzyme technology and biological remediation. Enzym Microb Technol 38:291–316
Wick B, Kuhne RF, Vlek PLG (1998) Soil microbiological parameters as indicators of soil quality under improved fallow management systems in south-western Nigeria. Plant Soil 1998:97–107
Wilmes P, Bond PL (2009) Microbial community proteomics: elucidating the catalysts and metabolic mechanisms that drive the Earth's biogeochemical cycles. Curr Opin Microbiol 12:310–317

Yang M, Yang D, Xuan Y (2018) Soil microbial communities and enzyme activities in sea- buckthorn *(Hippophaerhamnoides)* plantation at different ages. Plos One 13(1):e0190959

Yoshida S (1998) Reaction of manganese peroxidase of Bjerkandera adusta with synthetic lignin in acetone solution. J Wood Sci 44(6):486–490

Yuan BC, Yue DX (2012) Soil microbial and enzymatic activities across a chronosequence of Chinese pine plantation development on the Loess Plateau of China. Pedosphere 22:1–12

Zaidi S, Usmani S, Singh BR, Musarrat J (2006) Significance of Bacillus subtilis strain SJ-101 as a bioinoculant for concurrent plant growth promotion and nickel accumulation in *Brassica juncea*. Chemosphere 64(6):991–997

Zehnder GW, Murphy JF, Sikora EJ, Kloepper JW (2001) Application to rhizobacteria for induced resistance. Eur J Plant Pathol 107:39–50

Zeman N, McCrea J (1985) Alpha-amylase production using a recombinant DNA organism. Cereal Food World 30:777–780

Zhang WW, Lu ZT, Yang K, Zhu JJ (2017) Impacts of conversion from secondary forests to larch plantations on the structure and function of microbial communities. Appl Soil Ecol 111:73–83

Zsolt K, Robert T, Eniko NN, Victor BS, Mariana T, Orsolya M, Zita F (2015) Arylsulfatase A: an important metabolic factor in pathophysiology of different diseases. Acta Medica Marisiensis 61(3):233–235

Chapter 6
Marine Microbes in Bioremediation: Current Status and Future Trends

Neetu Sharma, Abhinashi Singh, Sonu Bhatia, and Navneet Batra

Abstract As per evolutionary studies, life was believed to be originated in the marine environment. About 70% of earth's surface is covered with water which hosts a wide variety of life forms under extreme conditions. Vast research has been carried out to explore the terrestrial habitats for a variety of products, but the marine environment still remains little explored. The marine microorganisms have undergone great evolutionary changes due to variable and extreme marine conditions. Hence enzymes from marine microflora bear novel properties with wide applications in multidisciplinary areas. Due to highly challenging environmental conditions, the aquatic ecosystem inhabits microorganisms that produce enzymes having unique/novel characteristics such as thermostability, cold adaptability, high pressure, pH, and extreme salt tolerance. *Desulfurococcus* sp., *Pyrococcus* sp., *Thermococcus* sp., and *Geobacillus* sp. are aquatic producers of thermostable amylases, peptidases, and lipases. Cold-active enzymes such as beta-glycosidases and peptidases have been isolated from psychrophiles inhabiting in cold marine areas such as deep-sea muds. Other polysaccharide-degrading enzymes are also well studied in aquatic systems including chitinase, alginate lyases, agarases, carrageenans, and cellulose hydrolases. Halophilic microbes from waters of the Pacific Ocean, Black Sea, and Mediterranean Sea have been explored for enzymes like beta-D-galactosidase, alpha-D-galactosidase, etc. These enzymes have a wide range of applicability in pulp and paper, biofuel, food, and textile industry, replacing the conventional processes and making the process eco-friendly and cost-effective. Further fungal enzymes lignin peroxidase, manganese peroxidase, and laccase can be used in the treatment and bioremediation of industrial effluents and wastewater contaminants which escapes traditional treatment processes. This chapter deals with the review of the research work associated with the present scenario of marine microbes in bioremediation and their future trends.

Keywords Bioremediation · Oxygenases · Degradation · Pollutants · Marine enzymes

N. Sharma (✉) · A. Singh · S. Bhatia · N. Batra
GGDSD College, Chandigarh, India

A. Kumar, S. Sharma (eds.), *Microbes and Enzymes in Soil Health and Bioremediation*, Microorganisms for Sustainability 16,
https://doi.org/10.1007/978-981-13-9117-0_6

6.1 Introduction

Environmental quality has degraded globally due to large-scale production of various chemicals including xenobiotics such as polychlorinated biphenyls (PCBs), trichloroethylene (TCE), perchloroethylene (PCE), trinitrotoluene (TNT), etc. These compounds are highly toxic and are biomagnified via the food web. Bioremediation is considered as an efficient biotechnological approach which uses microbes to detoxify and degrade environmental contaminants. Microbial metabolism has the capability to transform organic contaminants into compounds which can be processed through natural biogeochemical cycles. A number of bacterial and fungal strains metabolizing environmental contaminants have been isolated from the natural environments, and the toxic chemical degradation-encoding genes have been analyzed. 71% of the earth's surface accounts for marine waters, generating 32% of the world's net primary production (Alexander 1999). Marine organisms are associated with a wide array of functions like antibiotic and enzyme production, marine light absorption (Stramski and Kiefer 1998), heavy metal bioremediation (Rainbow 1995), biosurfactant production (Maneerat and Phetrong 2007), biodegradation and bioremediation of hydrocarbons (Margesin and Schinner 2001), oil biodegradation, and bioremediation of diesel-contaminated soils (Gallego et al. 2001).

Bioremediation based on marine microorganisms holds promise in decreasing the level of harmful compounds and restoration of the environment by employing various metabolic techniques such as degradation, transformation, and accumulation of toxic compounds (Karigar and Rao 2011). Higher efficiency is exhibited by bioremediation processes as compared to physicochemical processes for removal of heavy metals even at a low metal concentration (Gadd and White 1993).

Marine microorganisms show better physiological and biochemical adaptations to the physical, chemical, and biological conditions of the ocean environment. There are a number of changes that occur sporadically in the marine environment including sea surface temperature, pH of the surroundings, changes in light and UV light patterns, sea level rise, tropical storms, etc. Microbial community structure and function can also be altered by factors like rainfall and river flood which adds pollutants and xenobiotics into the seawater. Extreme halophiles which have the ability to tolerate high salt concentrations strive in marine ecosystems (Bowers et al. 2009). Marine bacteria can also adapt to nutritionally deprived conditions and can survive through long starvation periods (Wai et al. 1999). Lower temperatures and high pressures of marine environment favor psychrophilic and barophilic microorganisms (Delong and Yayanos 1987). Marine microorganisms can adapt swiftly to the varying, noxious environments which make them suitable for bioremediations. Attempts have been made to hasten bioremediation by using bioaugmentation which involves the introduction of exogenous microorganisms into the polluted environments. These microorganisms are monitored for their role in pollutant degradation and to evaluate its influence on the ecosystem.

The marine microorganisms contribute to around 90% of the total biomass present in oceans and play an important role in the cycling of nutrients owing to their diverse metabolic activities. They are often classified as extremophiles owing to their continuous exposure to extreme pH, pressure, temperature, salinity, and oxygen concentrations at various depths (Dash et al. 2013). According to the studies carried by Dewapriya and Kim (2014), approximately 3.6×10^{29} bacterial cells constitute the ocean biomass. But the reported population is quite less due to the failure of cultivation techniques used to isolate them. Most of the marine bacteria cannot be isolated by conventional culture-based methods, so there is a need to have culture-independent techniques like metagenomic approach to identify these novel groups of bacteria which still remains unexplored under the depths of oceans. More research is being focused on exploring fungal- and bacterial-based whole cells due to the flexibility of manipulation in them by using principles of genetic engineering (Zhang and Kim 2012).

6.2 Characteristic Features of Marine Bacteria

The characteristic features present in marine bacteria are different from their terrestrial counterparts and requires sodium and potassium ions which helps to maintain their cytoplasmic osmolarity in high salt conditions. The other important function of sodium ion is to facilitate the uptake of substrates for the growth of cells (Hase et al. 2001). The other similar component reported by Robertson et al. (1990) was β-glutamate utilized by marine bacteria to maintain osmotic balance. The oligotrophy is another feature of marine microbes which is associated with the low availability of nutrients in the marine environment. But most of the marine bacteria are reported to carry out degradation of complex organic components to release simpler organic components to be utilized as potential raw materials for growth (Purushothaman 1998). This feature can be utilized as a potent bioremediation agent. The population and diversity of bacterial members varies with the depth and local conditions. The different diversity is reported in different water bodies across the world. Piskorska et al. (2007) documented the presence of alpha-, beta-, gamma-proteobacteria, flavobacteria, actinobacteria, and bacilli in the Indian Ocean. The other such similar study reported the presences of *Desulfococcus*, *Desulfitobacterium*, and *Syntrophus* as dominant member species in the Pacific Ocean (Inagaki et al. 2006).

6.3 Approaches for Studying Marine Bacterial Diversity

Conventional culture-based methods can be employed to isolate approximately 1% of the microbes present in the environment. The potential isolation sites for the bacteria vary from marine sediments, coral reefs, mangrove sites to deep sea vents.

The metagenomics approach is the most studied approach in the last few years to explore uncultured microbes from different environmental samples including marine area. The process is based on direct extraction of DNA and if required amplification followed by making metagenomic libraries (Felczykowska et al. 2015). This is followed by cloning in selected hosts like *E. coli* and genera *Bacillus*, *Pseudomonas*, and *Streptomyces* and their screening for transformants for required characteristics based on functional or sequence-based approach (Aakvik et al. 2009). Based on the aim of the study, sequence-based study involves the use of various techniques like PCR, microarray, probe hybridization, high-throughput sequencing, etc. With the increasing advancement, high-throughput sequencing technique is popularly employed for the identification of isolates. The results of sequencing are further processed using different bioinformatics tools and compared with the available sequences in the genebank and assessed for their novelty. Function-based technique involves lengthy procedure, but the proficiency of obtaining novel organisms if any is high as compared to sequence-based approach. The major problems associated with the above approach is the selection of expression host to carry out psychrophilic expression (Pulicherla and Rao 2013). The expression of a catabolic gene is evaluated either using plate assays based on the formation of clear zones due to the presence of enzymatic activity. The other two available methods are SIGEX (substrate-induced gene expression) and PIGEX (product-induced gene expression) screening. The former is based on substrate-induced expression of gene under study, while the latter is based on the enzymatic product-based evaluation (Uchiyama and Watanabe 2008; Uchiyama and Miyazaki 2010). The list of some native and genetically engineered marine bacteria is discussed in Tables 6.1 and 6.2.

Table 6.1 List of important marine bacteria employed in bioremediation

Bacteria	Genes involved	Uses
Pseudomonas putida G7	nahAc gene	PAH (naphthalene, phenanthrene, or pyrene) (Lee et al. 2018)
Pseudomonas putida	cad A, cad B	Cadmium degradation (Chellaiah 2018)
Pseudomonas aeruginosa	mer A, mer B	Inorganic mercury degrdadtion (Dash and Das 2012)
Alcaligenes eutrophus CH34	czc gene cluster	Cd, Zn, Co degradation (Nies and Silver 1995)
Achromobacter sp. HZ01	Omp A, Omp H	Degradation of hydrocarbons, bioemulsifier production (Hong et al. 2016)
Staphylococccus aureus	chr B	Chromate reduction (Aguilar-Barajas et al. 2008)
Bacillus sp.	czc D	Cobalt, zinc, cadmium degradation (Abdelatey et al. 2011)

Table 6.2 List of genetically engineered marine bacteria employed in bioremediation

Bacteria	Contaminant	References
P. putida X3 strain	Methyl parathion and cadmium	Zhang et al. (2016)
P. putida MC4-5222	1,2,3-Trichloropropane degradation	Samin et al. (2014)
E. coli, *Pseudomonas* sp.	Degradation of plastic	Slater et al. (1988)
Alcaligenes eutrophus H850 Lr.	PCB	Van Dyke et al. (1996)
Pseudomonas fluorescens 4 K44	Naphthalene, anthracene	Sayler and Ripp (2000)
Pseudomonas fluorescens 105865	BTEX	Sousa et al. (1998)
Thalassospira Lucentensis	Hydrocarbon degradation	Sutiknowati (2010)

6.4 Other Organisms

Viruses are found to be dominant organisms in the marine environment. They are reported to play an important role in the cycling of important organic components including carbon and sulfur. They contributed to the addition of carbon levels in the oceanic waters by lysing bacteria. Apart from this their lytic action on phytoplanktons has been found to be associated with the release of DMS (dimethyl sulfide) gas into the atmosphere. DMS is produced by the breakdown of dimethyl sulphoniopropionate produced by the phytoplanktons (Reid and Edwards 2001). Actinomycetes, the close associates of prokaryotic members, were considered as potential sources of novel enzymes and metabolites. Around 32 genera were reported by Lecavalier and Lechevalier (1970) from marine sources including *Streptosporangium*, *Actinopolyspora*, *Micropolyspora*, *Rhodococcus*, *Nocardia*, etc. The widely studied trait of actinomycetes is its potential to produce a wide range of antibiotics. They found to harbor a wide variety of enzymes which aids in the cycling of organic components in the marine environment (Goodfellow and Haynes 1984).

6.5 Microbial Enzymes in Bioremediation

There is a rapid increase in the concentration of different pollutants in terrestrial and aquatic environments across the world. The various physicochemical approaches were utilized for removal of these contaminants but all in vein. The last decade saw a major change in the trend of treatment techniques and thus started exploring a new dimension of bioremediation. The term bioremediation refers to the use of living organisms including plants, microbes, and their products for the removal of contaminants. The main advantage of this strategy is that it is quite safe and delivers promising results as compared to the traditional physicochemical approaches. There are a series of enzymes involved in decontamination of particular site ranging from hydrolase, oxidoreductase, oxygenases, laccase, cellulase, protease, kinase, etc. Depending on the nature of existence in particular microbial system, enzymes are classified into extra- and intracellular enzymes. The key enzymes in bioremediation are reported to be belonging to the extracellular category. The metagenomic studies by Ferrer et al. (2016) reported the diversity of various enzymes in marine water in 211 screened samples found to be oxidoreductases (1%) and glycosidases (96%) with a maximum percentage of 77% of esterases and lipases. Hydrolases are known to play a key role in the initiation of degradation of various toxic compounds like carbamate, organophosphate, organochlorine insecticides like DDT, etc. by attacking major chemical bond on the compound and thus facilitating its further breakdown (Vasileva-Tonkova and Galabova 2003; Lal and Saxena 1982). The various classes of hydrolytic enzymes such as cellulase, hemicellulase, lipases, amylases, pullulanases, proteases, etc. have wide application in various fields ranging from food, pharmaceutical, to chemical industry and decontamination of terrestrial and aquatic bodies (Sanchez-Porro et al. 2003).

- Broad substrate specificity is exhibited by laccases as they show ability to oxidize range of xenobiotic compounds including polycyclic aromatic hydrocarbons, pesticides, and chlorinated phenols. Removal of aromatic amines and phenols from water can be achieved by laccase mainly by oxidation of the pollutant to free radicals or quinones which can be further polymerized and precipitate (Niku-Paavola and Viikari 2000). Laccases are also being employed in degradation of variety of chemical dyes used in textile industry (Couto et al. 2005). Synthetic dyes such as Azure B and Brilliant Blue R have been decolorized by laccases from *Flavodon flavus* (Soares et al. 2001). Laccases from *Pycnoprus sanguineus* showed complete decolorization of bromophenols blue and malachite green (Rodríguez et al. 1999). Resistant xenoestrogen nonylphenols can be degraded by aquatic hyphomycete, *Clavariopsis aquatic* (Muller et al. 2012).

Lipases are another important class of enzymes which can carry out the breakdown of triacylglycerols to glycerol and free fatty acids. They also have wide application in various industries. Lipase activity was reported to be the most important factor for studying the degradation of hydrocarbons in contaminated sites (Riffaldi et al. 2006). They are capable of carrying out a variety of reactions like aminolysis, alcoholysis, hydrolysis, and esterification (Prasad and Manjunath 2011). The mentioned catalytic reactions can be used to employ these enzymes in the degradation of oil. Proteases carried out the breakdown of proteinaceous material by breaking peptide bonds. Based on the position of action on peptide chains, proteases have been categorized into endopeptidases and exopeptidases. Due to this they have been widely studied and used in the industrial sector like proteases in the food industry, pharmaceuticals, and leather industry. These characteristics made them suitable candidates for bioremediation. Oxidoreductase enzymes played a crucial role in the breakdown of various hydrocarbons ranging from aliphatic to aromatic compounds prevailing in the environment (Fig. 6.1). The enzymes carry out oxidation-reduction process and carry out the transformation of toxic forms such as phenolic compounds to nontoxic residues by the aid of the oxidation-reduction process (Park et al. 2006). Different studies on degradation of phenolic compounds were carried out using

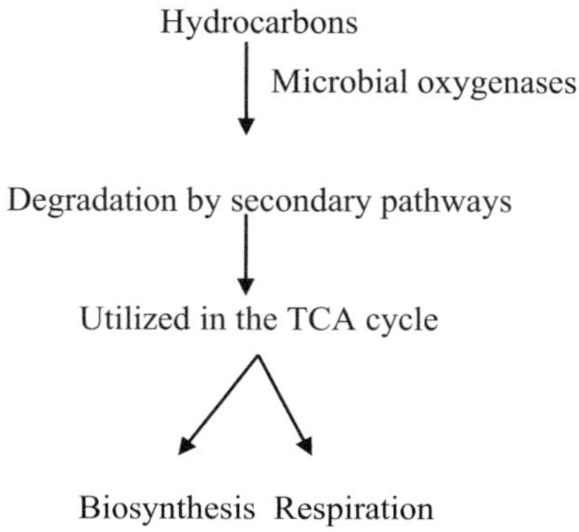

Fig. 6.1 Aerobic degradation of hydrocarbons catalyzed by oxygenases. (Fritsche and Hofrichter 2000)

fungi which showed successful removal as a result of enzymatic action of a cluster of extracellular enzymes like reductase and peroxidase (Rubilar et al. 2008). Another important class of enzymes belonging to oxidoreductases are oxygenases which are further categorized into mono and dioxygenases based on their ability to add oxygen to the substrate. The former added one atom of oxygen while the lateral can add two atoms of oxygen for the oxidation of the compound. They are reported to play a crucial role in the breakdown of a variety of organic components ranging from simpler aliphatic compounds to complex substituted aromatic rings (Arora et al. 2009). Most of the recalcitrant compounds comprising a range of insecticides, pesticides, and herbicides are made up of halogen-substituted organic compounds; thus oxygenases can be used for degradation of such compounds. Depending on the type of cofactor required, monooxygenases are categorized as flavin dependent which require flavin as a prosthetic group and NADP or NADPH as coenzymes and P450 monooxygenases containing heme as a prosthetic group. The former was reported to be present in both prokaryotes and eukaryotes and was successfully employed in the breakdown of a variety of aliphatic and aromatic compounds (Van Berkel et al. 2006; Urlacher et al. 2004). The halogenated aromatic ring containing compounds are the most persistent type of compounds and also pose a major risk to the environment and living beings including humans. Some such examples are dichloropheyltrichloroethane (DDT), trichlorophenol (TCP), polychlorinated biphenyls (PCBs), dioxin, 2,4-dichlorophenoxyacetic acid (2,4-D), etc. The bioremediation involving bacterial action requires the removal of the halogen group from the ring as the initiating step. The dehalogenation can be carried out by various mechanisms including reductive, hydrolytic, and oxygen-reliant dehalogenation (Copley 1997). The reductive dehalogenation involves use of hydrogen atom to remove halogen; hydrolytic dehalogenation is based on replacement with a hydroxyl group. The oxygenolytic dehalogenation is carried out by either mono or dioxygenases which involves replacing a halogen atom by oxygen-derived hydroxyl group (Copley 1997). Arora et al. (2009) reported the use of monooxygenases in a number of degradation and transformation processes like hydroxylation, dehalogenation, desulfurization, and denitrification of various compounds. Some of the important monooxygenase type dehalogenases are chlorophenol 4-monooxygenase, pentachlorophenol 4-monooxygenase, and 2,4,6-trichlorophenol monooxygenase, desulfurization process involving monooxygenases are DBT sulfone monooxygenases (DszA) and DBT monooxygenase (DszC), denitrification carrying enzymes are 2-nitrophenol 2-monooxygenase, 4-nitrophenol 2-monooxygenase, and 4-nitrophenol 4-monooxygenase (Arora et al. 2010; Li et al. 1996; Zhang et al. 2009).

Like monooxygenases, dioxygenases aids in the degradation of different classes of hydrocarbons by introducing molecular oxygen. They are known to play an important role in the oxidation of compounds belonging to the aromatic class. The catechol dioxygenases are reported to carry out degradation of various types of aromatic molecules into simpler degradable forms of aliphatic hydrocarbons in the soil (Que and Ho 1996). The dioxygenases responsible for hydroxylation of aromatic rings require NADP/NADPH as an electron donor. These are transferred by reductase enzyme to the terminal oxygenase component to form different products like catechol and protocatechuic acid which act as important precursors for cleavage of aromatic ring (Parales and Resnick 2006; Guzik et al. 2011). The further degradation can be mediated either by ortho or meta cleavage of the ring. Intradiol dioxygenases are known to carry out the intradiol cleavage by adding two atoms of oxygen at 1,2-position of catechol or its derivatives leading to the production of cis,cis-muconic acid or its derivatives which are further transformed into simpler hydrocarbons like pyruvic acid and acetaldehyde utilizing catechol pathway or pyruvic acid and oxaloacetic acid utilizing protocatechuate pathway (Latus et al. 1995; Guzik et al. 2011; Mohiuddin and Fakhruddin 2012).

Peroxidases are unique enzymes which are known to carry out the oxidation of various contaminants present in soil including phenolic compounds. Peroxidases can be classified into two types: heme and nonheme. The former type has been reported in prokaryotes, animals, and plants. The lateral type has been subdivided into three categories. In bacterial system, Class I enzymes include catalase peroxidases, lignin peroxidase belonging to Class II peroxidase, and Class III mainly belonging to plant peroxidases such as horseradish (Hiner et al. 2002).

6.6 Bioremediation Potential of Fungi

It has been well reported that both bacteria and fungi have tremendous bioremediation ability and target a wide range of environmental contaminants including petroleum products, heavy metals, and toxic industrial effluents. Fungi are considered to be a better bioremediating agent due to diverse morphology, large biomass, and specialized enzymatic metabolism for degradation of a large variety of pollutants. Fungus' ability to extend their mycelia networks and utilizing toxic pollutants as a growth substrate even at a low concentration makes them a promising tool for bioremediation. Studies have reported that fungi metabolize the pollutants to a much large extent in comparison to bacteria. For instance, bacteria transform polychlorinated dibenzo-*p*-dioxins and polychlorinated dibenzofurans less efficiently whereas white rot fungi mineralize these toxic compounds in a short time period (Harms et al. 2011). In recent years focus has shifted toward exploring the application of marine fungi and derived enzymes in bioremediation. Marine microbes show better adaptability and tolerance toward extreme environments such as high temperature, high pressure, variable pH, and high salt-containing environments than their terrestrial counterparts.

6.7 Applications of Marine Fungal Enzymes for Bioremediation

Pollutants from various industrial sources, heavy metals, petroleum hydrocarbons, radioactive materials, effluents from pharmaceutical industries, etc., contaminate and deteriorate the terrestrial and aquatic environment. Conventional physical and chemical treatment processes are not able to remove these pollutants, and they escape into the environment and persist for a longer time period. Degradation of environmental pollutants using biotechnological processes proved to be a better, safe, economical, and eco-friendly way toward cleaning of the environment. For this purpose, microorganisms like bacteria and algae have been used to bioremediate polluted environment (Harms et al. 2011; Barnes et al. 2018). In recent years, research focus has shifted toward exploring the bioremediation potential of fungi. In this regard, marine fungi due to great ecological diversity, better growth and survival adaptability in the extreme environment than bacteria, and synthesis of enzymes with novel characteristics show enormous potential for the efficient degradation and removal of organic and inorganic pollutants (Velmurugan and Lee 2012; Bonugli-Santos et al. 2015). Different species of marine fungi and their enzymes can be employed to degrade pollutants in terrestrial and aquatic environments (Table 6.3).

Table 6.3 List of marine fungi employed in bioremediation

Fungal species	Pollutant	References
Phialophora sp.	Dye	Torres et al. (2011)
Penicillium sp.	Dye	Torres et al. (2011)
Cladosporium sp.	Dye and PAHs	Torres et al. (2011) and Birolli et al. (2018)
Aspergillus sclerotiorum	PAHs	Passarini et al. (2011)
Mucor racemosus	PAHs	Passarini et al. (2011)
Phanerochaete chrysosporium	Dye	Cripps et al. (1990) and Sarma (2018)
Pleurotus ostreatus	PAH	Ghosal et al. (2016)
Trematophoma sp.	Aliphatic hydrocarbons and PAH	Moghimi et al. (2017)

6.8 Biodegradation of Heavy Metals

A large number of studies have been conducted throughout the globe describing the presence and persistence of heavy metals in both terrestrial and aquatic environment causing adverse effects to human health such as cancer. Heavy metals like chromium, cadmium, copper, nickel, arsenic, etc. are released into the environment by different anthropogenic activities and lead to their accumulation in soil and aquatic systems. Several bioremediation methods are being applied to either control or remove these metals. In this direction, marine fungi and their derived enzymes have shown the ability to biodegrade such metals and remove them from the environment.

6.9 Biodegradation of Polycyclic Aromatic Hydrocarbon Pollutants

Polycyclic aromatic hydrocarbons (PAHs) are organic molecules composed of two or more benzene rings. These are released into the environment by both natural and anthropogenic sources and can have large-scale implications due to their persistent nature. A large amount of PAHs is released by volcanic eruptions, burning of fossil fuels, oil spills, etc. Most commonly present PAHs in the environment include naphthalene, anthracene, pyrene, benzopyrene, etc. Due to PAHs' toxicity, carcinogenicity, and mutagenicity, they pose a great risk not only to human health but also to the complete ecosystem. In this view, a large number of studies have been conducted tapping the potential of fungi for biodegradation of PAHs. Passarini et al. (2011) investigated the potential of different fungal strains for biodegradation of PAHs. Results obtained showed *Aspergillus sclerotiorum* CBMAI 849 significantly reduced the pyrene and benzopyrene followed by *Mucor racemosus* which reduces benzopyrene CBMAI 847. PAH degradation has also been reported by Birolli et al. (2018) using *Cladosporium* sp. CBMAI 1237. Results obtained showed significant degradation of PAHs like anthracene, anthraquinone, anthrone, phenanthrene, fluoranthene, pyrene, etc.

Fungi use two types of enzymatic mechanisms to degrade PAHs, ligninolytic and non-ligninolytic (Fig. 6.2). The ligninolytic enzymatic mechanism involves the secretion of lignin-degrading enzymes, lignin peroxidase, manganese peroxidase, and laccase, and non-ligninolytic enzymatic mechanism involves the use of cytochrome P450 monooxygenase enzyme. Due to the efficient and broad spectrum degradation of PAHs, significant importance has been given to the ligninolytic enzyme system. Lignin peroxidise degradation mechanism involves the oxidation of aromatic ring of PAHs by the formation of radicals which further produces cations resulting in the destabilization of bonds. Laccases are copper-containing enzymes that catalyze the oxidation of molecular oxygen to water and oxidize a wide range of phenolic compounds (Kadri et al. 2017). D'Souza-Ticlo et al. (2009)

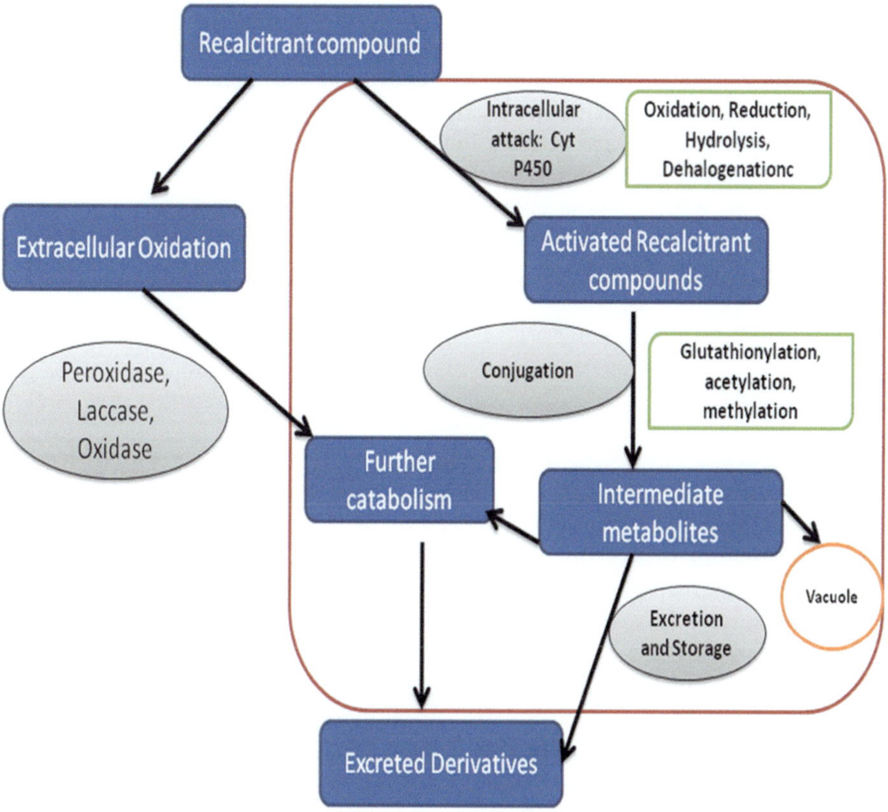

Fig. 6.2 Fungal mechanism (ligninolytic and non-ligninolytic) for bioremediation of toxic compounds. (Adapted from Deshmukh et al. 2016)

studied the bioremediation potential of a thermostable laccase from marine fungi *Cerrena unicolor*.

6.10 Degradation of Dyes and Textile Effluents

Besides several norms and regulations framed by different governmental agencies, a large amount of untreated dye-containing wastewater is released from textile industries around the globe causing severe water pollution and directly affecting both freshwater and marine ecosystem. Untreated effluent from textile industries commonly consists of synthetic dyes belonging to the class of anthraquinone, triphenylmethane, and azo dyes (Diwaniyan et al. 2010; Bonugli-Santos et al. 2015). These compounds are toxic, and being synthetic in nature, they are recalcitrant and persist for a longer time period in the ecosystem, thus adversely affecting the aquatic habitat. Various physicochemical methods are available for the removal or control of the release of dyes into the water bodies but are not of much applicability due to the involvement of chemicals, high cost, and limited efficiency (Bonugli-Santos et al. 2015). Enzymes from marine microbes can prove to be an efficient alternative for the treatment of dyes and textile effluents. Several marine fungi and their enzymatic system have been explored for their bioremediation potential. One such system comprises of ligninolytic enzymes having great bioremediation importance which can convert toxic dye components into nontoxic derivatives via oxidative mechanisms (Ciullini et al. 2008; Arun et al. 2008).

6.11 Conclusion

The studies related to marine enzymes lead to the exploration of the diversity of novel enzymes that have different properties than terrestrial habitat. Due to the extreme conditions like low light, high salinity and pressure, and varying temperatures of the marine ecosystem, it contributes toward the synthesis of enzymes with different characteristics having huge bioremediation potential. The available literature shows marine microorganism-derived enzymes have enormous capability to bioremediate toxic chemicals, polycyclic aromatic hydrocarbons, heavy metals, etc. But studies related to their long-term efficiency, gene modifications for large-scale production, and economic viability are still in their infancy, and there biotechnological methods for production and application must be studied in much detail.

Acknowledgments The authors are thankful to the Goswami Ganesh Dutta Sanatan Dharma College, Chandigarh for providing the facilities.

References

Aakvik T, Degnes KF, Dahlsrud R, Schmidt F, Dam R, Yu L, Völker U, Ellingsen TE, Valla S (2009) A plasmid RK2-based broad-host-range cloning vector useful for transfer of metagenomic libraries to a variety of bacterial species. FEMS Microbiol Lett 296(2):149–158

Abdelatey LM, Khalil WK, Ali TH, Mahrous KF (2011) Heavy metal resistance and gene expression analysis of metal resistance genes in gram-positive and gram-negative bacteria present in Egyptian soils. J Appl Sci Environ Sanitat 6(2)

Aguilar-Barajas E, Paluscio E, Cervantes C, Rensing C (2008) Expression of chromate resistance genes from Shewanella sp. strain ANA-3 in Escherichia coli. FEMS Microbiol Lett 285(1):97–100

Alexander DE (1999) Encyclopedia of environmental science. Springer, New York, 0-412-74050-8

Arora PK, Kumar M, Chauhan A, Raghava GP, Jain RK (2009) OxDBase: a database of oxygenases involved in biodegradation. BMC Res Not 2(1):67

Arun A, Raja PP, Arthi R, Ananthi M, Kumar KS, Eyini M (2008) Polycyclic aromatic hydrocarbons (PAHs) biodegradation by basidiomycetes fungi, Pseudomonas isolate, and their cocultures: comparative in vivo and in silico approach. Appl Biochem Biotechnol 151(2–3):132–142

Barnes NM, Khodse VB, Lotlikar NP, Meena RM, Damare SR (2018) Bioremediation potential of hydrocarbon-utilizing fungi from select marine niches of India. 3 Biotech *8*(1):21

Birolli WG, Santos DDA, Alvarenga N, Garcia AC, Romão LP, Porto AL (2018) Biodegradation of anthracene and several PAHs by the marine-derived fungus Cladosporium sp. CBMAI 1237. Mar Pollut Bull 129(2):525–533

Bonugli-Santos RC, dos Santos Vasconcelos MR, Passarini MR, Vieira GA, Lopes VC, Mainardi PH, Dos Santos JA, de Azevedo Duarte L, Otero IV, da Silva Yoshida AM, Feitosa VA (2015) Marine-derived fungi: diversity of enzymes and biotechnological applications. Front Microbiol 6:269

Bowers KJ, Mesbah NM, Wiegel J (2009) Biodiversity of poly-extremophilic Bacteria: does combining the extremes of high salt, alkaline pH and elevated temperature approach a physicochemical boundary for life? Saline Syst *5*(1):9

Chellaiah, E.R., 2018. Cadmium (heavy metals) bioremediation by Pseudomonas aeruginosa: a minireview. Applied water science, 8(6), p.154

Ciullini I, Tilli S, Scozzafava A, Briganti F (2008) Fungal laccase, cellobiose dehydrogenase, and chemical mediators: combined actions for the decolorization of different classes of textile dyes. Bioresour Technol 99(15):7003–7010

Copley SD (1997) Diverse mechanistic approaches to difficult chemical transformations: microbial dehalogenation of chlorinated aromatic compounds. Chem Biol 4(3):169–174

Couto SR, Sanromán MA, Gübitz GM (2005) Influence of redox mediators and metal ions on synthetic acid dye decolourization by crude laccase from Trametes hirsuta. Chemosphere 58(4):417–422

Cripps C, Bumpus JA, Aust SD (1990) Biodegradation of azo and heterocyclic dyes by Phanerochaete chrysosporium. Appl Environ Microbiol 56(4):1114–1118

D'Souza-Ticlo D, Sharma D, Raghukumar C (2009) A thermostable metal-tolerant laccase with bioremediation potential from a marine-derived fungus. Mar Biotechnol 11(6):725–737

Dash HR, Das S (2012) Bioremediation of mercury and the importance of bacterial mer genes. Intl Biodeterior Biodegrad 75:207–213

Dash HR, Mangwani N, Chakraborty J, Kumari S, Das S (2013) Marine bacteria: potential candidates for enhanced bioremediation. Appl Microbiol Biotechnol 97(2):561–571

Delong EF, Yayanos AA (1987) Properties of the glucose-transport system in some deep-sea bacteria. Appl Environ Microbiol 53:527–532

Deshmukh R, Khardenavis AA, Purohit HJ (2016) Diverse metabolic capacities of fungi for bioremediation. Indian J Microbiol 56(3):247–264

Dewapriya P, Kim SK (2014) Marine microorganisms: an emerging avenue in modern nutraceuticals and functional foods. Food Res Int 56:115–125

Diwaniyan S, Kharb D, Raghukumar C, Kuhad RC (2010) Decolorization of synthetic dyes and textile effluents by basidiomycetous fungi. Water Air Soil Pollut 210(1–4):409–419

Felczykowska A, Krajewska A, Zielińska S, Łoś JM, Bloch SK, Nejman-Faleńczyk B (2015) The most widespread problems in the function-based microbial metagenomics. Acta Biochim Pol 62(1):161

Ferrer M, Martínez-Martínez M, Bargiela R, Streit WR, Golyshina OV, Golyshin PN (2016) Estimating the success of enzyme bioprospecting through metagenomics: current status and future trends. Microb Biotechnol 9(1):22–34

Fritsche W, Hofrichter M (2000) Aerobic degradation by microorganisms. Biotechnology 11:146–164

Gadd GM, White C (1993) Microbial treatment of metal pollution: working biotechnology? Trends Biotechnol 11:353–359

Gallego JLR, Loredo J, Lamas JF, Azquez FV, Anchez JS (2001) Bioremediation of diesel-contaminated soils: evaluation of potential in situ techniques by the study of bacterial degradation. Biodegradation 12:325–335

Ghosal D, Ghosh S, Dutta TK, Ahn Y (2016) The current state of knowledge in microbial degradation of polycyclic aromatic hydrocarbons (PAHs): a review. Front Microbiol 7:1369

Goodfellow M, Haynes JA (1984) Actinomycetes in marine sediments. In: Oritz-Oritz L et al (eds) Biological, biochemical and biomedical aspects of actinomycetes. Academic, New York/ London, pp 453–472

Guzik U, Greń I, Hupert-Kocurek K, Wojcieszyńska D (2011) Catechol 1, 2-dioxygenase from the new aromatic compounds–degrading Pseudomonas putida strain N6. Int Biodeterior Biodegradation 65(3):504–512

Harms H, Schlosser D, Wick LY (2011) Untapped potential: exploiting fungi in bioremediation of hazardous chemicals. Nat Rev Microbiol *9*(3):177

Hase CC, Fedorova ND, Galperin MY, Dibrov PA (2001) Sodium ion cycle in bacterial pathogens: evidence from cross-genome comparisons. Microbiol Mol Biol Rev 65(3):353–370

Hiner AN, Ruiz JH, López JNR, Cánovas FG, Brisset NC, Smith AT, Arnao MB, Acosta M (2002) Reactions of the class II peroxidases, lignin peroxidase andArthromyces ramosus peroxidase, with hydrogen peroxide catalase-like activity, compound III formation, and enzyme inactivation. J Biol Chem 277(30):26879–26885

Hong YH, Deng MC, Xu XM, Wu CF, Xiao X, Zhu Q, Sun XX, Zhou QZ, Peng J, Yuan JP, Wang JH (2016) Characterization of the transcriptome of Achromobacter sp. HZ01 with the outstanding hydrocarbon-degrading ability. Gene 584(2):185–194

Inagaki F, Nunoura T, Nakagawa S, Teske A, Lever M, Lauer A, Suzuki M, Takai K, Delwiche M, Colwell FS, Nealson KH (2006) Biogeographical distribution and diversity of microbes in methane hydrate-bearing deep marine sediments on the Pacific Ocean Margin. Proc Natl Acad Sci 103(8):2815–2820

Kadri T, Rouissi T, Brar SK, Cledon M, Sarma S, Verma M (2017) Biodegradation of polycyclic aromatic hydrocarbons (PAHs) by fungal enzymes: a review. J Environ Sci 51:52–74

Karigar CS, Rao SS (2011) Role of microbial enzymes in the bioremediation of pollutants: a review. Enzyme Res 2011:1. https://doi.org/10.4061/2011/805187

Lal R, Saxena DM (1982) Accumulation, metabolism, and effects of organochlorine insecticides on microorganisms. Microbiol Rev 46(1):95

Latus M, Seitz H, Eberspacher J, Lingens F (1995) Purification and characterization of Hydroxyquinol 1, 2-dioxygenase from Azotobacter sp. strain GP1. Appl Environ Microbiol 61(7):2453–2460

Lecavalier MP, Lechevalier H (1970) Chemical composition as a criterion in the classification of aerobic actinomycetes. Int J Syst Evol Microbiol 20(4):435–443

Lee DW, Lee H, Lee AH, Kwon BO, Khim JS, Yim UH, Kim BS, Kim JJ (2018) Microbial community composition and PAHs removal potential of indigenous bacteria in oil contaminated sediment of Taean coast, Korea. Environ Pollut 234:503–512

Li MZ, Squires CH, Monticello DJ, Childs JD (1996) Genetic analysis of the dsz promoter and associated regulatory regions of *Rhodococcus erythropolis* IGTS8. J Bacteriol 178(22):6409–6418

Maneerat S, Phetrong K (2007) Isolation of biosurfactant-producing marine bacteria and characteristics of selected biosurfactant. Songklanakarin J Sci Technol 29:781–791

Margesin R, Schinner F (2001) Biodegradation and bioremediation of hydrocarbons in extreme environments. Appl Microbiol Biotechnol 56:650–663

Moghimi H, Tabar RH, Hamedi J (2017) Assessing the biodegradation of polycyclic aromatic hydrocarbons and laccase production by new fungus Trematophoma sp. UTMC 5003. World J Microbiol Biotechnol *33*(7):136

Mohiuddin M, Fakhruddin ANM (2012) Degradation of phenol via meta-cleavage pathway by Pseudomonas fluorescens PU1. ISRN Microbiol 2012:1–6

Nies DH, Silver S (1995) Ion efflux systems involved in bacterial metal resistances. J Ind Microbiol 14(2):186–199

Niku-Paavola ML, Viikari L (2000) Enzymatic oxidation of alkenes. J Mol Catal B Enzym 10(4):435–444

Parales RE, Resnick SM (2006) Aromatic ring-hydroxylating dioxygenases. In Pseudomonas. Springer, Boston, pp 287–340

Passarini MR, Rodrigues MV, da Silva M, Sette LD (2011) Marine-derived filamentous fungi and their potential application for polycyclic aromatic hydrocarbon bioremediation. Mar Pollut Bull 62(2):364–370

Piskorska M, Smith G, Weil E (2007) Bacteria associated with the coral *Echinopora lamellosa* (Esper 1795) in the Indian Ocean-Zanzibar region. Afr J Environ Sci Technol 1(5):93–98

Prasad MP, Manjunath K (2011) Comparative study on biodegradation of lipid-rich wastewater using lipase producing bacterial species. Indian J Biotechnol 10:121–124

Pulicherla KK, Rao KS (2013) Marine biocatalysts and their stability: a molecular approach. In: Marine enzymes for biocatalysis. Woodhead Publishing, Cambridge, pp 71–87

Purushothaman A (1998) Microbial diversity. In: Proceedings of the technical workshop on biodiversity of Gulf of Mannar Marine Biosphere Reserve, M. S. Swaminathan Research Foundation, Chennai, pp 86–91

Que L Jr, Ho RY (1996) Dioxygen activation by enzymes with mononuclear non-heme iron active sites. Chem Rev 96(7):2607–2624

Rainbow PS (1995) Bio monitoring of heavy metal availability in the marine environment. Mar Poll Bull 31:183–192

Reid PC, Edwards M (2001) Plankton and climate. In Steele JH et al (eds) Encyclopedia of ocean sciences, vol. 4. Academic, New York, pp 2194–2200

Riffaldi R, Levi-Minzi R, Cardelli R, Palumbo S, Saviozzi A (2006) Soil biological activities in monitoring the bioremediation of diesel oil-contaminated soil. Water Air Soil Pollut 170(1–4):3–15

Rubilar O, Diez MC, Gianfreda L (2008) Transformation of chlorinated phenolic compounds by white rot fungi. Crit Rev Environ Sci Technol 38(4):227–268

Samin G, Pavlova M, Arif MI, Postema CP, Damborsky J, Janssen DB (2014) A Pseudomonas putida strain genetically engineered for 1, 2, 3-trichloropropane bioremediation. Appl Environ Microbiol 80(17):5467–5476

Sarma VV (2018) Obligate marine fungi and bioremediation. In: Mycoremediation and environmental sustainability. Springer, Cham, pp 307–323

Sayler GS, Ripp S (2000) Field applications of genetically engineered microorganisms for bioremediation processes. Curr Opin Biotechnol 11(3):286–289

Slater SC, Voige WH, Dennis DE (1988) Cloning and expression in *Escherichia coli* of the Alcaligenes eutrophus H16 poly-beta-hydroxybutyrate biosynthetic pathway. J Bacteriol 170(10):4431–4436

Soares GM, Costa-Ferreira M, de Amorim MP (2001) Decolorization of an anthraquinone-type dye using a laccase formulation. Bioresour Technol 79(2):171–177

Sousa S, Duffy C, Weitz H, Glover LA, Bär E, Henkler R, Killham K (1998) Use of a lux-modified bacterial biosensor to identify constraints to bioremediation of btexcontaminated sites. Environ Toxicol Chem 17(6):1039–1045

Stramski D, Kiefer DA (1998) Can heterotrophic bacteria be important to marine light absorption? J Plankton Res 20:1489–1500

Sutiknowati LI (2010) Hydrocarbon-degrading bacteria: isolation and identification. Makara J Sci 11(2007):98–103

Torres JMO, Cardenas CV, Moron LS, Guzman APA, dela Cruz TEE (2011) Dye decolorization activities of marine-derived fungi isolated from Manila Bay and Calatagan Bay, Philippines. Philippine J Sci 140(2):133–143

Uchiyama T, Miyazaki K (2010) Product-induced gene expression, a product-responsive reporter assay used to screen metagenomic libraries for enzyme-encoding genes. Appl Environ Microbiol 76(21):7029–7035

UchiyamaT, Watanabe K (2008) Substrate-induced gene expression (SIGEX) screening of metagenome libraries. Nature Protocols 3(7):1202

Urlacher VB, Lutz-Wahl S, Schmid RD (2004) Microbial P450 enzymes in biotechnology. Appl Microbiol Biotechnol 64(3):317–325

Van Berkel WJH, Kamerbeek NM, Fraaije MW (2006) Flavoprotein monooxygenases, a diverse class of oxidative biocatalysts. J Biotechnol 124(4):670–689

Van Dyke MI, Lee H, Trevors JT (1996) Survival of luxAB-marked Alcaligenes eutrophus H850 in PCB-contaminated soil and sediment. J Chem Technol Biotechnol Intl Res Process Environ Clean Technol 65(2):115–122

Vasileva-Tonkova E, Galabova D (2003) Hydrolytic enzymes and surfactants of bacterial isolates from lubricant-contaminated wastewater. Zeitschrift für Naturforschung C 58(1–2):87–92

Velmurugan N, Lee YS (2012) Enzymes from marine fungi: current research and future prospects. Marine Fungi and Fungal-like Organisms (Marine and Freshwater Botany), pp 441–474

Wai SN, Mizunoe Y, Yoshida S (1999) How *Vibrio cholerae* survive during starvation. FEMS Microbiol Lett 180:123–131

Zhang C, Kim SK (2012) Application of marine microbial enzymes in the food and pharmaceutical industries. Advances in food and nutrition research, vol 65. Academic, New York, pp 423–435

Zhang JJ, Liu H, Xiao Y, Zhang XE, Zhou Zhang JJ, Liu H, Xiao Y, Zhang XE, Zhou NY (2009) Identification and characterization of catabolic para-nitrophenol 4-monooxygenase and para-benzoquinone reductase from Pseudomonas sp. strain WBC-3. J Bacteriol 191(8):2703–2710

Zhang R, Xu X, Chen W, Huang Q (2016) Genetically engineered Pseudomonas putida X3 strain and its potential ability to bioremediate soil microcosms contaminated with methyl parathion and cadmium. Appl Microbiol Biotechnol 100(4):1987–1997

Chapter 7
Role of Microbial Hydrolases in Bioremediation

Abhishek Sharma, Taruna Sharma, Tanvi Sharma, Shweta Sharma, and Shamsher Singh Kanwar

Abstract Bioremediation deals with the utilization of microorganisms to degrade environmental pollutants. Bioremediation principally depends upon those microorganisms which enzymatically attack the pollutants and convert them to less toxic or innocuous products. A large number of enzymes from bacteria, fungi, and plants degrade perilous organic pollutants to compounds like CO_2, CH_4, H_2O, and biomass without harmfully disturbing the environment. Bioremediation is an economically and environmentally pleasant biotechnological approach empowered by microbial enzymes. The knowledge of the mechanisms of bioremediation-related enzymes like hydrolases has been extensively studied in the present review. Microbial breakdown and environmental reactions like hydrolysis, a peculiar feature of lipases and esterases, can renovate toxic compounds into less toxic compounds. Bioremediation using these hydrolytic enzymes is a usually safe, cheaper, and eco-friendly system in eliminating the pollutants from the environment. The present review attempts to afford descriptive information on the lipases/esterases sourced from a number of microorganisms involved in the biodegradation of a broad series of pollutants.

Keywords Bioremediation · Enzymes · Environment · Lipases · Pollutants

A. Sharma · T. Sharma · S. S. Kanwar (✉)
Department of Biotechnology, Himachal Pradesh University, Shimla, India

T. Sharma
Department of Biotechnology and Bioinformatics, Jaypee University of Information Technology, Waknaghat, Solan, Himachal Pradesh, India

S. Sharma
Directorate of Mushroom Research (DMR), Waknaghat, Solan, Himachal Pradesh, India

A. Kumar, S. Sharma (eds.), *Microbes and Enzymes in Soil Health and Bioremediation*, Microorganisms for Sustainability 16,
https://doi.org/10.1007/978-981-13-9117-0_7

7.1 Introduction

Nature has always been self-dependent and self-sufficient to sustain its existence. To every issue that comes along, there lies a solution nearby. The self-cleaning capacity of natural sources has been one of the factors that influence the varied life on the planet. As the flowing stream of water has the capacity to carry the waste once mixed in it, to its side, so is the case with soil. The soil has the ability to degrade the waste after a certain period of time. The things have always been in a balanced form in nature, but as the balance of human population has exceeded to an irreversible extent, the same sources have now reached to such levels of pollution that now they are at the verge of harming the life instead of supporting it. Microorganisms have been one of the promising tools that maintain the self-cleaning capacity of natural sources. Thus, the organisms are being exploited to obtain the most efficient products to reverse the phase of exploitation. Therefore, the word "Bioremediation" means to utilize living organisms as a source to solve the definite environmental issues like contaminated soil or groundwater or to prevent pollution (Sasikumar and Papinazath 2003; Mohammed et al. 2011; Ron and Rosenberg 2014; Krzmarzick et al. 2018; Hu et al. 2018; Sahoo et al. 2018; Das et al. 2018; Chen et al. 2019).

Microorganisms are appropriate to the task of contaminant destruction because they possess enzymes that allow them to use environmental contaminants as a food and energy source. Bioremediation using microorganisms, which enzymatically degrade the hazardous organic pollutants and convert them to CO_2, CH_4, and H_2O without adversely affecting the environment (Ron and Rosenberg 2014; Yuniati 2018). Microbial hydrolytic enzymes (lipases or esterases) can be used in the management of waste produced during food processing, degradation of plastics and insecticides, treatment of biofilm deposits and oil-contaminated soils, etc. (Sood et al. 2016; Gangola et al. 2018; Austin et al. 2018; Hu et al. 2018; Sahoo et al. 2018; Luo et al. 2018; Islam and Roy 2018; Chen et al. 2019; Khan et al. 2019; Srivastava et al. 2019). Till date, these enzymes have been used for commercial production of various chemicals and synthetic reaction but these have very high potential to convert the harmful waste and use that as substrate. Also, bioremediation using hydrolytic enzymes like lipases and esterases is usually a least harmful, safe, cheaper, and eco-friendly method in removing the toxic products from the environment.

7.2 Various Microbial Enzymes Involved in Bioremediation

Enzymes are well-known to human beings since ancient era, and plenty of advantageous bioprocesses are practicable simply because of the enzymes (Sharma et al. 2016; Sood et al. 2016; Patel et al. 2017; Kumar et al. 2018a, b, c, d). Today more than 4000 enzymes are distinguished, and among them about 200 are in commercial use (Kumar and Kanwar 2011a, b; Sharma et al. 2017a, b, 2018a, b, c; Thakur et al. 2018a). Among microbes, bacteria are considered better sources of these enzymes than other higher organisms because of the ease with which bacterial cells may be cultured and genetically manipulated (Kumar and Kanwar 2012a, b, c, d; Sharma et al. 2017c, d; Jamwal et al. 2017).

Environmental pollution by natural and xenobiotic compounds has poor health and environmental effects, raising imperative concerns (Wang et al. 2017; Das et al. 2018). Microbial enzymes sufficiently degrade various industrial waste containing phenols, aromatic amines, nitriles, etc. to innocuous products (Sood et al. 2016; Gangola et al. 2018; Islam and Roy 2018). Several microbial enzymes such as oxygenases, lipases, esterases, amidases, amylases, amyloglucosidases, cellulases, nitrile hydratases, pectinases, proteases, etc. are used (Margesin et al. 1999; Riffaldi et al 2006; Karigar and Rao 2011) for waste treatment (Table 7.1). These microbial enzymes catalyze the elimination of chlorinated phenolic compounds from industrial wastes (Gianfreda et al. 1999; Mai et al. 2000; Have and Teunissen 2001; Piontek et al. 2001; Sood et al. 2016; Krzmarzick et al. 2018). Basically, these enzymes are incorporated to salvage various wastes or pollutants to formulate them adequately for reprocess, e.g., to recover extra oil from oilseeds, to change starch to sugar, and to change whey to a variety of valuable products (Hu et al. 2018; Sahoo et al. 2018; Luo et al. 2018). Microbial oxygenases have broad substrate specificity for a wide range of compounds, including the chlorinated aliphatic compounds (Fetzner and Lingens 1994; Arora et al. 2009). Hence, these microbial enzymes are functioning to degrade the halogenated contaminants also (Fetzner and Lingens 1994; Karigar and Rao 2011; Yuniati 2018; Kumar et al. 2018a, b, c, d; Khan et al. 2019; Chen et al. 2019).

7.3 Microbial Bioremediation by Hydrolytic Lipases/Esterases

The environmental pollution by industrial waste, household kitchen waste, oil spillage, and petroleum hydrocarbons is solemn trouble to the world in the present scenario. These environmental contaminants drastically affect the aquatic as well as terrestrial ecosystems. So, in order to save the environment for the better future, the "bioremediation technologies" using microbial hydrolytic lipases/esterases provide a secure and profitable substitute to frequently used remediation approaches. Bacterial enzymatic activity is a generally imperative process involved in the hydrolysis of organic pollutants. Extracellular microbial hydrolytic activity is a foremost step in degradation and consumption of organic polymers as only those compounds can pass through cell pores which are of *Mr* 600 Da (Williams 1977; Vasileva-Tonkova and Galabova 2003). Hydrolytic microbial enzymes (esterases/ lipases) may split the most important chemical bonds like ester bond of chemical pollutant to change their toxic behavior. This feature of these peculiar enzymes is mainly

Table 7.1 Major microbial enzymes in bioremediation

Enzyme	Role in bioremediation	Industrial use	References
Oxygenases	Degrade organic compounds by escalating their water solubility, can split aromatic molecules, and can also perform dehalogenation reactions of polyhalogenated compounds	Biosensors, organic synthesis, and biofuels	Fetzner and Lingens (1994), Arora et al. (2009), and Yuniati (2018)
Monooxygenases	Degrade hydrocarbons like substituted methanes, alkanes, cycloalkanes, alkenes, haloalkenes, and aromatic heterocyclic hydrocarbons	Involved in bio-desulfurization, dehalogenation, denitrification, and hydroxylation of compounds	Arora et al. (2009) and Kumar et al. (2018a, b, c, d)
Dioxygenases	Degrade aromatic compounds into aliphatic products		Kim et al. (2002) and Krzmarzick et al. (2018)
Laccase	Depolymerization of lignin to an array of phenols and degradation of bisphenol A	Cleaning agents for certain water purification systems and catalysts for the manufacture of anticancer drugs	Kim et al. (2002), Gangola et al. (2018), and Das et al. (2018)
Esterases	To degrade man-made pollutants, such as plastics, polyurethane, and polyesters	Used in cosmetics, paper and pulp, feed processing, detergents, synthesis of carbohydrate derivatives, food additives, etc.	Bhardwaj et al. (2012), Sood et al. (2016), Sharma et al. (2016), Yoshida et al (2016), Gangola et al. (2018), Austin et al. (2018), Luo et al. (2018), Lopes et al. (2018), Bhatt et al. (2019), and Chen et al. (2019)
Lipases	To degrade cooking waste	Food, detergents, pharmaceutical, leather, textile, cosmetics, and paper industries	Hermansyah et al. (2007), Yoshida et al. (2016), Okino-Delgado et al. (2017), Hu et al. (2018), Sahoo et al. (2018), Liu et al. (2018), Khan et al. (2019), and Srivastava et al. (2019)
Cellulases	Convert waste cellulosic material into foods	The textile industry, paper and pulp industry, and manufacture of detergents	Karmakar Ray (2011) and Islam and Roy (2018)

(continued)

Table 7.1 (continued)

Enzyme	Role in bioremediation	Industrial use	References
Proteases	Hydrolyze peptide bonds	Pharmaceutical, manufacture of cheese and detergents	Beena et al. (2010) and Tavano et al. (2018)
Nitrilases	Remediate cyanide-polluted waste and noxious nitriles	Hydrolysis of nitrile compounds, synthesis of important carboxylic acids, and treatment of cyanide and toxic nitriles	Park et al. (2017)
Peroxidases	Degrade lignin and also oxidize Mn^{2+}, methoxybenzenes, and phenolic aromatic substrates	Treatment of industrial wastewaters. Development of cosmetic and dermatological products	Wong (2009), Tsukihara et al. (2006), Have and Teunissen (2001)

effective for the biodegradation of oil spill, food waste, plastic waste, organophosphate, and insecticides (Table 7.2).

Hydrolases catalyze reactions like condensations and alcoholysis, and the most important advantages of hydrolases are their ease of use, lack of cofactor stereoselectivity, and ability to tolerate the addition of water-miscible solvents. Extracellular hydrolytic enzymes like lipases and esterases have quite diverse potential usages in different areas such as food industry, feed additive, biomedical sciences, and chemical industries (Kumar and Kanwar 2011a, b; Kumar and Kanwar 2012a, b, c, d; Sood et al. 2016; Sharma et al. 2017a, b, c, d, e). These enzymes belong to the α/β hydrolase super family of enzymes that catalyze the hydrolysis and synthesis of ester bonds (Sharma et al. 2017a, 2018a, b, c). Esterases (EC 3.1.1.1) can hydrolyze short-chain fatty esters (≤C8), whereas lipases (EC3.1.1.3) hydrolyze long-chain acylglycerols (≥C8). These enzymes are excellent biocatalysts for various reactions like esterification, transesterification, aminolysis, alcoholysis, etc. Hence, their peculiar characteristics make them significant enzymes for various applications (Table 7.2) and also an imperative group of biocatalysts in organic chemistry as well as an indispensable tool in bioremediation (Jaeger and Eggert 2002; Sood et al. 2016; Sharma et al. 2017a, b, c, d, e).

Hydrolytic lipases can degrade lipid molecules derived from microorganisms, animals, and plants. Lipolytic activity is answerable for the severe diminution of the whole hydrocarbon in the contaminated area. Research undertaken in this field is probably to develop the understanding of the bioremediation of oil spills (Margesin et al. 1999; Riffaldi et al. 2006; Yoshida et al. 2016; Okino-Delgado et al. 2017; Hu et al. 2018; Sahoo et al. 2018; Liu et al. 2018; Chen et al. 2019). Microbial lipases are versatile biocatalyst because of their effective applications in the industries. Microbial lipases are ubiquitous enzymes which catalyze the lipolytic reactions at the lipid-water interface, where lipolytic moiety frequently forms stability among monomeric, micellar, and emulsified states.

Table 7.2 Major microbial lipases/esterases in bioremediation

Enzyme	Role in bioremediation	References
Pseudomonas sp. strain D2D3 lipase	Biodegradation of kitchen waste	Shon et al. (2002)
Rhizopus delemar lipase	Biodegradation of polylactic acid (PLA)	Masaki et al. (2005)
Comamonas acidovorans esterase	Biodegradation of polyurethane	Masaki et al. (2005)
Rhizopus delemar, *R. arrhizus*, and *Penicillium* lipase	Biodegradation of plastic waste like polycaprolactone, polybutylene succinate, and polyethylene adipate	Tokiwa et al. (2009)
Penicillium chrysogenum lipase	High lipid content wastes in used cooking oil	Kumar et al. (2012)
Agrobacterium radiobacter phosphotriesterases	Degradation of insecticides	Riya and Jagatpati (2012)
Ralstonia sp. DI-3 lipases	Degradation of diazinon an organophosphorus insecticide	Wang and Liu (2016)
Aspergillus ibericus and *Aspergillus uvarum* lipase	Bioremediation of olive oil extraction wastes	Salgado et al. (2016)
Synechocystis esterase	Biodegradation of dimethyl phthalate	Zhang et al. (2016)
Citrus sinensis (Orange plant waste) lipase	Bioremediation of cooking oil waste	Okino-Delgado et al. (2017)
Ideonella sakaiensis polyesterase	Degradation of plastic waste like poly(ethylene terephthalate) and mono(2-hydroxyethyl) terephthalic acid	Yoshida et al. (2016) and Austin et al. (2018)
Clostridium botulinum esterase	Degradation of plastic waste like poly(ethylene terephthalate)	Biundo et al. (2016)
Acinetobacter sp. strain LMB-5 esterase	Biodegradation of phthalate esters	Yue et al. (2017)
M. ruber, *M. sanguineus*, and *Monascus* sp. esterase	Polyurethane a plastic product biodegradation	El-Morsy et al. (2017)
Recombinant esterase (*Mucor*)	Bioremediation of pyrethroid-contaminated vegetables	Fan et al. (2017)
Bacillus licheniformis lipase	Biodegradation in kitchen waste oil	Sahoo et al. (2018)
Pseudomonas aeruginosa HFE733 lipase	Biodegradation in food wastewater treatment	Hu et al. (2018)
Porcine pancreatic lipase	Biodegradation of mycotoxin ptaulin and carbon dots	Liu et al. (2018), and Srivastava et al. (2019)
Bacillus subtilis esterase	Biodegradation and detoxification of cypermethrin a soil contaminant	Gangola et al. (2018)
Candida antarctica lipase	Degradation of aliphatic polyesters and oxidation of fatty acid methyl esters derived from unsaturated vegetable oils	Kundys et al. (2017)
Yarrowia lipolytica W29 lipase	Degrade waste cooking oil	Lopes et al. (2018)
Rhodopseudomonas palustris PSB-S esterase	Decomposes pyrethroid pesticide	Luo et al. (2018)

(continued)

Table 7.2 (continued)

Enzyme	Role in bioremediation	References
Bacillus spp. esterase	Biodegradation of pesticides like cypermethrin, sulfosulfuron, and fipronil	Bhatt et al. (2019)
Sphingobium sp. strain C3 esterase AppH	Hydrolytic degradation of 2-(4-aryloxyphenoxy) propionate herbicides	Chen et al. 2019
Lactobacillus plantar lipase	Degradation of poly(ε-caprolactone)	Khan et al. (2019)

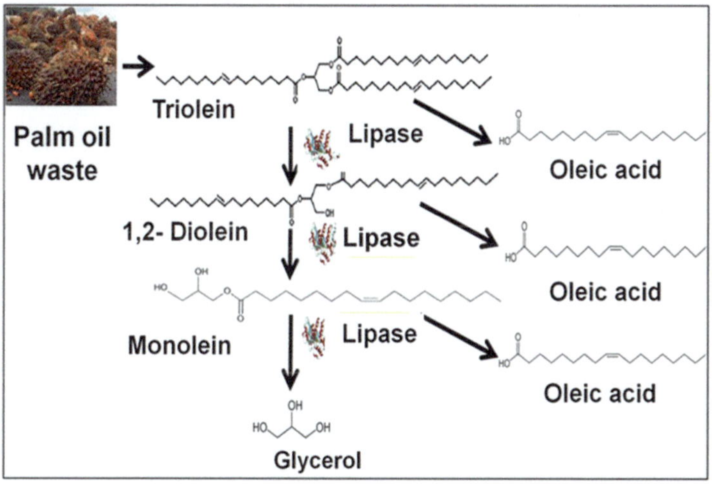

Fig. 7.1 Possible mechanism to degrade triolein, a palm oil waste by microbial lipases. (Hermansyah et al. 2007)

Triglyceride is the chief constituents of natural oil or fat which can be successively hydrolyzed by microbial lipases and esterases to diacylglycerol, monoacylglycerol, glycerol, and fatty acids. These products which further can be extensively used as raw resources in emulsification processes. The work on triolein (a palm oil waste) degradation by *Candida rugosa* lipase in the two-phase system (water/oil) was confirmed to be successful (Fig. 7.1). The fungal lipase breaks the ester bonds of triolein to produce successively diolein, monoolein, and glycerol. Throughout the catalytic degradation of triolein, oleic acid is produced at every successive reaction step and finally the glycerol formed which is generally hydrophilic and therefore easily dissolves into the water phase (Hermansyah et al. 2007).

Microbial hydrolytic lipases or esterases can hydrolyze polylactic acid (PLA), i.e., a plastic waste obtained from renewable resources. *Rhizopus delemar* lipase and polyurethane esterase from *Comamonas acidovorans* have been studied (Masaki et al. 2005) for the degradation of low molecular weight PLA and high molecular weight poly (ethylene terephthalate) (PET). PET is one of the main polyester plastic artificials manufactured in the world. A large number of applications make use of PET like beverage bottles, clothing, packaging, and carpeting. PET is a recalcitrant to catalytic or biological degradation due to the inadequate convenience of the ester linkages. Recently, a study showed that *Ideonella sakaiensis* polyesterase can degrade poly (ethylene terephthalate), mono (2-hydroxyethyl) terephthalic acid (Yoshida et al. 2016) to less toxic form such as ethylene glycol and terephthalic acid (Fig. 7.2).

Microbial lipolytic activity was observed to be the most valuable sign for testing hydrocarbon degradation (Margesin et al. 1999; Riffaldi et al. 2006). Besides their role in bioremediation, these lipolytic enzymes have a lot of prospective applications in chemical, food, cosmetic, detergent, and paper industries, but its manufacture rate has been limited for industrial utilization (Sharma et al. 2011; Joseph et al. 2006; Kumar et al. 2016; Sharma et al. 2018a, b, c).

The major wastes usually produced by food industry are lipids which cause harsh reparation to the atmosphere due to the formation of oily films on aquatic surfaces which leads to interruption in oxygen diffusion and eventually affects the greenhouse effect (Nwobi et al. 2006; Kumar et al. 2012; Hocevar et al. 2012; Brown et al. 2016). Biodegradation of kitchen waste like fat, oil, and grease (FOGs) plays an imperative part in wastewater management. Although several industrial food processing and food restaurants produce FOG wastewaters, however, there is little

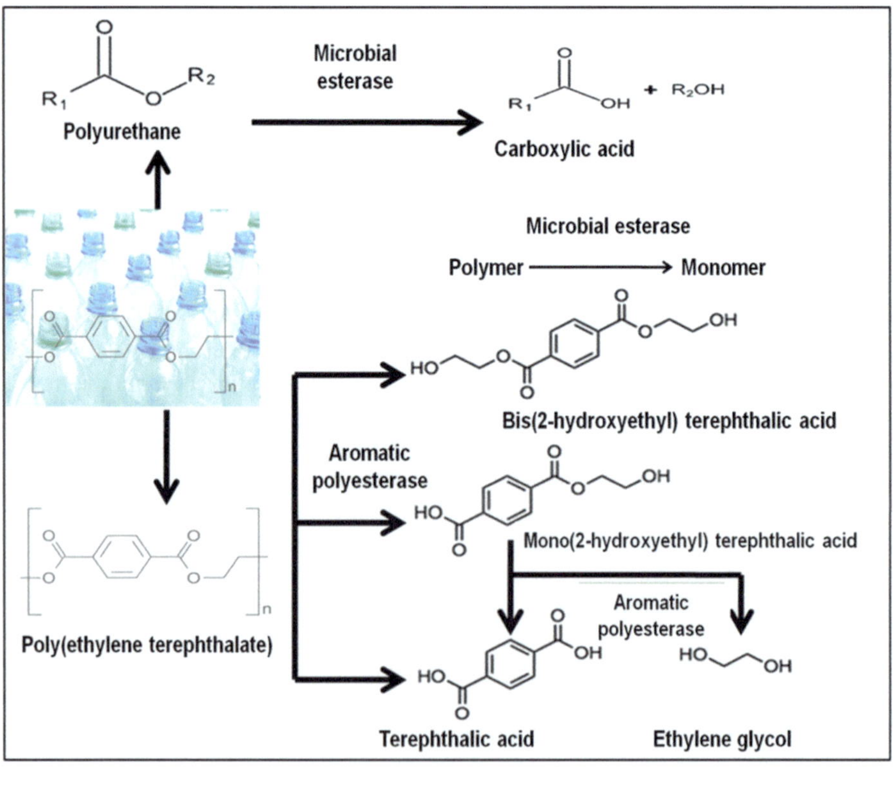

Fig. 7.2 Degradation of plastic waste by microbial lipases/esterases enzymes. (Yoshida et al. 2016)

expertise for their pretreatment (Koh et al. 1992). FOG elimination via microorganisms has been accredited by many researchers (Tan and Gill 1985; Han et al. 1988; Koh et al. 1992; Kyong et al. 2002; Rigo et al. 2008; Kumar et al. 2012; Basso et al. 2012). Prior literature in this field suggests that the preliminary attack on triglycerides by microorganisms is extracellular which generally involves the hydrolysis of the ester bonds by lipases or esterases, which easily eliminate the fatty acids from the triglycerides. These microbial enzymes are extremely specific toward their substrate (Hur et al. 1999; Nunn 1986; Park et al. 1991; Shimada et al. 1992; Mita et al. 2010; Sharma et al. 2017a, b). These hydrolytic enzymes are also very much cooperative in the treatment of biofilm deposits and oil-contaminated soils (Bhardwaj et al. 2012).

Some earlier studies showed why organophosphorus insecticides like pyrethroids and malathion are found to be unsafe to human beings (Wang et al. 2009; Goda et al. 2010; Khan et al. 2016). Harmful organochlorine insecticides like dichloro diphenyl trichloroethane (DDT) and heptachlor are quite stable in well-aerated soil and can only be freely degraded in the anaerobic atmosphere (Williams 1977; Lal and Saxena 1982; Vasilevea-Tonkova and Galabova 2003). Chemically, these insecticides contain a carboxylic ester which can be easily hydrolyzed by microbial carboxylesterases (Galego et al. 2006; Baffi et al. 2008). Microbial carboxylesterases esterases are very important in the treatment of xenobiotics, and its mechanism is related to the mass assembly of versatile microbial carboxylesterases (Galego et al. 2006; Baffi et al. 2008; Sood et al. 2016; Sharma et al. 2016). The degradation of malathion by *Alicyclobacillus tengchogenesis*, *Brevibacillus* sp., *Bacillus licheniformis*, and *Bacillus cereus* has also been observed (Littlechild 2015). Malathion was observed to be a carbon source for many bacteria like *Bacillus licheniformis*, and hence hydrolytic enzymes of *B. licheniformis* can also help in the bioremediation of malathion-containing soil (Singh et al. 2012; Xie et al. 2013). Till date numerous soil inhabitant microorganisms and microbial genes related to these hydrolytic enzymes having the capability to degrade harmful insecticides have been isolated, cloned, and characterized (Yang and Jian 2010; Gangola et al. 2018). These microbial enzymes can be used in the management of waste produced during fat or food processing, degradation of plastic and insecticides, and treatment of biofilm deposits and oil-contaminated soils (Bhardwaj et al. 2012; Yoshida et al. 2016; Okino-Delgado et al. 2017; Sharma et al. 2017e, 2018b; Gangola et al. 2018; Austin et al. 2018; Hu et al. 2018; Sahoo et al. 2018). Therefore, the widespread uses of these hydrolytic enzymes with purposeful values appropriated well for their industrial applications as well as their function in bioremediation.

7.4 Advantages of Bioremediation Using Microbial Enzymes

Universal demands for recycling of any polymeric materials are reasonably noteworthy from environmental friendly perspective. However, examples of chemical recycling are limited, unsafe, and costly. For this reason, development of environ-

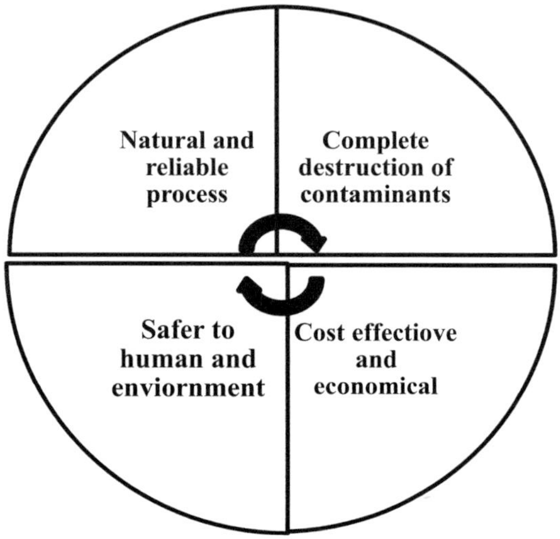

Fig. 7.3 Advantages of bioremediation using microbial enzymes

mentally hospitable approach of recycling is robustly needed which microbial hydrolytic enzyme strongly fulfilled (Hu et al. 2018; Sahoo et al. 2018; Liu et al. 2018). With the growing demands of products, the expectations of growth in the harmful factors also increase hand in hand. The bioremediation using microbial lipases or esterases as a tool to reduce these effects is highly recommended (Fig. 7.3). Enzymatic bioremediation using lipases and esterases is usually a least harmful, safe, cheaper, eco-friendly, and finest method in removing the toxic products in the environment. Bioremediation through microbial enzymes is a suitable waste treatment process and their ability to degrade the contaminant increase in numbers in the presence of a contaminant. The processes involved in bioremediation are valuable for the complete degradation of contaminants. Various harmful compounds mainly xenobiotics can be altered to nontoxic products using enzymatic bioremediation (Kumar et al. 2011; Prajapati et al. 2018; Gangola et al. 2018; Islam and Roy 2018; Austin et al. 2018; Hu et al. 2018; Sahoo et al. 2018; Bhatt et al. 2019; Chen et al. 2019).

7.5 Disadvantages of Bioremediation Using Microbial Enzymes

As every concept or process has its own pros and cons, so is the case of bioremediation. Being the best method in comparison to other processes it leaves one in doubt before being brought in practice. Bioremediation through microbial hydrolases is also restricted to nonbiodegradable products maybe due to the specific behavior of these microbial enzymes (Fig. 7.4). Bioremediation using lipases or esterases habitually takes longer time than other treatment processes (Margesin et al. 1999; Riffaldi

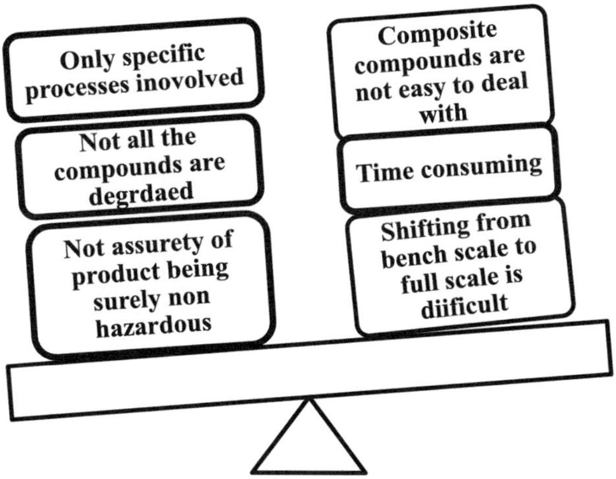

Fig. 7.4 Disadvantages of bioremediation using microbial enzymes

et al. 2006; Kumar et al. 2011; Singh et al. 2012; Xie et al. 2013; Chen et al. 2019). Sometimes, the products of biodegradation using microbial lipases or esterases turn out to be more harmful than the original compound which is simply due to a mutation in genes hydrolyzing these hazardous compounds. So, greater research inputs are required to perform bioremediation using microbial enzymes so as to make green and healthy earth.

7.6 Conclusion

The microbial hydrolases can be the best alternative for the depletion of pollution-causing agents in the environment, but like any other technique, a complete knowledge and understanding about the organism being used and the purpose of it is being used are necessary to avoid further issues and to obtain the most efficient products to reverse the phase of exploitation. Further, it will be remarkable to deal with the improvised structural activity of these enzymes, which preferably has evolved with the surroundings to degrade structurally varied ecological pollutants.

Acknowledgments The financial support in the form of fellowship (DST/INSPIRE Fellowship/2013/1036; IF 131077) to one of the authors (AS) by Department of Science and Technology, New Delhi (India) is thankfully acknowledged.

Conflict of Interest The authors declare no conflict of interests publishing this article in this book.

References

Arora PK, Kumar M, Chauhan A, Raghava GP, Jain RK (2009) A database of oxygenases involved in biodegradation. BMC Res Notes 2:67

Austin HP, Allena MD et al (2018) Characterization and engineering of a plastic-degrading aromatic polyesterase. www.pnas.org/cgi/doi/10.1073/pnas.1718804115

Baffi MA, Rocha LG, Soares de Sousa C, Ceron CR, Bonetti AM (2008) Esterase enzymes involved in pyrethroid and organophosphate resistance in a Brazilian population of *Riphicephallus* (*Boophilus*) *microplus* (Acari, Ixodidae). Mol Biochem Parasitol 160:70–73

Basso RC, Meirelles AJAA, Batista EAC (2012) Liquid-liquid equilibrium of pseudoternary systems containing glycerol + ethanol + ethylic biodiesel from crambe oil (*Crambe abyssinica*) at T/K = (289.2, 318.2, 338.2) and thermodynamic modeling. Fluid Phase Equilib 333:55–62

Beena AK, Geevarghese PI, Jayavardanan KK (2010) Detergent potential of a spoilage protease enzyme liberated by a psychrotrophic spore former isolated from sterilized skim milk. Am J Food Technol 7:89–95

Bhardwaj H, Gupta R, Tiwari A (2012) Microbial population associated with plastic degradation open access scientific reports 1:1–4

Bhatt P, Gangola S, Chaudhary P, Khati P, Kumar G, Sharma A, Srivastava A (2019) Pesticide induced up-regulation of esterase and aldehyde dehydrogenase in indigenous *Bacillus* spp. Bioremed J 23:42. https://doi.org/10.1080/10889868.2019.1569586

Biundo A, Ribitsch D, Steinkellner G, Gruber K, Guebitz GM (2016) Polyester hydrolysis is enhanced by a truncated esterase: less is more. Biotechnol J 10:1308–1322

Brown LD, Cologgi KF, Ulrich AC (2016) Bioremediation of oil spills on land. In: Fingas M (ed) Oil spill science and technology. Elsevier Inc, Amsterdam, pp 699–729

Chen D, Kong X, Wu S, Huang J, Cheng X, Jiang J (2019) An esterase AppH for the hydrolysis of 2-(4-aryloxyphenoxy) propionate herbicides in *Sphingobium* sp. strain C3. Int Biodeterior Biodegrad 136:34–40

Das R, Li G, Mai B, An T (2018) Spore cells from BPA degrading bacteria *Bacillus* sp. GZB displaying high laccase activity and stability for BPA degradation 1:640–641:798–806

El-Morsy EM, Hassan HM, Ahmed E (2017) Biodegradative activities of fungal isolates from plastic contaminated soils. Mycosphere 8:1071–1087

Fan X, Liang W, Li Y, Li H, Liu X (2017) Identification and immobilization of a novel cold-adapted esterase, and its potential for bioremediation of pyrethroid-contaminated vegetables. Microb Cell Factories 16:1–12

Fetzner S, Lingens F (1984) Bacterial dehalogenases: biochemistry, genetics, and biotechnological applications. Microbiol Rev 58:641–685

Fetzner S, Lingens F (1994) Bacterial dehalogenases: biochemistry, genetics, and biotechnological applications. F Microbiol Rev 58:641–685

Galego LG, Ceron CR, Carareto MA (2006) Characterization of esterases in a Brazilian population of *Zaprionus indianus* (Diptera: Drosophilidae). Genetica 126:89–99

Gangola S, Sharma A, Bhatt P, Khati P, Chaudhary P (2018) Presence of esterase and laccase in *Bacillus subtilis* facilitates biodegradation and detoxification of cypermethrin. Scientific Rep. https://doi.org/10.1038/s41598-018-31082-5

Gianfreda L, Xu F, Bollag JM (1999) Laccases: a useful group of oxidoreductive enzymes. Biorem J 3:1–25

Goda SK, Elsayed IE, Khodair TA, Elayed W, Mohamed ME (2010) Screening for and isolation and identification of malathion-degrading bacteria: cloning and sequencing a gene that potentially encodes the malathion-degrading enzyme, carboxylesterase in soil bacteria. Biodegradation 21:903–913

Han DS, Kwon DY, Rhee JS (1988) Lipase-catalyzed reactions for fats and oils in a non-polar solvent. Kor J Appl Microbiol Bioeng 16:250–258

Have RT, Teunissen PJM (2001) Oxidative mechanisms involved in lignin degradation by white-rot fungi. Chem Rev 101:3397–3413

Hermansyah H, Wijanarko A, Gozan M et al (2007) Consecutive reaction model for triglyceride hydrolysis using lipase. J Teknologi 2:151–157

Hocevar L, Soares VRB, Oliveira FS, Korn MGA, Teixeira LSG (2012) Application of multivariate analysis in mid-infrared spectroscopy as a tool for the evaluation of waste frying oil blends. J Am Oil Chemists Soc 89:781–786

Hu J, Cai W, Wang C, Du X, Lin J, Cai J (2018) Purification and characterization of alkaline lipase production by *Pseudomonas aeruginosa* HFE733 and application for biodegradation in food wastewater treatment. Biotechnol Biotechnol Equip 32:583–590

Hur BK, Woo DJ, Kim CB (1999) Hydrolysis mechanisms of fish oil by lipase-100T. J Microbiol Biotechnol 9:624–630

Islam F, Roy N (2018) Screening, purification and characterization of cellulase from cellulase producing bacteria in molasses. BMC Res Notes. https://doi.org/10.1186/s13104-018-3558-4

Jaeger KE, Eggert T (2002) Lipases for biotechnology. Curr Opin Biotechnol 13:390–397

Jamwal S, Kumar R, Sharma A, Kanwar SS (2017) Response surface methodology (RSM) approach for improved extracellular RNase production *by a Bacillus* sp. J Adv Microbiol 3:131–144

Joseph B, Ramteke PW, Kumar PA (2006) Studies on the enhanced production of extracellular lipase by *Staphylococcus epidermidis*. J Gen Appl Microbiol 52:315–320

Karigar CS, Rao S (2011) Role of microbial enzymes in the bioremediation of pollutants: a review. Enzyme Res 2011:1–11

Karmakar Ray RR (2011) Current trends in research and application of microbial cellulases. Res J Microbiol 6:41–53

Khan S, Zaffar H, Irshad U, Ahmad R, Khan AR et al (2016) Biodegradation of malathion by *Bacillus licheniformis* strain ML-1. Arch Biol Sci Belgrade 68:51–59

Khan I, Nagarjuna R, Dutta JR, Ganesan R (2019) Enzyme-embedded degradation of poly(ε-caprolactone) using lipase-derived from probiotic *Lactobacillus plantarum*. ACS Omega 4:2844–2852. https://doi.org/10.1021/acsomega.8b02642

Kim JS, Park JW, Lee SE, Kim JE (2002) Formation of bound residues of hydroxybentazon by oxidoreductive catalysts in soil. J Agricult Food Chem 50:3507–3511

Koh JS, Ko YH, Kim GS (1992) Degradation of fats, oils and hydrocarbons by *Acinetobacter calcoaceticus*. Kor J Appl Microbiol Biotechnol 20:477–482

Krzmarzick MJ, Taylor DK, Fu X, McCutchan AL (2018) Diversity and niche of archaea in bioremediation. Archaea 2018:1–17

Kumar A, Kanwar SS (2011a) Synthesis of ethyl ferulate in organic medium using celite-immobilized lipase. Bioresour Technol 102:2162–2167

Kumar A, Kanwar SS (2011b) Synthesis of isopropyl ferulate using silica immobilized lipase in an organic medium. Enzyme Res 2011:1–8

Kumar A, Kanwar SS (2012a) Catalytic potential of a nitrocellulose membrane- immobilized lipase in aqueous and organic media. J Appl Polymers Sci 124:37–44

Kumar A, Kanwar SS (2012b) An innovative approach to immobilize lipase onto the natural fiber and its application for the synthesis of 2-octyl ferulate in an organic medium. Curr Biotechnol 1:240–248

Kumar A, Kanwar SS (2012c) Lipase production in solid state fermentation (SSF): recent developments and biotechnological applications. Dyn Biochem Process Biotechnol Mol Biol 6:13–27

Kumar A, Kanwar SS (2012d) An efficient immobilization of *Streptomyces* sp. STL-D8 lipase onto photo-chemically modified cellulose-based natural fibers and its application in ethyl ferulate synthesis. Trends Carbohydrates Res 4:13–23

Kumar A, Bisht BS, Joshi VD, Dhewa T (2011) Review on bioremediation of polluted environment: a management tool. Indian J Environ 1:1079–1093

Kumar S, Mathur A, Singh V, Nandy S, Kumar K, Negi S (2012) Bioremediation of waste cooking oil using a novel lipase produced by *Penicillium chrysogenum* SNP5 grown in solid medium containing waste grease. Bioresour Technol 120:300–304

Kumar A, Dhar K, Kanwar SS, Arora PK (2016) Lipase catalysis in organic solvents: advantages and applications. Biol Proced Online 2016:1–18

Kumar A, Wu G, Wu Z, Liu Z (2018a) Improved catalytic properties of a serine hydroxymethyl transferase from *Idiomarina loihiensis* by site-directed mutagenesis. Intl J Biol Macromol 117:1216–1223

Kumar A, Patel SKS, Madan B, Poglu R et al (2018b) Immobilization of xylanase using protein inorganic hybrid system. J Microbiol Biotechnol. https://doi.org/10.1007/s12088-017-0700-2

Kumar A, Wu G, Liu Z (2018c) Synthesis and characterization of cross-linked enzyme aggregates of serine hydroxyl methyltransferase from *Idiomerina leihiensis*. Intl J Biol Macromol 117:683–690

Kumar A, Patel SKS, Madan B, Poglu R, Listeri R, Jeong SH, Kim T, Haw JR, Kim SY, Kim IW, Lee JK (2018d) Immobilization of xylanase using protein inorganic hybrid system. J Microbiol Biotechnol. https://doi.org/10.1007/s12088-017-0700-2

Kundys A, Bialecka-Florjanczyk E, Fabiszewska A, Malajowicz J (2017) *Candida antarctica* lipase B as catalyst for cyclic esters synthesis, their polymerization and degradation of aliphatic polyesters. J Poly Environ 26:396–407

Lal R, Saxena DM (1982) Accumulation, metabolism and effects of organochlorine insecticides on microorganisms. Microbiol Rev 46:95–127

Littlechild JA (2015) Archaeal enzymes and applications in industrial biocatalysts. Archaea 2015:1–10

Liu B, Peng X, Meng X (2018) Effective biodegradation of mycotoxin patulin by porcine pancreatic lipase. Front Microbiol. https://doi.org/10.3389/fmicb.2018.00615

Lopes M, Miranda SM, Alves JM, Pereira AS, Belo I (2018) Waste cooking oils as feedstock for lipase and lipid-rich biomass production. Eur J Lipid Sci Technol 121:1–9

Luo X, Zhang D, Zhou X, Du J, Zhang S, Liu Y (2018) Cloning and characterization of a pyrethroid pesticide decomposing esterase gene, *Est3385*, from *Rhodopseudomonas palustris* PSB-S. Sci Rep 8:1–8

Mai C, Schormann W, Milstein O, Huttermann A (2000) Enhanced stability of laccase in the presence of phenolic compounds. Appl Microbiol Biotechnol 54:510–514

Margesin R, Zimmerbauer A, Schinner F (1999) Soil lipase activity-a useful indicator of oil biodegradation. Biotechnol Tech 13:859–863

Masaki K, Kamini NR, Ikeda H, Iefuji H (2005) Cutinase-like enzyme from the yeast *Cryptococcus* sp. strain S-2 hydrolyzes polylactic acid and other biodegradable plastics. Appl Environ Microbiol 71:7548–7550

Mita L, Sicac V, Guidad M, Nicoluccia C, Grimaldia T, Caputoe L, Biancoa M, Rossie S, Bencivengae U, Eldinf MSM, Tufano MA, Mita DG (2010) Employment of immobilised lipase from *Candida rugosa* for the bioremediation of waters polluted by dimethyl phthalate, as a model of endocrine disruptors. J Mol Cata B: Enzymatic 62:133–141

Mohammed AS, Kapri A, Goel R (2011) Heavy metal pollution: source, impact, and remedies. Biomanagement of metal-contaminated soils. https://doi.org/10.1007/978-94-007-1914-9

Nunn WD (1986) A molecular view of fatty acid catabolism in *Escherichia coli*. Microbiology 50:179–192

Nwobi BE, Ofoegbu O, Adesina OB (2006) Extraction and qualitative assessment of African sweet orange seed oil. Afr J Food Agricult Nutr 6:1–11

Okino-Delgado CH, Prado DZD, Facanali R, Marques MMO, Nascimento AS, Fernandes CJDC, et al (2017) Bioremediation of cooking oil waste using lipases from wastes. PLoS ONE https://doi.org/10.1371/journal.pone.0186246

Park CH, Kim YK, Yu HJ et al (1991) Microbial degradation of fats and oils in industrial wastewater. Kor J Appl Microbiol Biotechnol 19:88–93

Park JM, Sewell T, Benedik MJ (2017) Cyanide bioremediation: the potential of engineered nitrilases. Appl Microbiol Biotechnol 101:3029–3042

Patel SKS, Anwar MZ, Kumar A, Otari SV, Pagolu RT, Kim SY, Kim IW, Lee JK (2017) Fe2O3 yolk-shell particle-based laccase biosensor for efficient detection of 2,6-methoxyphenyl. Biochem Eng J 132:1. https://doi.org/10.1016/j.bej.2017.12.013

Piontek K, Smith AT, Blodig W (2001) Lignin peroxidase structure and function. Biochem Soc Trans 29:111–116

Riffaldi R, Levi-Minzi R, Cardelli R, Palumbo S, Saviozzi A (2006) Soil biological activities in monitoring the bioremediation of diesel oil-contaminated soil. Water Air Soil Pollut 170:3–15

Rigo E, Rigoni RE, Lodea P, Oliveira D, Freire DMG, Luccio MD (2008) Application of different lipases as pretreatment in the anaerobic treatment of wastewater. Environ Eng Sci 25:1243–1248

Riya P, Jagatpati T (2012) Biodegradation and bioremediation of pesticides in soil: its objectives, classification of pesticides, factors and recent developments. World J Sci Technol 2:36–41

Ron EZ, Rosenberg E (2014) Enhanced bioremediation of oil spills in the sea. Curr Opin Biotechnol 27:191–194

Sahoo RK, Kumar M, Mohanty S, Sawyer M et al (2018) Statistical optimization for lipase production from solid waste of vegetable oil industry. Preparat Biochem Biotechnol 48:321–326

Salgado JM, Abrunhosa L, Venancio A, Domínguez JM, Belo I (2016) Combined bioremediation and enzyme production by *Aspergillus* sp. in the olive mill and winery wastewaters. Int Biodeterior Biodegrad 110:16–23

Sasikumar CS, Papinazath T (2003) Environmental management: bioremediation of polluted environment. In: Bunch MJ, Suresh VM, Kumaran TV (eds) Proceedings of the third international conference on environment and health, Chennai, India, 15–17 December 2003

Sharma D, Sharma B, Shukla AK (2011) Biotechnological approach of microbial lipase: a review. Biotechnology 10:23–40

Sharma T, Sharma A, Kanwar SS (2016) Purification and characterization of an esterase from *Bacillus pumilus*. J Adv Biotechnol Bioeng 4:9–16

Sharma T, Sharma A, Kanwar SS (2017a) An overview on esterases: structure, classification, sources and their application. In: Rai V (ed) Recent advances in biotechnology, vol 2. Shree Publishers, New Delhi, pp 216–229

Sharma A, Meena KR, Kanwar SS (2017b) Molecular characterization and bioinformatics studies of a lipase from Bacillus thermoamylovorans BHK67. Intl J Biol Macromol 107:2131–2140

Sharma A, Meena KR, Rana S, Singh M, Kanwar SS (2017c) Fabrication and functionalization of magnesium nanoparticle for lipase immobilization in *n* -propyl gallate synthesis. J King Saud University- Sci 29:536–546

Sharma A, Sharma T, Meena KR, Kanwar SS (2017d) Physical adsorption of lipase onto mesoporous silica. Intl J Curr Adv Res 6:3836–3841

Sharma A, Sharma T, Gupta R, Kanwar SS (2017e) Correlation of enzyme activity and stabilities with extreme environments: complexities and flexibilities in adaptations. In Rai V (ed) Recent advances in biotechnology, vol 1. Shree Publishers New Delhi, pp 17–31

Sharma A, Sharma T, Meena KR, Kumar A, Kanwar SS (2018a) High throughput synthesis of ethyl pyruvate by employing superparamagnetic iron nanoparticles-bound esterase. Process Biochem 71:109–117

Sharma A, Sharma T, Kanwar SS (2018b) Entrapment of *Streptomyces* sp. lipase in cupric-beads for methyl salicylate synthesis. Trends in Carbo Res 10:21–29

Sharma A, Meena KR, Kanwar SS (2018c) Microbial thermo-tolerant and solvent- tolerant lipases and their applications. Curr Res Microbiol 2:1–15

Shimada Y, Sugihara A, Nagao T, Tominaga Y (1992) Induction of *Geotrichum candidum* lipase by long-chain fatty acids. J Ferment Bioeng 74:77–80

Shon HO, Kyong HO, Tian D et al (2002) Degradation of fat, oil and grease (FOGs) by the lipase-producing bacterium *Pseudomonas* sp. strain D2D3. J Microbiol Biotechnol 12:583–591

Singh B, Kaur J, Singh K (2012) Biodegradation of malathion by *Brevibacillus* sp. strain KB2 and *Bacillus cereus* strain PU. World J Microbiol Biotechnol 28:1133–1141

Sood S, Sharma A, Sharma N, Kanwar SS (2016) Carboxylesterases: sources, characterization, and broader. Insight Enzyme Res 1:1–11

Srivastava I, Sar D, Mukherjee P, Schwartz-Duval AS, Huang Z, Jaramillo C, Civantos IA et al (2019) Enzyme-catalysed biodegradation of carbon dots follow sequential oxidation in a time dependent manner. Nanoscale. https://doi.org/10.1039/C9NR00194H

SureshKumar P, Thomas J, Poornima V (2018) Structural insights on bioremediation of polycyclic aromatic hydrocarbons using microalgae: a modeling-based computational study. Environ Monit Assess. https://doi.org/10.1007/s10661-017-6459-4

Tan KH, Gill CO (1985) Batch growth of *Saccharomycopsis lipolytica* on animal fats. Appl Microb Biotechnol 21:292–298

Tavano OL, Berenguer-Murcia A, Secundo F, Fernandez-Lafuente R (2018) Biotechnological applications of proteases in food technology. Comprehen Rev Food Sci food Saf 17:412–436

Thakur N, Kumar A, Sharma A, Bhalla TC, Kumar D (2018a) Purification and characterization of alkaline, thermostable and organic solvent stable protease from a mutant of *Bacillus* sp. Biocatal Agricult Biotechnol 16:217–224

Thakur R, Kumar A, Kumar S (2018b) Synthesis and characterization of cross-linked enzyme aggregates (CLEAs) of thermostable xylanase from *Geobacillus thermodenitrificans* X1. Process Biochem 80:72. https://doi.org/10.1016/j.procbio.2019.01.019

Tokiwa Y, Calabia BP, Ugwu CU, Aiba S (2009) Biodegradability of plastics. Int J Mol Sci 10:3722–3742

Tsukihara T, Honda Y, Sakai R, Watanabe T (2006) Exclusive overproduction of recombinant versatile peroxidase MnP2 by genetically modified white rot fungus, *Pleurotus ostreatus*. J Biotechnol 126:431–439

Vasileva-Tonkova E, Galabova D (2003) Hydrolytic enzymes and surfactants of bacterial isolates from lubricant contaminated wastewater. Z Naturforsch 58:87–92

Wang G, Liu Y (2016) Diazinon degradation by a novel strain *Ralstonia* sp. DI-3 and X-ray crystal structure determination of the metabolite of diazinon. J Biosci 41:359–366

Wang BZ, Gou P, Hang BJ, Li L, He J et al (2009) Cloning of a novel pyrethroid hydrolyzing carboxylesterase gene from *Sphingobium sp*. Appl Environ Microbiol 75:5496–5500

Wang Q, Chen M, Shan G, Chen P, Cui S, Yi S, Zhu L (2017) Bioaccumulation and biomagnification of emerging bisphenol analogs in aquatic organisms from Taihu Lake, China. Sci Total Environ 598:814–820

Williams PP (1977) Metabolism of synthetic organic pesticides by anaerobic microorganisms. Residue Rev 66:63–135

Wong DWS (2009) Structure and action mechanism of ligninolytic enzymes. Appl Biochem Biotechnol 157:174–209

Xie Z, Xu B, Ding J, Liu L, Zhang X et al (2013) Heterologous expression and characterization of a malathion-hydrolyzing carboxylesterase from a thermophilic bacterium, *Alicyclobacillus tengchogensis*. Biotechnol Lett 35:1283–1289

Yang CR, Jian JZ (2010) Recent advance in biodegradation in China: new microorganism and pathway, biodegradation engineering, and bioenergy from pollutant biodegradation. Process Biochem 45:1937–1943

Yoshida S, Hiraga K, Takehana T et al (2016) A bacterium that degrades and assimilates poly(ethylene terephthalate). Science 351:1196–1199

Yue F, Lishuang Z, Jing W, Ying Z, Bangce YE (2017) Biodegradation of phthalate esters by a newly isolated *Acinetobacter* sp. strain LMB-5 and characteristics of its esterase. Pedosphere 27:606–615

Yuniati MD (2018) Bioremediation of petroleum contaminated soil: a review. Eath Environ Sci 118:1–7

Zhang X, Liu L, Zhang S et al (2016) Biodegradation of dimethyl phthalate by freshwater unicellular cyanobacteria. Biomed Res Int 2016:1–8

Chapter 8
Laccases for Soil Bioremediation

María Pilar Guauque-Torres and Ana Yanina Bustos

Abstract Bioremediation tool, by diminishing noxiousness or promoting pollutant mineralization to CO_2 and water, is one of the most efficient, cost-effective, and eco-friendly approaches for the rehabilitation of polluted soils. Bioremediation process is mainly based on the ability of different enzymes or complex enzymes to act on various substrates. Laccases are ligninolytic enzymes, classified as benzenediol oxygen reductase (EC 1.10.3.2) and also known as multicopper oxidases. They are widely distributed in insects, plants, archaea, fungi, and bacteria. Industrially, laccases coming from fungi are the most commonly used; however, recently bacterial laccases have attracted attention because of their versatility, which includes higher thermostability, better tolerance to different concentrations of Cu^{2+}, and higher resistance to changes regarding pH and halo and high chloride. The versatility of laccases allows its use on the soil to polymerize pollutants, and it also permits the bioaugmentation with immobilized laccases to degrade pollutants. Taking into account that laccase is one of the oldest enzymes ever described and it is relevant to the decomposition of xenobiotics, the present chapter will be dedicated to the exploration of laccase as an invaluable tool for soil bioremediation. We will review the main aspects related to the structure of laccases, substrates, and mechanisms of action. Additionally, we will also focus on two main topics: the production and the immobilization techniques to enhance the availability and stability of laccases. We highlighted some of the successful strategies employed to enhance laccase

M. P. Guauque-Torres (✉)
GIAM-Z Research Group, Faculty of Environmental Engineering, Universidad Libre, Campus Majavita, Socorro, Santander, Colombia
e-mail: pilar.guauque@unilibre.edu.co

A. Y. Bustos
Universidad San Pablo – Tucumán, San Miguel de Tucumán, Tucumán, Argentina

Centro de Investigación en Biofísica, Aplicada y Alimentos (CIBAAL-CONICET), Santiago del Estero, Argentina

Facultad de Agronomía y Agroindustrias (FAyA), Universidad Nacional de Santiago del Estero (UNSE), Santiago del Estero, Argentina
e-mail: abustos@uspt.edu.ar

A. Kumar, S. Sharma (eds.), *Microbes and Enzymes in Soil Health and Bioremediation*, Microorganisms for Sustainability 16,
https://doi.org/10.1007/978-981-13-9117-0_8

production, including the screening of new promising laccase, recombinant laccase production, as well as different immobilization strategies applied to increase the enzyme's stability.

Keywords Soil bioremediation · Laccases · Recombinant laccase production · Immobilization laccase techniques

8.1 Introduction

Soil is essential for ecosystems and human life for its central role on maintaining balance to ensure biodiversity (microorganism, vegetables, and animal species) and also for assuring a source of food, fuel, minerals, and materials for life on earth as we know it so far. In addition, soil helps to regulate emissions of greenhouse gases and to clean water, functioning as a filter if the physicochemical characteristics of every place are conserved. However, the industrial growth, the population increase, and the consumer lifestyle have altered earth's natural balance, affecting the physicochemical characteristics of soil and causing erosion and chemical pollution among other associated problems that diminish biodiversity and fertile surface to crops.

Multiple strategies have been used to slow down the progress of soil pollution, and bioremediation is one of the most efficient, cost-effective, and eco-friendly approaches. Bioremediation exploits the ability of microorganisms of adapting their metabolism, using different molecules as a source of carbon and nutrients. In this sense, adapted microorganisms and their metabolic machinery can degrade or modify pollutants to diminish the soil contamination (Shiomi 2015; Baldantoni et al. 2017; De Lima et al. 2018; Speight and El-Gendy 2018).

In the present chapter, soil bioremediation will be focused on laccases as one of the most important enzymes present in many bacteria and fungi species related to the decontaminating process. An overview of the structure of laccases will help to understand the mechanism of reactions and the type of substrates they can catalyze, such as humic acids, xenobiotics, polycyclic aromatic hydrocarbons (PAHs), pesticides, and polyphenol compounds. The reaction mechanism can operate by direct or mediator oxidation depending on substrate reduction potential.

In addition, the new natural hypersecretory strains will be reviewed as well as the improvement strategies of yielding focusing on the production of recombinant proteins, using the heterologous or homologous expression system.

In the soil bioremediation context, immobilization will be addressed in two ways. The first one will be immobilizing laccases on different supports through different techniques, which will help us to find an optimum combination "carrier–enzyme–target reaction" that diminishes the pollutants in the soil. The second one will be using laccases to catalyze the polymerization of pollutants and reduce their mobility and bioavailability.

8.2 Soil Pollutant and Bioremediation

Soil is essential for ecosystems and human life as a provider of food and fuel, but also as a major storehouse for carbon, as it regulates emissions of greenhouse gases which are vital for regulating climate. Moreover, soil filters and cleans the water for human consumption around the world, ensuring biodiversity and multiple socioeconomic benefits. Since the past century, industrial growth has had a direct relationship with the level of pollution, and soils, acting as sinks for this complex contamination, have been affected. Thirty-three percent of the land suffers erosion, compaction, salinization, and/or chemical pollution, leaving the loss of productive soils and attracting the attention of the scientific community and different governments to protect this biome, not only from an ecosystem perspective but also because of economic losses. In the European Union countries, 250000 sites need rehabilitation, and it is expected that this number will increase to about 50% percent by 2025 (Mougin et al. 2009). Mining and oil industries are in the first and second place as soil contaminants, followed by organic pollutants such as aromatic hydrocarbons and polycyclic aromatic hydrocarbons (Mougin et al. 2009). The costs for cleaning up contaminated sites are in the range of 425–500 USD billion (Megharaj and Naidu 2017).

Sustainable soil management (SSM) is a tool established by FAO that seeks to prevent and repair damaged soils, using scientific and local knowledge as well as proven technologies that improve the production and safeguarding of ecosystems and biodiversity (FAO 2017). Different strategies have been developed based on biological, physicochemical, and thermal treatment for the dismissal of soil pollution. In this section of the book, the focus will be on biological ones, especially bioremediation (Surridge et al. 2009; De Lima et al. 2018).

Bioremediation tool is one of the most efficient, cost-effective, and eco-friendly approaches for the rehabilitation of polluted soils, by diminishing noxiousness or promoting pollutant mineralization to CO_2 and water (Baldantoni et al. 2017; De Lima et al. 2018). Bioremediation applies microorganisms (fungi and bacteria) in contaminated places, which use its pollutants like carbon and nutrient source while cleaning up the surface (Shiomi 2015; De Lima et al. 2018; Speight and El-Gendy 2018). The treated soil can be reused if the degradation of targeted pollutants is acceptable and there is no xenobiotic as a degradation product (Mougin et al. 2009). The main disadvantage is the long time required for achieving the acceptable pollution thresholds.

Biochemistry of bioremediation proceeds by biodegradation, co-metabolism, and/or synthesis. Biodegradation implies mineralization of xenobiotic into carbon dioxide and other inorganic compounds (Food and Agricultural Organization of the United States (FAO) and Intergovernmental Technical Panel on Soils (ITPS) 2015; Speight and El-Gendy 2018). Especially in hydrocarbons remediation, the most common biochemical xenobiotic process is co-metabolism, where organisms grow consuming xenobiotic, namely co-substrate, without taking advantage of any nutrient or energy for this reaction (García-Rivero and Peralta-Perez 2008). Finally, the synthesis process refers to the conjugation and oligomerization of low molecular weight molecules forming complex compounds which could worsen the environmental impact because of the limited bioavailability of the final product. One or several interacting organisms can be involved in the general biochemical process with different reactions and metabolic cycles, which evidence the complexity of this topic (Mougin et al. 2009).

From the biological point of view, bioremediation techniques involve three possibilities, natural attenuation, bioaugmentation, or biostimulation, used alone or in combination to optimize the pollutant degradation (Šašek et al. 2003; Mougin et al. 2009; Speight and El-Gendy 2018). Furthermore, from an engineering perspective, bioslurry reactors, biopile, and landfarming are the main methods used for bioremediation. The first one has been used for the treatment of non-halogenated semi-volatile organic compounds (SVOCs), chlorpyrifos, volatile organic compounds (VOCs), petroleum hydrocarbons and explosive compounds in soil (Luthy et al. 1995; Pant and Rai 2018). The second technique, biopile treatment, has been applied to the treatment of fuel-contaminated soil and non-chlorinated VOCs (Siracusa et al. 2017; Llorens-Blanch et al. 2018). Landfarming has been used to remediate refinery petroleum sludges and is useful when oil spills affect marine beach sand and sediments (Van De Vijver et al. 2015; Nikolopoulou and Kalogerakis 2016).

8.2.1 Enzymes for Bioremediation

Bioremediation process is based on the ability of different enzymes or complex enzymes to act in different substrates using middle operational conditions such as temperature, pH, and salinity range among others. However, it is possible to find enzymes prepared to operate in concentrations of a high contaminant, as recalcitrant xenobiotics (Gianfreda and Bollag 1994). The bioremediation process can occur by intracellular oxidation process (Quintero Díaz 2011) or using extracellular enzymes (EE) especially adapted (glycosylation or disulfide bonds to improve thermostability and pH/proteases resistance) to assure their activity in pollutant soil (Burns et al. 2013).

Ligninolytic enzymes biodegradation process is known as an enzymatic extracellular system composed by lignin peroxidase (LiP), laccase, and manganese peroxidase (MnP) typically found in white-rot fungi which degrades xenobiotics such as pesticides, dyes, and explosives (Bourbonnais and Paice 1990; Theuerl and Buscot 2010; Quintero Díaz 2011; Saptarshi et al. 2013). Lignin peroxidase (LiP) and manganese peroxidase (MnP) have a high redox potential (1.15–1.25 V) for oxidizing lignin in the presence of H_2O_2. LiP degrades nonphenolic lignin units, whereas MnP oxidizes phenolic or nonphenolic compounds (Mougin et al. 2009; Shiomi 2015).

Laccases complete the ligninolytic system as low redox potential enzymes (0.5–0.8 V) which reduce their ability to degrade nonphenolic aromatic compounds. However, in the presence of mediators with redox potential higher than 0.9 V, laccases expand their substrates possibilities (Shiomi 2015).

Interesting and feasible strategies for soil bioremediation with the ligninolytic system have been developed using the white-rot fungus *Phanerochaete chrysosporium* for the degradation of alachlor (2-chloro-2,6-diethyl-N-[methoxymethyl]-acetanilide; AL) and atrazine (2-chloro-4-ethylamino-6-isopropylamino-S-triazine; AT) in contaminated soil. AL was the most susceptible substrate, with a percentage of the transformation of 54%, while AT suffers a 32%. The difference is observed mainly in the first stages of the degradation process (Chirnside et al. 2011).

However, the complexity of the ligninolytic system can result in difficulties to explain interactions and fate of the degradation products leaving some doubts about the bioremediation process. For this reason, a lot of studies have been developed to focus on the degradation process catalyzed by one type of enzymes only. Taking into account that laccase is the oldest enzyme recognized and the relevance of its role in xenobiotics decomposition, the next sections will be dedicated to laccase as an invaluable tool for soil bioremediation. Although the vast majority of the literature has been oriented on wastewater treatment, our main focus was reviewing soil bioremediation mediated by laccases.

8.3 Laccases for Soil Bioremediation

8.3.1 Sources of Laccases

Laccases were first detected in the varnish tree *Rhus vernicifera* (Yoshida 1883), but they have also been identified into the cell wall of different plant species as *Rhus succedanea*, *Lactarius piperatus*, and *Prunus persica*, where lignin biosynthesis occurs. Some insects genus such as *Diploptera*, *Bombyx*, *Drosophila*, *Calliphora*, *Musca*, *Phormia*, *Lucilia*, *Schistocerca*, and *Manduca* also expressed this enzyme, probably because it is related to cuticle formation (Madhavi and Lele 2009). Bacterial laccases in different genres such as *Bacillus (B.) subtilis*, *Marinomonas mediterranea*, *Azospirillum lipoferum*, *Streptomyces ipomoea*, and *Streptomyces griseus* are important for morphogenesis, spore formation, and copper homeostasis (Fernández-Fernández et al. 2013).

However, the highest production of laccases is obtained from fungi species such as ascomycetes, deuteromycetes, and basidiomycetes, especially in saprotrophic and ligninolytic fungi. The most common producers are *Trametes versicolor*, *Trametes ochracea*, *Trametes hirsuta*, *Trametes gallica*, *Trametes villosa*, *Coriolopsis polizona*, *Cerena maxima*, *Pleurotus enryngii*, and *Lentinus tigrinus* (Madhavi and Lele 2009; Fernández-Fernández et al. 2013). The widespread distribution of laccases in earth depends on their role in pigments formation, lignin, and toxic compounds degradation (Morozova et al. 2007). These three principal functions have directed their ability to oxidize a broad range of aromatic substrates like diphenols, methoxy-substituted monophenols, diamines, and aromatic amines. It can even oxidize polymeric molecules such as polyphenols and polycyclic aromatic hydrocarbons (Eichlerová et al. 2012; Shiomi 2015; Košnář et al. 2018).

8.3.2 Structure of Laccase

Laccases are benzenediol oxygen reductase (EC 1.10.3.2) also known as multicopper oxidases (MCOs); they have a molecular weight between 60 and 90 kDa and possess a high glycosylation grade (15–50% weight of carbohydrates) which is a common feature in extracellular enzymes. The isoelectric point (pI) of fungal laccases is around 3 and 7 while in plant laccases is close to 9. Optimum pH for fungal laccases is 3.5–5.2, while the optimum pH in bacterial laccases is near to 6.8–7.4. However, it is important to say that the same organism can produce different isozymes with different characteristics (Piontek et al. 2002; Shraddha et al. 2011; Fernández 2013). Fungal laccases are more tolerant to low moisture and support lower pH in the soil than bacterial laccases, but can be pathogenic to animals and plants (Mougin et al. 2009). Although blue multicopper oxidases structure can be arranged with two (2D), three (3D), or six (6D) domains, fungal laccases usually have 3D, while bacterial laccases have 2D (Ausec et al. 2011a).

The structure of an active holoenzyme laccase molecule has three redox sites (T1, T2, and T3), each one containing four copper atoms per monomer. The copper atom at T1 redox site has a strong absorption at 600 nm (which gives the enzyme a blue color), and its principal function is to abstract one electron from reducing substrates (organic compounds). On the other hand, T2 and T3 form a trinuclear metallic cluster that is in charge of reducing the molecular oxygen to produce water. The mononuclear T2 site is coordinated by two histidine (His) residues and exhibits its characteristic signals in electronic paramagnetic resonance, (EPR) whereas T3 site has two coppers atoms with an absorption band on 330 nm, the Cu couple being stabilized by six His residues. Both T3 Cu atoms do not have EPR signals because of the diamagnetic basal state given by a strong antiferromagnetic coupling between the two Cu (II) ions, maintained by a hydroxyl bridge. Highly conserved metal-binding motifs can be a tool to recognize MCO, in spite of sequence diversity (Fig. 8.1) (Piontek et al. 2002; Román et al. 2010; Lawton and Rosenzweig 2011; Shiomi 2015).

Piontek et al. (2002) reported a crystal structure for laccase of *Trametes versicolor* (TvL) at 1.9 Å containing four copper atoms and seven N-acetyl glucosamine moieties (Fig. 8.1. PDB file 1GYC). TvL is the smallest of the family of blue MCOs (bMCOs) with almost 500 amino acids organized in three sequential domains with a β-barrel-type architecture configured in the space of $65*55*45Å^3$. The structure presents some 3_{10} helices and β-sheets forming a multicopper active site. There is one disulfide bridge in domain 1 near the C-terminal portion (Cys 85–Cys 488) attending the stability need of an extracellular enzyme and a second disulfide bridge connecting domains 1 and 2 (Cys 117–Cys 488). The trinuclear copper cluster is stabilized by His residues of domains 1 and 3 and hydrogen bonding networks, providing rigidity to the enzyme especially to the crystal structure elucidation on N-terminal and C-terminal regions (Piontek et al. 2002).

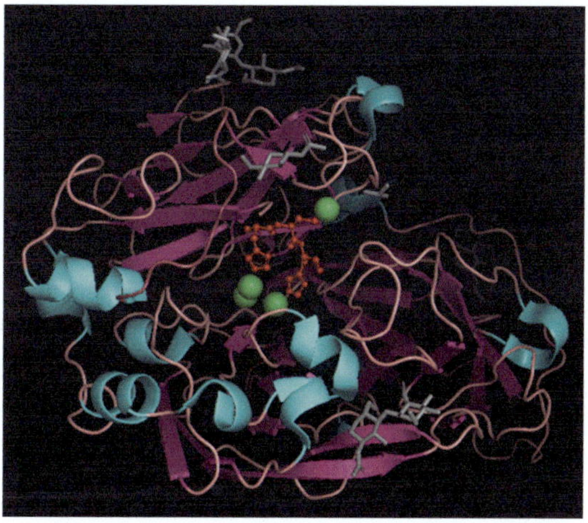

Fig. 8.1 Laccase of *Trametes versicolor* (PDB: 1GYC—Piontek et al. 2002) created with executable Open-Source PyMol 2.0. The image shows copper atoms (green) and trinucleotide His-Cys-His linking copper at T1 site with trinuclear T2/T3 site (red). For a better quality of image, amino acids IDFHLEA (456) were hidden

The same report discloses the crystallization of the enzyme at pH 5.6, showing a potential distribution of its electrostatic surface, which is predominantly negative as a consequence of laccases' acidic pI (3.5). The authors found a small negatively charged cavity that can bind the substrate near site T1, explaining that radical cation products of the reaction can be stabilized in this way (Piontek et al. 2002).

The water molecules found forming two channels evidenced communication of T2 and T3 sites with solvent. The comparison between PDB files for other MCOs showed that these channels are highly conserved structures in terms of amino acids residues and water molecules involved. These features can be related to fast access to molecular oxygen from the T2/T3 cluster and water release as a reaction product. These features are in concord with the proposed laccase reaction mechanism "two-site ping-pong bi-bi" where the binding of new substrates occurs after the products are released (Piontek et al. 2002).

Laccases are classified according to their reduction potential. Laccases typically found in fungi have high potential (HPLs 0.6–0.8 V), whereas laccases found in bacteria have low potential (LPLs with 0.4–0.6 V). The reduction potential of the T1 site has been determined for most laccases while T2/T3 potential reduction sites remain ignored. Although laccase has a smaller reduction potential in a ligninolytic system, when compared to lignin peroxidases and manganese peroxidase, it is known as one of the multicopper oxidases with the highest reduction potential (Lawton and Rosenzweig 2011; Abdel-Hamid et al. 2013; Shiomi 2015). Changes at reduction potential in the T1 site (up to 200 mV) can be related to the chemistry identity of amino acid residues in an axial position of copper, which changes the geometry coordination of the site. Another hypothesis is the stretching of the bond between the metal and the ligating amino acid, which decreases the dense contribution of electrons to copper (Piontek et al. 2002).

8.3.3 Substrate for Laccases

Laccases have *o*-diphenol and *p*-diphenol activity; they are also capable of oxidizing aryl diamines, aminophenols, polyamines, polyphenols, and lignins as well as some inorganic ions. Their possible target substrates range from humic acids to xenobiotics going through fluoroquinolones antibiotics among others. Due to their unspecificity, laccases have a broad action spectrum on different xenobiotics molecules, putting them on the center of attention of applied biotechnology for remediation. For this reason, laccases have been used to test oxidation on polycyclic aromatic hydrocarbons (PAHs), pesticides, and polyphenol compounds (Fernández-Fernández et al. 2013; De Lima et al. 2018).

Natural substrates of laccases (fungal or bacterial) are humic acids; in fact, laccases are recognized as one of the major driving forces in humification. Laccases catalyze the oxidation of phenols in the presence of oxygen to produce phenoxy radicals and quinones, although biomass oxidation is also catalyzed by peroxidases and abiotic factors. These laccases can either couple with a smaller molecule in a polymerization process or degrade to a larger one, which reduces the bioavailability of pollutants (Lisov et al. 2018).

Nevertheless, laccases can also catalyze the oxidation of bigger molecules if mediators are present. In this way, laccases expand the range of substrates over which they can act such as xenobiotics, known by their toxicity and recalcitrant behavior in terrestrial and aquatic environments.

Some xenobiotics such as PAHs are of major interest because of their prevalence or critic existence in the environment. As mentioned before, PAHs are the third main soil contaminants after heavy metals and mineral oil (Mougin et al. 2009) making PAHs one of the principal targets to test the bioremediation process. PAHs can derive from volcanoes' eruption, incomplete combustion of fossil fuel, and agricultural activities. Their hydrophobic and benzene-related structure makes them toxic, mutagenic, and/or carcinogenic for humans but also for animals and plants. For this reason, their remediation has been fixed as a priority by the US Environmental Protection Agency (EPA), to asses environmental clean-up and to take care of human health (Surridge et al. 2009). Xenobiotics toxicity increases as long as the molecular weight does the same, which means that anthracene (AnT), naphthalene (NaT), and phenanthrene (PhT) (two or three fused aromatic rings) are not carcinogenic; but benzo[α]anthracene (BaA), benzo[α]pyrene (BaP), and pyrene (Pyr) (four or more rings) are recognized as toxic compounds (Dodor et al. 2004)..

On the other hand, the increasing population in the world is accompanied by the growing needs for food. Large areas of land are being used for different cash crops, usually monocultures, where plagues attack. For this reason, many pesticides have been developed to control them, with different biochemistry mechanisms that can affect human and ecosystem health. It is calculated that three million tons of pesticides are applied around the world per year and one-third of them remain in agricultural soils, reaching dangerous accumulation levels. Despite all this, physicochemical and microbiologic properties in soil can promote some strategies for decontamination such as mineralization of pollutants into CO_2 and water, inclusion of pollutants as biomass—decreasing the environmental impact—and adsorption on clay minerals or organic matter (Gianfreda and Bollag 1994; Shraddha et al. 2011; Botterweck et al. 2014).

The use of enzymes as catalysts for the degradation of pesticides or for derivating them into safer metabolites could be a suitable alternative for reducing soil pollution. Some immobilized enzymes could provide better stabilizing conditions and a faster and convenient process compared to that with free enzymes and strains (Zhang et al. 2013). Research has been conducted to test the enzyme degradation of chlorpyrifos and carbofuran. Chlorpyrifos is an organophosphate pesticide widely used in the world for insects control on cotton, corn, almond, orange, bananas, and apple cultures. In the same way, carbofuran is a broad-spectrum pesticide used for nematodes, insects, and acarids control. The World Health Organization (WHO) classifies chlorpyrifos as moderately hazardous, while carbofuran is highly unsafe. Detailed information about these studies can be found in upcoming sections (Wang et al. 2016, 2017b).

Phenolic compounds are another major family of molecules responsible for environmental pollution. Laccase from *Cerrena unicolor* reached a complete transformation of 2,4-dichlorophenol (2,4-DCP) and catechol in soil colloids, whereas TvL transformed an urea derivate (N,N-dimethyl-N-(hydroxyphenyl)-urea) into a pH-dependent reaction (Durán and Esposito 2000).

In another approach about the versatility of laccases on bioremediation, they have been applied to nanobiotechnology as a biosensor for different reactions in the presence of electron transference (Shraddha et al. 2011). Independently of substrates or application, the understanding of the mechanism of oxidation mediated by laccases is an interesting way to find new uses that could help exploit their potential as biocatalyst on different processes.

8.3.4 Mechanism of Laccase-Catalyzed Reactions

The general reaction implies the oxidation of four electrons of the reducing agent with the concomitant reduction of four electrons of molecular oxygen to produce water (Fig. 8.2). Crystal studies on *Trametes versicolor* laccase (TvL) show that the substrate-binding domain surrounds the copper in T1 site, about 6.5 Å under the enzyme surfaces, indicating that T1 site is responsible for accepting the organic compounds (phenol, aromatic amine, etc.) as the primary electron acceptor site. The electron transfer reaction to trinuclear cluster T2/T3, 12 Å away of T1, occurs by a His-Cys-His tripeptide that is well conserved in bMCOs. The reaction mechanism on TvL is proposed as an electron transfer between Cu1 and sulfur of Cys453 and then to the carbonyl oxygen of the same residue and via a hydrogen bond to nitrogen of His 452 ligated to one of the copper atoms of the dinuclear T3 site (distance 2.16 Å). Copper atoms of T3 site act as electron acceptors where an oxygen

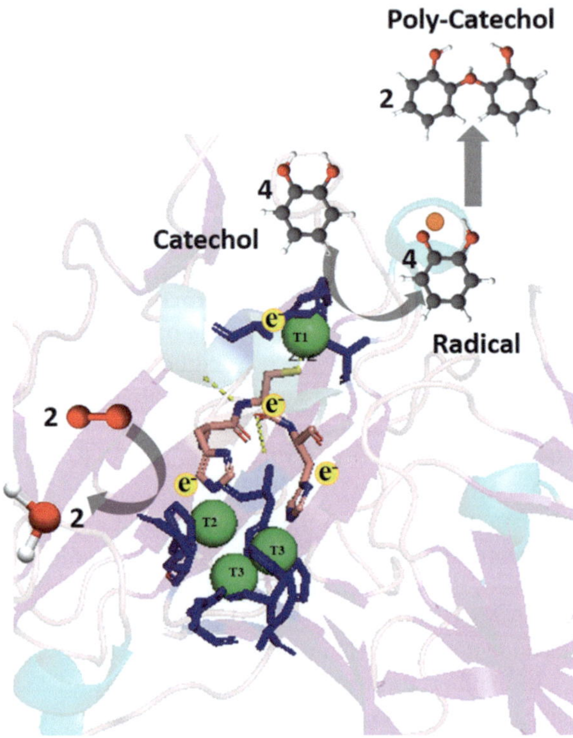

Fig. 8.2 Mechanism of *Trametes versicolor* laccase (PDB: 1GYC—Piontek et al. 2002). The image shows copper atoms (green) and trinucleotide His-Cys-His linking copper at T1 site with trinuclear T2/T3 site (pink/blue). Some His are coordinated symmetrically to Cu atoms with a mean distance of 2.16 Å (Blue). Electron transfer reaction is exemplified by four electrons (yellow), oxidation depicts near T1 (radical in orange), and reduction reaction appears on T2/T3 site. Reaction of polymerization of pollutants. (Adapted from Fernández et al. 2013)

molecule (aerobic oxidation) is reduced to water. This final reaction has a peroxide intermediate which is stabilized by the T2 site (Piontek et al. 2002; Fernández 2013).

The reduction potential of the T1 site is the limiting factor for the reaction to proceed. Laccases oxidize organic compounds with a lower or equal reduction potential as that of the T1 site, which determines a catalytic efficiency on the oxidation of different substrates. This would indicate that the higher the potential reduction of laccases, the more biotechnology uses will be possible (Fernández 2013).

As we mentioned before, laccases have the smaller redox potential in the ligninolytic system, decreasing their possibilities for degrading nonphenolic substrates. Moreover, as the active site of the enzyme (T1 copper) is small, some substrates with higher molecular weight cannot pass. Nonetheless, the presence of small chemical compounds known as mediators can surpass this behavior. Mediators expanded the substrate range of laccases because they possess a redox potential higher than 0.9 V, so they can include nonphenolic compounds such as complex dye compounds and lignin (Shiomi 2015). Small molecules such as 1-hydroxybenzotriazole (HBT), 2,2-azinobis-(3-ethylbenthiazoline-6-sulfonate) (ABTS), catechol, or syringaldazine are oxidized continuously by laccase, generating an intermediate molecule with higher redox potential able to develop chemical oxidation on the second substrate (nonphenolic). This intermediate molecule (cation) reduces again to regenerate the initial mediator, closing the oxidation cycle. In this way, the mediator acts as an electron carrier between laccases and nonphenolic substrates overcoming redox potential limits and steric hindrances (Fig. 8.3) (Eichlerová et al. 2012).

Nine different mediators typically used on laccase reactions were studied. The main conclusion was that ABTS is the most sensible mediator, as it can detect 13.5 pg/mL or $0.199*10^{-12}$ mol/mL of TvL (Eichlerová et al. 2012). The critical point is the formation of an intermediate cation that produces a radical in the nonphenolic substrate, encouraging its own degradation. Interestingly, plant laccases use the same mechanism but for lignin polymerization. In the end, phenol radicals produced by laccases polymerize to produce higher weight molecules (Fernández-Fernández 2013).

Oxidation with laccases or laccase-mediator systems improves the possibilities of degrading various recalcitrant compounds such as synthetic dyes,

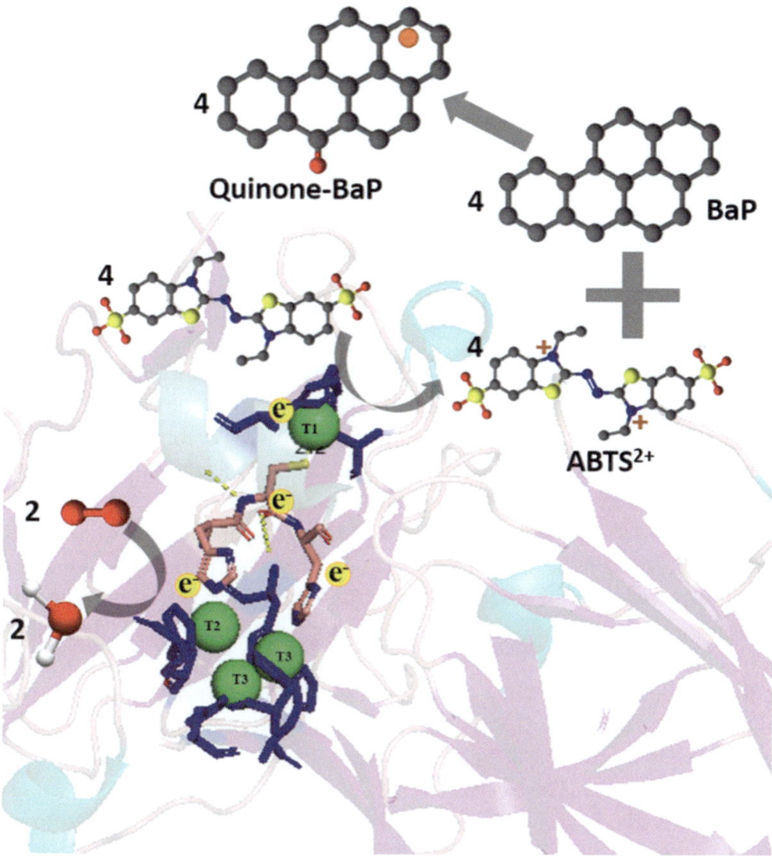

Fig. 8.3 Oxidation mechanism of laccase-mediator systems (PDB: 1GYC Piontek et al. 2002). The color convention is the same than Fig. 8.2. Reaction of degradation of PAHs. (Adapted from Fernández et al. 2013)

organophosphorus compounds, phenols, chlorophenols, lignin-related structures, and polyaromatic hydrocarbons (PAHs). The advantages of working with this enzyme are the use of oxygen as the final acceptor, and the scarce specificity to substrates oxidation, which expands the application possibilities. For example, laccases and laccases-mediator system have been evaluated for textile dye degradation, organic synthesis, delignification, battery cells, and biosensors (Shiomi 2015).

Laccases are versatile enzymes for bioremediation due to their different mechanisms for oxidation and their unspecificity for binding substrates. Three mechanisms are important to take advantage of laccase bioremediation. First is the direct oxidation on substrates with redox potential below 0.8 V. Second is the laccase-mediator system which extends the target substrates depending on the redox potential of mediators used to reaction (ABTS 1.09 V and HBT 1.2 V among others). In this case, the enzyme acts on the first substrate or mediator to produce radicals that oxidize a second substrate, typically alkene to ketone or aldehyde, which implies major susceptibility of this second substrate to the subsequent degradation reactions. Finally, laccases can also immobilize pollutants by the polymerization or the coupling of hazardous molecules to clay minerals or soil humic substances, decreasing the biological activity, the bioavailability, and the toxicity in microcosm's soil. Xenobiotics are susceptible to this immobilization laccase mediated, which are phenolic compounds, including anilines and chlorinated phenols such as 2,4,6-trinitrotoluene or 3,4-dichloroaniline (Viswanath et al. 2014).

8.3.5 Laccases as Indicator of Bioremediation Process

Bioremediation using complete cells, free or immobilized, has been used for removing or decreasing pollution on soil because of their differential selectivity. At the present section, some examples of bioremediation with complete cells have been reviewed, focusing on the importance of laccase activity as an indicator of bioremediation (Abatenh et al. 2017; Megharaj and Naidu 2017).

Dayi et al. (2018) evaluated dye discoloration of four fungal strains in soil and liquid medium. Treatment of 10 mg/L of methyl red with *Trametes versicolor* on soil medium led to 91% of discoloration with 76% of biodegradation and 15% of biosorption in 10 days, while the same conditions with the RB220 dye led to 80% of discoloration divided in 62% of biodegradation and 18% of biosorption. The higher dye concentration has the lower discoloration because of the toxicity of this type of molecules. Laccase (Lac) and manganese peroxidase (MnP) were detected on the fourth day of treatment, indicating that they play a fundamental role in the bioremediation process (Dayi et al. 2018).

Some studies about complete cell immobilization of different microorganisms have been done to prove their effectiveness on bioremediation processes. For instance, Compart et al. (2007) developed a ceramic support from slate powder as sintering hollow spheres to immobilize the fungus *Psilocybe castanella*. Laccase activity was measured by ABTS oxidation after incubation on sand for 45 min at

75 rpm resulting in an 80% protection of the inoculum against loss of enzyme activity. Results showed that the supports of slate-spheres protect the inoculum when it is used as a mixture with soil (Compart et al. 2007).

Similarly, Baldantoni et al. (2017) used untreated soil, soil added with a fungal consortium, and soil amended with compost to probe degradation of anthracene (AnT) and Benzo[α]pyrene (BaP) for validation of polycyclic aromatic hydrocarbons (PAHs) bioremediation. Laccase activity was measured as catechol oxidation. AnT content, with lower molecular weight, decreased in all conditions until 5% of the initial values after 154 days, but the higher rate of degradation was achieved with fungal consortium soil, especially in the early stages of the experiment. BaP content reached a 50% residual content after nine months due to its higher hydrophobicity and its molecular weight. Its use can be suggested in PAH bioremediation (Baldantoni et al. 2017), considering the ameliorating effects of different evaluated conditions. In a similar work, *Ganoderma lucidum* mycelia pellets and corn-cobs were immobilized on Ca-alginate modified by polycaprolactone. Anthracene removal results in 96% after 20 days at pH 5.0 and 45 °C (Xie et al. 2015).

Indeed, some patents have been dedicated to the bioremediation of pesticides like dichlorodiphenyltrichloroethane (DDT), proposing the application of a mixture of soil, white-rot fungus, and laccase obtained with the same method in the polluted area. According to the authors, this method guarantees the degradation of DDT in soils up to over 50% with the advantages of convenience for operation, low cost, and no secondary pollution (Zhao and Ma 2011).

8.4 Production of Laccases

Laccases were first found in the tree *Rhus vernicifera* (Yoshida 1883). Henceforth they have been isolated and characterized from a variety of organisms, including plants, insects, fungi, bacteria, and archaea and have also been employed in different applications (Couto and Herrera 2006; Antošová et al. 2016). However, most purified laccases have been found in low yields and, consequently, their potential large-scale applications are limited (Osma et al. 2010; Antošová et al. 2016). Yield is a major limiting factor for the enzyme applications, and therefore numerous strategies have been employed to attempt producing larger quantities of laccases at lower prices. The main approaches employed for overcoming this obstacle are the screening of natural hypersecretory strains and the use of recombinant organisms using heterologous or homologous expression in a microbial system. Additionally, the stability of the enzyme, as well as some enzymatic parameters such as the specific activity, can be modified using immobilization and protein engineering (Lettera et al. 2016; Upadhyay et al. 2016). In this section, we reviewed some of the successful strategies employed to enhance laccase production, including screening of new promising laccase and recombinant laccase production, highlighting the recombinant systems (fungal and bacterial) that could contribute to bioremediation by developing robust tools.

8.4.1 Natural Laccases Hypersecretory Strains

Reports show that the highest enzyme yields achieved by laccases in the industry and for bioremediation come from fungi (Majeau et al. 2010). In fact, the screening and selection of laccase-producing strains still constitute an effective approach. Additionally, the optimization of the culture conditions does not only allow the increase of synthesis capacity of the producer organisms but also reduces the costs (Martínez-Morales et al. 2015; Neifar et al. 2016; Yang et al. 2017). Many works focus on the selection of the house of natural origin with optimal performance and desired properties (Si et al. 2013; Fang et al. 2015; Kandasamy et al. 2016; Iracheta-Cárdenas et al. 2016; Olajuyigbe and Fatokun 2017). Table 8.1 shows the natural laccase-producing strains recently described. After this, a novel laccase from *Sporothrix carnis* was isolated and characterized. The enzyme exhibited high thermostability and pH-versatility indicating its potential to be applied in numerous processes (Olajuyigbe and Fatokun 2017).

On the other hand, an extracellular laccase enzyme produced from the green-blue microalga *Spirulina platensis* CFTRI was purified. The laccase activity was increased by divalent cations such as Cu^{+2}, Zn^{+2}, and Mn^{+2}. Due to its high stability,

Table 8.1 New natural laccases hypersecretory strains

Source	Yield	Molecular mass (kDa)	Optimum activity and stability	References
Sporothrix carnis	Yield, 3.9%, and purification fold, 2.84	56 kDa	OT: 50 °C. retained more than 50% of activity at 80 °C after 180 min of incubation pH: Neutral. Stability over pH of 3.0–11.0	Olajuyigbe and Fatokun (2017)
Spirulina platensis	Recovery, 51.5%, and purification fold, 5.8	66 kDa	OT: 30 °C. retained 80% activity for 90 min at 50 °C. pH: 3. Retained activity during 1 hour of exposure to pH 8	Afreen et al. (2017)
Aquisalibacillus elongates	Yield, 68.2%, and purification fold, 99.8	75 kDa	OT: 40 °C. pH profiles depending on the substrate. For ABTS, 6.0; syringaldazine (SGZ), 7.0; 2,6-DMP, 8.0	Rezaei et al. (2017)
Pseudomona extremorientalis	6,980 U/L	NA	OT: 40–50 °C, pH 8.0. 100% of its initial activity at pH values between 7 and 10 (24 h)	Neifar et al. (2016)
Trametes versicolor	Lcc1, 27%; Lcc2, 11%	Lcc1: 60 kDaLcc2: 100 kDa	pH 3 for both, NA	Martínez-Morales et al. (2015)

OT optimum temperature, *2,6-DM* 2,6-dimethoxyphenol

this cyanobacterial laccase was proposed to be used in wastewater treatment (Afreen et al. 2017). Also, a halophilic bacterial laccase from *Aquisalibacillus elongates* showed delignification activity in extreme conditions. The enzyme showed specificity against a broad spectrum of substrates and resistance against a wide range of potentially inhibitory agents such as solvents, salts, and others (Rezaei et al. 2017) (Table 8.1).

Remarkably, in an attempt to simultaneously detect laccase's activity quickly and efficiently, an instrument-free assay was developed. The method, consisting of dried paper discs previously impregnated with a substrate, was successfully applied for laccase detection and allowed the reduction of unnecessary purification steps (Dias et al. 2017).

However, laccase production varies according to the species and strain evaluated, and most of the native species produce laccase at low concentrations. Besides, some of the natural producers (such as ligninolytic fungi) often provide several isoenzymes with similar physicochemical properties, hindering the purification of specific enzymes (Desai and Nityanand 2011). For this reason, in most cases, the production of laccase from natural sources does not satisfactorily respond to the demands of higher yields at low cost, laccases growth conditions required for many microorganisms, and present low compatibility with the industrial standard fermentation processes (Piscitelli et al. 2010). The generation of recombinant enzymes makes it possible to overcome, at least in part, these drawbacks and could simplify the handling and purification of enzymes.

8.4.2 Recombinant Laccase Production

Heterologous expression is a valuable tool to produce proteins when only metagenomic sequences are available and also for isolating an enzyme of interest when many isoforms are present. This tool is also helpful when the enzyme is expressed at low concentrations, or it is silent (Fang et al. 2011, 2012). This approach permits the production of proteins using a rational design or directed evolution to increase the expression, catalytic activity, stability, etc. In fact, enzyme yield can be improved by the use of multiple gene copies and strong and inducible promoters. Additionally, when appropriate signal sequences are designed, proteins can be secreted outside the cell, simplifying the downstream processing. Besides, heterologous expression eases biochemical and structural characterization of recombinant laccases (Rosano and Ceccarelli 2014; Wang et al. 2017a).

Expression of recombinant protein in hosts that are easy to grow and manage allowed higher productivity in less time and reduced production costs. The versatility and the possibilities of expanding the production of recombinant proteins enabled their successful application both in bioremediation and new commercial developments (Ferrer-Miralles et al. 2009; Piscitelli et al. 2010).

Eukaryotic and prokaryotic hosts can be used, being the yeast, *Pichia pastoris* the most employed one (Antošová et al. 2016; Ergün et al. 2016). Other unicellular hosts commonly employed are *Escherichia (E.) coli* and *B. subtilis*, *Saccharomyces cerevisiae*, and also some plants including *Oryza sativa* and *Nicotiana tabacum*. Filamentous fungi hosts such as *Aspergillus niger* and *Trichoderma reesei* have resulted in high recombinant laccase yields and have also proved to be efficient in protein expression (Saloheimo and Niku-Paavola 1991; Record et al. 2002; Kiiskinen 2004). In bacteria, activity yields of secreted laccases reached 5,600 U/L (Durao et al. 2008; Ihssen et al. 2015), and the protein yields up to 350 mg/L (Koschorreck et al. 2008; Dubé et al. 2008), while the production of laccases using filamentous fungi as heterologous hosts brings activity levels between 600 and 780 U/L approximately (Téllez-Jurado et al. 2006) and maximum yields of 920 mg/L (Hatamoto et al. 1999; Couto and Toca-Herrera 2007). However, they are not used massively because genetic tools have been developed more efficiently for yeasts and prokaryotes (Antošová et al. 2016). Remarkably, laccases, as other ligninolytic enzymes, are particularly hard to express in a heterologous way (Gu et al. 2014; Ergün et al. 2016). Many of the laccases expressed in heterologous hosts reached even lower levels than in their natural environment (below 10 U/mL) (Yang et al. 2017).

As mentioned above, the main limitation is the low yield obtained in these guests. Additionally, recombinant enzymes can form aggregates, which complicates their purification (Suzuki et al. 2003).

Some strategies applied to enhance enzymes production in heterologous hosts include promoter and signal peptide selection, codon optimization, protein engineering, as well as optimization of the culture medium composition. Some research suggests that employing a host with similar codon usage and GC content as the original gene could improve the protein expression (Nishibori et al. 2013). However, the different codon preferences between the expression host and the native producer are not enough to explain the difference in production yields observed in homologous expression systems (Piscitelli et al. 2010). In this sense, and taking into account the different and even contradictory results that have been reported, it is difficult to predict the most suitable combination of parameters that would allow the optimization of laccase production (Piscitelli et al. 2010; Antošová et al. 2016).

In addition, as it was mentioned, protein engineering is a useful tool to modify the stability of an enzyme as well as its specific activity. In this sense, tailor-made laccases can be produced to satisfy the demand for a particular application (Bornscheuer et al. 2012). Using directed or random mutagenesis, it is possible to obtain enzymes with higher thermal stability, substrate specificity, and optimum pH. It is also possible to improve the catalytic parameters (Vmax, KM, and Ki) of the enzymes. Additionally, "DNA shuffling" method, which consists of a recombination of homologous genes created in vitro that allow sensible improvements in the enzymatic properties in short periods, can also be used. The most significant advances in this research topic are summarized in very comprehensive reviews (Mate and Alcalde 2015; Pardo and Camarero 2015).

8.4.3 Laccases from Fungal Origin

Among fungi, *T. versicolor* can be considered the model organism for laccase studies. In recent years, *Cerrena* genus has received special attention due to its high yields of laccase production and the promising characteristics of enzymes that in some cases exceed commercial ones (Chen et al. 2012). However, there is still little information available on *Cerrena* species producing laccase.

8.4.3.1 *Trametes versicolor*

Laccase enzymes make up the most important part of the extracellular lignin-degradation system of *T. versicolor*, being laccase III (CVL3) the isoenzyme secreted at the highest level (Iimura et al. 1995) (Table 8.2). The enzyme is highly glycosylated, and it has been shown that carbohydrate chains play an important role in the proteolysis and thermal resistance of the enzyme (Yoshitake et al. 1993). In an earlier study, Kajita et al. (2004) overexpressed the laccase III (*cvl*3) gene by self-cloning it in *T. versicolor*. One of the transformant strain obtained was able to overproduce extracellular laccase, whose activity exceeded that of the wild type, especially when copper (II) was added (Kajita et al. 2004). In a later study, the same gene was introduced into tobacco plants, obtaining a transgenic strain that produced laccase in the rhizosphere, and was able to remove environmental pollutants, being more efficient than the control (Sonoki et al. 2005).

Another study (Nishibori et al. 2013), *Cryptococcus* (*Cryp.*) was tested as an alternative host for *P. pastoris* for the expression of the laccase gene from *T. versicolor* and *Gaeumannomyces* (*G.*) *graminis*. Results showed that the activity of *T. versicolor* and *G. graminis* laccases was 143 times and 60 times higher, respectively, in *Cryp.* compared with *Pichia* (Table 8.2). The authors suggest that using a host such as *Cryp.* with comparable GC content and codon usage as *T. versicolor* and *G. graminis* may improve the protein expression. However, recent studies did not arrive at the same conclusion (Piscitelli et al. 2010). In addition, the proposed host offers similar performances as *P. pastoris* regarding proteins yields, achieving high fermentation efficiency with minimal growth requirement, and also the capability for post-translational modifications. The authors proposed *Cryp.* as an alternative host to produce enzymes at high levels when other hosts do not work (Nishibori et al. 2013).

In a recent study (Iimura et al. 2018), laccase III from *T. versicolor* was cloned and expressed in Saccharomyces cerevisiae. In this study, the essential requirements for the efficient production of laccase in the host, as well as some factors that affect its expression, were clarified (Table 8.2). Purified enzyme results in a hyper-mannosylated isoform mixture that lowered the substrate affinity without modifying thermal resistance and optimum pH.

Table 8.2 Remarkably recombinant laccase production

Origin	Source	Gene	Host	Activity	Optimum activity and stability	References
Fungal origin	*Trametes versicolor (before Coriolus versicolor)*	Laccase III (CVL3)	*T. versicolor*	NA	NA	Kajita et al. (2004)
		Laccase III (CVL3)	*Tobacco plants*	NA	NA	Sonoki et al. (2005)
		Laccase III (CVL3)	*Pichia (P.) pastoris*	0.014 U/ml (ABTS)	NA	Nishibori et al. (2013)
		Laccase III (CVL3)	*Cryptococcus*	2.0 U/mL (ABTS)	NA	Nishibori et al. (2013)
		Laccase III (CVL3)	*Saccharomyces cerevisiae*	170 U/mg (ABTS)	NA	Iimura et al. (2018)
	Cerrena sp. strain HYB07	Laccase 1	*P. pastoris*	6.3 U∗mL/L (laccase 1)	OT: 55 °C pH 3.5. Remained stable between pH 4–10 and 20–60 °C	Yang et al. (2015)
		Laccase 3	*P. pastoris*	0.2 U/mL	OT, 55; pH, 3	Yang et al. (2016b)
		Laccase 6	*P. pastoris*	2.9 U/mL	NA	Yang et al. (2016b)
		Laccase 7	*P. pastoris*	0.5 U/mL	NA	Yang et al. (2016b)
		Laccase 8	*P. pastoris*	0.7 U/mL	NA	Yang et al. (2016b)

Bacterial origin	*Escherichia (E.) coli*	CueO (yacK)	*E. coli*	4 U/mg	pH, 5; OT, NA. half-life of 5 h at 70 °C	Kim et al. (2001)
	E. coli	yacK	*P. pastoris*	41 U/mL	pH, 3.0; OT, 55 °C. half-life at 70 °C for 60 min	Ma et al. (2017)
	Bacillus (B.) subtilis	cotA	*E. coli*	Specific activity: 1.28 mol/min/mg protein	pH, 3.0 (ABTS), and pH, 7.0 (SGZ). OT: 75 °C. half-life of 2 h at 80 °C	Martins et al. (2002)
	B. subtilis	cotA	*E. coli*	1.26 U/mL	pH, 4.4; OT, 60 °C. remained 21% of the initial activity after 10 h at 80 °C and 163% at pH 9.0 (after 10 days)	Guan et al. (2014)
	B. subtilis	cotA	*P. pastoris*	1.65 U/mL	64% and 95% of its initial activity after a 10-day incubation at pH 9.0 and 10.0	Wang et al. (2015)
	B. coagulans	cotA	*E. coli*	0.72 mol/min	Optimal pH 4. Retained full activity for 30 min at 70 °C and more than 60% of the activity at pH 11	Ihssen et al. (2015)
	B. pumilus	cotA	*E. coli*	4.2 mol/min	Optimal pH 4. Retained full activity for 30 min at 70 °C	Ihssen et al. (2015)
	Streptomyces (S.) coelicolor	SLAC	*E. coli*	NA	pH: 9.4	Machczynski et al. (2004)
	S. anulatus	SaSL	*E. coli*	NA	OT: 85 °C. pH optimum of 3.6 (ABTS) and 9 (2,6-DMP). Retained 80% activity at pH 11	Lisov et al. (2018)
	S. griseus	EpoA	*E. coli*	NA	OT, 40 °C; pH, 6.5. Retained 40% activity after preincubation at 70 °C for 60 min	Endo et al. (2003)

NA not available, *2,6-DMP* 2,6-dimethoxyphenol, *OT* optimum temperature, *SGZ* syringaldazine

8.4.3.2 *Cerrena* Genus

In the last years, many researchers have been interested in strains of *Cerrena* genre due to its ability to produce laccases with various applications. However, our current knowledge of laccase genes belonging to *Cerrena* remains scarce, since only a few have been cloned and molecularly characterized.

Yang and his co-workers have previously isolated and characterized a *Cerrena* sp. strain that was able to produce laccase in significant quantities (Yang et al. 2014). The laccase family of *Cerrena* HYB07 contains nine members, some of which were heterologously expressed in *P. pastoris* and characterized (Table 8.2) (Yang et al. 2015; Yang et al. 2016b).

Identity analysis of the eight laccase genes among themselves as well as with other fungal laccases showed moderate values. On the other hand, a high redox potential for all enzymes was predicted, except for Lac 6. Moreover, they showed different enzymatic properties and expression profiles in liquid and solid state fermentation, being Lac7 the main isozyme produced by HYB07 as indicated by transcription analysis by RT-PCR and LC-MS/MS analyses (Yang et al. 2016a, b). In fact, in the submerged fermentation, Lac7 and Lac2 were the most significantly expressed genes during six days, representing more than 95% of the transcripts found, while in a solid state, differences were less evident (Yang et al. 2016b).

8.4.4 Laccases from Bacterial Origin

As was referenced, yeasts are commonly used for many applications including industrial production of various proteins such as laccases. However, in some cases, fungal laccases have little stability, especially at high temperatures and pH values, and, in consequence, its activity usually decays rapidly (Ihssen et al. 2015).

In contrast, due to the widespread occurrence and versatility of prokaryotic organisms, their laccases could have certain advantages over fungal ones due to their higher thermostability, Cu^{2+} resistance, and pH and halo- and high-chloride concentrations tolerance. It has been reported that bacterial laccases are involved in several protective processes that include pigmentation, UV safeguard, oxidation of metals, and degradation of xenobiotic substances (Upadhyay et al. 2016; Nunes and Kunamneni 2018). Remarkably, bacterial laccases do not require glycosylation and can be expressed more easily in *E. coli*, probably the host of choice for expression system for recombinant proteins in many applications. As mentioned, the expression in prokaryotic systems generates low yields, so several approaches for increasing the yield have been described.

Moreover, catalytic properties, as well as the expression level and stability, could be improved by direct evolution (Pardo and Camarero 2015). Despite presenting low redox potentials, bacterial enzymes can be used as functional complements of their fungal homologs in valuable processes such as bioremediation as well as several other industrial and biotechnological applications (Singh et al. 2011; Chandra and Chowdhary 2015; Martins et al. 2015).

The first bacterial laccase reported was from the *Azospirillum lipoferum* (Givaudan et al. 1993). A bioinformatic analysis by Ausec et al. (2011a, b) reveals a high diversity of genes for putative genes of laccase enzymes in bacteria. Indeed, more than 1,200 genes for laccase-like enzymes were detected in 2,200 complete bacterial genomes of chromosomes and plasmids of various bacteria. Interestingly, in ≈75% of the genes, signal peptides were predicted, suggesting that some bacterial laccases can be exported outside the cells (Ausec et al. 2011b), in contrast to the first intracellular bacterial laccases described. Additionally, some strains of *Streptomyces* (*S.*) spp., which produce extracellular laccases, have been reported (Beppu et al. 2002; Endo et al. 2003; Niladevi et al. 2009; Molina-Guijarro et al. 2009). In addition, bacterial laccases were described in *E. coli* (Ma et al. 2017), *B.subtilis*, *B. halodurans*, *B. licheniformis* (Martins et al. 2002; Koschorreck et al. 2009; Guan et al. 2014), *Streptomyces coelicolor* (Machczynski et al. 2004), and *Thermus thermophilus* (Miyazaki 2005) among others.

8.4.4.1 *Escherichia coli*

The *yacK* gene coding a putative multicopper oxidase (called CueO), from *E. coli* K 12, was first heterologously cloned and expressed in other *E. coli* strains. The enzyme exhibited phenoloxidase and ferroxidase activities. Copper addition stimulated the enzyme activity (Kim et al. 2001) as well as a decrease in the thermal stability of the enzyme. The authors suggested that the mechanism confers a high level of protection against copper in the bacteria and leaves the most soluble form of Fe II available for absorption. It has been reported that a crude enzymatic extract from *E. coli* containing CueO was able to oxidize two important HPAs (such as benzo[α] pyrene and anthracene) similarly as laccase from *T. versicolor* (Zeng et al. 2011). In a more recent study, the *yacK* gene was cloned and expressed in a *P. pastoris* strain as a host showing high thermostability (Ma et al. 2017) (Table 8.2). Together, these findings indicate the potential of *E. coli*–CueO to be used in bioremediation strategies.

8.4.4.2 *Bacillus* Genus

Until now, the most studied bacterial laccase is the CotA from *(B.) subtilis* consisting in an endospore coat component. The CotA is a 65-kDa proteinacious shell that encases the spore and plays a significant role in spore survival. The enzyme participates in the biosynthetic pathway of the brown spore pigment, which is believed to be a product similar to melanin with a protective effect against UV light and hydrogen peroxide (Driks 1999). CotA protein was purified and characterized from an overproducing *E. coli* strain showing laccase activity with intrinsic high thermal stability (Table 8.2) and similarities with multicopper oxidases (Martins et al. 2002).

Remarkably, the multicopper oxidases require a sufficient supply of copper for catalysis. However, *E. coli* cells have inducible efflux pumps to prevent the toxic accumulation of copper in their cytoplasm (Outten et al. 2001). In copper-supplemented media, a switch to anoxic conditions leads to the synthesis of a recombinant CotA holoenzyme; thus, cells grown under microaerobic conditions accumulate up to 80-fold more copper than aerobically grown cells. To achieve successful production of recombinant CotA, culture conditions were adapted by introducing a static culture stage after induction that allows intracellular copper increase and ensures Cu supply for the enzyme levels (Durao et al. 2008). The same procedure was applied for the efficient expression of other *Bacillus* strains in *E. coli* (Reiss et al. 2011). In other work, Gupta and Farinas (2010) proposed the directed evolution of CotA laccase from *B. subtilis* to increase substrate specificity. For that, a CotA genes library was expressed in the spore layer. It was found that a mutant CotA was 120 times more specific against the substrate of choice (ABTS). These results demonstrate that spores of *B. subtilis* can be a useful strategy for screening protein libraries (Gupta and Farinas 2010).

Recently, new bacterial strains of *B. subtilis* exhibiting high laccase activity were identified, cloned, and expressed (Guan et al. 2014; Wang et al. 2015). The CotA laccases resulted in a biologically active enzyme that has activity in a wide range of pHs and high stability in the presence of alkaline pHs and high temperatures (Table 8.2).

In a very comprehensive study (Ihssen et al. 2015), the activity yields and biochemical properties of diverse origin multicopper oxidases of bacterial origin, including *B. subtilis*, *B. pumilus*, *B. coagulans*, and *B. licheniformis*, were determined. In almost all cases, a switch to oxygen-limited growth conditions after induction significantly increased the activity. *Bacillus* enzymes showed better performances than those observed in *Streptomyce*s homologs and some Gram-negative strains in terms of yields achieved using *E. coli* as a host. Also, they showed relevant biochemical properties for potential applications. Remarkably, a novel *B. coagulans* laccase showed significant yields and also exceptional activity at high pHs and significant storage stability (Table 8.2). These properties make it an interesting candidate for numerous applications (Ihssen et al. 2015).

8.4.4.3 *Streptomyces* Genus

As mentioned before, laccases and other four-copper oxidases are commonly organized by three cupredoxin domains, and the active center of those contains four copper atoms. Besides, a two-domain (2D) laccase was described in bacteria. In fact, Machczynski et al. (2004) reported and characterized an enzyme from *Streptomyces coelicolor* that represented a new family of laccases that lacked the second domain and yet exhibited significant activity. The enzyme, called SLAC, was recombinantly expressed in *E. coli* and showed higher stability since it retained its activity and its dimeric nature after being boiled and exposed to detergents such as SDS (Machczynski et al. 2004).

Recently, the interaction of 2D laccase from *Streptomyces anulatus* (called SaSL) with humic substances under alkaline pH was studied. The recombinant SaSL showed a significant thermo-resistance and was able to metabolize HA at alkaline pH values (Table 8.2). Based on these findings, the authors proposed a putative role of the 2D laccases in the humification processes of alkaline soils (Lisov et al. 2018).

On the other hand, EpoA from *S. griseus* has been characterized physicochemically and biochemically and then expressed as recombinant in *E. coli* (Endo et al. 2003). The enzyme has a relatively close substrate specificity since it does not metabolize some classic laccase substrates. The enzyme seems to participate in morphogenesis of *Streptomyces* spp. (Beppu et al. 2002).

Moreover, laccases from *S. psammoticus* and *S. ipomoea* showed unusual high activity at pH values of 7–8 as well as tolerance to osmotic conditions (NaCl >1 M), conditions commonly found in wastewater and others (Niladevi et al. 2009; Molina-Guijarro et al. 2009). Additionally, significant laccase activity was also observed in the culture supernatant of *S. psammoticus* and S. cyaneus (Arias et al. 2003; Niladevi et al. 2009), suggesting the potential application of these strains for bioremediation applications.

Taking into account the versatility of reactions catalyzed by laccases and the increasing production possibilities aforementioned, the next section will be dedicated to exploring the increase of immobilization strategies to improve stability to different extreme conditions and especially the reuse possibilities.

8.5 Laccase Immobilization

Industrial applications for enzymes have been increasing in the last century on diverse processes associated with nutraceutical, pharmaceutical, and cosmetic products, beverages and food production, or soil and water decontamination. The mid-temperatures and pH, the aqueous system, the biodegradability, and the catalytic efficiency in the reaction are interesting features to take into account in the bioprocess, as an alternative to reduce the environmental impact and sustain the industrial production. However, the lack of stability in some environments, the expensive production, and the capacity of reuse of free enzymes have retarded their use at an industrial scale. For this purpose, different immobilization procedures have been developed for trying to change the surface microenvironment and degree of multipoint attachment to a synthetic or natural support. Various immobilization strategies increase their stability to temperature and pH changes or presence of organic solvents while improving the recovery and reuse of the biocatalyst and dismiss costs associated with the process (Guauque-Torres et al. 2013).

Immobilization techniques can be classified as physical adsorption, entrapment, cross-linking, or covalent attachment (Fig. 8.4). Physical adsorption takes advantage of differential charges between protein surface and supports to create an electrostatic interaction that insolubilizes target enzymes without losing their activity, but

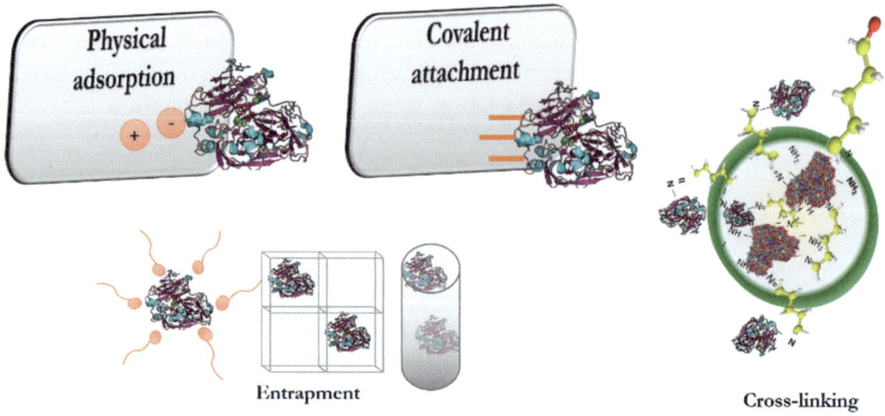

Fig. 8.4 Immobilization techniques. TvL (PDB: 1GYC—Piontek et al. 2002). Glutaraldehyde (yellow) and bovine serum albumin (BSA-PDB 3 V03—Majorek et al. 2012) are used as additives for cross-linking

the weakness of this interaction can increase leaching rate leaving losses of enzymatic material and decrease the reuse cycles. Entrapment improves mechanical stabilization through confining enzyme into fibers, networks, or polymeric matrices that permit the substrate and products to pass through but retain the enzyme. However, mass transfer limitations and deactivation during immobilization are the major disadvantages in this technique.

The third strategy is also known as carrier-free immobilization or self-immobilization. It uses bi-functional reactive as glutaraldehyde, carbodiimides, or dextran to link different functional groups on side chains of amino acid, acting as a hinge between different enzyme molecules. Even though cross-linking can avoid dilution of activity with consequent economical savings due to suppression of support, it is possible too that its inter or intra-linking could change the active structure, leaving inactive enzymes. The covalent attachment focuses on functional groups of amino acid residues (lysine, cysteine, aspartic or glutamic acid, imidazole, and phenolic groups) bonding to the support with the cross-linkers aforementioned (glutaraldehyde or dextran). This technique is the most used for irreversible immobilization due to its better stability in different environments and the fact that it reduces the leaching of the enzyme. However, it is essential to know the distribution of nucleophiles groups on the protein surface, to avoid the inactivation by the interaction with the amino acid residues involved in the active site or binding domain (Mohamad et al. 2015). The main disadvantages of covalent immobilization are the expensive modifications required to perform the carrier material to improve the interaction with the enzyme without modifying their activity and selectivity.

Remarkable new approaches mix immobilization techniques to afford the best performance of immobilized enzymes. Kumar and Kanwar (2012) prepared an immobilized lipase by adsorption followed by entrapment. Firstly, the enzyme was adsorbed on lignocellulosic coconut fibers containing 32.8% of lignin. The reactivity between

OH groups of natural fiber and NH_2 groups of glutaraldehyde (2% v/v) was employed to achieve subsequent physical entrapment. The final fiber-immobilized biocatalyst showed better stability on temperature (55–60 °C), pH (alkaline), resistance to detergent, and metals presence, and also, it retained more than 50% of its original activity after fourth reuse cycle (Kumar and Kanwar 2012).

A growing field of research at nanoscale provided different structures to be considered as strategies to choose where the traditional immobilization techniques have restrained industrial or therapeutic implementations. We reviewed three innovative designs of tailored strategies to improve the performance of the biocatalyst that has been proved in different enzymes and that can be surely used to enhance the performance of catalyst based on laccases extending their potential applications.

The first strategy exploits the ability of the protein to self-assemble or to be aggregated with transition metals as a precipitating agent to produce nanohybrids or nanoflowers which can be used as biosensors or biocatalysts due to their improved catalytic and physicochemical properties (Kumar et al. 2018; Correa et al. 2019). This technique uses the affinity of surface-exposed histidines and other side chains residues to metal ions by coordinating unpaired electrons (Iyer and Przybycien 1995). The aggregates obtained are centrifuged, washed, and stored for further experiments. Reusability of nanoflowers should be improved because the centrifugation alters the intricate hierarchical shaped petals. By modifying the method, glutaraldehyde can be added to improve reusability of the final biocatalyst (70% of initial activity after four reuses) which could be verified by morphology conservation after centrifugation and reuse (Lee et al. 2017). Another strategy to maintain the morphology of the structure in reuse cycles is to prepare laccase-loaded magnetic nanoflowers (MNFs) by attaching amino-functionalized magnetic nanoparticles (Fu et al. 2019).

In the second approach, carbon nanotubes were used to make mesoporous particles of SiO_2. The technique implies the mixture of the components, spray pyrolysis to ensure porousness and enhance the area available to immobilization, and the stage of posttreatment with glutaraldehyde to provide anchoring points for enzyme coupling. The size of porous was 12 and 8 nm before and after immobilization, whereas the Brunauer–Emmett–Teller (BET) surface area was of 494 and 171 m^2 /g showing the ability of the technique to create microparticles with controlled size porous and functionalization required for binding enzymes. Besides, the mesoporous SiO_2-immobilized enzyme proved to have achieved better specific activities, robustness, reusability, stability, and sensitivity in bio-sensing applications (Kumar et al. 2019).

Finally, SnO_2 hollow nanotubes were synthesized by electrospinning because of simplicity, briefness, and economic benefits. The technique involved the use of electrospinning with a mixture of polyvinylpyrrolidone, $SnCl_2$, dimethylformamide (DMF), and ethanol with ulterior calcination to assure SnO_2 hollow nanotubes in the range of micrometers with diameters between 200 and 300 nm. Changes at glutaraldehyde functionalization were evaluated to improve physicochemical stability of biocatalyst of lipase, horseradish peroxidase (HRP), and glucose oxidase (GOx). Comparison of SnO_2 hollow nanotubes with other nanoparticles reported previously showed improved mechanical properties such as larger surface areas, higher porosity and consequently better loading capacity, making this support as a promising candidate for another type of enzymes (Anwar et al. 2017).

8.5.1 Materials for Immobilization

The requirements on material for immobilization depend on the complexity of the reaction media and matrix (soil, water) where it will be applied. Some of the most important characteristics for choosing a carrier are: Mechanical strength, resistance to decomposition by chemical or microbial agents, low affinity to reaction products and availability of surface reactive groups to bind target enzyme. Finding the correct combination between carrier–enzyme–target reaction is a bigger challenge, and it is crucial for their application at industrial scale (Verma et al. 2019).

There are many materials for immobilization with different physicochemical and mechanical properties, from organic materials such as biopolymers or synthetic polymers and resins to inorganic materials such as silica, alumina, and carbon nanotubes. Each one of them offers different reactive functional groups on the surface, affinities, stability, and cost-effective characteristics. In a green perspective about materials for immobilization, recent studies have proved the success of the immobilization process for different enzymes on natural fibers of dried coconut, *Bombax ceiba*, pinewood nanobiochar, or even bacterial cellulose. The advantages of this type of support are low cost, availability, tensile strength, recycling, and biodegradability (Kumar et al. 2014; Chen et al. 2015; Naghdi et al. 2018).

Immobilization of extracellular enzyme may be useful for extending their activity and improve stability. Laccase has been assayed with a great quantity of them to produce a biocatalyst with enhanced activity in specific reactions and mediums. At the present section, we will focus on laccases immobilized and optimized for soil matrix (De Lima et al. 2018; Zdarta et al. 2018).

For instance, Koyani and Vazquez-Duhalt (2016) reported the laccase encapsulation on chitosan-tripolyphosphate nanoparticles that maintain kinetic parameters to syringaldazine oxidation but enhance stability for microbial degradation. After a 24-h exposure to wastewater, the encapsulated form retained 82.8% of its activity in contrast with 7.8% for the free enzyme. In compost extract, after 36 h of incubation, the difference was 72.4% to 0% for immobilized and free enzyme, whereas soil extract led to 57.9% and 17.3% of initial activity for immobilized and free laccase, respectively (Koyani and Vazquez-Duhalt 2016).

In a recent study about self-immobilization strategy, *Cerrena* laccase was immobilized using the cross-linked enzyme aggregates technique. Ammonium sulfate and glutaraldehyde (30 mM) were used as a precipitant and a cross-linking agent, respectively. The recovery activity rate was 68.1% at pH 8 after immobilization for 3 h at 25 °C. The CLEA-laccase retained activity when exposed to organic solvents, different NaCl concentrations, and metal ions presence (Yang et al. 2016c).

There are multiple test reactions to probe successfully the immobilization strategies using different substrates. Flavonoids as phenolic structures are part of the natural substrates of laccases. Polymerization of flavonoids is a concept-proof for one of the main strategies of laccases to soil bioremediation, which is immobilization of pollutants. Song et al. (2018) immobilized *Myceliophthora thermophila* laccase onto bacterial cellulose (BC) membranes with residual activity of 88% and a color absorption increase that corresponds to oligomers formation which might be produced through the nucleophilic attack of the laccase-catalyzed reaction (Song et al. 2018). Polyethylene glycol has been added to an epoxy resin–laccase system with a threefold activity increase to improve the polymerization and removal of phenolic compounds from wastewater. Indeed, the immobilized PEGylated laccase of *Myceliophthora thermophila* shows different phenol polymerization behavior depending on inner interactions of support, enzyme, and PEG additive into biocatalyst. Su et al. (2018) probed that degree polymerization of catechol changed from 7 with free enzyme to 14 when laccase is immobilized on epoxy resin with PEG as a junction between them (Su et al. 2018).

Examples of enzymes immobilized on biodegradable polymers or natural supports are an excellent option for soil bioremediation. In this way, the bioaugmentation strategy is possible and feasible, since it provides stability and high volumetric activity to a structure that protects the enzyme of an adverse environment without the addition of exogenous compounds that can introduce adverse effects.

On the same direction, different ceramic materials typically founded on soil can be evaluated as supports since enzymes are naturally adsorbed on clay minerals by different interaction forces such as cation exchange, electrostatic adsorption, and hydrophobic binding. Special attention must be directed to clay–enzymes interactions taking into account that upon immobilization, the enzymatic structure is exposed to microenvironment governed by the chemical nature of the support, leading to conformational changes or steric hindrance that influences the overall behavior of the biocatalyst (Naidja et al. 2000).

For example, Ruggiero et al. (1989) immobilized TvL on kaolinite, silt loam soil, and montmorillonite (MMT) with an efficient removal of 2,4-DCP of 95% for the first two mentioned. These biocatalysts showed stability for 15 cycles of reaction, and they had better proteolysis resistance (Ruggiero et al. 1989). Laccase of *Trametes versicolor* was also immobilized on aluminum hydroxide, finding a strong affinity and similar kinetic parameters with its free counterpart using oxygen as substrate. Despite immobilized laccase showing similar stability in pH and temperature than free enzyme, the improved resistance to inhibition by humic acids makes the immobilization process a viable alternative to enhance pollutant immobilization on soils avoiding the spread of soil contamination (Ahn et al. 2007). In a similar work, Wu et al. (2014) studied the interaction of TvL with iron and aluminum. Despite the immobilized laccase having lower substrate affinity (kinetic parameters), the adsorbed laccase exhibited better catalytic activities for acidic pH and improved resistance to proteolysis and extended the lifespan of laccase (Wu et al. 2014). Laccase was crosslinked to bentonite and used on a wastewater sample to reach a 90.13% of removal of phenolic compounds in 1h of reaction. (Alsoufi 2018). On the same direction, a hybrid layered double hydroxide Mg/Al was used as support for the laccase of *Myceliophthora thermophila* with 97% of the recovered activity compared to free enzyme using syringaldazine as substrate (Camacho-Cordova and Morales-Borges 2009).

Moreover, new approaches can be probed to optimize degradation rates in immobilized enzymes. To illustrate this, we reviewed a couple of studies. Nonionic surfactant-modified clay combines high adsorption surface with hydrophobicity to improve PAHs degradation (NaT/PhT). The remaining relative activity of laccase after 14 days of incubation with PAHs compounds was greater for the immobilized enzyme (66.32%) than for the free counterpart (53.41%). Degradation variances were attributed to the molecular weight and stability differences between them. The best results were found with Triton X-100 before and after critical micelle concentration (CMC) with removal percentages above 80% for both PAHs (Chang et al. 2016). Recent studies try to mimic conditions of wastewater using malachite green and Cd(II) as typical compounds present on textile effluents. An immobilized TvL on kaolinite with 88.22% of efficiency was able to decolorize a solution after 300 min of incubation. The immobilization process improved pH stability and reuse for five cycles (Wen et al. 2019).

Immobilized laccase has also been used for the oxidation of high toxicity PAHs (higher molecular weight). To probe the oxidative potential of the laccase-mediator system, kaolinite-immobilized laccase on AnT and BaP in ABTS presence was evaluated in contaminated waters. Results indicated that the mediator system oxidized more than 80% of the PAHs in solution, whereas, in ABTS absence, percentages were below 20%. In spite of free laccase having similar oxidation behavior, the immobilized one had stability on reuse for four cycles retaining 60% of the oxidation potential (Dodor et al. 2004). For a better explanation of the system on realistic conditions, an aqueous mixture of PAHs in equimolar concentration (AnT, BaP, BaA, and Pyr as binary, ternary and quaternary mixtures) has been evaluated too. Soil and kaolinite were used as immobilization support for TvL while ABTS and HBT were used as mediators. Soil laccase has lower K_m than free laccase and is slightly acidic than the optimum pH of the ABTS system (165 μM < 261 μM and pH 4 < 4.5). Taking into account the oxidation of PAHs alone, soil laccase–HBT system demonstrated the best behavior in the reaction for the molecules with higher redox potential assayed with approximately 70% and 30% for BaA and Pyr, respectively. Potential oxidation of kaolinite laccase was evaluated on most complex mixtures. Due to its higher HBT redox potential (1.2 V compared with 1.09 V of ABTS), this system was more favorable to oxidation of PAHs mixtures. In general, AnT with the lowest molecular weight inhibited the oxidative transformation of the other PAHs, and quaternary mixtures had the lower oxidation results because of the complexity of their system (Dodor et al. 2018). Similar results were previously found on biotransformation of BaP using laccase of *Coriolopsis gallica* when the addition of HBT increased the oxidation rate ninefold compared with the same reaction with ABTS alone (Šašek et al. 2003).

As previously mentioned, different materials have been used as supports with multiples applications. Every biocatalyst for a special application can be optimized to obtain the maximum efficiency. Zdarta has excellently reviewed the complete information about the advantages and disadvantages of materials for support in 2018 (Zdarta et al. 2018).

8.5.2 Bacterial Laccases Immobilized

Bacterial laccases also have been studied to probe oxidation behavior, in spite of structural differences (two-domain structure compared with a three-domain fungal structure). Bacterial sources present biotechnological advantages because of the scaling process previously optimized for other enzymes and bacterial laccases. Particularly, they have a broad substrate spectrum, a wide pH range, high thermal stability, and tolerance to alkaline soil environments (Guan et al. 2018).

The role of *Streptomyces anulatus* laccase on humic acid transformation was evaluated in soils. The results indicated selectivity pH-oriented for the oxidation of electron donors ($K_4[Fe(CN)_6]$, ABTS) at acidic pH and phenolic substrates (2-methoxyphenol, 2,6-dimethixyphenol) at alkaline pH values. The last one is an interesting result for suggesting a mechanism in the humification process in alkaline soils and contributing to further expand the range of applications of laccases on soil bioremediation (Lisov et al. 2018). Environmental pH has played an essential role in polymerization rate, so bacterial laccases open the door to different approaches on calcareous and alkaline soils.

Interesting approaches had been patented with immobilized bacterial laccases for the chemical treatment of phenolic and/or aromatic amines. The textile industry is the target of this invention, where bacterial spores of *B. subtilis* are immobilized in different supports like cellulose, alumina, natural, and synthetic polymers among others using adsorption, entrapment, or covalent binding and spacers such as polyethylene glycol, cellulose, and dextrins for bleaching process (Guebitz et al. 2003).

Fungal and bacterial laccases have been immobilized on silica, cationic exchange resins, glass ceramic, graphite, clay minerals, carbon fiber, chitosan, and alginate among other material supports, for different applications in wastewater treatment, biofuel cells, biosensors, dyes discoloration, paper and textile industries, and xenobiotics degradation, driven specifically by their aqueous medium demand. For a deeper understanding of this laccase applications, excellent reviews can be found in the literature (Durán and Esposito 2000; Theuerl and Buscot 2010; Shraddha et al. 2011; Burns et al. 2013; Fernández-Fernández et al. 2013).

8.5.3 Laccases for Soil Bioremediation

The versatility of laccases permits their use from a catalyst on the soil to polymerization of pollutants and for bioaugmentation, with immobilized laccases for degradation of contaminants. We will mention some of the researches developed in those topics focusing the first subsection on the biochemical synthesis process understood as oligomerization of pollutants to dismiss their mobility and bioavailability. The second subsection will refer to laccases immobilized on different materials to bioaugmentation with immobilized enzymes to degrade pollutants (Table 8.3).

Table 8.3 Laccase immobilization focusing on soil applications

Laccase source	Immobilization type	Application/test reaction	Conditions (support/additives)	Optimus (specific activity/ kinetic parameters/stability/ reuse)	Reference
Lactuca serriola L.	Covalent	Phenol remediation/ABTS oxidation	Support: Bentonite Additives: 3-APTES[c] solution (10% v/v; 1 h); glutaraldehyde solution (10% v/v; 1 h); ABTS Room temperature	Immobilization yield: 91% pH activity = 6 Temperature–activity profile (°C): 40[a]; 45[b] Stability = 60 °C[a] (15 min) Reuse: 30 cycles at 100%	Alsoufi (2018)
Myceliophthora thermophila	Adsorption	Soil remediation/ syringaldazine oxidation	Support: Mg/Al-layered double hydroxide modified with glutamic acid Agitation: 400 rpm for 12 h. Temperature = 5 °C	Specific activity retained: 97.2% Specific activity retained (lyophilized): 92.3% Decompose temperature (°C): 230[a]; 885 (lost 51.5% of its original mass)[b]	Camacho-Cordova and Morales-Borges (2009)
Trametes versicolor	Covalent	Bioremediation/ABTS oxidation—PAH remediation (surfactants)	Support: Nonionic surfactant-modified clay. Ca-montmorillonite (SAz-1) Additives: 3-APTES, glutaraldehyde; ABTS; Tritòn X-100 and Brij 35	Removal percentage (14 days): 60.87% (PhT); 53.68% (NaT)[a]; TX-100 presence 80% (PhT and NaT)[b]	Chang et al. (2016)
Trametes versicolor	Covalent	Bioremediation/ABTS oxidation—PAH remediation	Support: Kaolinite silanized pellet. Additives: 3-APTES; ABTS	K_m (mM): 0.262[a]; 0.165[b] pH activity = 4.5 Activity on temperature stability at 80 °C (°C): 0%[a]; 85%[b] Oxidation percentage BaP and NaT (%): 85[a]; 80[b]	Dodor et al. (2004)

Trametes versicolor	Adsorption	Soil phenol remediation/ ABTS oxidation	Support: Iron/aluminum (aluminum hydroxide, gibbsite, lepidocrocite, and goethite) Additives: ABTS Agitation: 180 rpm (2 h) Temperature = 30 °C	Specific activity retained[d]: 64% K_m (mmol/L): 0.31[a]; 0.46[b, d] pH activity profile[d] = 3.6 Temperature activity profile[d] = 25 °C Activation energy (kJ/Mol)[d] = 26.73 Residual activity after 16 h in protease and humic acids presence: 55%[a]; 90%[b d]	Wu et al. (2014)
Trametes versicolor	Covalent	Bioremediation/ABTS and HBT oxidation—PAH remediation	Support: Soil, kaolinite Additives: 3-APTES; tween 80; ABTS, HBT	K_m (mM): 261.4[a]; 119[b, e] Oxidation percentage BaP and AnT (%): 100[a]; 70–80[b, e] Oxidation of PAHs in a quaternary system (AnT + BaP + BAA + PyR): 55, 42, 5, and 5% for every one Reuse: 4 cycles at 80% (BaP/ AnT)	Dodor et al. (2018)
Geotrichum candidum	Adsorption	Soil bioremediation/14C-labeled 4-methylphenol (4MP); 2,4-dichlorophenol (2,4-DCP)	Support: Lakewood sand Additives: Silica columns	Pollutants polymerization (retention percentage 2 weeks): 38.3[a]; 88.4[b] Immobilizing percentage 4MP (%): 10[uT]; 90[iT]	Shannon and Barta (1988)

(continued)

Table 8.3 (continued)

Laccase source	Immobilization type	Application/test reaction	Conditions (support/additives)	Optimus (specific activity/ kinetic parameters/stability/ reuse)	Reference
Trametes villosa	Covalent	Bioremediation/oxidation 14C-labeled-2,4-DCP	Support: Montmorillonite (MMT) Additives: 3-APTES; glutaraldehyde	Specific activity (%) 14 days on soil: Less than 25%^{a}; more than 60%^{b} Removal percentage (14 days): 10%^{a}; 100%^{b} (2.4-DCP)	Ahn et al. (2002)
Trametes versicolor	Covalent	Bioremediation (phenolic compounds)/oxidation 2,4-dichlorophenol (2,4-DCP)	Support: MMT, kaolinite, glass beads and soil Additives: 3-APTES; glutaraldehyde	Immobilization yield (%):71^{MMT} K_m (mM): 0.08^{a}; 0.07$^{b\ MMT}$ Residual specific activity percentage (%): 118^{MMT} Storage stability after 4 months = 100%	Gianfreda and Bollag (1994)
Lenzites betulinus	Adsorption/ cross-linking	Soil bioremediation/ABTS oxidation—PAH remediation	Support: Nylon, chitosan Additives: Glutaraldehyde	Immobilization pH: 4 Immobilization temperature (°C) = 40 BaP degradation rate at 40 °C (%): 40^{a}; 70^{b} Pyr degradation rate at pH 4 (%): 80^{a}; 90^{b}	Wang et al. (2018)
Trametes versicolor	Entrapment (electro-spinning technique—LCEFM)	Soil bioremediation/ABTS oxidation—PAH remediation	Support: Poly(D,L-lactide) (PDLLA), poly(D,L-lactide-co-glycolide) (PDLGA), methoxypolyethylene glycol poly(lactide-co-glycolide) (MPEG-PLGA) Additives: Triblock copolymer PEO-PPO-PEO (F108); CH_2Cl_2	Remotion efficiency PaT-FaT-BaA-BaP (%): 95.1, 93.2, 79.1, and 72.5 Degradation half-lives; MPEG immobilized PaT-FaT-BaA-BaP (h): 17.9^{a}; 0.115^{b}; 0.048^{b}; 0.428^{b}	Dai et al. (2011)

Lenzites betulinus	Entrapment (embedding cross-linking)	Soil bioremediation/ABTS oxidation—Chlorpyrifos	Support: Sodium alginate; $CaCl_2$ Additives: Glutaraldehyde	Ratio: Support/cross-linked/enzyme: Alginate 3%/glutaraldehyde 1%/laccase 60 mL Immobilization time: 4 h Cross-linking time: 6 h Degradation rate of chlorpyrifos 48 h (%): 46[a]; 70[b]. Degradation rate on alkali pH (%): 15[a]; 65[b]. Degradation rate at 35 °C (%): 80[a]; 40[b].	Wang et al. (2016)
Lenzites betulinus	Adsorption–entrapment	Soil bioremediation/ABTS oxidation—Carbofuran	Support: Sodium alginate; powder-activated carbon (PAC); $CaCl_2$	Ratio: Support/cross-linked/enzyme: PAC 0.6 g/alginate 0.4 g/$CaCl_2$ 15%/laccase 80 mL Immobilization time: 4 h Cross-linking time: 6 h Degradation rate of carbofuran 48 h (%): 50[a]; 95[b]. Degradation rate on alkali pH (%): 30[a]; 70[b]. Degradation rate at 35 °C (%): 45[a]; 85[b].	Wang et al. (2017b)

[MMT] Montmorillonite, [uT] Untreated column, [iT] Enzymatically column
[a]Free laccase
[b]Immobilized laccase
[c]3-aminopropyltriethoxysilane
[d]Goethite-laccase
[e]Kaolinite immobilized

8.5.3.1 Immobilization of Pollutants Catalyzed by Laccases to Reduce Pollution

At the first approach, soil bioremediation takes advantage of the ability of laccases to catalyze polymerization and immobilize pollutants, leading to the direct incorporation of xenobiotic residues into the soil organic matter as molecules with higher molecular weight, reducing mobility and avoiding leaching to wider areas or groundwater.

Enzymatic treated and untreated columns with Lakewood sand were packed to evaluate the immobilization and polymerization capacities of *Geotrichum candidum* laccase. Results of radiolabeled substrates show that 90.9% of 14C-labeled 4-methylphenol (4MP) was retained on the enzymatically treated column after two retention days, while only 10% was retained on the untreated column. In the same way, percentages for treated and untreated columns of 2,4-DCP were 51.5 and 20.8% after 1 retention day, respectively. However, 2 weeks later, 88.4% of the 2,4-DCP was retained compared to 38.3% from the control column. Due to elongated spots of radioactivity on the column plates, it was established that polymerization of radiolabeled substrates is one of the most important processes in the enzyme–substrate interaction followed by absorption to packed material (Shannon and Bartha 1988).

Even though free extracellular enzymes can adsorb soil colloids and improve their performance, their activity can be reduced or inhibited by extreme pH and temperature, proteases biodegradation, and mass diffusional transfer, among other biological and non-biological factors (Ahn et al. 2002). For this reason, it is important to probe the immobilization as a tool to improve the stability of enzymes in the soil making cost-effective the bioremediation process.

Concerning the phenolic pollutants on soil, a *Trametes villosa* laccase–TvL (free or immobilized on MMT) was used to degrade 14C-labeled-2,4-DCP in soils with 2.8 and 7.4% organic matter percentage. The higher soil organic matter (SOM) was the most difficult to remove from 2,4-DCP by the free laccase in different moisture conditions, while immobilized laccase was able to remove 95% regardless of water content. Free and immobilized enzymes degrade 100% of the substrate in the lowest SOM content. Different removal mechanisms were present, such as polymerization, covalent binding, SOM, and unaltered adsorption on the soil after 14 days of incubation. However, it is necessary to establish the equilibrium between the loss of enzyme activity during immobilization and free laccase additions to reach equal remediation level, as the immobilization cost increases (Ahn et al. 2002).

Remobilization of xenobiotics pollutants during humus degradation is a concerning problem to be solved for avoiding new contamination of soil or groundwater. Inorganic material on soils provide surfaces to anchoring some of these hazardous molecules, but dynamics on soil biotics/nonbiotics factors remained as a black box in environmental risk assessment. Many pesticides identified as health dangerous compounds can be immobilized or sequestered on soil (known as non-extractable residues (NER)), where they can be enzymatically degraded in the presence of extracellular enzymes such as phenol oxidases and peroxidases, making them a promissory tool for danger decrease. TvL immobilized on copper alginate beads was mixed with sterile soil containing 14C-metalaxyl and compared with sterile soil without the enzyme. Results showed that interactions between the radiolabeled probe molecule and the soil in the control treatment were ester linkages, while the same in the enzymatically treated one had the prevalence of more stable covalent linkages such as ether and C–C bonds, especially in humic acids. In spite of alginate–enzyme beads disaggregating in 10 days, the addition of 130 mU/g of TvL beads to soil enhanced radiolabeled NER fractions by twofold, demonstrating the oxidative coupling as an alternative tool for strengthening the bioremediation process (Botterweck et al. 2014). Previous studies also support the immobilization process as an alternative to xenobiotic bioremediation on the soil. The pesticide, 3,4-dichloroaniline was spread on the soil, and on evaluation 2 years after treatment, 46% of the compound remained bound, while 83% of another pesticide, 14C-labeled atrazine, remained on the soil after 9 years with 50% of the pesticide bound to organic matter. In general, the microbial release of bound xenobiotics can occur very slowly, and they are susceptible to mineralization or re-bounded to another organic matter as the same way that occurs in the first immobilization (Bollag 1992).

8.5.3.2 Immobilization of Laccases for Pollutants Remotion

In the second approach, laccases can be immobilized on different supports to catalyze xenobiotic degradation. Focusing on phenolic compounds, Gianfreda and Bollag (1994) immobilized *T. versicolor* laccase on MMT, kaolinite, glass beads, and soil in the presence of soil–sand mixtures. The measured activity was determined polarographically, using a biological oxygen monitor in the oxidation of 2,4-dichlorophenol. MMT was the best support, immobilizing 71% of the laccase with 118% of specific activity in the test reaction. At the same time, only 56% of laccase was adsorbed on glass beads but retained 236% of residual activity. Michaelis constant values remained unchanged compared to free laccase, but an inhibitory effect was observed in the presence of soil–sand mixture especially with higher contents of SOM. Reusability in presence and absence of soil for different biocatalysts decreased at 50% approximately at fourth incubation cycle (24 h each one). Storage stability was 100% for 4 months (Gianfreda and Bollag 1994).

Given the growing interest on bioremediation for PAHs, Wang et al. (2018) compared free laccase with their counterpart immobilized on nylon and chitosan to test the degradation of Pyr and BaP under different pH and temperatures conditions in a polluted soil made with 3:1 volume ratio of soil to PAH solution. At a 72-h degradation, rates of both Pyr and BaP exceeded 80 and 50% independently of temperature. The immobilized laccase increased degradation rates of Pyr and BaP in 10–30% with chitosan as best support in front of free laccase. Optimum conditions were fixed as 40 °C and pH 4 (Wang et al. 2018).

In another approach, Chinese researchers developed polymeric fibrous membranes obtained by electro-spinning technique (LCEFM) containing laccases in the core to improve PAHs degradation and pores on the shell to facilitate the accessibility of toxic compounds to active center. This LCEFM increased the removal efficiencies through two synergistic processes: adsorption of the hydrophobic fibrous shell that extracts PAHs of the soil particles and degradation with laccases inside them. Phenanthrene (PaT), fluoranthene (FaT), BaA, and BaP were removed with an efficiency of 95.1%, 93.2%, 79.1%, and 72.5%, respectively, while removal half-time was one order magnitude faster than free laccase in a sample of soil of Yangtze River with variable PAHs concentrations between 180 and 315 μg/kg. Kinetic parameters for adsorption and degradation process showed that the first step was a limiting factor, so the higher degradation efficiencies of PHAs may be due to the pre-concentration effect of the fibrous membranes. Storage stability of LCEFM was 70% of the initial activity while it was only 30% for free laccase. In the same way, reusing cycles for LCFEM was ten batches with more than 70% of their initial activities (Dai et al. 2011).

Finally, the immobilization of laccases to improve pesticides degradation was also addressed. In the chlorpyrifos assessment, a mixed immobilization technique has been tested for the first time based on encapsulation and cross-linking methods using a fungal laccase to improve degradation rate of this compound. The best ratio support/cross-linked/enzyme was alginate 3%/glutaraldehyde 1%/laccase 60 mL with a cross-linking time of 6 h and a support–enzyme contact time for 4 h. A slurry was prepared in a volume ratio soil/water (1:3) at 25 °C and 120 rpm to test chlorpyrifos degradation rate with immobilized and free laccase. The best conditions to evaluate chlorpyrifos degradation were neutral pH with a temperature of 30 °C and concentrations of pollutants under 200 mg/mL. Free laccase was able to degrade 46% of chlorpyrifos in the medium while immobilized laccase showed 70% of degradation (Wang et al. 2016).

For the carbofuran degradation test, a mixed immobilization technique was also used, combining adsorption and entrapment to improve the activity and stability of the immobilized enzyme. The appropriate conditions to succeed on the embedding–adsorption method were 0.6 g of powdered active carbon as a surface to immobilization with 0.4 g of sodium alginate, and 15% of $CaCl_2$ used to consolidate the support, and 80 mL of crude laccase put in contact for 6 h. For carbofuran degradation, a slurry containing soil–carbofuran polluted was made with water (1:3 ratio), and the conditions with best-evaluated results were in a concentration of 100 mg/mL, pH 6, 30°C and 48h. Authors reported degradation rates by carrier and laccase (free and immobilized) of 10.4%, 52%, and 83.2%, respectively (Wang et al. 2017b).

8.6 Conclusions and Perspectives

The versatility of laccases allows its use on soil bioremediation to polymerize pollutants or its bioaugmentation with immobilized enzymes that degrade pollutants. Increasing literature about production and immobilization of laccases focusing on soil matrix—two main topics of this review—shows the growing interest to enhance production, stability, and cost-effective process at an industrial scale. The improvement of production has been focused on the screening of natural hypersecretory strains and the use of recombinant organisms using the heterologous or homologous expression of a microbial system. The immobilization strategies aim to increase the specific activity, the physicochemical stability, and the mechanical reusability through different techniques, emphasizing in an optimum combination "carrier–enzyme–target reaction" that will attend soil pollution. As a result, enzymes with improved properties were obtained, which allow their application in entirely new areas where laccases have not been previously used, extending the possibilities to exploit them as one of the most important tools for soil bioremediation.

Acknowledgments This work was supported by Universidad Libre de Colombia, Universidad San Pablo (Tucumán Argentina), Centro de Investigación en Biofísica, Aplicada y Alimentos (CIBAAL-CONICET) and Universidad Nacional de Santiago del Estero (UNSE) (Santiago del Estero, Argentina).

The authors acknowledge Drs. Moldes-Moreira, Sanromán-Braga, and Fernández-Fernández for permission to adapt material reprinted in this chapter.

References

Abatenh E, Gizaw B, Tsegaye Z, Wassie M (2017) The role of microorganisms in bioremediation – a review. Open J Env Biol 2:38–46

Abdel-Hamid AM, Solbiati JO, Cann IKO (2013) Insights into lignin degradation and its potential industrial applications. In: Advances in applied microbiology. Elsevier, San Diego, pp 1–28

Afreen S, Shamsi TN, Baig MA et al (2017) A novel multicopper oxidase (laccase) from cyanobacteria: purification, characterization with potential in the decolorization of anthraquinonic dye. PLoS One 12:e0175144

Ahn M, Dec J, Kim JE, Bollag J (2002) Treatment of 2,4-Dichlorophenol polluted soil with free and immobilized laccase. Environ Qual 31:1509–1515

Ahn M, Zimmerman AR, Martinez CE et al (2007) Characteristics of Trametes villosa laccase adsorbed on aluminum hydroxide. Enzym Microb Technol 41:141–148. https://doi.org/10.1016/j.enzmictec.2006.12.014

Alsoufi MA (2018) Use of immobilized laccase in bioremediation of phenolic compounds which causes environmental pollution. J Biodivers Environ Sci 12:370–377

Antošová Z, Sychrova H, Sychrová H, Sychrova H (2016) Yeast hosts for the production of recombinant laccases: a review. Mol Biotechnol 58:93–116. https://doi.org/10.1007/s12033-015-9910-1

Anwar MZ, Kim DJ, Kumar A et al (2017) SnO 2 hollow nanotubes: a novel and efficient support matrix for enzyme immobilization. Sci Rep 7:1–12. https://doi.org/10.1038/s41598-017-15550-y

Arias ME, Rodrı J, Soliveri J et al (2003) Kraft pulp biobleaching and mediated oxidation of a nonphenolic substrate by Laccase from Streptomyces cyaneus CECT 3335, vol 69, pp 1953–1958. https://doi.org/10.1128/AEM.69.4.1953

Ausec L, van Elsas JD, Mandic-Mulec I (2011a) Two- and three-domain bacterial laccase-like genes are present in drained peat soils. Soil Biol Biochem 43:975–983. https://doi.org/10.1016/j.soilbio.2011.01.013

Ausec L, Zakrzewski M, Goesmann A et al (2011b) Bioinformatic analysis reveals high diversity of bacterial genes for laccase-like enzymes. PLoS One 6:e25724. https://doi.org/10.1371/journal.pone.0025724

Baldantoni D, Morelli R, Bellino A et al (2017) Anthracene and benzo(a)pyrene degradation in soil is favoured by compost amendment: perspectives for a bioremediation approach. J Hazard Mater 339:395–400. https://doi.org/10.1016/j.jhazmat.2017.06.043

Beppu T, Endo K, Hosono K, Ueda K (2002) A novel extracytoplasmic phenol oxidase of Streptomyces: its possible involvement in the onset of morphogenesis. Microbiology 148:1767–1776. https://doi.org/10.1099/00221287-148-6-1767

Bollag J (1992) Decontaminating soil with enzymes. An in situ method using phenolic and anilinic compounds. Environ Sci Technol 26:1876–1881

Bornscheuer UT, Huisman GW, Kazlauskas RJ et al (2012) Engineering the third wave of biocatalysis. Nature 485:185. https://doi.org/10.1038/nature11117

Botterweck J, Schmidt B, Schwarzbauer J, et al (2014) Enhanced non-extractable residue formation of 14 C-metalaxyl catalyzed by an immobilized laccase. Biol Fertil Soils 1015–1024. https://doi.org/10.1007/s00374-014-0923-x

Bourbonnais R, Paice MG (1990) Oxidation of non-phenolic substrates. An expanded role for laccase in lignin biodegradation. FEBS Lett 267:99–102. https://doi.org/10.1016/0014-5793(90)80298-W

Burns RG, DeForest JL, Marxsen J et al (2013) Soil enzymes in a changing environment: current knowledge and future directions. Soil Biol Biochem 58:216–234. https://doi.org/10.1016/j.soilbio.2012.11.009

Camacho-Cordova D, Morales-Borges R (2009) Immobilization of laccase on hybrid layered double hydroxide. Quim Nova 32:1495–1499

Chandra R, Chowdhary P (2015) Properties of bacterial laccases and their application in bioremediation of industrial wastes. Environ Sci Process Impacts 17:326–342. https://doi.org/10.1039/C4EM00627E

Chang Y-T, Lee J-F, Liu K-H et al (2016) Immobilization of fungal laccase onto a nonionic surfactant-modified clay material: application to PAH degradation. Environ Sci Pollut Res 23:4024–4035. https://doi.org/10.1007/s11356-015-4248-6

Chen Q, Marshall MN, Geib SM et al (2012) Effects of laccase on lignin depolymerization and enzymatic hydrolysis of ensiled corn Stover. Bioresour Technol 117:186–192. https://doi.org/10.1016/j.biortech.2012.04.085

Chen L, Zou M, Hong FF (2015) Evaluation of fungal laccase immobilized on natural nanostructured bacterial cellulose. Front Microbiol 6:1245. https://doi.org/10.3389/fmicb.2015.01245

Chirnside AEM, Ritter WF, Radosevich M (2011) Biodegradation of aged residues of atrazine and alachlor in a mix-load site soil by fungal enzymes. Appl Environ Soil Sci 2011:1–10. https://doi.org/10.1155/2011/658569

Compart LCA, Machado KMG, Matheus DR, Cardoso AV (2007) Immobilization of Psilocybe castanella on ceramic (slate) supports and its potential for soil bioremediation. World J Microbiol Biotechnol 23:1479–1483. https://doi.org/10.1007/s11274-007-9393-8

Correa S, Puertas S, Gutiérrez L et al (2019) Design of stable magnetic hybrid nanoparticles of Si-entrapped HRP. PLoS One 14:e0214004. https://doi.org/10.1371/journal.pone.0214004

Couto SR, Herrera JLT (2006) Industrial and biotechnological applications of laccases: a review. Biotechnol Adv 24:500–513

Couto SR, Toca-Herrera JL (2007) Laccase production at reactor scale by filamentous fungi. Biotechnol Adv 25:558–569

Dai Y, Yin L, Niu J (2011) Laccase-carrying electrospun fibrous membranes for adsorption and degradation of PAHs in shoal soils. Environ Sci Technol 45:10611–10618. https://doi.org/10.1021/es203286e

Dayi B, Duishemambet Kyzy A, Akdogan HA (2018) Characterization of recuperating talent of white-rot fungi cells to dye-contaminated soil/water. Chin J Chem Eng 27:634. https://doi.org/10.1016/j.cjche.2018.05.004

De Lima DP, Dos Santos E dos A, Marques MR et al (2018) Fungal bioremediation of pollutant aromatic amines. Curr Opin Green Sustain Chem 11:34–44. https://doi.org/10.1016/j.cogsc.2018.03.012

Desai SS, Nityanand C (2011) Microbial laccases and their applications: a review. Asian J Biotechnol 3:98–124. https://doi.org/10.3923/ajbkr.2011.98.124
Dias AA, Matos AJS, Fraga I et al (2017) An easy method for screening and detection of laccase activity. Open Biotechnol J 11:89
Dodor DE, Hwang H, Ekunwe SI (2004) Oxidation of anthracene and benzo[a]pyrene by immobilized laccase from Trametes versicolor. Enzym Microb Technol 35:210–217. https://doi.org/10.1016/j.enzmictec.2004.04.007
Dodor DE, Miyittah M, Ahiabor BDK (2018) Immobilized laccase mediator-catalyzed oxidation of aqueous mixtures of polycyclic aromatic hydrocarbons. Polycycl Aromat Compd 0:1–11. doi: https://doi.org/10.1080/10406638.2018.1462210
Driks A (1999) Bacillus subtilis spore coat. Microbiol Mol Biol Rev 63:1–20
Dubé E, Shareck F, Hurtubise Y et al (2008) Homologous cloning, expression, and characterisation of a laccase from Streptomyces coelicolor and enzymatic decolourisation of an indigo dye. Appl Microbiol Biotechnol 79:597–603. https://doi.org/10.1007/s00253-008-1475-5
Durán N, Esposito E (2000) Potential applications of oxidative enzymes and phenoloxidase-like compounds in wastewater and soil treatment: a review. Appl Catal B Environ 28:83–99. https://doi.org/10.1016/S0926-3373(00)00168-5
Durao P, Chen Z, Fernandes AT et al (2008) Copper incorporation into recombinant CotA laccase from Bacillus subtilis: characterization of fully copper loaded enzymes. JBIC J Biol Inorg Chem 13:183–193
Eichlerová I, Šnajdr J, Baldrian P (2012) Laccase activity in soils: considerations for the measurement of enzyme activity. Chemosphere 88:1154–1160. https://doi.org/10.1016/j.chemosphere.2012.03.019
Endo K, Hayashi Y, Hibi T et al (2003) Enzymological characterization of EpoA, a laccase-like phenol oxidase produced by Streptomyces griseus. J Biochem 133:671–677
Ergün BG, Çalık P, Gündüz Ergün B, Çalık P (2016) Lignocellulose degrading extremozymes produced by Pichia pastoris: current status and future prospects. Bioprocess Biosyst Eng 39:1–36. https://doi.org/10.1007/s00449-015-1476-6
Fang Z, Li T, Wang Q et al (2011) A bacterial laccase from marine microbial metagenome exhibiting chloride tolerance and dye decolorization ability. Appl Microbiol Biotechnol 89:1103–1110. https://doi.org/10.1007/s00253-010-2934-3
Fang Z-M, Li T-L, Chang F et al (2012) A new marine bacterial laccase with chloride-enhancing, alkaline-dependent activity and dye decolorization ability. Bioresour Technol 111:36–41. https://doi.org/10.1016/j.biortech.2012.01.172
Fang Z, Liu X, Chen L et al (2015) Identification of a laccase Glac15 from Ganoderma lucidum 77002 and its application in bioethanol production. Biotechnol Biofuels 8:54. https://doi.org/10.1186/s13068-015-0235-x
FAO (2017) Voluntary guidelines for sustainable soil management. 155th Session of the FAO Council, Rome, Italy, 5th December 2016 15
Fernández M, Moldes D, Sanromán MA (2013) Inmovilización de Lacasa : Métodos y potenciales aplicaciones industriales. http://www.investigo.biblioteca.uvigo.es/xmlui/handle/11093/75. Accessed 07 Aug 2019
Fernández-Fernández M, Sanromán MÁ, Moldes D (2013) Recent developments and applications of immobilized laccase. Biotechnol Adv 31:1808–1825. https://doi.org/10.1016/j.biotechadv.2012.02.013
Ferrer-Miralles N, Domingo-Espín J, Corchero JL et al (2009) Microbial factories for recombinant pharmaceuticals. Microb Cell Factories 8:17
Food and Agricultural Organization of the United States (FAO), Intergovernmental Technical Panel on Soils (ITPS) (2015) Status of the world ' s soil resources (SWSR) – main report. FAO, Rome
Fu M, Xing J, Ge Z (2019) Preparation of laccase-loaded magnetic nanoflowers and their recycling for efficient degradation of bisphenol A. Sci Total Environ 651:2857–2865. https://doi.org/10.1016/j.scitotenv.2018.10.145
García-Rivero M, Peralta-Perez M (2008) Cometabolismo en la biodegradación de hidrocarburos cometabolism in the biodegradation of hydrocarbons. Rev Mex Ing Quim 7:1–12

Gianfreda L, Bollag J-MJ-M (1994) Effect of soils on the behavior of immobilized enzymes. Soil Sci Soc Am J 58:1672–1681. https://doi.org/10.2136/sssaj1994.03615995005800060014x
Givaudan A, Effosse A, Faure D et al (1993) Polyphenol oxidase in Azospirillum lipoferum isolated from rice rhizosphere: evidence for laccase activity in non-motile strains of Azospirillum lipoferum. FEMS Microbiol Lett 108:205–210. https://doi.org/10.1111/j.1574-6968.1993.tb06100.x
Gu C, Zheng F, Long L et al (2014) Engineering the expression and characterization of two novel laccase isoenzymes from Coprinus comatus in Pichia pastoris by fusing an additional ten amino acids tag at N-terminus. PLoS One 9:e93912. https://doi.org/10.1371/journal.pone.0093912
Guan Z-B, Zhang N, Song C-M et al (2014) Molecular cloning, characterization, and dye-decolorizing ability of a temperature-and pH-stable laccase from Bacillus subtilis X1. Appl Biochem Biotechnol 172:1147–1157
Guan ZB, Luo Q, Wang HR et al (2018) Bacterial laccases: promising biological green tools for industrial applications. Cell Mol Life Sci 75:3569–3592. https://doi.org/10.1007/s00018-018-2883-z
Guauque-Torres M, Foresti M, Ferreira M (2013) Cross-linked enzyme aggregates (CLEAs) of selected lipases: a procedure for the proper calculation of their recovered activity. AMB Express 3:25. https://doi.org/10.1186/2191-0855-3-25
Guebitz G, Cavaco PA, Kandelbauer A, et al (2003) Biocatalyst containing a laccase. Patent EP1468968 (A1)
Gupta N, Farinas ET (2010) Directed evolution of CotA laccase for increased substrate specificity using Bacillus subtilis spores. Protein Eng Des Sel 23:679–682. https://doi.org/10.1093/protein/gzq036
Hatamoto O, Sekine H, Nakano E, ABE K (1999) Cloning and expression of a cDNA encoding the laccase from Schizophyllum commune. Biosci Biotechnol Biochem 63:58–64
Ihssen J, Reiss R, Luchsinger R et al (2015) Biochemical properties and yields of diverse bacterial laccase-like multicopper oxidases expressed in Escherichia coli. Sci Rep 5:10465. https://doi.org/10.1038/srep10465
Iimura Y, Katayama Y, Kawai S et al (1995) Degradation and Solubilization of 13 C_, 14 C-side chain labeled synthetic lignin (Dehydrogenative Polymerizate) by laccase III of Coriolus versicolor. Biosci Biotechnol Biochem 59:903–905. https://doi.org/10.1271/bbb.59.903
Iimura Y, Sonoki T, Habe H (2018) Heterologous expression of Trametes versicolor laccase in Saccharomyces cerevisiae. Protein Expr Purif 141:39–43. https://doi.org/10.1016/j.pep.2017.09.004
Iracheta-Cárdenas MM, Rocha-Peña MA, Galán-Wong LJ et al (2016) A Pycnoporus sanguineus laccase for denim bleaching and its comparison with an enzymatic commercial formulation. J Environ Manag 177:93–100. https://doi.org/10.1016/j.jenvman.2016.04.008
Iyer HV, Przybycien TM (1995) Metal affinity protein precipitation: effects of mixing, protein concentration, and modifiers on protein fractionation. Biotechnol Bioeng 48:324–332. https://doi.org/10.1002/bit.260480405
Kajita S, Sugawara S, Miyazaki Y et al (2004) Overproduction of recombinant laccase using a homologous expression system in Coriolus versicolor. Appl Microbiol Biotechnol 66:194–199. https://doi.org/10.1007/s00253-004-1663-x
Kandasamy S, Muniraj IK, Purushothaman N et al (2016) High level secretion of laccase (LccH) from a newly isolated white-rot basidiomycete, Hexagonia hirta MSF2. Front Microbiol 7:707. https://doi.org/10.3389/fmicb.2016.00707
Kiiskinen L-L (2004) Expression of Melanocarpus albomyces laccase in Trichoderma reesei and characterization of the purified enzyme. Microbiology 150:3065–3074. https://doi.org/10.1099/mic.0.27147-0
Kim C, Lorenz WW, Hoopes JT, Dean JFD (2001) Oxidation of phenolate siderophores by the multicopper oxidase encoded by the Escherichia coli yacK gene. J Bacteriol 183:4866–4875
Koschorreck K, Richter SM, Ene AB et al (2008) Cloning and characterization of a new laccase from Bacillus licheniformis catalyzing dimerization of phenolic acids. Appl Microbiol Biotechnol 79:217–224

Koschorreck K, Schmid RD, Urlacher VB (2009) Improving the functional expression of a Bacillus licheniformis laccase by random and site-directed mutagenesis. BMC Biotechnol 9:12. https://doi.org/10.1186/1472-6750-9-12

Košnář Z, Částková T, Wiesnerová L et al (2018) Comparing the removal of polycyclic aromatic hydrocarbons in soil after different bioremediation approaches in relation to the extracellular enzyme activities. J Environ Sci (China) 76:1–10. https://doi.org/10.1016/j.jes.2018.05.007

Koyani RD, Vazquez-Duhalt R (2016) Laccase encapsulation in chitosan nanoparticles enhances the protein stability against microbial degradation. Environ Sci Pollut Res 23:18850–18857. https://doi.org/10.1007/s11356-016-7072-8

Kumar A, Kanwar SS (2012) An innovative approach to immobilize lipase onto natural fiber and its application for the synthesis of 2-Octyl Ferulate in an organic medium. Curr Biotechnol e 1:241–248. https://doi.org/10.2174/2211550111201030241

Kumar R, Kumar A, Kanwar S (2014) Synthesis of methyl succinate by natural-fibre immobilized lipase of Streptomyces sp. STL-D8. Curr Biotechnol 3:152–156. https://doi.org/10.2174/2211550102666140107232547

Kumar A, Kim IW, Patel SKS, Lee JK (2018) Synthesis of Protein-Inorganic Nanohybrids with Improved Catalytic Properties Using Co 3 (PO 4) 2. Indian J Microbiol 58:100–104. https://doi.org/10.1007/s12088-017-0700-2

Kumar A, Park GD, Patel SKS et al (2019) SiO2 microparticles with carbon nanotube-derived mesopores as an efficient support for enzyme immobilization. Chem Eng J 359:1252–1264. https://doi.org/10.1016/j.cej.2018.11.052

Lawton TJ, Rosenzweig AC (2011) Detection and characterization of a multicopper oxidase from nitrosomonas Europaea, 1st edn. Elsevier Inc, Amsterdam

Lee HR, Chung M, Kim MI, Ha SH (2017) Preparation of glutaraldehyde-treated lipase-inorganic hybrid nanoflowers and their catalytic performance as immobilized enzymes. Enzym Microb Technol 105:24–29. https://doi.org/10.1016/J.ENZMICTEC.2017.06.006

Lettera V, Pezzella C, Cicatiello P et al (2016) Efficient immobilization of a fungal laccase and its exploitation in fruit juice clarification. Food Chem 196:1272–1278. https://doi.org/10.1016/j.foodchem.2015.10.074

Lisov AV, Trubitsina LI, Lisova ZA et al (2018) Transformation of humic acids by two-domain laccase from Streptomyces anulatus. Process Biochem 76:128. https://doi.org/10.1016/j.procbio.2018.11.001

Llorens-Blanch G, Parladé E, Martinez-Alonso M et al (2018) A comparison between biostimulation and bioaugmentation in a solid treatment of anaerobic sludge: drug content and microbial evaluation. Waste Manag 72:206–217. https://doi.org/10.1016/J.WASMAN.2017.10.048

Luthy RG, Ghoshal S, Ramaswami A, Nakles DV (1995) Bioslurry treatment of NAPL-contaminated soil. In: Soil and groundwater pollution. Springer, Dordrecht, pp 36–40

Ma X, Liu L, Li Q et al (2017) High-level expression of a bacterial laccase, CueO from Escherichia coli K12 in Pichia pastoris GS115 and its application on the decolorization of synthetic dyes. Enzym Microb Technol 103:34–41. https://doi.org/10.1016/j.enzmictec.2017.04.004

Machczynski MC, Vijgenboom E, Samyn B, Canters GW (2004) Characterization of SLAC: a small laccase from Streptomyces coelicolor with unprecedented activity. Protein Sci 13:2388–2397

Madhavi V, Lele SS (2009) Laccase: properties and applications. BioResources 4:1694–1717. https://doi.org/10.15376/biores.4.4.1694-1717

Majeau J-A, Brar SK, Tyagi RD (2010) Laccases for removal of recalcitrant and emerging pollutants. Bioresour Technol 101:2331–2350

Majorek KA, Porebski PJ, Dayal A, Zimmerman MD, Jablonska K, Stewart AJ, Chruszcz M, Minor W (2012) Structural and immunologic characterization of bovine, horse, and rabbit serum albumins. Mol Immunol 52(3–4):174–182

Martínez-Morales F, Bertrand B, Pasión Nava AA et al (2015) Production, purification and biochemical characterization of two laccase isoforms produced by Trametes versicolor grown on oak sawdust. Biotechnol Lett 37:391–396. https://doi.org/10.1007/s10529-014-1679-y

Martins LO, Soares CM, Pereira MM et al (2002) Molecular and biochemical characterization of a highly stable bacterial laccase that occurs as a structural component of the Bacillus subtilis endospore coat. J Biol Chem 277:18849–18859

Martins LO, Durão P, Brissos V, Lindley PF (2015) Laccases of prokaryotic origin: enzymes at the interface of protein science and protein technology. Cell Mol Life Sci 72:911–922. https://doi.org/10.1007/s00018-014-1822-x

Mate DM, Alcalde M (2015) Laccase engineering: from rational design to directed evolution. Biotechnol Adv 33:25–40

Megharaj M, Naidu R (2017) Soil and brownfield bioremediation. Microb Biotechnol 10:1244–1249. https://doi.org/10.1111/1751-7915.12840

Miyazaki K (2005) A hyperthermophilic laccase from Thermus thermophilus HB27. Extremophiles 9:415–425. https://doi.org/10.1007/s00792-005-0458-z

Mohamad NR, Marzuki NHC, Buang NA et al (2015) An overview of technologies for immobilization of enzymes and surface analysis techniques for immobilized enzymes. Biotechnol Biotechnol Equip 29:205–220. https://doi.org/10.1080/13102818.2015.1008192

Molina-Guijarro JM, Pérez J, Muñoz-Dorado J et al (2009) Detoxification of azo dyes by a novel pH-versatile, salt-resistant laccase from Streptomyces ipomoea. Int Microbiol 12:13–21

Morozova OV, Shumakovich GP, Shleev SV, Yaropolov YI (2007) Laccase-mediator systems and their applications: a review. Appl Biochem Microbiol 43:523–535. https://doi.org/10.1134/S0003683807050055

Mougin C, Boukcim H, Jolivat C (2009) Soil bioremediation strategies based on the use of fungal enzymes. In: Singh A, Kuhad RC, Ward OP (eds) Advances in applied bioremediation. Springer, Berlin/Heidelberg, pp 123–150

Naghdi M, Taheran M, Brar SK et al (2018) Pinewood nanobiochar: a unique carrier for the immobilization of crude laccase by covalent bonding. Int J Biol Macromol 115:563–571. https://doi.org/10.1016/J.IJBIOMAC.2018.04.105

Naidja A, Huang PM, Bollag J-M (2000) Enzyme-clay interactions and their impact on transformations of natural and anthropogenic organic compounds in soil. J Environ Qual 29:677. https://doi.org/10.2134/jeq2000.00472425002900030002x

Neifar M, Chouchane H, Mahjoubi M et al (2016) Pseudomonas extremorientalis BU118: a new salt-tolerant laccase-secreting bacterium with biotechnological potential in textile azo dye decolourization. 3 Biotech 6:1–9. https://doi.org/10.1007/s13205-016-0425-7

Nikolopoulou M, Kalogerakis N (2016) Ex situ bioremediation treatment (Landfarming). In: Hydrocarbon and lipid microbiology protocols. Springer protocols handbooks. Springer, Berlin/Heidelberg, pp 195–220

Niladevi KN, Sukumaran RK, Jacob N et al (2009) Optimization of laccase production from a novel strain—Streptomyces psammoticus using response surface methodology. Microbiol Res 164:105–113. https://doi.org/10.1016/j.micres.2006.10.006

Nishibori N, Masaki K, Tsuchioka H et al (2013) Comparison of laccase production levels in Pichia pastoris and Cryptococcus sp. S-2. J Biosci Bioeng 115:394–399. https://doi.org/10.1016/j.jbiosc.2012.10.025

Nunes CS, Kunamneni A (2018) Laccases—properties and applications. In: Enzymes in human and animal nutrition. Elsevier, Amsterdam, pp 133–161

Olajuyigbe FM, Fatokun CO (2017) Biochemical characterization of an extremely stable pH-versatile laccase from Sporothrix carnis CPF-05. Int J Biol Macromol 94:535–543

Osma JF, Toca-Herrera JL, Rodríguez-Couto S (2010) Uses of laccases in the food industry. Enzyme Res 2010:1–8. https://doi.org/10.4061/2010/918761

Outten FW, Huffman DL, Hale JA, O'Halloran TV (2001) The independent cue and cus systems confer copper tolerance during aerobic and anaerobic growth in Escherichia coli. J Biol Chem 276:30670–30677

Pant A, Rai JPN (2018) Bioremediation of chlorpyrifos contaminated soil by two phase bioslurry reactor: processes evaluation and optimization by Taguchi's design of experimental (DOE) methodology. Ecotoxicol Environ Saf 150:305–311. https://doi.org/10.1016/J.ECOENV.2017.12.052

Pardo I, Camarero S (2015) Laccase engineering by rational and evolutionary design. Cell Mol Life Sci 72:897–910

Piontek K, Antorini M, Choinowski T (2002) Crystal structure of a laccase from the fungus Trametes versicolor at 1.90-?? Resolution containing a full complement of coppers. J Biol Chem 277:37663–37669. https://doi.org/10.1074/jbc.M204571200

Piscitelli A, Pezzella C, Giardina P et al (2010) Heterologous laccase production and its role in industrial applications. Bioeng Bugs 1:254–264. https://doi.org/10.4161/bbug.1.4.11438

Quintero Díaz JC (2011) Revisión: degradación de plaguicidas mediante hongos de la pudrición blanca de la madera. Rev Fac Nac Agron Medellín 64:5867–5882

Record E, Punt PJ, Chamkha M et al (2002) Expression of the Pycnoporus cinnabarinus laccase gene in Aspergillus Niger and characterization of the recombinant enzyme. Eur J Biochem 269:602–609

Reiss R, Ihssen J, Thöny-Meyer L (2011) Bacillus pumilus laccase: a heat stable enzyme with a wide substrate spectrum. BMC Biotechnol 11:9

Rezaei S, Shahverdi AR, Faramarzi MA (2017) Isolation, one-step affinity purification, and characterization of a polyextremotolerant laccase from the halophilic bacterium Aquisalibacillus elongatus and its application in the delignification of sugar beet pulp. Bioresour Technol 230:67–75

Román R, Torres-Duarte C, Ayala M, Vázquez-Duhalt R (2010) Producción a escala piloto de lacasa de Coriolopsis gallica. Rev Mex Micol 32:1.2

Rosano GL, Ceccarelli EA (2014) Recombinant protein expression in Escherichia coli: advances and challenges. Front Microbiol 5:172. https://doi.org/10.3389/fmicb.2014.00172

Ruggiero P, Sarkar J, Bollag J (1989) Detoxification of 2,4 dichlorophenol by a laccase immobilized on soil or clay. Soil Sci Soc Am J 147:361–370

Saloheimo M, Niku-Paavola M-L (1991) Heterologous production of a Ligninolytic enzyme: expression of the Phlebia Radiata laccase gene in Trichoderma Reesei. Bio/Technology 9:987–990. https://doi.org/10.1038/nbt1091-987

Saptarshi SR, Duschl A, Lopata AL (2013) Interaction of nanoparticles with proteins: relation to bio-reactivity of the nanoparticle. J Nanobiotechnology 11:26. https://doi.org/10.1186/1477-3155-11-26

Šašek V, Glaser JA, Baveye P (2003) The utilization of bioremediation to reduce soil contamination: problems and solutions. Springer, Dordrecht

Shannon MJR, Bartha R (1988) Immobilization of leachable toxic soil pollutants by using oxidative enzymes. Appl Environ Microbiol 54:1719–1723

Shiomi N (2015) Advances in bioremediation of wastewater and polluted soil. AvE4EvA

Shraddha SR, Sehgal S et al (2011) Laccase: microbial sources, production, purification, and potential biotechnological applications. Enzyme Res 2011:217861. https://doi.org/10.4061/2011/217861

Si J, Peng F, Cui B (2013) Purification, biochemical characterization and dye decolorization capacity of an alkali-resistant and metal-tolerant laccase from Trametes pubescens. Bioresour Technol 128:49–57. https://doi.org/10.1016/j.biortech.2012.10.085

Singh G, Bhalla A, Kaur P et al (2011) Laccase from prokaryotes: a new source for an old enzyme. Rev Environ Sci Bio/Technol 10:309–326. https://doi.org/10.1007/s11157-011-9257-4

Siracusa G, Becarelli S, Lorenzi R et al (2017) PCB in the environment: bio-based processes for soil decontamination and management of waste from the industrial production of Pleurotus ostreatus. New Biotechnol 39:232–239. https://doi.org/10.1016/J.NBT.2017.08.011

Song JE, Su J, Noro J et al (2018) Bio-coloration of bacterial cellulose assisted by immobilized laccase. AMB Express:8. https://doi.org/10.1186/s13568-018-0552-0

Sonoki T, Kajita S, Ikeda S et al (2005) Transgenic tobacco expressing fungal laccase promotes the detoxification of environmental pollutants. Appl Microbiol Biotechnol 67:138–142. https://doi.org/10.1007/s00253-004-1770-8

Speight JG, El-Gendy NS (2018) Bioremediation of contaminated soil. In: Introduction to petroleum biotechnology. Elsevier, Amsterdam, pp 361–417

Su J, Noro J, Fu J et al (2018) Exploring PEGylated and immobilized laccases for catechol polymerization. AMB Express 8. https://doi.org/10.1186/s13568-018-0665-5

Surridge AKJ, Wehner FC, Cloete TE (2009) Bioremediation of polluted soil. In: Singh A, Kuhad RC, Ward OP (eds) Advances in applied bioremediation. Springer, Berlin/Heidelberg

Suzuki T, Endo K, Ito M et al (2003) A thermostable laccase from Streptomyces lavendulae REN-7: purification, characterization, nucleotide sequence, and expression. Biosci Biotechnol Biochem 67:2167–2175

Téllez-Jurado A, Arana-Cuenca A, Becerra AEG et al (2006) Expression of a heterologous laccase by Aspergillus niger cultured by solid-state and submerged fermentations. Enzym Microb Technol 38:665–669

Theuerl S, Buscot F (2010) Laccases: toward disentangling their diversity and functions in relation to soil organic matter cycling. Biol Fertil Soils 46:215–225. https://doi.org/10.1007/s00374-010-0440-5

Upadhyay P, Shrivastava R, Agrawal PK (2016) Bioprospecting and biotechnological applications of fungal laccase. 3. Biotech 6:15. https://doi.org/10.1007/s13205-015-0316-3

Van De Vijver E, Van Meirvenne M, Seuntjens P (2015) Investigating bioremediation of petroleum hydrocarbons through landfarming using apparent electrical conductivity measurements. EGU Gen Assem 2015, held 12-17 April 2015 Vienna, Austria id9304 17

Verma R, Kumar A, Kumar S (2019) Synthesis and characterization of cross-linked enzyme aggregates (CLEAs) of thermostable xylanase from Geobacillus thermodenitrificans X1. Process Biochem 80:72–79. https://doi.org/10.1016/j.procbio.2019.01.019

Viswanath B, Rajesh B, Janardhan A et al (2014) Fungal laccases and their applications in bioremediation. Enzyme Res 2014:1–21. https://doi.org/10.1155/2014/163242

Wang T-N, Lu L, Wang J-Y, Xu T-F, Li J, Zhao M (2015) Enhanced expression of an industry applicable CotA laccase from Bacillus subtilis in Pichia pastoris by non-repressing carbon sources together with pH adjustment: recombinant enzyme characterization and dye decolorization. Process Biochem 50(1):97–103

Wang X, Yao M, Liu L et al (2016) Degradation of chlorpyrifos in contaminated soil by immobilized laccase. J Serbian Chem Soc 81:1215–1224. https://doi.org/10.2298/JSC160128066W

Wang J, Lu L, Feng F (2017a) Combined strategies for improving production of a thermo-alkali stable laccase in Pichia pastoris. Electron J Biotechnol 28:7–13. https://doi.org/10.1016/j.ejbt.2017.04.002

Wang X, Liu L, Yao M, et al (2017b) Degradation of carbofuran in contaminated soil by immobilized laccase. Polish J Environ Stud 26:1305–1312. https://doi.org/10.15244/pjoes/68428

Wang X, Shi Y, Ni Z et al (2018) Degradation of polycyclic aromatic hydrocarbons in contaminated soil by immobilized laccase. J Serbian Chem Soc 83:549–559. https://doi.org/10.2298/JSC171004022W

Wen X, Du C, Wan J et al (2019) Immobilizing laccase on kaolinite and its application in treatment of malachite green effluent with the coexistence of cd (II). Chemosphere 217:843–850. https://doi.org/10.1016/j.chemosphere.2018.11.073

Wu Y, Jiang Y, Jiao J et al (2014) Adsorption of Trametes versicolor laccase to soil iron and aluminum minerals: enzyme activity, kinetics and stability studies. Colloids Surf B Biointerf 114:342–348. https://doi.org/10.1016/j.colsurfb.2013.10.016

Xie H, Liu H, Xie Y et al (2015) Fabrication of a novel immobilization system and its application for removal of anthracene from soil. Biochem Eng J 97:8–16. https://doi.org/10.1016/j.bej.2015.01.011

Yang J, Ng TB, Lin J, Ye X (2015) A novel laccase from basidiomycete Cerrena sp.: cloning, heterologous expression, and characterization. Int J Biol Macromol 77:344–349

Yang J, Lin Q, Ng TB, Ye X, Lin J, Rito-Palomares M (2014) Purification and characterization of a novel Laccase from Cerrena sp. HYB07 with dye decolorizing ability. PLoS One 9(10):e110834

Yang J, Wang G, Ng TB, et al (2016a) Laccase production and differential transcription of laccase genes in Cerrena sp. in response to metal ions, aromatic compounds, and nutrients. Front Microbiol 6:1558

Yang J, Xu X, Ng TB et al (2016b) Laccase gene family in Cerrena sp. HYB07: sequences, heterologous expression and transcriptional analysis. Molecules 21:1017
Yang J, Xu X, Yang X et al (2016c) Cross-linked enzyme aggregates of Cerrena laccase: preparation, enhanced NaCl tolerance and decolorization of Remazol brilliant blue reactive. J Taiwan Inst Chem Eng 65:1–7. https://doi.org/10.1016/J.JTICE.2016.04.025
Yang J, Li W, Bun Ng T et al (2017) Laccases: production, expression regulation, and applications in pharmaceutical biodegradation. Front Microbiol 8:832. https://doi.org/10.3389/fmicb.2017.00832
Yoshida H (1883) LXIII.—chemistry of lacquer (Urushi). Part I. communication from the chemical society of Tokyo. J Chem Soc Trans 43:472–486
Yoshitake A, Katayama Y, Nakamura M et al (1993) N-linked carbohydrate chains protect laccase III from proteolysis in Coriolus versicolor. J Gen Microbiol 139:179–185. https://doi.org/10.1099/00221287-139-1-179
Zdarta J, Meyer AS, Jesionowski T, Pinelo M (2018) Developments in support materials for immobilization of oxidoreductases: a comprehensive review. Adv Colloid Interf Sci 258:1–20. https://doi.org/10.1016/j.cis.2018.07.004
Zeng J, Lin X, Zhang J et al (2011) Oxidation of polycyclic aromatic hydrocarbons by the bacterial laccase CueO from E. coli. Appl Microbiol Biotechnol 89:1841–1849
Zhang Y, Dong X, Jiang Z et al (2013) Assessment of the ecological security of immobilized enzyme remediation process with biological indicators of soil health. Environ Sci Pollut Res 20:5773–5780. https://doi.org/10.1007/s11356-012-1455-2
Zhao Y, Ma W (2011) Method for repairing soil with DDT (Dichloro-diphenyl-Trichloroethane) pollution by fungus and enzyme combination. Patent CN102091717A

Chapter 9
Environmental Fate of Organophosphate Residues from Agricultural Soils to Fresh Farm Produce: Microbial Interventions for Sustainable Bioremediation Strategies

Rishi Mahajan, Shalini Chandel, and Subhankar Chatterjee

Abstract The dependency of the growing population for the requirement of food has put an immense pressure on agriculture. As a direct consequence, different stakeholders especially associated with the agri-ecosystem are making concentrated efforts to enhance crop productivity. This has resulted in indiscriminate use of chemical pesticides/insecticides in agricultural fields. Pesticides are mainly used to control unwanted growth of plants (weeds) and also to control the population of pests, so that the agricultural and industrial products remain safe. In modern agriculture, several pesticides including organochlorine, organophosphate, carbamate, fungicides, herbicides, and synthetic pyrethroids are well effective in this regard. Because of their low cost of manufacturing, organophosphate pesticide (OPP) is the preferred one among them. The worldwide use of organophosphate pesticides (OPPs) in natural agri-ecosystems is now a well-documented fact. Out of five billion pounds of pesticides which are used worldwide every year, organophosphate pesticides (mostly insecticides) constitute 20–38%, and the main candidates are chlorpyrifos, dichlorvos, diazinon, dimethoate, fenitrothion, methyl parathion, monocrotophos, malathion, and profenophos. Regular use of these pesticides results in an increase in environmental and occupational exposures. During the last few decades, there is a growing concern among consumers as well as among farmers

R. Mahajan
Department of Microbiology, College of Basic Sciences, CSK Himachal Pradesh Agricultural University, Palampur, Himachal Pradesh, India

S. Chandel
Department of Plant Sciences, School of Life Sciences, Central University of Himachal Pradesh, TAB—Shahpur, Kangra, Himachal Pradesh, India

S. Chatterjee (✉)
Bioremediation and Metabolomics Research Group: Department of Environmental Sciences and Department of Chemistry and Chemical Sciences, Central University of Himachal Pradesh, TAB—Shahpur, Kangra, Himachal Pradesh, India
e-mail: subhankar@cuhimachal.ac.in

A. Kumar, S. Sharma (eds.), *Microbes and Enzymes in Soil Health and Bioremediation*, Microorganisms for Sustainability 16,
https://doi.org/10.1007/978-981-13-9117-0_9

about their negative effect in human and environmental health. In spite of the efforts to shift toward organic farming practices, the residual levels of OPs in soil and water bodies are still posing a threat to environment. To eliminate the OP pesticides or reduce their concentration from the environment, development of sustainable microbial-based bioremediation strategies has been initiated in the early 1970s, and the enzymatic degradation of OPs by organophosphorus hydrolase enzymes has been well studied in this regard. Modern biotechnological inventions and recently developed *omics*-based techniques have further increased the effectiveness of this process.

Keywords Organophosphate pesticides · Acetylcholinesterase · Organophosphorus hydrolase · Microbial bioremediation

9.1 Introduction

Organophosphate (OP) compounds are a group of human-made chemicals which were developed originally (1930–1940) as human nerve gas agents, but later they were extensively used as insecticides in agriculture. OP compounds are the most widely used insecticides today. They are used in agriculture, in the home, in gardens, and in veterinary practices. Statistics from the Food and Agriculture Organization suggested that the OP pesticide usage in agriculture is highest in the Asian countries, accounting for 29,554 metric tons during 2010–2015, with India dominating the charts in terms of OP usage (FAO 2019). The commonly used OP pesticides include parathion, malathion, methyl parathion, chlorpyrifos, diazinon, dichlorvos, phosmet, fenitrothion, tetrachlorvinphos, azamethiphos, azinphos methyl, and terbufos.

Inhibition of acetylcholinesterase (AChE) is the primary mechanism of OP's mode of action. OP compounds contain carbon and phosphorous acid derivatives which help them to absorb through the skin, lungs, and gastrointestinal tract (Narang et al. 2015). Therefore, they can easily bind to acetylcholinesterase (AChE) molecules by posing mimicking effect (Kazemi et al. 2012) which results in inhibition of enzymatic activity of AChE. At the end, ACh accumulates in the synaptic clefts of muscles and nerves leading to overstimulation of cholinergic receptors (Antonijevic and Stojiljkovic 2007). Molecular mechanism of OP activity is shown in Fig. 9.1. The inhibition reaction takes place through phosphorylation, dealkylation, and reactivation by hydroxylation of serine at the active site of AChE. The reaction mainly proceeds through a proton transfer from the nucleophilic serine in the active site of AChE to the leaving group of the OP. Covalent bond is formed between the OP molecule and the serine residue (OPAChE adduct) which further eliminate one of the functional groups associated with OP. Direct and indirect proton transfer reactions are involved in the proton-shifting mechanism.

To ensure high crop yields, farmers are greatly depending upon the chemical input into the soil. This kind of practice has led to indiscriminate use of OPs in

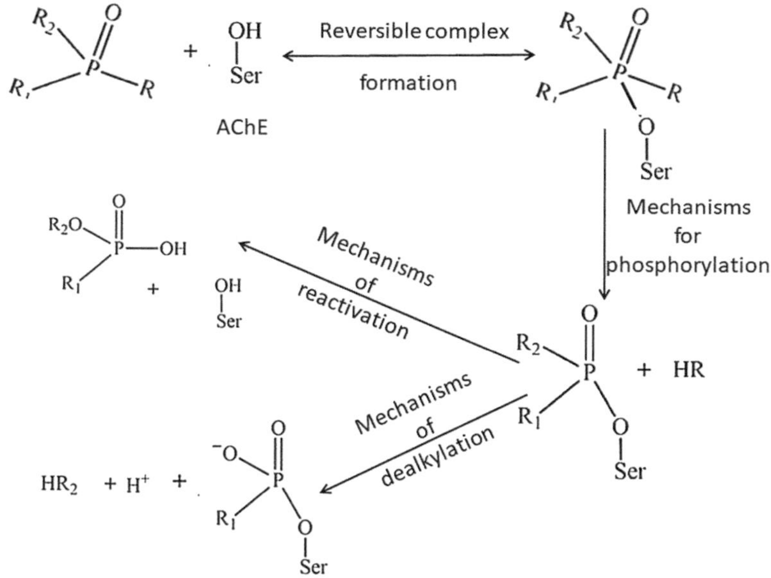

Fig. 9.1 Molecular mechanism involved in the action of organophosphate pesticides on the active site of acetylcholinesterase (Adopted from Rathnayake and Northrup 2016; Kumar et al. 2018; Field and Wymore 2014)

agricultural fields, and unfortunately, the soil acts as a sink for all the chemical inputs. The OP residue can either persist on the agricultural produce or, from the soil, can further leach into water systems and thus eventually enter into the human food chain. Both acute and chronic toxic effects of OP residues have been observed in humans and led to serious environmental concerns over the use of OP. This observation has motivated the scientific community to evaluate microbial interventions for sustainable bioremediation strategies. This chapter focuses on (a) the environmental fate of OP (after they enter into the environment), (b) their exposure to human population, (c) adverse health effects, and (d) microbial interventions for sustainable bioremediation strategies.

9.2 Organophosphate Residues: Environmental Fate

Due to the inevitable dependency on OP pesticides (OPPs) for increased agricultural output, accumulation of OPPs in diverse environmental compartments as well as in some non-targeted niche poses a serious threat to the environment. It has often been observed that OP exposure pathways overlap for many targeted and non-targeted living organisms which include wildlife species and humans; for example, the application of OPPs in agricultural crops can cause pesticide drift to nearby communities. Similarly, OPPs can run off into water bodies (surface and groundwater) and can cause detrimental effects on aquatic and terrestrial species that feed around that water bodies and ultimately harm the human populations of the nearby locality. In many occasions, wildlife species are used as sentinels for the risk assessment of OPPs to human health. Many environmental species including humans possess not only the active site of action of OPs (acetylcholinesterase) but also the cholinergic receptors (found throughout most taxonomic groups of animals) (Vermeire et al. 2003).

The possible routes of OP exposure to the environment are represented schematically in Fig. 9.2. The OPs enter into the environment (target soil or air or surface water) through different modes after their application onto crops (by spraying) or into the soil (as seed treatment). These modes include mainly emissions, leaching, drainage, and volatilization. From the soil, the OP residues can enter into either surface water or groundwater via drainage or leaching. Sedimentation results when OP residues are transported from surface waters to the different sediments. After spraying of OPs onto crops, these compounds can also enter into sewage treatment plants from where the OP residue can enter into the non-target soil. As a result of the movement of the OP residues in different environmental components, a number of non-target species are adversely affected. These non-targeted species range from microorganisms in the soil to aquatic organisms and terrestrial organisms. A detailed schematic representation of possible routes of OP exposure to human and wildlife is presented in Fig. 9.3.

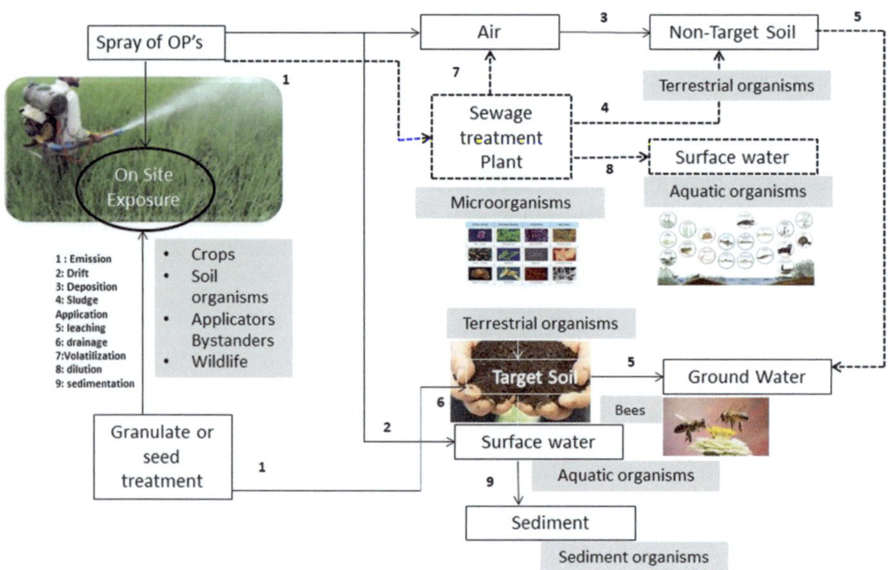

Fig. 9.2 Distribution routes of OPP from agricultural field to different parts of the environment. Gray boxes contain the receptor organisms; dotted lines and boxes are used when the release is via the sewage treatment plant

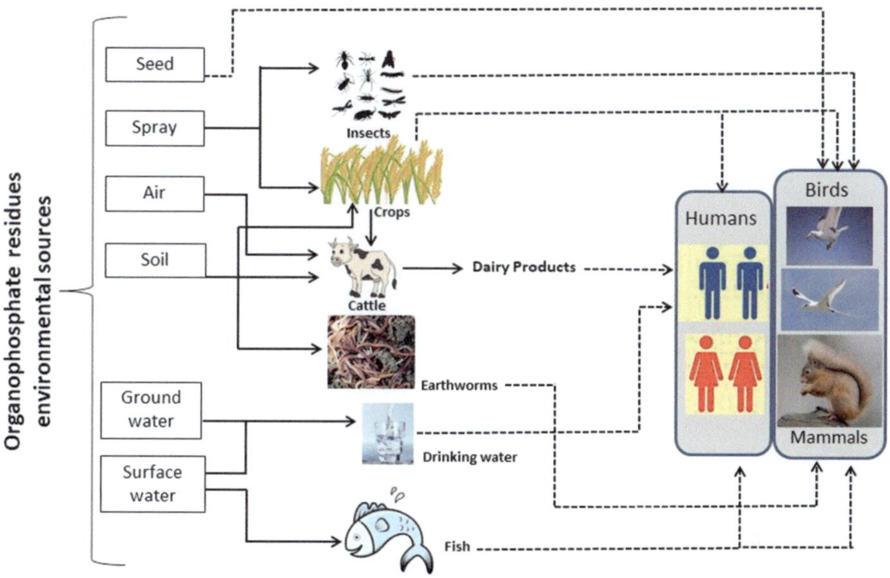

Fig. 9.3 A detailed schematic representation of possible routes of OP exposure to humans and wildlife

9.2.1 Organophosphate Residues: From Agricultural Soils to Fresh Farm Produce

In today's scenario, where there is an emphasis on modernization of farming techniques for high crop productivity, the use of OP pesticides in agricultural fields is on an all-time rise, especially in agricultural-predominant countries (Tilman et al. 2002). Fresh farm produce (vegetables and fruits) are an essential component of healthy lifestyle and are often recommended for good physical, mental, and gut health as they are important source of vitamins, nutrients, polyphenols, and dietary fibers (Fenik et al. 2011; Mahajan et al. 2018). Though recommended for good health, often the consumption of OP (or OP residues) contaminated vegetables and fruits are leading to serious ill effects on human health (Muñoz-Quezada et al. 2012, Swarnam and Velmurugan 2013; Sapbamrer and Hongsibsong 2014). A number of reviews have revealed that the majority of the vegetables and fruits grown worldwide contain significant OP residues (Tables 9.1 and 9.2).

The significant concern is about the fact that the OP residues can persist in soil for a long period of time and when crops are grown in such soil, OP residues can still be detected in fresh produce in spite of no application of OPs. Once such observation was made by Essumang and co-workers in 2013 while growing okra in Ghana, where they observed significantly higher concentrations of chlorpyrifos (1321 ng g^{-1}), diazinon (6.10 ng g^{-1}), and malathion (7.0 ng g^{-1}), in spite of the fact that no OPs were used during the growing cycle.

Table 9.1 Widely used organophosphate pesticide residues observed in common vegetables

Organophosphate pesticide	Vegetables	Maximum residue levels (ng g^{-1})	References
Ethoprophos	Tomato Potato Sweet potato Carrot Cucumber Zucchini Kohlrabi Eggplant	20–50	Darko and Akoto (2008), Szala and Szponik (2012), Essumang et al. (2013), Yu et al. (2016), and Witczak et al. (2018)
Diazinon		10–200	
Chlorpyrifos-methyl		50–500	
Parathion-methyl		0–10	
Fenchlorphos		0–10	
Chlorpyrifos		10–400	

Table 9.2 Widely used organophosphate pesticide residues observed in common fruits

Organophosphate pesticide	Fruits	Maximum residue levels (ng g^{-1})	References
Ethoprophos	Kiwi Avocado Mango Olive Red banana Green apple Orange Grapefruit Melon	0–20	Quijano et al. (2016), Fenik et al. (2011), and Witczak et al. (2018)
Diazinon		0–10	
Chlorpyrifos-methyl		50–500	
Parathion-methyl		0–10	
Fenchlorphos		0–10	
Chlorpyrifos		10–300	

The concept of organic farming is of much significance in today's context since substantial awareness is growing among consumers and producers. However, the matter of concern is about the fact that our soils are already contaminated because of the overuse of OPPs. In countries where organic farming practices were initiated almost three to four decades back, the agricultural produce obtained today is still bearing OP residues. Once such report by Gnusowski and co-workers, in 2011, indicated detection of OP residues of chlorpyrifos in organic produce of cucumber grown over a period of 2004–2010. Recent reports have further strengthened this viewpoint when Fuhrimann and co-workers in 2019 reported comparative studies between conventional and organic agricultural farming in Costa Rica; Farina et al. (2018) reported that edible tissues from vegetables (cauliflower, spinach, celery, cabbage, broccoli, lettuce, and mustard) grown in organic farms of Cameron Highlands in Malaysia accounted for significant OP residues.

9.3 Organophosphate Residues: Affecting Human Health

Humans are consistently exposed to OPs starting from OPs production until their leaching into the environment. Primarily, people who are working in OP-producing industries might be exposed to OPs if proper handling measures are not put into practice. Further exposure can occur if there is accidental ingestion or if OPs come in direct contact with eyes. Secondary sources of OP exposure include eating contaminated food or get infected through contaminated soils or contaminated runoff water or groundwater.

A toxicity report published by Robb and Baker 2018, suggested that an estimated more than three million people are exposed to OP, which results in approximate death of 300,000 people worldwide. OP binds irreversibly to acetylcholinesterase in the cholinergic synapses in the central nervous system and peripheral nervous system, which results in high concentrations of acetylcholine in the synaptic clefts that eventually cause initial excessive stimulation and, later, blockade of synaptic transmission (Namba 1971). Peter et al. (2014) suggested that, after the OP exposure (within minutes to hours), symptoms like salivation, lacrimation, urination, defecation, gastric cramps, and emesis (sludge) may occur. Long-term exposure to organophosphates can cause confusion, anxiety, loss of memory, loss of appetite, disorientation, depression, and personality changes.

Analysis of human blood and urine samples has revealed that human populations are highly exposed to OPs across the world, especially in countries where organic farming is widely practiced (Roca et al. 2014; Croes et al. 2015; Spaan et al. 2015; Cartier et al. 2015). The most alarming fact is that, the OP metabolites in urine samples have been found in higher concentrations among children as compared to adults, indicating that the children are more vulnerable to OP exposure (Mie et al. 2017). Studies have also provided significant evidences regarding OP exposure to pregnant women. The results showed OP (or OP residues) could significantly affect the brain functions of children within the age group of 2–11 years (Young et al. 2005; Eskenazi et al. 2007; Engel et al. 2011). On the basis of these results, OPP can be called as endocrine disrupting chemical (EDC) (Mie et al. 2017). As a consequence of such reports, the commonly used OPP, i.e., chlorpyrifos, has been enlisted as human developmental neurotoxicant (Grandjean and Landrigan 2014). A general outline of the possible health effects of organophosphate exposure in humans is presented in Fig. 9.4.

9.4 Organophosphate Residues: Microbial Interventions for Sustainable Bioremediation Strategies

In developing countries, where agriculture holds the major stake, multidimensional uses of OPs in agricultural crops, fruits, and vegetables have resulted in its high residual levels in the environment. In spite of the efforts to shift towards organic

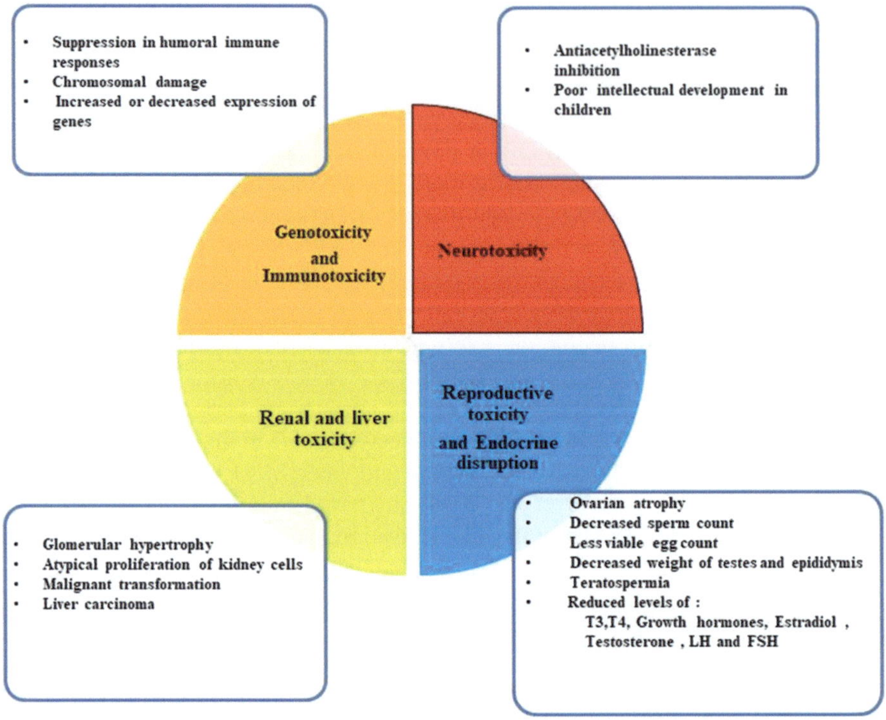

Fig. 9.4 A general outline of possible health effects of OPP exposure in human

farming practices, the residual levels of OPs in soil and water bodies (groundwater, surface water) has further compounded the problem. In this regard, detection of OP's in soil and water bodies is very important and therefore there is an urgent need to develop simple and easy to use analytical techniques for OP detection. Recent work in our laboratory on 'a simple and rapid analytical method for detection of two OPs (profenofos and fenthoin) using HPLC-DAD method' has given a consice idea about the method of OPP detection in the environmental samples (Mahajan and Chatterjee 2018). To detoxify and decontaminate the environment from OP pollution, use of microbial-based bioremediation strategy has started in the early 1970s and has emerged as the most viable and eco-friendly technique. In the year 1973, the first microorganism capable of degrading organophosphorus compounds was isolated and identified as *Flavobacterium* sp. (Singh and Walker 2006).

Microbial bioremediation is a cost-effective and environmentally friendly sustainable methodology that can be used to remove the toxic pollutants from the environment. Various bacterial and fungal species are being used that are having a potential enzymatic mechanism for degradation of the specific organic pollutants (Chatterjee and Dutta 2003; Chatterjee and Karlovsky 2010; Mahajan et al. 2019). A few important organophosphate pesticides which have been degraded by this technique are presented in Table 9.3.

Table 9.3 List of few important organophosphate pesticides which are degraded through microbial interventions

Organophosphate pesticides	Microbial cultures associated
Chlorpyrifos	*Pseudomonas pseudoalcaligenes* *Pseudomonas aeruginosa* *Lactobacillus plantarum* *Bacillus aryabhattai* *Brucella melitensis* *Bacillus subtilis* *Bacillus cereus* *Trichosporon* sp.
Diazinon	*Serratia marcescens* *Arthrobacter* sp. *Leuconostoc mesenteroides* *Serratia liquefaciens*
Fenitrothion	*Burkholderia* sp. *Corynebacterium* sp. *Lactobacillus brevis* *Cupriavidus* sp.
Profenofos	*Pseudomonas putida* *Burkholderia gladioli*
Parathion-methyl	*Leuconostoc mesenteroides* *Pseudomonas pseudoalcaligenes* *Stenotrophomonas* sp.
Monocrotophos	*Arthrobacter atrocyaneus* *Bacillus megaterium* *Aspergillus fumigatus*
Fenamiphos	*Pseudomonas putida* *Acinetobacter rhizosphaerae* *Microbacterium esteraromaticum* *Caulobacter crescentus*

Adsorption is the first major step adopted by the microorganisms which restricts the movement of OP in the soil matrix. The eventual degradation depends on several factors such as solubility of the OP, its volatility in soil, charge, polarity, molecular structure, and the size. **Photodegradation** is another means by which rapid degradation of OPs can be achieved. The possible mechanisms include oxidation of the P=S bonds or the isomerization. The mechanism of **hydrolysis** plays a critical role in the overall degradation process of OPs, which accounts for the cleavage of P–S or P–O bonds. A detailed introspection on the modes of microbial degradation of OP has been provided in a recent review article by Kumar et al. (2018).

Further, the **enzymatic hydrolysis** of OPs, especially by using enzymes derived from microbial origin, plays a crucial role in developing cost-effective and sustainable biotechnologies for OP residue management in the environment. A significant advantage of using microbial enzymes is its border substrate specificity.

Organophosphate-degrading (*opd*) gene which encodes for enzyme organophosphorus hydrolase (OPH; EC. 3·1·8·1) has been well studied in *Pseudomonas diminuta* and *Flavobacterium* spp. (strain ATCC27551). The enzyme has been reported to cleave phosphorus–ester bonds, viz., P–O, P–CN, P–F, and P–S, although the efficiencies of enzyme-mediated cleavage differ significantly. Many reports suggested that the enzyme catalytic efficiency needs to be improved especially for the catalysis of P–S bond for effective bioremediation of OPP. To improve organophosphorus hydrolase (OPH) efficiency, site-directed mutagenesis approach was followed by Schofield and DiNovo (2010). By using this technique, they were able to change a specific amino acid in OPH and successfully prepared eight mutants. All of these mutant cell lysates resulted in variation in their protein specific activities when tested against OP. Gotthard et al. (2013) carried out the structural and enzymatic characterization of OPHC2 from *Pseudomonas pseudoalcaligenes* (GenBank ID: AJ605330) and reported its esterase and phosphotriesterase activities.

Among the several OPPs, the two most widely studied OPPs in respect to microbial biodegradation studies include parathion and malathion. Parathion (O,O-diethyl-O-p-nitrophenyl phosphorothioate) was used widely as a broad-spectrum insecticide, acaricide, fumigant, and nematocide, until it was designated as restricted-use pesticide. Similarly, malathion (diethyl-2-di-methoxy phosphinothioyl-sulfanyl-butane-dioate) is a neurotoxin that belongs to OPP and is mostly used in food crops and vegetables worldwide.

The predominant biodegradation pathway of malathion involves formation of mono- and diacid metabolites which were generated by carboxylesterase activity. Complete mineralization of the pesticide was done through oxidative desulfurization and demethylation reactions. Other minor routes of metabolism include initial oxidation to remove sulfur and methyl groups. A detailed account of the degradation pathway has been outlined by Gao et al. (2010) in the University of Minnesota Biocatalysis/Biodegradation Database. This database also reported a number of microorganisms (e.g., *Rhizobium* spp., *Pseudomonas aeruginosa* AA112, *Arthrobacter* sp., *Trichoderma viride*, *Aspergillus niger*, *Pseudomonas* sp., and *Penicillium notatum*) which can efficiently degrade malathion through different enzymatic pathways (Gao et al. 2010).

In one of the possible pathways, *Rhizobium* spp., *Arthrobacter* sp., *Trichoderma viride*, and *Pseudomonas* sp. have biotransformed malathion into malathion monocarboxylate via the action of malathion esterase. This compound was then broken down to malathion dicarboxylate by the action of malathion monocarboxylate esterases. At this stage, an oxidoreductase enzyme catalyzes the breakdown of malathion dicarboxylate to dimethyldithiophosphate. At the same time, formation of oxaloacetate was also evidenced which further entered into an intermediate metabolic pathway. Dimethyldithiophosphate was further metabolized to dimethylthiophosphate by oxidoreductase enzyme. Various enzymes like dimethylthiophosphate phosphodiesterase, methylthiophosphate phosphomonoesterase, and thiophosphate oxidoreductase have eventually helped the pathway molecule to break down to methanol and phosphate. Methanol ultimately enters into the C1 cycle (Gao et al. 2010; database).

Parathion biodegradation is mainly achieved through the involvement of microorganisms, including *Flavobacterium* sp. ATCC 27551 and *Brevundimonas diminuta* MG. The detailed pathway has been explained by Gao et al. (2010) in the University of Minnesota Biocatalysis/Biodegradation Database. Briefly, in the aerobic degradation pathway, parathion was initially hydrolyzed to para-nitrophenol and diethylthio phosphoric acid. In another pathway, parathion was oxidized to paraoxon through the enzymatic action of parathion oxidoreductase. Further breakdown of paraoxon could be either mediated by enzyme aryldialkyl-phosphatase or sometimes the involvement of enzyme may not be the case. However, eventually metabolized products could be either p-nitrophenol or diethyl phosphoric acid, which would enter the p-nitrophenol or the phorate pathway. Under anaerobic conditions, parathion could be reduced to aminoparathion (mediated through NAD(P)H nitroreductase), which was further hydrolyzed to p-aminophenol and diethylthiophosphoric acid. The enzyme aryldialkyl-phosphatase was involved in this step.

The practical applicability of the microbial enzymes further lies in the process development of biosensors for the rapid detection of OPs in the environment. The OPH enzyme reaction can be combined with a variety of transduction schemes, which would eventually lead to a direct and rapid determination of OPs in the environment. Combining OPH enzyme with pH electrode (a potentiometric transducer) results in the development of **potentiometric OPH-based enzyme electrode biosensor**. In a similar manner, the chromophoric products produced during OP hydrolysis can be detected by creating **optical OPH-based biosensors**. Finally, monitoring of oxidation–reduction current (generated as a result of the hydrolysis products of OPs) can also be achieved by integrating OPH with an amperometric transducer. Recent developments in the field of nanotechnology can also be used in developing biosensors using immobilized OPH enzymes. Significant research work is being carried out in this respect, details of which can be accessed by evaluating the works of Beleno Cabarcas et al. (2018), who have developed chitosan nanocomposite-modified OPH-based amperometric sensor for organophosphorus pesticide determination, and Hondred et al. (2018), who have printed graphene electrochemical biosensors fabricated by inkjet maskless lithography for rapid and sensitive detection of organophosphates.

9.5 Future Prospects

Considering the present scenario, where there is a considerable risk associated with the OP exposure to humans, the need of the hour is to primarily limit the every possible loopholes in the system that results in the entry of the OPs into the environment. This calls in for stringent government interventions by the way of policy-making for systematic phasing out of the different OPs used in agriculture. Nongovernmental organizations can play a pivotal role by ensuring that the farming community is made aware of the deleterious health effects of OP usage. If farmers can feed the world, they can certainly bring the change in the way we can prevent the contamination of food through OPs. The scientific community, on the other hand, has a much critical role in developing biotechnologies for mitigating the impact that OP has already made in our environment.

Acknowledgments The authors are thankful to Dr. Ashok Nadda, Department of Biotechnology and Bioinformatics, Jaypee University of Information Technology, HP, who has kindly invited us to contribute this chapter and also for his encouragement and advice since the invitation. Research in SC lab is supported by the Council of Scientific and Industrial Research (CSIR), India.

References

Antonijevic B, Stojiljkovic MP (2007) Oxtime efficacy against organophosphates. Clin Med Res 5(1):71–82

Beleno Cabarcas MT, Stoytcheva M, Zlatev R, Montero G, Velkova Z, Gochev V (2018) Chitosan nanocomposite modified OPH-based Amperometric sensor for organophosphorus pesticides determination. Curr Anal Chem 14(1):75–82

Cartier C, Warembourg C, Le Maner-Idrissi G, Lacroix A, Rouget F, Monfort C, Limon G, Durand G, Saint-Amour D, Cordier S, Chevrier C (2015) Organophosphate insecticide metabolites in prenatal and childhood urine samples and intelligence scores at 6 years of age: results from the mother–child PELAGIE cohort (France). Environ Health Perspect 124(5):674–680

Chatterjee S, Dutta TK (2003) Metabolism of butyl benzyl phthalate by Gordonia sp. strain MTCC 4818. Biochem Biophys Res Commun 309(1):36–43

Chatterjee S, Karlovsky P (2010) Removal of the endocrine disrupter butyl benzyl phthalate from the environment. Appl Microbiol Biotechnol 87(1):61–73

Croes K, Den Hond E, Bruckers L, Govarts E, Schoeters G, Covaci A, Loots I, Morrens B, Nelen V, Sioen I, Van Larebeke N (2015) Endocrine actions of pesticides measured in the Flemish environment and health studies (FLEHS I and II). Environ Sci Pollut Res 22(19):14589–14599

Darko G, Akoto O (2008) Dietary intake of organophosphorus pesticide residues through vegetables from Kumasi, Ghana. Food Chem Toxicol 46(12):3703–3706

Engel SM, Wetmur J, Chen J, Zhu C, Barr DB, Canfield RL, Wolff MS (2011) Prenatal exposure to organophosphates, paraoxonase 1, and cognitive development in childhood. Environ Health Perspect 119(8):1182–1188

Eskenazi B, Marks AR, Bradman A, Harley K, Barr DB, Johnson C, Morga N, Jewell NP (2007) Organophosphate pesticide exposure and neurodevelopment in young Mexican-American children. Environ Health Perspect 115(5):792–798

Essumang DK, Asare EA, Dodoo DK (2013) Pesticides residues in okra (non-target crop) grown close to a watermelon farm in Ghana. Environ Monit Assess 185(9):7617–7625

FAO- Pesticides Use Data [Internet] (2019) [cited January 28, 2019]. Available from: http://www.fao.org/faostat/en/#data/RP

Farina Y, Munawar N, Abdullah MP, Yaqoob M, Nabi A (2018) Fate, distribution, and bioconcentration of pesticides impact on the organic farms of Cameron Highlands, Malaysia. Environ Monit Assess 190(7):386

Fenik J, Tankiewicz M, Biziuk M (2011) Properties and determination of pesticides in fruits and vegetables. TrAC Trends Anal Chem 30(6):814–826

Field MJ, Wymore TW (2014) Multiscale modeling of nerve agent hydrolysis mechanisms: a tale of two Nobel prizes. Phys Scr 89(10):108004

Fuhrimann S, Winkler MS, Staudacher P, Weiss FT, Stamm C, Eggen RI, Lindh CH, Menezes-Filho JA, Baker JM, Ramírez-Muñoz F, Gutiérrez-Vargas R (2019) Exposure to pesticides and health effects on farm owners and workers from conventional and organic agricultural farms in Costa Rica: protocol for a cross-sectional study. JMIR Res Protocols 8(1):e10914

Gao J, Ellis LBM, Wackett LP (2010) The University of Minnesota Biocatalysis/biodegradation database: improving public access. Nucleic Acids Res 38:D488–D491

Gnusowski B, Nowacka A, Łozowicka B, Szpyrka E, Walorczyk S (2011) Pesticide residues in organic food and feed of plant origin. J Res Appl Agricult Eng 56(3):102–107

Gotthard G, Hiblot J, Gonzalez D, Chabriere E, Elias M (2013) Crystallization and preliminary X-ray diffraction analysis of the organophosphorus hydrolase OPHC2 from Pseudomonas pseudoalcaligenes. Acta Crystallogr Sect F: Struct Biol Cryst Commun 69(1):73–76

Grandjean P, Landrigan PJ (2014) Neurobehavioural effects of developmental toxicity. Lancet Neurol 13(3):330–338

Hondred JA, Breger JC, Alves NJ, Trammell SA, Walper SA, Medintz IL, Claussen JC (2018) Printed graphene electrochemical biosensors fabricated by inkjet maskless lithography for rapid and sensitive detection of organophosphates. ACS Appl Mater Interfaces 10(13):11125–11134

Kazemi M, Tahmasbi AM, Valizadeh R, Naserian AA, Soni A (2012) Organophosphate pesticides. Agricult Sci Res J 2(9):512–522

Kumar S, Kaushik G, Dar MA, Nimesh S, Lopez-Chuken UJ, Villarreal-Chiu JF (2018) Microbial degradation of organophosphate pesticides: a review. Pedosphere 28(2):190–208

Mahajan R, Attri S, Mehta V, Udayabanu M, Goel G (2018) Microbe-bio-chemical insight: reviewing interactions between dietary polyphenols and gut microbiota. Mini Rev Med Chem 18(15):1253–1264

Mahajan R, Verma S, Kushwaha M, Singh D, Akhter Y, Chatterjee S (2019) Biodegradation of di-n-butyl phthalate by psychrotolerant Sphingobium yanoikuyae strain P4 and protein structural analysis of carboxylesterase involved in the pathway. Int J Biol Macromol 122:806–816

Mahajan R, Chatterjee S (2018) A simple HPLC–DAD method for simultaneous detection of two organophosphates, profenofos and fenthion, and validation by soil microcosm experiment. Environ Monit Assess 190(6)

Mie A, Andersen HR, Gunnarsson S, Kahl J, Kesse-Guyot E, Rembiałkowska E, Quaglio G, Grandjean P (2017) Human health implications of organic food and organic agriculture: a comprehensive review. Environ Health 16(1):111

Muñoz-Quezada MT, Iglesias V, Lucero B, Steenland K, Barr DB, Levy K, Ryan PB, Alvarado S, Concha C (2012) Predictors of exposure to organophosphate pesticides in schoolchildren in the province of Talca, Chile. Environ Int 47:28–36

Namba T (1971) Cholinesterase inhibition by organophosphorus compounds and its clinical effects. Bull World Health Organ 44(1–3):289

Narang U, Narang P, Gupta OP (2015) Organophosphorus poisoning: a social calamity. J Mahatma Gandhi Institute of Medical Sci 20(1):46–51

Peter JV, Sudarsan TI, Moran JL (2014) Clinical features of organophosphate poisoning: a review of different classification systems and approaches. Indian J Critic Care Med 18(11):735

Quijano L, Yusà V, Font G, Pardo O (2016) Chronic cumulative risk assessment of the exposure to organophosphorus, carbamate and pyrethroid and pyrethrin pesticides through fruit and vegetables consumption in the region of Valencia (Spain). Food Chem Toxicol 89:39–46

Rathnayake LK, Northrup SH (2016) Structure and mode of action of organophosphate pesticides: a computational study. Computational and Theoretical Chemistry 1088:9–23

Robb EL, Baker MB (2018) Organophosphate toxicity. In Stat pearls [internet]. StatPearls Publishing

Roca M, Miralles-Marco A, Ferré J, Pérez R, Yusà V (2014) Biomonitoring exposure assessment to contemporary pesticides in a school children population of Spain. Environ Res 131:77–85

Sapbamrer R, Hongsibsong S (2014) Organophosphorus pesticide residues in vegetables from farms, markets, and a supermarket around Kwan Phayao Lake of northern Thailand. Arch Environ Contam Toxicol 67(1):60–67

Schofield DA, DiNovo AA (2010) Generation of a mutagenized organophosphorus hydrolase for the biodegradation of the organophosphate pesticides malathion and demeton-S. J Appl Microbiol 109(2):548–557

Singh BK, Walker A (2006) Microbial degradation of organophosphorus compounds. FEMS Microbiol Rev 30:428–471

Spaan S, Pronk A, Koch HM, Jusko TA, Jaddoe VW, Shaw PA, Tiemeier HM, Hofman A, Pierik FH, Longnecker MP (2015) Reliability of concentrations of organophosphate pesticide metabolites in serial urine specimens from pregnancy in the Generation R Study. J Expos Sci Environ Epidemiol 25(3):286

Swarnam TP, Velmurugan A (2013) Pesticide residues in vegetable samples from the Andaman Islands, India. Environ Monit Assess 185(7):6119–6127

Szala J, Szponik M (2012) Dynamics of chlorpyrifos residues in cauliflower cultivation fad-eDynamika zanikania pozostałości chloropiryfosu w uprawie kalafiora. Prog Plant Protect 52(4):1117–1119

Tilman D, Cassman KG, Matson PA, Naylor R, Polasky S (2002) Agricultural sustainability and intensive production practices. Nature 418(6898):671

Vermeire T, MacPhail R, Waters M (2003) Integrated human and ecological risk assessment: a case study of Organophosphorous pesticides in the environment. Hum Ecol Risk Assess Int J 9(1):343–357

Witczak A, Pohoryło A, Abdel-Gawad H, Cybulski J (2018) Residues of some organophosphorus pesticides on and in fruits and vegetables available in Poland, an assessment based on the European union regulations and health assessment for human populations. Phosphorus Sulfur Silicon Relat Elem:1–10

Young JG, Eskenazi B, Gladstone EA, Bradman A, Pedersen L, Johnson C, Barr DB, Furlong CE, Holland NT (2005) Association between in utero organophosphate pesticide exposure and abnormal reflexes in neonates. Neurotoxicology 26(2):199–209

Yu R, Liu Q, Liu J, Wang Q, Wang Y (2016) Concentrations of organophosphorus pesticides in fresh vegetables and related human health risk assessment in Changchun, Northeast China. Food Control 60:353–360

Chapter 10
Secreted Microbial Enzymes for Organic Compound Degradation

Lauren Sara McKee and Annie Rebekah Inman

Abstract Microbes in the belowground environment draw nutrition from the complex organic biomass found in the soil. Their primary means of interaction with the molecular components of their environment is via extracellular enzymes that deconstruct high molecular weight organic compounds in the soil. These include the natural carbohydrates and polyaromatic compounds of decaying plant and microbial biomass, predominantly cellulose, lignin, and chitin. Also important are several classes of organic xenobiotics of anthropogenic origin, such as polyaromatic hydrocarbons, polychlorinated biphenyls, and diverse synthetic fertilisers and pesticides. Many biotechnological processes have now been established that exploit this natural toolbox of biomass-degrading enzymes for the industrial production of biofuels or biomaterials. However, our understanding of the natural role these enzyme systems play within the soil remains limited. It is well accepted, for example, that an active microbiota is vital for productive agriculture, but the impacts of soil management regimes on the microbiota remain opaque. In this chapter, we review current knowledge on microbial enzyme secretion and activity in the soil and explore current research into the regulation of enzyme production. We summarise the range of enzyme activities found in the soil environment and their contribution to the recycling and degradation of organic compounds, a vital elemental turnover that may be impacted by a warming climate. The methods employed by microbes to maintain an effective level of enzyme activity in the extracellular environment are described. Finally, we discuss the ways in which we might make use of microbial enzymes to improve the sustainability of agriculture and industry.

L. S. McKee (✉)
Division of Glycoscience, Department of Chemistry, KTH Royal Institute of Technology, Stockholm, Sweden

Wallenberg Wood Science Centre, KTH Royal Institute of Technology, Stockholm, Sweden
e-mail: mckee@kth.se

A. R. Inman
Division of Glycoscience, Department of Chemistry, KTH Royal Institute of Technology, Stockholm, Sweden

A. Kumar, S. Sharma (eds.), *Microbes and Enzymes in Soil Health and Bioremediation*, Microorganisms for Sustainability 16,
https://doi.org/10.1007/978-981-13-9117-0_10

Keywords Biomass recycling · Enzyme activity · Industrial biotechnology · Secreted enzymes · Soil microbiota · Sustainable agriculture

10.1 Introduction

The secretion of enzymes into the extracellular environment by soil-dwelling fungi and bacteria is vital for global elemental turnover and may be the rate-determining step in the recycling of natural plant and microbial biomass (Moorhead et al. 2013). The efficiency of the overall cycling process is governed by factors including the species composition of the microbiota, the overall functional capacity of the community, and the biophysical properties of the enzymes themselves. Particularly important considerations include the pH, salt, and temperature tolerance of each enzyme, and the kinetic parameters of the catalysed reactions under natural conditions. These biophysical characteristics also strongly influence the extent to which soil organisms and their enzymes can be put to use in industrial biotechnology processes.

The functional enzymatic capacity of the soil microbiota has been studied in a wide variety of contexts. For example, composting and bioreactor communities have been sequenced and characterised (Ortseifen et al. 2016; Jensen et al. 2018; Zhou et al. 2018), as have the natural microbiota of soils in forest, agricultural, and industrial landscapes (Katz et al. 2016; Berini et al. 2017; Bedoya et al. 2019). Secreted enzymes discovered from these communities have been put to use in the generation of pulp, paper, biofuels, and biomaterials, and are considered vital to the implementation of economically viable biorefinery technologies (Carvalho et al. 2019; Nitsos et al. 2019). The enzymes secreted by phytopathogens and root colonisers in the soil have also been studied (Kubicek et al. 2014) and are relevant in the context of crop protection and biological control of plant disease.

Soil organic matter (SOM) in natural and managed land is a long-term carbon sink, largely comprising aggregated polysaccharides and polyphenolic compounds, bound together by microbial exopolysaccharides and a fungal mycelial network. It is vital for land cohesion, water retention, and the accessibility of nutrients and water to plants. It is also host to the microbial diversity and biochemical processes that define healthy or fertile soil. Unfortunately, human activities can negatively impact soil functions. The application of fertilisers and pesticides to agricultural soils contributes to substantial short-term increases in farm productivity, but in the longer term these interventions cause great harm to agricultural soil and the wider ecosystem (Tilman et al. 2002). Degradation and loss of fertility in the soil inevitably lead to further greenhouse gas emissions from land conversion and the chemical synthesis of pesticides and fertilisers, compounds that eventually leach into groundwater and become elements to be remediated at wastewater treatment sites. 'Ecological' or 'organic' practices are often vaguely defined, but for the purposes of this chapter we will consider low-input agricultural systems as those that aim to restrict the use of synthetic fertilisers and pesticides, using manure and slurry instead. While there can be short-term losses due to lower yields or higher disease rates, it is thought that long-term productivity will be higher in low-input systems, as soil fertility is maintained for longer, especially where diverse mixtures of crops are grown.

While loss of biodiversity of the macro- and micro-flora and fauna is evident where blanket coverage of pesticides is used, most conventional farming techniques actually use small, targeted applications of specific and relevant plant protection agents, which have a relatively small overall impact on the microbiota. In fact, the nature of the fertilising agent used is a vastly more important determinant of microbiota composition than the choice of crop cultivated or the type of pest control utilised (Hartmann et al. 2015). Soil from low-input farms seems to contain more carbon than that from conventionally farmed land (Gattinger et al. 2012) and to have a more diverse and active microbiota (Tilman et al. 2002).

Further complicating matters, residuals from human industry are globally ubiquitous in soils and include industrial effluents, petrohydrocarbons, and many long-banned chemicals such as pesticides that persist through physical association to soil particles. These harmful and recalcitrant substances can be difficult to remove, and we are increasingly turning to members of the soil microbiota to assist with our clean-up efforts. Bioremediation and bioaugmentation approaches to the cleaning of contaminated land are becoming more sophisticated, and at the same time there are efforts to harness the power of the soil microbiota to limit the use of synthetic agricultural additives in the first place. The biocontrol industry is beginning to boom, but large-scale implementation is still lacking, and a mechanistic understanding of the processes involved has still not been fully established. A final ever-pressing issue is the matter of climate change and whether a warming climate and more variable and extreme weather patterns may limit the ability of the soil to store carbon. This has profound impacts for our atmosphere, as well as our ability to feed ourselves as the global population continues to soar.

10.2 Organic Compound Degradation in the Soil

The complex and highly networked structure of organic biomass requires that a large consortium of degradative enzymes be deployed for effective depolymerisation. The plant cell wall alone contains four major classes of enmeshed polymers: cellulose, lignin, pectin, and the hemicelluloses. Soil microbes secrete synergistic enzyme mixtures to deconstruct this network, allowing nutrition to be drawn from complex organic material. This enzymatically catalysed turnover of dead biomass recycles C, N, and other elements, and feeds a healthy microbiota. At the same time, organic compounds deriving from agriculture and other industries are abundant and persist in soil. In many cases, soil microbes have adapted to the presence of these and are able to deconstruct them using extracellular enzymes.

Two broad categories of enzymes are utilised by soil bacteria and fungi to deconstruct organic macromolecules into short, soluble oligomers that can be taken up into the cytoplasm and used for nutrition. Hydrolytic enzymes such as cellulases generally target a specific chemical bond (e.g. a glycosidic bond connecting two glucose monomers), whereas oxidative enzymes such as laccases can act more broadly, and potentially deconstruct a wide range of substrates that contain similar chemical structures, using oxygen or hydrogen peroxide as electron donors.

The success of secreted enzymes is not guaranteed. Enzymes deployed into the extracellular space must be stable and active in that environment. However, biophysical and biochemical characterisation of such enzymes in vitro typically reveals them to be active in a narrow range of conditions such as temperature, pH, salt content, and redox state. The question of enzymatic durability in the soil therefore arises, as there is an apparently high potential for enzyme 'wastage' in the environment due to degradation, denaturation, loss of activity, or unproductive binding. The production and secretion of enzymes is energetically very taxing for microorganisms, and so many species have developed elegant means of limiting energy expenditure on enzyme secretion.

10.2.1 Cell Wall Glycans

The vast majority of organic carbon entering the soil is derived from growing plants, which contribute via the aboveground plant and leaf litter and the belowground root litter. Small organic molecules such as simple sugars, sugar alcohols, and amino acids leach directly from these litter types and can be rapidly taken up and utilised for microbial respiration within hours or days (Gougoulias et al. 2014). But the bulk of plant biomass comprises complex polymers of high molecular weight (Fig. 10.1). These compounds require deconstruction by extracellular enzymes prior to being used for respiration, and this can take from days to weeks.

Of particular significance in the biodegradation of complex biomass are the mycorrhizal fungi, including obligate symbionts (arbuscular mycorrhizal fungi, AMF) that exclusively recycle carbon from their host, and more generalist facultative symbionts (ectomycorrhizal fungi, ECM). Bacteria and fungi in forest soils have been shown to play clearly distinct roles in carbon processing: oxidative processes and chitin degradation dominate in microbiota rich in saprophytic fungi, whereas hydrolytic processes for cellulose and heteroglycan degradation dominate in microbiota rich in AMF and bacteria (You et al. 2014).

The most abundant component of plant cell walls is cellulose, a β-1,4-linked polymer of D-glucose monomers forming linear glucan chains that exhibit a rigid crystalline microfibril structure, held by an extensive H-bonding network (Srivastava et al. 2017). The recalcitrant nature of the cellulose structure necessitates the use of multiple enzymes for deconstruction. The classical view of cellulose degradation involves *endo*-β-1,4-glucanases, cellobiohydrolases, and β-glucosidases (Lopes et al. 2018). This view has recently been expanded to include the newly discovered lytic polysaccharide monooxygenases (LPMOs, discussed in detail below), which are thought to help overcome the problem of cellulose crystallinity by forming nicks

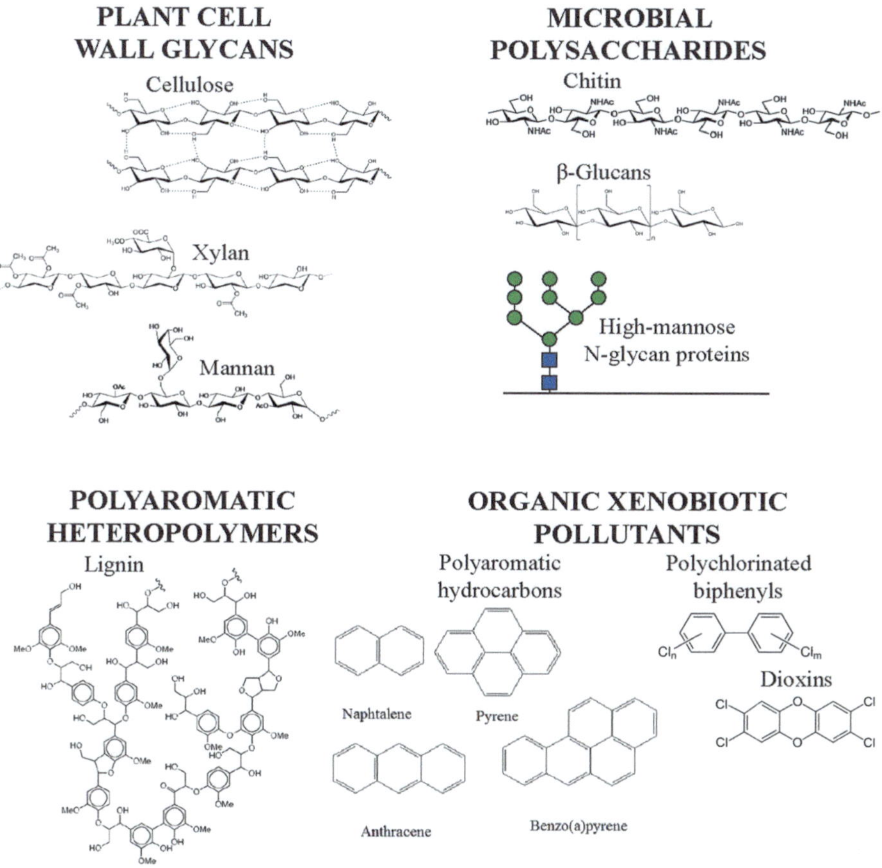

Fig. 10.1 Natural and synthetic organic compounds found in the soil that are enzymatically degraded by soil microbes. The graphic depicts the abundant crystalline cellulose found in plant cell walls, as well as representative structures of two common hemicelluloses, xylan and mannan. Also shown is lignin, a highly variable polyaromatic compound. Lignin is abundant in wood due to its location in the secondary cell walls of plant tissue. Also depicted are common components of microbial cell walls and exopolysaccharides, including chitin, β-1,3-glucans, and high mannose glycoproteins (blue squares, *N*-acetylglucosamine; green circles, mannose). Finally, some examples of common organic xenobiotic pollutants are also shown; these are of anthropogenic origin

in the cellulose surface to permit access to other enzymes (Bissaro et al. 2018; Tandrup et al. 2018). Many fungi secrete all of these activities, often with some apparent redundancy. For example, *Trichoderma reesei* secretes at least 30 cellulose-targeting enzymes, including seven endo-glucanases (Strakowska et al. 2014).

Of course, in the plant cell wall cellulose is intimately associated with other glycans such as the hemicelluloses and pectins, as well as lignin, which is highly abundant in wood (Srivastava et al. 2017). Lignin is a randomly cross-linked polymeric phenylpropanoid that is degraded by oxidative enzymes such as laccases, manganese peroxidases, and lignin peroxidases (Pollegioni et al. 2015; Kamimura et al.

2018). These degradative biocatalysts rely on yet other enzymes to produce the necessary cations and hydrogen peroxide co-factors. The removal of lignin is vital for efficient cellulose degradation by various classes of fungi, as it protects the cellulose from physicochemical and enzymatic attack. But it is probable that lignin itself is rarely if ever the primary carbon source for lignolytic organisms, which typically couple lignin degradation with cellulolysis: the energy obtained via metabolism of the high glucose yields taken from cellulose compensates for energy expended on lignin degradation.

While the aforementioned bulky and structurally recalcitrant polymers are to a large extent deconstructed and metabolised by fungi, this leaves a great deal of carbohydrate material in the form of more amorphous heteroglycans from the cell wall matrix. This includes the pectins and hemicelluloses from plants, which comprise multiple sugars and multiple linkages, necessitating the use of large numbers of degradative enzymes to hydrolyse just one complex carbohydrate (Larsbrink et al. 2014a, b). These relatively more soluble glycans may be degraded before the crystalline cell wall components, as they diffuse more readily out of the networked structure. Some studies have shown that Gram-positive bacteria are enriched in areas of complex polysaccharide degradation (Waldrop et al. 2000; Bell et al. 2009), whereas Gram-negative bacteria are more likely to compete for simpler organic compounds and to secrete enzymes that produce very short oligosaccharides that contribute to a large shared pool of nutrient carbon (You et al. 2014).

Additionally, as the soil microbes take up and metabolise so much of the plant-derived carbon in the soil, it is logical that they themselves contribute to the total SOM. Fungal and bacterial exopolysaccharides and necromass, for example, comprise substantial amounts of accessible organic carbon. Indeed, an important but often overlooked source of nutrition for scavenging microbes in the soil and litter environment is dead microbial cells (Burns 2013). The chitin, varied β-glucans, and glycoproteins found in decaying fungal mycelium and the sticky polysaccharides secreted by bacteria constitute a rich source of C, N, and P. There is increasing evidence that these microbial materials serve a specialised niche of highly active bacteria (Larsbrink et al. 2017). Indeed, decaying fungal mycelium represents a hotspot of enzymatic activity, a microenvironment characterised by specific and highly active genera of bacteria (Brabcova et al. 2016).

To utilise all of these varied glycan substrates in the soil, microbes deploy a vast arsenal of enzymes with catalytic activity (Fig. 10.2). Fungi often secrete hemicellulose- and lignin-degrading cocktails, and many bacteria can secrete synergistic cocktails of enzymes targeting one very complex glycan. The recognition of complex polysaccharides is often facilitated by accessory domains appended to hydrolytic enzymes, the most important of which are the carbohydrate-binding modules (CBMs). A CBM may bind a simple sugar, a free oligo- or polysaccharide, or the surface of a crystalline polysaccharide (Gilbert et al. 2013), and it is thought that this binding helps to promote enzyme activity in several ways. It can increase the concentration of an enzyme on the surface of a polysaccharide and also prolong the attachment time of an enzyme and its substrate (Herve et al. 2010), leading to greater acquisition of metabolisable sugars.

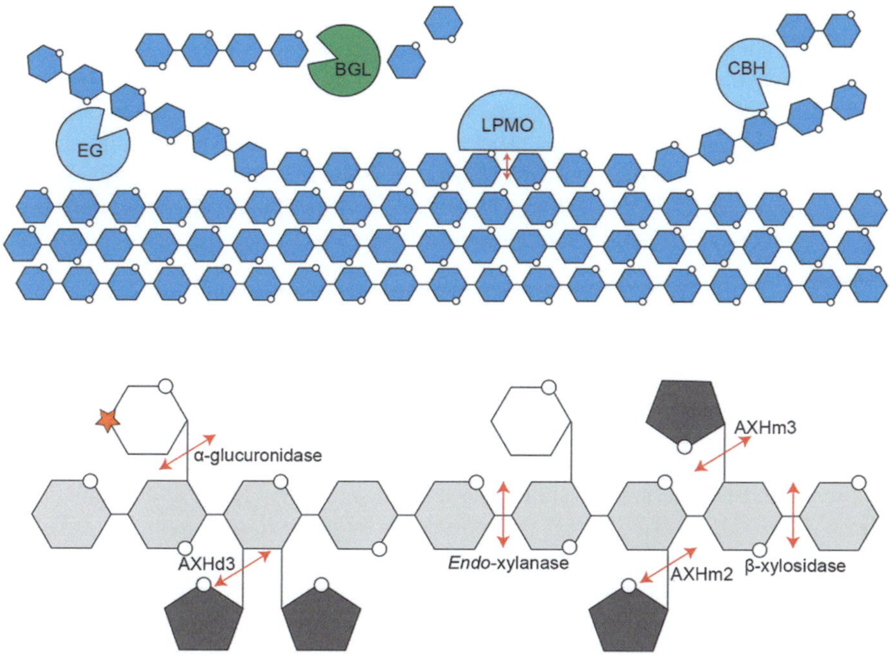

Fig. 10.2 Simplified schematic models showing synergistic enzyme degradation of complex cell wall glycans. Top: the deconstruction of crystalline cellulose, comprising solely glucose monomers (blue hexagons), requires the concerted action of lytic polysaccharide monooxygenases (LPMOs), cellobiohydrolases (CBHs), *endo*-glucanases (EGs), and β-glucosidases for full enzymatic conversion to glucose. The LPMO creates nicks in the crystalline surface, causing chain loosening and providing access to EG and CBH. BGL converts the oligosaccharide glucan chains into glucose. Bottom: an example structure of heteroxylan comprising a backbone of β-1,4-linked xylose (light grey) decorated with α-L-arabinose (Ara) sugars (dark grey) connected to the backbone at the O2 and/or O3 positions, and α-glucuronic acid (GlcA) sugars (white) connected at O2. An *endo*-xylanase is required to deconstruct the xylan backbone into oligosaccharides that are hydrolysed by a β-xylosidase. An α-glucuronidase is required to remove the GlcA, which may or may not be methylated (orange star). Finally, multiple types of α-L-arabinofuranosidase are required to remove Ara residues at different positions: AXHm2 and AXHm3 enzymes remove singly substituted Ara, while AXHd3 removes one of the Ara residues from a double substitution

10.2.2 Organic Xenobiotic Pollutants

The soil is a significant sink of persistent hydrophobic organic contaminants (HOCs), which adhere to the soil matrix by sorption to the large surface area of soil particles, by diffusing into pores in the SOM matrix, or by interacting with lipids in the soil (James et al. 2011, Cachada et al. 2014). These HOCs include industrial effluents such as polyaromatic hydrocarbons (PAHs), dioxins, and polychlorinated biphenyls (PCBs), all of which persist in the environment and accumulate in food chains, leading to widespread toxicity in animals and humans.

PAHs are likely the organic pollutants that have received the most attention in the literature. These are found ubiquitously in the global soil environment, although most studies focus on hotspots of contamination at specific industrial sites. Hotspots are in fact relatively rare and are characterised by high PAH concentrations deriving from a single identifiable source. The vast majority of soil PAH contamination is diffuse, low concentration, and derived from multiple sources (Johnsen and Karlson 2007; Okere et al. 2017). PAHs are classed as priority pollutants by regulatory bodies as they can be genotoxic, mutagenic, and carcinogenic (Rogers et al. 2002; Yan et al. 2009).

Unfortunately, due to the lipophilic nature of both PAHs and the outer surface (cuticle) of plant leaves, atmospheric PAHs settle readily on vegetation, and when this plant material is deposited as litter, the PAHs are transferred to the soil. The specific fate of PAHs in the soil depends on the ability of the microbes present to access and metabolise them, but the level of microbial PAH transformation is high. The relatively low molecular weight 3- and 4-ring PAHs such as anthracene, phenanthrene, and pyrene are metabolisable by soil microbes (Dries and Smets 2002; Hadibarata et al. 2013), whereas the more toxic 5- and 6-ring structures seem to be more resistant to degradation and therefore accumulate to proportionally higher levels (Johnsen et al. 2006). There is evidence that some species can co-metabolise PAH compounds when mixtures are present, indicating that the metabolism of a low molecular weight PAH might synergistically enhance the metabolism of a high molecular weight PAH, leading to higher removal rates (Barret et al. 2010). The mechanisms of pollutant co-metabolism are highly complex, and it remains unclear whether most regulation occurs at the level of protein production or microbial growth rate (Ghosal et al. 2016), but it has been shown that the microbial decomposition of natural litter is sometimes enhanced in the presence of PAH contaminants (Qasemian et al. 2012), indicating an overall increase in degradative behaviours.

In some bacteria, PAHs can be metabolised and serve as the sole carbon source: the process typically begins with oxidation of one of the rings of the PAH by a ring hydroxylating dioxygenase (RHD) enzyme, followed by cleavage of this product into metabolisable aliphatic products (Ghosal et al. 2016). While bacterial transformation relies most heavily on dioxygenase enzymes, in fungi the process typically uses monooxygenase enzymes (Ghosal et al. 2016), and PAHs cannot be used as a carbon source. Fungi generally rely on secreted proteins, utilising enzymes similar to the laccases, lignin peroxidases, and manganese peroxidases that are involved in lignin degradation (Li et al. 2010; Wang et al. 2018).

This natural ability of the microbiota to adapt to the presence of PAHs by gaining the ability to transform and even metabolise them has led to the emergence of natural bioremediation as a major method used to decontaminate toxic environs. *In situ* bioremediation of contaminated sites remains a highly variable process, and optimisation requires more detailed knowledge of the function of the microbiota and how it responds to the presence of organic contaminants. Indeed, the presence of PAHs has a profound impact on the microbial community in the soil. The *Bacteroidetes* and *γ-Proteobacteria* appear to be less abundant in PAH-contaminated soils than in clean soils, while *Actinobacteria* dominate in contaminated soils, particularly when

4-ring PAHs are present (Sawulski et al. 2014). In order to predict microbial activity and properly manage a site, it is essential to understand long-term changes in the microbial community in the presence of pollutants such as PAH. Often, a single microbe is not capable of transforming or degrading a PAH, but relies on the broader microbial consortium for assistance. For example, fungi are generally unable to degrade the high molecular weight 5- and 6-ring PAHs, but some white-rot species can use extracellular enzymes to transform them into more polar metabolites that can be degraded by co-resident bacterial species (Meulenberg et al. 1997). Despite their ability to degrade some of the more abundant PAHs, the actual rate of removal of PAHs by soil microbes can be slow due to the very low bioavailability of the pollutants. Consequently, bioremediation processes often require some human intervention to speed up the natural process, such as by introducing specific plants, nutrients, or exogenous microbial species (Ye et al. 2014).

10.3 Enzymology-Ecology of the Soil Microbiota

The physicochemical nature of SOM poses multiple challenges to the microorganisms that depend upon it for nutrition. The high molecular weight and structural complexity of the dominant carbohydrates require a two-phase metabolic approach: hydrolysis of organic polymers by extracellular enzymes and subsequent uptake of the lower molecular weight reaction products. The depolymerisation of natural compounds by bacteria and fungi in the soil is a rate-limiting step in carbon (C) cycling in the rhizosphere and litter layers (Schimel and Schaeffer 2012), and consequently the secretion of enzymes by these organisms is critical for a functional ecosystem.

The energy dynamics of enzyme secretion in the soil are complex and determined by a wide range of biotic and abiotic factors. The modelling of these types of community behaviours is inherently challenging due to the natural variability of ecosystems and diversity of the soil microbiota. Our understanding remains limited by the relevance and availability of accurate data, but there is a clear consensus that an active microbiota is vital for biomass recycling and for the maintenance of a robust, healthy ecosystem that includes plants, animals, and insects.

10.3.1 Energy Acquisition and Expenditure: A Delicate Balancing Act

The substrate field represented by SOM poses challenges to scavenging microbes due to its complexity and low concentrations. To gain enough nutrition from this environment to survive and thrive, the energy gained from biomass deconstruction must outweigh the energy expended on enzyme production and secretion. This requires efficient utilisation of extracellular enzymes (Traving et al. 2015). Extracellular enzymes may be freely released into the environment, or they may be tethered to the outer surface of the producing cell (Wallenstein and Burns 2011). Each strategy has strengths and limitations in environments defined by the movement and solubility of substrates, as well as physical factors such as temperature and hydration. For microbes in the soil, enzyme tethering to the cell surface may be the most cost-efficient strategy, as it limits the chances of enzyme loss to the environment and allows reaction products to be kept close to the cell, minimising the risk that 'cheating' organisms will steal oligosaccharides liberated by the enzyme producer (Folse and Allison 2012). Free extracellular enzymes are 'public goods' in the soil microbiota, as they can benefit organisms other than the producer (Allison 2005). Enzymes that are produced and not kept within close proximity to the producer are especially likely to be exploited by cheaters at the expense of the enzyme producers. Social models of microbiota in biomass-rich environments indicate that, in the absence of cheating, nutrient depolymerisation rates are high as enzyme production is high, leading to a rapid turnover of organic material. However, in diverse microbial communities, conditions may explicitly favour cheaters and collaborative scavengers, who compete with generalist enzyme producers: ironically, this can lead to a slowdown in the overall rate of nutrient deconstruction (Allison 2005).

The enzyme-producing microbe must be able to optimise both the rate of substrate encounter and the energy cost of maintaining a population of active enzymes. Clearly, a stable enzyme with a long active half-life is preferable as this allows the enzyme to catalyse more productive events in its lifetime, resulting in more nutrients being taken up by the cell. However, long-lived free enzymes may generate products far from a motile cell, resulting in a modest energy gain. Tethering enzymes to the cell surface is a way of prolonging the active half-life of an enzyme while also maintaining cellular proximity to the reaction products.

One means of tethering enzymes to the cell surface is via a cellulosome, a multiprotein complex of plant cell wall-degrading enzymes attached to the bacterial cell surface that has been dubbed a 'microbial nanomachine' due to its degradative power and efficiency (Gilbert 2007) (Fig. 10.3). This approach is utilised by *Clostridium thermocellum* and related bacterial species. Cellulosomes assemble and anchor multiple enzymes to the cell surface via tight and specific interactions with a scaffold protein called CipA scaffoldin. Cellulosomes are coordinated via specific protein-protein binding interactions between cohesins and dockerins (Carvalho et al. 2003). These are calcium-dependent interactions with high binding affinity. Type I cohesin-dockerin interactions tether enzymes to the scaffoldin, while type II interactions tether the cellulosome to the cell surface (Fig. 10.3). Interactions tend to be non-specific within a species, allowing one cohesin to recruit multiple potential dockerins, but specific between species (i.e. dockerins from *C. thermocellum* will not interact with cohesins from a different *Clostridium* species). Key to the success of the cellulosome is the synergistic nature of the enzymes recruited to the complex, but the cellulosome is a highly dynamic structure: different enzymes and different complexes can be recruited to the cell surface as needed (Blanvillain et al. 2007). This flexibility in organisation allows maximum active lifespan for

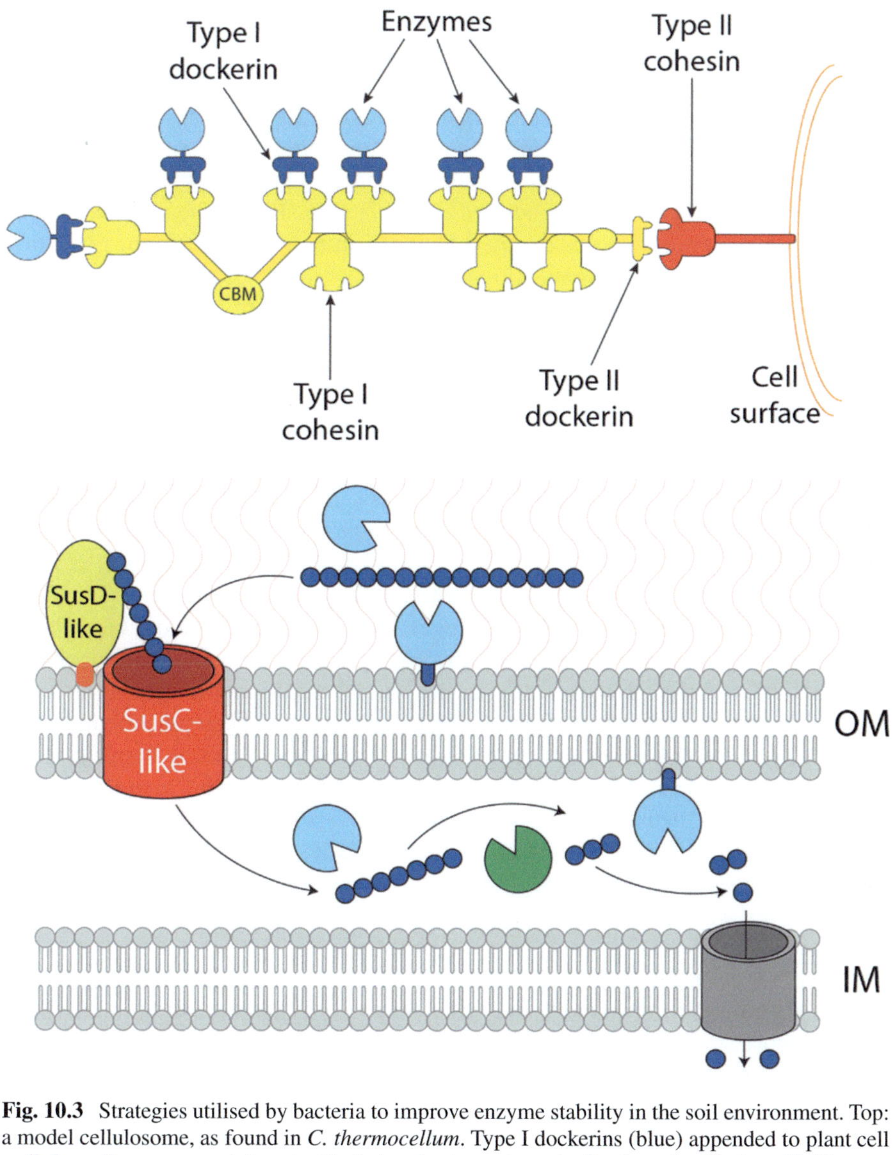

Fig. 10.3 Strategies utilised by bacteria to improve enzyme stability in the soil environment. Top: a model cellulosome, as found in *C. thermocellum*. Type I dockerins (blue) appended to plant cell wall-degrading enzymes interact with their cohesin partners (yellow) on the CipA scaffoldin protein (yellow). This forms the active core of the multi-protein complex. The entire cellulosome is tethered to the cell surface via a type II dockerin that interacts with its cohesin partner (red) appended to a protein (green) bound to the outer membrane. CBM: carbohydrate-binding module. Bottom: a schematic of an archetypal polysaccharide utilisation locus (PUL). The SusD outer-membrane protein (yellow) recognises a specific glycan; this is transported into the periplasm via the SusC pore (red). This induces a transcriptional response, leading to increased levels of production of the SusC/D proteins and all enzymes encoded by the PUL. The enzymes may be *endo*- or *exo*-acting (depicted in blue and green, respectively), they may be secreted to the periplasm or extracellular space, and they may also be tethered to the outer membrane. OM, outer membrane; IM, inner membrane. The arrows indicate the path of a complex glycan through the system as it is degraded

cellulolytic enzymes, while minimising the time and energy required for cellulosome assembly.

An energy-saving strategy utilised by other bacteria is to produce enzymes at a high level only when they are needed to degrade a nearby substrate. Bacteria of the phylum *Bacteroidetes*, which dominate complex microhabitats from the forest floor to the human gut, utilise an elegant system of genetic operons termed polysaccharide utilisation loci (PULs) (Grondin et al. 2017) (Fig. 10.3). The protein components of a PUL have names that correspond to their functional homologue in the first such operon described, the Starch Utilisation System (SUS) of *Bacteroides thetaiotaomicron*, uncovered by seminal work undertaken in the Abigail Salyers lab at the University of Illinois (D'Elia and Salyers 1996; Reeves et al. 1997; Shipman et al. 2000). A PUL is identified by the presence of two outer-membrane proteins: a SusD-like protein that recognises and binds a specific glycan structure and a SusC-like protein that translocates that glycan into the periplasm. In the periplasm, the glycan is sensed by a transcriptional regulator (termed SusR-like), which induces upregulated transcription of the PUL. Importantly, the PUL also encodes glycoside hydrolases and other enzymes that can specifically and efficiently deconstruct the activating glycan into metabolisable monosaccharides. The enzymes may be secreted into the extracellular space, attached to the outer membrane, or located in the periplasm (Martens et al. 2009). Upregulating enzyme production only when a specific activity is explicitly required can compensate for enzyme loss to the environment and permits growth spurts in the spatial and temporal proximity of a favoured carbon source. Many examples of PULs have now been functionally characterised, and they are known to target a vast array of complex and simple glycans from various sources (Ndeh and Gilbert 2018). Although most characterisation efforts have so far focussed on human gut symbiont species (Larsbrink et al. 2014a, b; Cuskin et al. 2015) rather than environmental species (Larsbrink et al. 2016), it is likely that the same processes dominate in the external environment, where *Bacteroidetes* with the same genetic markers of PULs are found (McBride et al. 2009; McKee et al. 2019).

Recent models of enzyme-based SOM decomposition strongly support observations from fundamental microbiology that enzyme production can be induced by specific substrates. When fresh litter is added to a microcosm, there is a delay of several days before enzyme activity reaches its peak, when the overall amount of microbial biomass decreases but enzyme activity increases drastically due to a shift towards a more active population (Moorhead et al. 2013). This supports data showing that isolated organisms have a preference in mixed-substrate media for simpler sugars and start to degrade more complex polysaccharides only when necessary. These models also support studies showing that different taxa of microbes are active at different stages of litter decomposition, with concurrent shifts in the dominant groups of microorganisms led by changes in the available substrate material (Allison and Martiny 2008; Rinkes et al. 2011).

10.3.2 Ecological Impacts: Carbon Cycling and Plant Growth

Natural and managed soils serve as a net carbon sink, with a steady flux of CO_2 between the soil and the atmosphere, termed 'soil respiration', accounted for primarily by the activity of soil microorganisms (Gougoulias et al. 2014). Global carbon cycling levels are defined by the balance between carbon fixing by photosynthetic organisms and respiration by autotrophs and heterotrophs. The aforementioned processes of organic compound degradation in the soil permit soil microbial respiration, by which CO_2 is eventually returned to the atmosphere.

In the active soil layers of the rhizosphere and the detritus, the rate of microbial turnover of decomposing biomass is rate-limiting for C cycling and is determined by the level of extracellular enzyme activity in the soil. By contrast, in the mineral soil layers, access to substrates is limited by aggregation and sorption, so the rate-limiting step is gaining access to occluded substrates (Dungait et al. 2012). Microbes in the mineral layer do still play an important role in determining the extent of C sequestration, as they both produce and deconstruct the necromass and sticky polysaccharides that can hold aggregated particles together. Microaggregates of soil often form around elements of microbially derived macromolecules, including viscous exopolysaccharides and the AMF glycoprotein glomalin (Wang et al. 2017). The true level of microbial involvement in the soil's ability to act as a carbon sink is complex and difficult to model or quantify. Due to alterations in the carbon cycle that are impacting human society via climate change, there is a pressing need to better understand the relationships between soil microbiota function and the global climate system. This has led to increasing integration of microbial ecology and evolutionary ecological theory. This scientific movement has to a large extent been driven by the evergrowing availability of massive sequence-based descriptions of microbial communities, although there is a need to move towards functional characterisation, rather than simple phylogenetic characterisation, of the soil microbiota (You et al. 2014).

What we do know is that the biogeochemically relevant processes that involve the soil microbiota fall into two major categories. First are the 'narrow' processes that are of critical importance to one specific ecosystem but involve a very small fraction of the total microbial community and only cycle a tiny amount of a certain element. Examples include specific plant-microbe interactions such as pathogenicity, mycorrhizal interactions, and nitrogen fixing rhizobia. Many of these specific interactions involve secreted proteins and enzymes for recognition and deconstruction of molecules such as cell wall components. By contrast, most soil microbes are generalists engaged in 'broad' processes that are responsible for the large flux of carbon through soil (Schimel and Schaeffer 2012). There is a great deal of functional redundancy among the species performing these broad processes. In fact, changes to the soil microbiota at the phylogeny level appear not to significantly impact elemental cycling at seasonal or annual scale, where the dominating factors are the rates of plant growth and death.

Most models estimate that more atmospheric CO_2 is still being fixed by photosynthesis every year than is released by soil respiration (Gougoulias et al. 2014). However, the degradation of soil through industrial activity, deforestation, and overly intensive agriculture has caused several gigatonnes of extra carbon to be released to the atmosphere. Soil carbon stocks on agricultural land are experiencing a steady decline, which reduces biodiversity and soil fertility, as well as limiting the ability of soil and its microbial inhabitants to adapt to a warming climate (Gattinger et al. 2012).

Microbes in the soil depend on photosynthetic plants for carbon and energy sources, but there exists a positive feedback loop, as plants are equally dependent on soil microbes for their own healthy growth. Indeed, careful stewardship of the soil underpins effective agricultural practice. There is an increasing body of evidence on the importance of the so-called plant growth-promoting rhizobacteria (PGPR) and how important they are in healthy fertile agricultural soil (Bhattacharyya and Jha 2012). Plant-microbe and microbe-microbe cross talk in the soil and particularly the rhizosphere, whether beneficial, symbiotic, or pathogenic, is highly species-specific. Communication is generally mediated via the secretion of secondary metabolites (Saraf et al. 2014; Deveau et al. 2016) and in some cases may also involve the secreted proteins that permit bacteria to attach to or glide over solid surfaces (Kolton et al. 2014). Additionally, there is mounting evidence supporting the importance of proteins secreted by both plants and microbes in mediating the nature and specificity of interactions (De-la-Peña et al. 2008; Nelson and Sadowsky 2015). The symbiosis between leguminous plants and rhizobacteria is the best known plant-microbe interaction, but there is also communication between plants and parasites. Studies have confirmed that when plants and microbes recognise each other, each changes their profile of secreted proteins in a way that may boost nutrient scavenging by a bacterium and/or may stimulate a symbiotic or defence response in the plant (De-la-Peña et al. 2008). Furthermore, recognition of plant or microbial material by a specific bacterium can be communicated to the surrounding microbial community via quorum sensing. Quorum sensing is when bacteria use externally secreted chemicals as a means of communication with other bacteria in an effort to coordinate their activities (Whiteley et al. 2017). The above-described interactions can lead to disease states and the development of symbiosis or have no obvious impact, depending upon the specific partners involved.

10.4 Applications of Secreted Enzymes in the Sustainability Context

Fundamental investigations into the enzymology of the soil environment are of relevance in the ongoing study of global biogeochemistry and can help us to design tailored in situ bioremediation and biocontrol efforts at environmental improvement. Using data on soil microbiota function, biochemists and structural biologists have uncovered new reaction mechanisms, new modes of substrate interaction, and new synergistic modes of enzyme activity (Gilbert 2010), especially since the widespread availability of large-scale sequencing technologies. Novel enzymes have been discovered from metagenomic, metaproteomic, metatranscriptomic, and functional metagenomic studies on the soil microbiota.

The development of efficient and economic enzyme-catalysed industrial processes is key to moving away from traditional chemical processes and a reliance on non-renewable resources. Furthermore, fundamental investigations into the enzymology of the soil microbiota can uncover powerful tools for use in biotechnology at an industrial scale, such as biomass degradation for fuel or materials production (Hemsworth et al. 2016). The use of synergistic enzymes (McKee et al. 2016; von Freiesleben et al. 2018) or enzymes targeting cross-linking structures (Arnling Bååth et al. 2018) is of particular benefit towards the total saccharification of biomass. More targeted applications of microbial enzymes can be used to produce oligosaccharides for applications such as in food (Raveendran et al. 2018), materials (Martínez-Abad et al. 2015), and medicine (Anbu et al. 2017) production. And due to their high specificity of action, enzymes and CBMs can also be used as research tools to directly study the structure of cell walls (McCartney et al. 2004) by probing complex biomass samples for the presence of specific target structures (Verhertbruggen et al. 2009). The targeted enzymatic removal of cell wall components has permitted gene expression analysis with spatial resolution in plant tissue (Giacomello et al. 2017) and enabled the detailed molecular characterisation of plant cell wall glycan structures (Gunl et al. 2011, Martínez-Abad et al. 2017).

10.4.1 Extracellular Enzymes for Industrial Biotechnology

There is a broad consensus among scientists and policy makers that we need to move away from our reliance on fossil resources for materials and especially fuels production, to address problems of environmental destruction and social injustice. The majority of fossil fuel consumption is for energy in the transport sector, accounting for huge quantities of CO_2 emissions (von Blottnitz and Curran 2007). The use of plant biomass for fuel production instead of non-renewable fossil resources has long been proposed to offer a more favourable carbon emissions profile. However, in the case of the so-called first-generation bioethanol fuels, produced from easily degraded starch in edible food crops such as corn, life cycle analyses show that the overall environmental footprint is complex and that the use of such fuels is not always more sustainable than burning gasoline. Perhaps more importantly, the diversion of food crops away from the food supply chain has serious consequences for global food security, and so the focus has shifted to second-generation bioethanol fuels, produced from lignocellulosic waste (LCW).

Theoretically, the sugar content and fermentability of LCW should permit bioethanol yields comparable with those of first-generation feedstocks (Taha et al. 2016). The major difference between the saccharification of first- and second-generation feedstocks is the complexity of the saccharification process. Biomass such as corn that has a high starch content can give extremely high glucose yields with simple amylase treatment. By contrast, as we have previously described, the saccharification of LCW requires a broad consortium of enzymes with complementary activities that can work synergistically to break down the different polymers within the plant cell wall. A lack of access for enzymes to the cell wall glycans is also a major problem, and the saccharification of LCW presents serious technical challenges not posed by edible biomass feedstocks. The recalcitrance of the plant cell wall requires significant inputs of energy, time, and other resources that have so far largely prevented the broad adoption of second-generation biofuel production (Robak and Balcerek 2018). A huge body of research is building up describing pretreatment technologies aimed at increasing efficiency. These include mechanical and chemical techniques to break open the cell wall, increase pore size in biomass, and partly depolymerise cell wall components. Many processes proposed to date have low efficiency, lead to major loss of sugar, require significant amounts of energy input, and produce high volumes of hazardous chemical waste (Taha et al. 2016). Biological pretreatments on the other hand utilise living cultures of microbes, often basidiomycete or ascomycete white-rot fungi, to disrupt the lignin moieties of the plant cell wall, leaving a large portion of the cellulose and hemicellulose fractions relatively intact and amenable to saccharification by a cocktail of mostly hydrolytic enzymes (Taha et al. 2016). These biological pretreatments are less wasteful and energy intensive than physicochemical approaches, but they are very slow and require careful control to reduce the amount of sugar lost to microbial metabolism.

Additional enzyme activities that shed new light on biomass recycling in nature and that may revolutionise the industrial exploitation of biomass continue to be uncovered (Nelson et al. 2017). An important recent example is the revelation of oxidative cellulose deconstruction by lytic polysaccharide monooxygenases (LPMOs) (Quinlan et al. 2011, Horn et al. 2012, Bissaro et al. 2018). LPMOs employ an oxidative mechanism to break glycosidic bonds at the cellulose surface, creating nicks that can grant access to *endo*-glucanase enzymes. The mechanism of action by LPMOs may involve electron donation from other cell wall components, hinting at potential synergy with lignin-degrading machinery (Westereng et al. 2015). It is now known that these enzymes can also act on non-crystalline carbohydrates (Agger et al. 2014), highlighting their previously unaccounted contributions to biomass deconstruction and their potential significance for a biomass-based industrial economy. Methods for LPMO characterisation are still in their infancy compared to the well-developed techniques utilised for glycoside hydrolases (Westereng et al. 2018), but the enzymes are now a major focus of research on cellulose saccharification (Eibinger et al. 2014), the enzymatic production of nanocellulose (Hu et al. 2018), and also the oxidative modification of the cellulose surface for materials development.

Soil bacteria may yet reveal another new mode of cellulose deconstruction. The cellulolytic soil bacterium *Cytophaga hutchinsonii* lacks the CBH proteins that are crucial to the classical mode of cellulose degradation described earlier in this chapter (Zhu and McBride 2017), but other proteins involved in the gliding motility of the species have been shown to be crucial for cellulose utilisation (Zhu and McBride 2014). It has been proposed that physical tethering to and gliding along the cellulose surface may be required for cellulose deconstruction by this species. Certainly, *C. hutchinsonii* uses a 'selfish' approach to cellulose degradation, releasing very little sugar into the external environment, suggesting close proximity between the active enzymes, the cell surface, and the substrate (Zhu and McBride 2017). Recent proteomic investigations of this cellulolytic system showed that members of a previously unexplored glycoside hydrolase family were abundant in cellulolytic proteomes, as were proteins that belong to no known classification, suggesting that there may be novel catalytic activities to be uncovered here (Taillefer et al. 2018).

An expansive enzymatic toolbox is therefore now available for use in LCW saccharification. The high cost of enzyme production remains the greatest stumbling block in the expansion of second-generation LCW bioethanol production, and therefore much research is focussed on improving the levels of enzyme produced by industrially relevant strains of fungi and bacteria. Many different soil microbes have been assessed as candidate enzyme producers for biotechnology, but here we will focus on just two important species: the fungus *Trichoderma reesei* and the bacterium *Bacillus subtilis*.

The ascomycete *T. reesei* is a true industrial workhorse for cellulose and biomass-degrading enzyme production (Schuster and Schmoll 2010). Trichoderma species are ubiquitous colonisers of cellulosic biomass in soil and the rhizosphere, where they deploy a synergistic mixture of secreted cellulases (cellobiohydrolases and *endo*-glucanases) for effective cellulose deconstruction (Henrissat et al. 1985). The species is also an effective degrader of lignin and the important hemicellulose xylan, underlining its powerful role in industrial biomass saccharification (Bischof et al. 2016). Annotation of the *T. reesei* genome revealed over 200 carbohydrate-active enzymes, a great many of which contribute to cellulose or xylan degradation (Hakkinen et al. 2012). Later studies made the important discovery that growing *T. reesei* on different carbohydrates or types of biomass led to changes in the profile of enzymes secreted, allowing the induced secretion of cocktails optimised for specific processes (Aro et al. 2005; Dos Santos Castro et al. 2014). Strain improvements have since increased the industrial standard of protein secretion to approximately 100 grams per litre during cultivation on cellulose or plant biomass, of which 60–80% may be cellulolytic enzymes (Schuster and Schmoll 2010). Metabolic engineering is now developing strains that can hyper-produce cellulolytic cocktails during growth on more simple carbon sources such as lactose (Li et al. 2017; de Paula et al. 2018).

Due to the huge body of research on gene expression and protein secretion in *Trichoderma* species (Conesa et al. 2001), *T. reesei* is now the most commonly used filamentous fungus for production of recombinant proteins, often utilising the native promoter of the cellobiohydrolase II (CBHII) (Meng et al. 2013). New tools are constantly being developed to allow the expression of recombinant genes in fungi (Cherry and Fidantsef 2003; Rantasalo et al. 2018) and very high yields are possible.

However, there is evidence that current techniques may have reached the upper limits of the super-producer, as no significant new increases in protein yield have been reported for some time (Nevalainen and Peterson 2014).

Gram-positive bacteria can also secrete very high concentrations of proteins into fermentation broth and are often used to produce industrial quantities of proteases, lipases, and carbohydrate-degrading enzymes (van Dijl and Hecker 2013). *B. subtilis* could be said to represent the greatest hope for 'next-generation' protein secreting cell factories. This soil microbe has the capacity to secrete some proteins in the grams per litre range (Earl et al. 2008), and the possibility for genetic engineering makes *B. subtilis* an ideal research and production host (Cui et al. 2018; Liu et al. 2019). Physiological, genetic, and proteomic investigations of the bacterium have shown that it is highly adaptable to changes in environmental conditions and substrate availability (Otto et al. 2010; Buescher et al. 2012; Nicolas et al. 2012). This adaptability underpins the microbe's success in its natural environment and in the industrial protein production context. The majority of the proteins naturally secreted by *B. subtilis* have been identified by integrated physiological-proteomic investigations. One of the most relevant outcomes of these efforts was the development of a library of *B. subtilis* signal peptides, allowing highly effective cloning strategies to be designed for maximal secretion of recombinant proteins in this host (Brockmeier et al. 2006). Systems biology and synthetic biology techniques are now permitting even greater improvements in protein secretion by *B. subtilis*. For example, choice or design of signal peptides (Frobel et al. 2012), optimisation and engineering of expression promoters (Marciniak et al. 2012; Toymentseva et al. 2012), and whole-cell metabolic engineering (Kouwen et al. 2008) have been used to further increase production efficiency of this super-secretor, improving the overall economic performance of the system.

10.4.2 Biocontrol for Sustainable Agriculture

Low-input farming systems tend to employ a mixed crop-livestock system: a portion of the crops grown are used to feed animals, whose manure is used to fertilise soil and improve crop growth. The result is a close to zero input of external carbon and very high retention of soil organic carbon (SOC) (Gattinger et al. 2012). Conventional agriculture, by contrast, has separated crop and livestock production, requiring the import of animal feed or synthetic fertiliser and involving the sale and export of any manure produced. The result is a much greater flux of carbon on- and off-site and an overall degradation of SOC stocks. For agriculture to be sustainable, soil management practices must consider the functioning of the microbiota and aim to preserve SOC and microbial abundance, activity, and diversity.

Microbial functionality is largely mediated via the proteins and enzymes secreted by soil organisms. Again, we will use the examples of *Trichoderma* fungi and *Bacillus* bacteria to summarise current technology in biocontrol research, as these have been the focus of the majority of scientific research and attempts at commercial exploitation. *Trichoderma* species are able to control the growth of certain ascomycetes, basidiomycetes, and oomycetes, and there are even reports suggesting they can inhibit the growth of nematodes (Sharma and Gothalwal 2017). *Trichoderma* spp. can utilise secreted lytic and proteolytic (Kredics et al. 2005; Chen et al. 2009) enzymes to attack other organisms, in addition to a range of secondary metabolites (Reino et al. 2008) that often impair pathogen growth by damaging cell wall structures or by interfering with membrane integrity (Kubicek et al. 2001; Benitez et al. 2004). Complementing their abilities to control pathogen growth, many *Trichoderma* species are also capable of colonising plant roots and interacting in such a way that systemic resistance is enhanced within the plant (Schuster and Schmoll 2010). Products are now commercially available that provide spores of *Trichoderma* spp. for use as soil amendments, but there is scant knowledge regarding the long-term effects on the microbial community composition or function, or how stable any changes might be.

In addition to the PGPR processes discussed above that allow bacteria to quite directly promote plant health, there are also ways that soil bacteria can reduce the incidence of fungal- or nematode-borne plant diseases. A major route to this protection is via the secretion of bacterial chitinase enzymes that attack the tough chitin-rich outer cell wall that protects fungal cells (Patil et al. 2000). It is also now quite well accepted that mixing chitin or chitosan into the soil can lead to at least a transient increase in soil suppressiveness, which is the ability of the soil microbiota to prevent plant disease. This suggests that the ability of soil bacteria to mitigate fungal attack on plants is enhanced by their sensing of fungal cell wall components. In laboratory conditions, supplementation with chitin or chitosan leads to increased chitinase ability in cultures of *B. cereus* (Liang et al. 2013). Adding chitin into soil samples may favour the growth of *B. cereus*, and the chitin-oligosaccharides then produced by the bacterium may lead to some additional diversity in the microbiota, contributing to a positive feedback loop that promotes soil suppressiveness and soil fertility. The impact of chitin amendments on soil microbiota function is only recently being explored in mechanistic detail. While the addition of chitin or chitosan can lead to a moderate reduction in the overall numbers of living bacteria within soil, it also causes profound alterations in the relative abundance of certain taxa, particularly the *Oxalobacteria* and *Actinobacteria* (Cretoiu et al. 2013; Cretoiu et al. 2014). This is sure to have an impact on the overall microbiota function, but, again, much research remains to be done for us to understand the long-term stability of this population shift.

10.4.3 The Need for Systems Thinking in Sustainable Industrial Development

Above we have described cases where soil microbes or their secreted enzymes are contributing to the development of novel industrial or agricultural practices. This is allowing processes to be designed with sustainability at their heart, notably by permitting more efficient use of natural resources. It is often considered to be inherently a good idea to move away from petroleum for fuel and materials production, and soil microbes are helping us to find new ways of using abundant renewable biomass instead. However, long-term considerations of the life cycle of a process are still necessary, even when an apparently sustainable raw material substitutes for petroleum.

The case of first-generation bioethanol fuels highlights how an apparent environmental good – moving away from dependence on fossil fuels – can in fact cause negative environmental and social impacts. Spurred by high ethanol yields and temporarily high oil prices, as well as by governmental subsidies that obfuscated economic reality, the corn ethanol industry was starting to thrive in the 1990s. At a certain point, the true impact of corn ethanol on food prices began to be felt, precipitating a closer look at the overall sustainability of the production process. The environmental impacts were also not as clear-cut as first believed, largely due to carbon emissions arising from land conversion to farm feedstock plants (von Blottnitz and Curran 2007). Furthermore, the impact on food security caused the UN to label biofuel a ‘crime against humanity’.

The move towards second-generation LCW fuels sidesteps the issue of food diversion, but here again the environmental impacts are complex and hard to accurately measure (Erakhrumen 2014). Life cycle assessments of the process show that there is significant variation in impact due to differing feedstocks, waste disposal routes, and regional variability in the resources used to generate electricity (Borrion et al. 2012). Global production of LCW is estimated at 10–50 billion tonnes per year (Das and Singh 2004; Taha et al. 2016), and its production is globally distributed, enabling a more equitable and socially sustainable production infrastructure than petroleum permits. The most abundant agricultural wastes globally are wheat and rice straw, but decentralised production will require that each geographical area utilises the most abundant local resource: this could be forestry residue in regions such as Scandinavia or parts of North America, and may be food waste in urban areas. Generally speaking, the carbon emissions profile of LCW bioethanol is almost always favourable as a replacement for fossil fuels for road transport, when only production and use of the fuels are included in analyses. However, other ecological factors complicate this issue. For example, the very definition of ‘waste’ biomass in agriculture is controversial. The removal of lignocellulosic harvest residues from the field represents a removal of carbon and nutrients that would otherwise be worked back into the soil, contributing to long-term fertility and crop productivity (Lal 2009). True sustainability of LCW bioethanol production demands that the stability of SOC and nutrient contents are maintained. This might be achievable by returning crop harvest residues to the soil, a practice that is common in North American agriculture but seems to be less so in other parts of the world. In many places, LCW is simply incinerated or disposed of as a solid waste and artificial fertilisers are then used to supplement nutrient depleted soils (Graham et al. 2002; Govaerts et al. 2008; Lal 2009). Of course, the production and transport of synthetic inorganic fertilisers depletes non-renewable resources such as phosphate ore and natural gas and has a substantial carbon footprint in addition to the problem of nutrient-rich runoff into local groundwater (Smil 2000; Dawson and Hilton 2011). In contrast, return of harvest residues to the soil contributes to organic carbon content, nutrient content, and important physical parameters such as water retention and resistance to erosion.

Studies on LCW ethanol production from corn and wheat residues indicate that around 25% of crop residue can be removed from the field in strong harvest years, without damaging the carbon and nutrient contents of the soil to the extent that crop productivity is affected (Blanco-Canqui and Lal 2009; Lemke et al. 2010). When residue is diverted for biofuel production, it has been proposed that residues of the biofuel production process itself may be returned to the soil in order to replenish SOC (Reijnders 2013). However, the development of biorefinery processes has tended to focus on circular production practices that make internal use of waste streams, utilising organic residues for energy generation or the production of high-value products such as biomaterials or metabolic compounds (Zhang 2011; Fahd et al. 2012). Due to the practicalities of infrastructure development, cellulosic waste from bioethanol production rather tends to be incinerated or transferred to landfill or wastewater streams.

Additionally, it is still not clear that returning biofuel production residues to the soil is an effective means of maintaining soil carbon and nutrient stocks. Residues tend to be enriched in lignin, which turns over much more rapidly in SOM than do carbohydrates (Smith 2012). Furthermore, residues are often enriched in phenols and heavy metals (Blanco-Canqui and Lal 2009) with antimicrobial activity (Dong et al. 2011) and display phytotoxicity related to the presence of fatty acids and phenolic compounds (Gell et al. 2011). All of these factors could alter the microbiota composition in ways that are not helpful to soil fertility. Finally, the issue is further complicated as production waste biomass deriving from genetically modified *T. reesei* strains often contributes to the production residue. The release of such material into the environment is legally complex and in any case may contain high levels of mycotoxin (Frisvad et al. 2018). This aspect of the bioethanol production system must be urgently addressed to ensure process sustainability at commercial scales.

10.5 Conclusions and Future Prospects

The role of the soil microbiota in maintaining soil health and fertility is well established. We rely on the microbiota for the degradation of both organic compounds and xenobiotic pollutants. Indeed, the degradation of organic material by the microbial community plays a crucial role in global carbon cycling. Additionally, microbial enzymes play important roles in agriculture, industry, and bioremediation. Given the inevitable intensification of agriculture, the accumulation of pollutants in the environment, and the need for environmentally friendly energy sources and materials, our dependence on the microbial community is sure to increase. Exploiting this resource requires recognition of both its potential and its limitations. There are many factors that limit microbial enzyme activity levels including pH, temperature, and substrate availability. Ultimately, the energy balance of enzyme production and secretion must be favourable for the microbe to ensure survival within the community.

While many microbes have been identified through large sequencing projects, full functional characterisation of communities remains rare. Additionally, the effects of different soil conditions on the microbiota as a whole remain poorly understood, in part due to the complexity of these communities. There is the potential to develop novel products that support the growth of phyla with desirable traits such as plant growth promotion or phytopathogen suppression, both of which rely on secreted microbial enzymes. Furthermore, there are examples, such as the addition of chitin to soil as a means to promote the growth of antifungal bacteria and thereby reduce fungal disease in plants, which support the promise of harnessing this resource (Silvia Cretoiu et al. 2013). However, to address these challenges, we must improve our understanding through both basic and applied research. A long-term perspective is essential if we hope to both fully utilise and protect the incredible soil microbial communities present around the globe.

Acknowledgements This chapter can be considered as a successor to Richard G. Burns's earlier contribution to the Springer book *Molecular Environmental Soil Science*. Richard's chapter, entitled Microbial Extracellular Enzymes and the Degradation of Natural and Synthetic Polymers in Soil (Burns 2013), is an excellent summary of the field. We are also grateful to Dr. Vaibhav Srivastava of KTH Glycoscience for providing helpful feedback on our manuscript.

References

Agger JW, Isaksen T, Varnai A, Vidal-Melgosa S, Willats WG, Ludwig R, Horn SJ, Eijsink VG, Westereng B (2014) Discovery of LPMO activity on hemicelluloses shows the importance of oxidative processes in plant cell wall degradation. Proc Natl Acad Sci U S A 111(17):6287–6292

Allison SD (2005) Cheaters, diffusion and nutrients constrain decomposition by microbial enzymes in spatially structured environments. Ecol Lett 8(6):626–635

Allison SD, Martiny JBH (2008) Resistance, resilience, and redundancy in microbial communities. Proc Natl Acad Sci USA 105(Suppl 1):11512

Anbu P, Gopinath SCB, Chaulagain BP, Lakshmipriya T (2017) Microbial enzymes and their applications in industries and medicine 2016. Biomed Res Int 2017:2195808–2195808

Arnling Bååth J, Mazurkewich S, Knudsen RM, Poulsen JN, Olsson L, Lo Leggio L, Larsbrink J (2018) Biochemical and structural features of diverse bacterial glucuronoyl esterases facilitating recalcitrant biomass conversion. Biotechnol Biofuels 11:213

Aro N, Pakula T, Penttila M (2005) Transcriptional regulation of plant cell wall degradation by filamentous fungi. FEMS Microbiol Rev 29(4):719–739

Barret M, Carrere H, Delgadillo L, Patureau D (2010) PAH fate during the anaerobic digestion of contaminated sludge: do bioavailability and/or cometabolism limit their biodegradation? Water Res 44(13):3797–3806

Bedoya K, Coltell-Simon O, Cabarcas F, Alzate JF (2019) Metagenomic assessment of the microbial community and methanogenic pathways in biosolids from a municipal wastewater treatment plant in Medellin, Colombia. Sci Total Environ 648:572–581

Bell CW, Acosta-Martinez V, McIntyre NE, Cox S, Tissue DT, Zak JC (2009) Linking microbial community structure and function to seasonal differences in soil moisture and temperature in a Chihuahuan desert grassland. Microb Ecol 58(4):827–842

Benitez T, Rincon AM, Limon MC, Codon AC (2004) Biocontrol mechanisms of *Trichoderma* strains. Int Microbiol 7(4):249–260

Berini F, Presti I, Beltrametti F, Pedroli M, Varum KM, Pollegioni L, Sjoling S, Marinelli F (2017) Production and characterization of a novel antifungal chitinase identified by functional screening of a suppressive-soil metagenome. Microb Cell Factories 16(1):16

Bhattacharyya PN, Jha DK (2012) Plant growth-promoting rhizobacteria (PGPR): emergence in agriculture. World J Microbiol Biotechnol 28(4):1327–1350

Bischof RH, Ramoni J, Seiboth B (2016) Cellulases and beyond: the first 70 years of the enzyme producer *Trichoderma reesei*. Microb Cell Factories 15(1):106

Bissaro B, Varnai A, Røhr ÅK, Eijsink VGH (2018) Oxidoreductases and reactive oxygen species in conversion of lignocellulosic biomass. Microbiol Mol Biol Rev 82(4)

Blanco-Canqui H, Lal R (2009) Crop residue removal impacts on soil productivity and environmental quality. Crit Rev Plant Sci 28(3):139–163

Blanvillain S, Meyer D, Boulanger A, Lautier M, Guynet C, Denancé N, Vasse J, Lauber E, Arlat M (2007) Plant carbohydrate scavenging through TonB-dependent receptors: a feature shared by phytopathogenic and aquatic bacteria. PLoS One 2(2):e224

Borrion AL, McManus MC, Hammond GP (2012) Environmental life cycle assessment of lignocellulosic conversion to ethanol: a review. Renew Sust Energ Rev 16(7):4638–4650

Brabcova V, Novakova M, Davidova A, Baldrian P (2016) Dead fungal mycelium in forest soil represents a decomposition hotspot and a habitat for a specific microbial community. New Phytol 210(4):1369–1381

Brockmeier U, Caspers M, Freudl R, Jockwer A, Noll T, Eggert T (2006) Systematic screening of all signal peptides from *Bacillus subtilis*: a powerful strategy in optimizing heterologous protein secretion in Gram-positive bacteria. J Mol Biol 362(3):393–402

Buescher JM, Liebermeister W, Jules M, Uhr M, Muntel J, Botella E, Hessling B, Kleijn RJ, Le Chat L, Lecointe F, Mader U, Nicolas P, Piersma S, Rugheimer F, Becher D, Bessieres P, Bidnenko E, Denham EL, Dervyn E, Devine KM, Doherty G, Drulhe S, Felicori L, Fogg MJ, Goelzer A, Hansen A, Harwood CR, Hecker M, Hubner S, Hultschig C, Jarmer H, Klipp E, Leduc A, Lewis P, Molina F, Noirot P, Peres S, Pigeonneau N, Pohl S, Rasmussen S, Rinn B, Schaffer M, Schnidder J, Schwikowski B, Van Dijl JM, Veiga P, Walsh S, Wilkinson AJ, Stelling J, Aymerich S, Sauer U (2012) Global network reorganization during dynamic adaptations of *Bacillus subtilis* metabolism. Science 335(6072):1099–1103

Burns RG (2013) Microbial extracellular enzymes and the degradation of natural and synthetic polymers in soil. In: Xu J, Sparks DL (eds) Molecular environmental soil science. Springer, Dordrecht

Cachada A, Pereira R, da Silva EF, Duarte AC (2014) The prediction of PAHs bioavailability in soils using chemical methods: state of the art and future challenges. Sci Total Environ 472:463–480

Carvalho AKF, Bento HBS, Reis CER, De Castro HF (2019) Sustainable enzymatic approaches in a fungal lipid biorefinery based in sugarcane bagasse hydrolysate as carbon source. Bioresour Technol 276:269–275

Carvalho AL, Dias FM, Prates JA, Nagy T, Gilbert HJ, Davies GJ, Ferreira LM, Romao MJ, Fontes CM (2003) Cellulosome assembly revealed by the crystal structure of the cohesin-dockerin complex. Proc Natl Acad Sci U S A 100(24):13809–13814

Chen LL, Liu LJ, Shi M, Song XY, Zheng CY, Chen XL, Zhang YZ (2009) Characterization and gene cloning of a novel serine protease with nematicidal activity from *Trichoderma pseudokoningii* SMF2. FEMS Microbiol Lett 299(2):135–142

Cherry JR, Fidantsef AL (2003) Directed evolution of industrial enzymes: an update. Curr Opin Biotechnol 14(4):438–443

Conesa A, Punt PJ, van Luijk N, van den Hondel CA (2001) The secretion pathway in filamentous fungi: a biotechnological view. Fungal Genet Biol 33(3):155–171

Cretoiu MS, Kielak AM, Schluter A, van Elsas JD (2014) Bacterial communities in chitin-amended soil as revealed by 16S rRNA gene based pyrosequencing. Soil Biol Biochem 76:5–11

Cretoiu MS, Korthals GW, Visser JHM, van Elsas JD (2013) Chitin amendment increases soil suppressiveness toward plant pathogens and modulates the Actinobacterial and Oxalobacteraceal communities in an experimental agricultural field. Appl Environ Microbiol 79(17):5291

Cui W, Han L, Suo F, Liu Z, Zhou L, Zhou Z (2018) Exploitation of *Bacillus subtilis* as a robust workhorse for production of heterologous proteins and beyond. World J Microbiol Biotechnol 34(10):145

Cuskin F, Lowe EC, Temple MJ, Zhu Y, Cameron E, Pudlo NA, Porter NT, Urs K, Thompson AJ, Cartmell A, Rogowski A, Hamilton BS, Chen R, Tolbert TJ, Piens K, Bracke D, Vervecken W, Hakki Z, Speciale G, Munoz-Munoz JL, Day A, Pena MJ, McLean R, Suits MD, Boraston AB, Atherly T, Ziemer CJ, Williams SJ, Davies GJ, Abbott DW, Martens EC, Gilbert HJ (2015) Human gut Bacteroidetes can utilize yeast mannan through a selfish mechanism. Nature 517(7533):165–169

Das H, Singh SK (2004) Useful byproducts from cellulosic wastes of agriculture and food industry--a critical appraisal. Crit Rev Food Sci Nutr 44(2):77–89

Dawson CJ, Hilton J (2011) Fertiliser availability in a resource-limited world: production and recycling of nitrogen and phosphorus. Food Policy 36:S14–S22

de Paula RG, Antoniêto ACC, Ribeiro LFC, Carraro CB, Nogueira KMV, Lopes DCB, Silva AC, Zerbini MT, Pedersoli WR, Costa MdN, Silva RN (2018) New genomic approaches to enhance biomass degradation by the industrial fungus *Trichoderma reesei*. Int J Genomics 2018:1974151–1974151

De-la-Peña C, Lei Z, Watson BS, Sumner LW, Vivanco JM (2008) Root-microbe communication through protein secretion. J Biol Chem 283(37):25247–25255

D'Elia JN, Salyers AA (1996) Contribution of a neopullulanase, a pullulanase, and an alpha-glucosidase to growth of *Bacteroides thetaiotaomicron* on starch. J Bacteriol 178(24):7173–7179

Deveau A, Gross H, Palin B, Mehnaz S, Schnepf M, Leblond P, Dorrestein PC, Aigle B (2016) Role of secondary metabolites in the interaction between *Pseudomonas fluorescens* and soil microorganisms under iron-limited conditions. FEMS Microbiol Ecol 92(8)

Dong X, Dong M, Lu Y, Turley A, Jin T, Wu C (2011) Antimicrobial and antioxidant activities of lignin from residue of corn Stover to ethanol production. Ind Crop Prod 34(3):1629–1634

Dos Santos Castro L, Pedersoli WR, Antonieto ACC, Steindorff AS, Silva-Rocha R, Martinez-Rossi NM, Rossi A, Brown NA, Goldman GH, Faca VM, Persinoti GF, Silva RN (2014) Comparative metabolism of cellulose, sophorose and glucose in *Trichoderma reesei* using high-throughput genomic and proteomic analyses. Biotechnol Biofuels 7(1):41

Dries J, Smets BF (2002) Transformation and mineralization of benzo[a]pyrene by microbial cultures enriched on mixtures of three- and four-ring polycyclic aromatic hydrocarbons. J Ind Microbiol Biotechnol 28(2):70–73

Dungait JAJ, Hopkins DW, Gregory AS, Whitmore AP (2012) Soil organic matter turnover is governed by accessibility not recalcitrance. Glob Chang Biol 18(6):1781–1796

Earl AM, Losick R, Kolter R (2008) Ecology and genomics of *Bacillus subtilis*. Trends Microbiol 16(6):269–275

Eibinger M, Ganner T, Bubner P, Rošker S, Kracher D, Haltrich D, Ludwig R, Plank H, Nidetzky B (2014) Cellulose surface degradation by a lytic polysaccharide monooxygenase and its effect on cellulase hydrolytic efficiency. J Biol Chem 289(52):35929–35938

Erakhrumen AA (2014) Growing pertinence of bioenergy in formal/informal global energy schemes: necessity for optimising awareness strategies and increased investments in renewable energy technologies. Renew Sust Energ Rev 31:305–311

Fahd S, Fiorentino G, Mellino S, Ulgiati S (2012) Cropping bioenergy and biomaterials in marginal land: the added value of the biorefinery concept. Energy 37(1):79–93

Folse HJ 3rd, Allison SD (2012) Cooperation, competition, and coalitions in enzyme-producing microbes: social evolution and nutrient depolymerization rates. Front Microbiol 3:338

Frisvad JC, Moller LLH, Larsen TO, Kumar R, Arnau J (2018) Safety of the fungal workhorses of industrial biotechnology: update on the mycotoxin and secondary metabolite potential of *Aspergillus niger, Aspergillus oryzae*, and *Trichoderma reesei*. Appl Microbiol Biotechnol 102(22):9481–9515

Frobel J, Rose P, Lausberg F, Blummel AS, Freudl R, Muller M (2012) Transmembrane insertion of twin-arginine signal peptides is driven by TatC and regulated by TatB. Nat Commun 3:1311

Gattinger A, Muller A, Haeni M, Skinner C, Fliessbach A, Buchmann N, Mader P, Stolze M, Smith P, Scialabba Nel H, Niggli U (2012) Enhanced top soil carbon stocks under organic farming. Proc Natl Acad Sci U S A 109(44):18226–18231

Gell K, van Groenigen J, Cayuela ML (2011) Residues of bioenergy production chains as soil amendments: immediate and temporal phytotoxicity. J Hazard Mater 186(2):2017–2025

Ghosal D, Ghosh S, Dutta TK, Ahn Y (2016) Current state of knowledge in microbial degradation of polycyclic aromatic hydrocarbons (PAHs): a review. Front Microbiol 7:1369

Giacomello S, Salmen F, Terebieniec BK, Vickovic S, Navarro JF, Alexeyenko A, Reimegård J, McKee LS, Mannapperuma C, Bulone V, Stahl PL, Sundstrom JF, Street NR, Lundeberg J (2017) Spatially resolved transcriptome profiling in model plant species. Nat Plants 3:17061

Gilbert HJ (2007) Cellulosomes: microbial nanomachines that display plasticity in quaternary structure. Mol Microbiol 63(6):1568–1576

Gilbert HJ (2010) The biochemistry and structural biology of plant cell wall deconstruction. Plant Physiol 153(2):444–455

Gilbert HJ, Knox JP, Boraston AB (2013) Advances in understanding the molecular basis of plant cell wall polysaccharide recognition by carbohydrate-binding modules. Curr Opin Struct Biol 23(5):669–677

Gougoulias C, Clark JM, Shaw LJ (2014) The role of soil microbes in the global carbon cycle: tracking the below-ground microbial processing of plant-derived carbon for manipulating carbon dynamics in agricultural systems. J Sci Food Agric 94(12):2362–2371

Govaerts B, Mezzalama M, Sayre KD, Crossa J, Lichter K, Troch V, Vanherck K, De Corte P, Deckers J (2008) Long-term consequences of tillage, residue management, and crop rotation on selected soil micro-flora groups in the subtropical highlands. Appl Soil Ecol 38(3):197–210

Graham MH, Haynes RJ, Meyer JH (2002) Soil organic matter content and quality: effects of fertilizer applications, burning and trash retention on a long-term sugarcane experiment in South Africa. Soil Biol Biochem 34(1):93–102

Grondin JM, Tamura K, Dejean G, Abbott DW, Brumer H (2017) Polysaccharide utilization loci: fueling microbial communities. J Bacteriol 199(15)

Gunl M, Kraemer F, Pauly M (2011) Oligosaccharide mass profiling (OLIMP) of cell wall polysaccharides by MALDI-TOF/MS. Methods Mol Biol 715:43–54

Hadibarata T, Zubir MM, Rubiyatno, Chuang TZ (2013) Microbial transformation and sorption of anthracene in liquid culture. Bioprocess Biosyst Eng 36(9):1229–1233

Hakkinen M, Arvas M, Oja M, Aro N, Penttila M, Saloheimo M, Pakula TM (2012) Re-annotation of the CAZy genes of *Trichoderma reesei* and transcription in the presence of lignocellulosic substrates. Microb Cell Factories 11:134

Hartmann M, Frey B, Mayer J, Mader P, Widmer F (2015) Distinct soil microbial diversity under long-term organic and conventional farming. ISME J 9(5):1177–1194

Hemsworth GR, Dejean G, Davies GJ, Brumer H (2016) Learning from microbial strategies for polysaccharide degradation. Biochem Soc Trans 44(1):94–108

Henrissat B, Driguez H, Viet C, Schülein M (1985) Synergism of Cellulases from *Trichoderma reesei* in the degradation of cellulose. Bio/Technology 3:722

Herve C, Rogowski A, Blake AW, Marcus SE, Gilbert HJ, Knox JP (2010) Carbohydrate-binding modules promote the enzymatic deconstruction of intact plant cell walls by targeting and proximity effects. Proc Natl Acad Sci U S A 107(34):15293–15298

Horn SJ, Vaaje-Kolstad G, Westereng B, Eijsink VG (2012) Novel enzymes for the degradation of cellulose. Biotechnol Biofuels 5(1):45

Hu J, Tian D, Renneckar S, Saddler JN (2018) Enzyme mediated nanofibrillation of cellulose by the synergistic actions of an endoglucanase, lytic polysaccharide monooxygenase (LPMO) and xylanase. Sci Rep 8(1):3195

James K, Peters RE, Laird BD, Ma WK, Wickstrom M, Stephenson GL, Siciliano SD (2011) Human exposure assessment: a case study of 8 PAH contaminated soils using in vitro Digestors and the juvenile swine model. Environ Sci Technol 45(10):4586–4593

Jensen MS, Fredriksen L, MacKenzie AK, Pope PB, Leiros I, Chylenski P, Williamson AK, Christopeit P, Østby H, Vaaje-Kolstad G, Eijsink VGH (2018) Discovery and characterization of a thermostable two-domain GH6 endoglucanase from a compost metagenome. PLoS One 13(5):e0197862

Johnsen AR, de Lipthay JR, Sorensen SJ, Ekelund F, Christensen P, Andersen O, Karlson U, Jacobsen CS (2006) Microbial degradation of street dust polycyclic aromatic hydrocarbons in microcosms simulating diffuse pollution of urban soil. Environ Microbiol 8(3):535–545

Johnsen AR, Karlson U (2007) Diffuse PAH contamination of surface soils: environmental occurrence, bioavailability, and microbial degradation. Appl Microbiol Biotechnol 76(3):533–543

Kamimura N, Sakamoto S, Mitsuda N, Masai E, Kajita S (2018) Advances in microbial lignin degradation and its applications. Curr Opin Biotechnol 56:179–186

Katz M, Hover BM, Brady SF (2016) Culture-independent discovery of natural products from soil metagenomes. J Ind Microbiol Biotechnol 43(2–3):129–141

Kolton M, Frenkel O, Elad Y, Cytryn E (2014) Potential role of Flavobacterial gliding-motility and type IX secretion system complex in root colonization and plant defense. Mol Plant-Microbe Interact 27(9):1005–1013

Kouwen TR, Dubois JY, Freudl R, Quax WJ, van Dijl JM (2008) Modulation of thiol-disulfide oxidoreductases for increased production of disulfide-bond-containing proteins in *Bacillus subtilis*. Appl Environ Microbiol 74(24):7536–7545

Kredics L, Antal Z, Szekeres A, Hatvani L, Manczinger L, Vagvolgyi C, Nagy E (2005) Extracellular proteases of *Trichoderma* species. A review. Acta Microbiol Immunol Hung 52(2):169–184

Kubicek CP, Mach RL, Peterbauer CK, Lorito M (2001) *Trichoderma*: from genes to biocontrol. Plant Pathol 83:11–23

Kubicek CP, Starr TL, Glass NL (2014) Plant cell wall-degrading enzymes and their secretion in plant-pathogenic fungi. Annu Rev Phytopathol 52:427–451

Lal R (2009) Soil quality impacts of residue removal for bioethanol production. Soil Till Res 102(2):233–241

Larsbrink J, Rogers TE, Hemsworth GR, McKee LS, Tauzin AS, Spadiut O, Klinter S, Pudlo NA, Urs K, Koropatkin NM, Creagh AL, Haynes CA, Kelly AG, Cederholm SN, Davies GJ, Martens EC, Brumer H (2014a) A discrete genetic locus confers xyloglucan metabolism in select human gut Bacteroidetes. Nature 506(7489):498–502

Larsbrink J, Thompson AJ, Lundqvist M, Gardner JG, Davies GJ, Brumer H (2014b) A complex gene locus enables xyloglucan utilization in the model saprophyte *Cellvibrio japonicus*. Mol Microbiol 94(2):418–433

Larsbrink J, Tuveng TR, Pope PB, Bulone V, Eijsink VG, Brumer H, McKee LS (2017) Proteomic insights into mannan degradation and protein secretion by the forest floor bacterium *Chitinophaga pinensis*. J Proteome 156:63–74

Larsbrink J, Zhu Y, Kharade SS, Kwiatkowski KJ, Eijsink VG, Koropatkin NM, McBride MJ, Pope PB (2016) A polysaccharide utilization locus from *Flavobacterium johnsoniae* enables conversion of recalcitrant chitin. Biotechnol Biofuels 9:260

Lemke RL, VandenBygaart AJ, Campbell CA, Lafond GP, Grant B (2010) Crop residue removal and fertilizer N: effects on soil organic carbon in a long-term crop rotation experiment on a Udic Boroll. Agric Ecosyst Environ 135(1):42–51

Li C, Lin F, Zhou L, Qin L, Li B, Zhou Z, Jin M, Chen Z (2017) Cellulase hyper-production by *Trichoderma reesei* mutant SEU-7 on lactose. Biotechnol Biofuels 10(1):228

Li X, Lin X, Zhang J, Wu Y, Yin R, Feng Y, Wang Y (2010) Degradation of polycyclic aromatic hydrocarbons by crude extracts from spent mushroom substrate and its possible mechanisms. Curr Microbiol 60(5):336–342
Liang TW, Liu CP, Wu C, Wang SL (2013) Applied development of crude enzyme from *Bacillus cereus* in prebiotics and microbial community changes in soil. Carbohydr Polym 92(2):2141–2148
Liu Y, Liu L, Li J, Du G, Chen J (2019) Synthetic biology toolbox and chassis development in *Bacillus subtilis*. Trends Biotechnol 37:548–562
Lopes AM, Ferreira Filho EX, Moreira LRS (2018) An update on enzymatic cocktails for lignocellulose breakdown. J Appl Microbiol 125(3):632–645
Marciniak BC, Trip H, van der Veek PJ, Kuipers OP (2012) Comparative transcriptional analysis of *Bacillus subtilis* cells overproducing either secreted proteins, lipoproteins or membrane proteins. Microb Cell Factories 11:66
Martens EC, Koropatkin NM, Smith TJ, Gordon JI (2009) Complex glycan catabolism by the human gut microbiota: the Bacteroidetes Sus-like paradigm. J Biol Chem 284(37):24673–24677
Martínez-Abad A, Berglund J, Toriz G, Gatenholm P, Henriksson G, Lindstrom M, Wohlert J, Vilaplana F (2017) Regular motifs in Xylan modulate molecular flexibility and interactions with cellulose surfaces. Plant Physiol 175(4):1579–1592
Martínez-Abad A, Ruthes AC, Vilaplana F (2015) Enzymatic-assisted extraction and modification of lignocellulosic plant polysaccharides for packaging applications. J Appl Polym Sci 133(2)
McBride MJ, Xie G, Martens EC, Lapidus A, Henrissat B, Rhodes RG, Goltsman E, Wang W, Xu J, Hunnicutt DW, Staroscik AM, Hoover TR, Cheng YQ, Stein JL (2009) Novel features of the polysaccharide-digesting gliding bacterium *Flavobacterium johnsoniae* as revealed by genome sequence analysis. Appl Environ Microbiol 75(21):6864–6875
McCartney L, Gilbert HJ, Bolam DN, Boraston AB, Knox JP (2004) Glycoside hydrolase carbohydrate-binding modules as molecular probes for the analysis of plant cell wall polymers. Anal Biochem 326(1):49–54
McKee LS, Martinez-Abad A, Ruthes AC, Vilaplana F, Brumer H (2019) Focussed metabolism of beta-glucans by the soil Bacteroidetes *Chitinophaga pinensis*. Appl Environ Microbiol 85(2):pii: e02231-18
McKee LS, Sunner H, Anasontzis GE, Toriz G, Gatenholm P, Bulone V, Vilaplana F, Olsson L (2016) A GH115 alpha-glucuronidase from *Schizophyllum* commune contributes to the synergistic enzymatic deconstruction of softwood glucuronoarabinoxylan. Biotechnol Biofuels 9:2
Meng F, Wei D, Wang W (2013) Heterologous protein expression in *Trichoderma reesei* using the cbhII promoter. Plasmid 70(2):272–276
Meulenberg R, Rijnaarts HH, Doddema HJ, Field JA (1997) Partially oxidized polycyclic aromatic hydrocarbons show an increased bioavailability and biodegradability. FEMS Microbiol Lett 152(1):45–49
Moorhead DL, Rinkes ZL, Sinsabaugh RL, Weintraub MN (2013) Dynamic relationships between microbial biomass, respiration, inorganic nutrients and enzyme activities: informing enzyme-based decomposition models. Front Microbiol 4:223
Ndeh D, Gilbert HJ (2018) Biochemistry of complex glycan depolymerisation by the human gut microbiota. FEMS Microbiol Rev 42(2):146–164
Nelson CE, Attia MA, Rogowski A, Morland C, Brumer H, Gardner JG (2017) Comprehensive functional characterization of the glycoside hydrolase family 3 enzymes from *Cellvibrio japonicus* reveals unique metabolic roles in biomass saccharification. Environ Microbiol 19(12):5025–5039
Nelson MS, Sadowsky MJ (2015) Secretion systems and signal exchange between nitrogen-fixing rhizobia and legumes. Front Plant Sci 6(491)
Nevalainen H, Peterson R (2014) Making recombinant proteins in filamentous fungi- are we expecting too much? Front Microbiol 5:75–75
Nicolas P, Mader U, Dervyn E, Rochat T, Leduc A, Pigeonneau N, Bidnenko E, Marchadier E, Hoebeke M, Aymerich S, Becher D, Bisicchia P, Botella E, Delumeau O, Doherty G, Denham

EL, Fogg MJ, Fromion V, Goelzer A, Hansen A, Hartig E, Harwood CR, Homuth G, Jarmer H, Jules M, Klipp E, Le Chat L, Lecointe F, Lewis P, Liebermeister W, March A, Mars RA, Nannapaneni P, Noone D, Pohl S, Rinn B, Rugheimer F, Sappa PK, Samson F, Schaffer M, Schwikowski B, Steil L, Stulke J, Wiegert T, Devine KM, Wilkinson AJ, van Dijl JM, Hecker M, Volker U, Bessieres P, Noirot P (2012) Condition-dependent transcriptome reveals high-level regulatory architecture in *Bacillus subtilis*. Science 335(6072):1103–1106

Nitsos C, Lazaridis P, Mach-Aigner A, Matis K, Triantafyllidis K (2019) Increasing the efficiency of lignocellulosic biomass enzymatic hydrolysis: hydrothermal pretreatment, extraction of surface lignin, wet milling and production of cellulolytic enzymes. ChemSusChem 12:1179

Okere UV, Schuster JK, Ogbonnaya UO, Jones KC, Semple KT (2017) Indigenous (14) C-phenanthrene biodegradation in "pristine" woodland and grassland soils from Norway and the United Kingdom. Environ Sci Process Impacts 19(11):1437–1444

Ortseifen V, Stolze Y, Maus I, Sczyrba A, Bremges A, Albaum SP, Jaenicke S, Fracowiak J, Puhler A, Schluter A (2016) An integrated metagenome and -proteome analysis of the microbial community residing in a biogas production plant. J Biotechnol 231:268–279

Otto A, Bernhardt J, Meyer H, Schaffer M, Herbst F-A, Siebourg J, Mäder U, Lalk M, Hecker M, Becher D (2010) Systems-wide temporal proteomic profiling in glucose-starved *Bacillus subtilis*. Nat Commun 1:137

Patil RS, Ghormade V, Deshpande MV (2000) Chitinolytic enzymes: an exploration. Enzyme Microb Technol 26(7):473–483

Pollegioni L, Tonin F, Rosini E (2015) Lignin-degrading enzymes. FEBS J 282(7):1190–1213

Qasemian L, Guiral D, Ziarelli F, Ruaudel F, Farnet AM (2012) Does anthracene affect microbial activities and organic matter decomposition? A comparative study in *Pinus halepensis* litters from Mediterranean coastal and inland areas. Chemosphere 89(5):548–555

Quinlan RJ, Sweeney MD, Lo Leggio L, Otten H, Poulsen JC, Johansen KS, Krogh KB, Jorgensen CI, Tovborg M, Anthonsen A, Tryfona T, Walter CP, Dupree P, Xu F, Davies GJ, Walton PH (2011) Insights into the oxidative degradation of cellulose by a copper metalloenzyme that exploits biomass components. Proc Natl Acad Sci U S A 108(37):15079–15084

Rantasalo A, Landowski CP, Kuivanen J, Korppoo A, Reuter L, Koivistoinen O, Valkonen M, Penttila M, Jantti J, Mojzita D (2018) A universal gene expression system for fungi. Nucleic Acids Res 46(18):e111

Raveendran S, Parameswaran B, Ummalyma SB, Abraham A, Mathew AK, Madhavan A, Rebello S, Pandey A (2018) Applications of microbial enzymes in food industry. Food Technol Biotech 56(1):16–30

Reeves AR, Wang GR, Salyers AA (1997) Characterization of four outer membrane proteins that play a role in utilization of starch by *Bacteroides thetaiotaomicron*. J Bacteriol 179(3):643–649

Reijnders L (2013) Sustainability of soil fertility and the use of lignocellulosic crop harvest residues for the production of biofuels: a literature review. Environ Technol 34(13–14):1725–1734

Reino JL, Guerrero RF, Hernández-Galán R, Collado IG (2008) Secondary metabolites from species of the biocontrol agent *Trichoderma*. Phytochem Rev 7(1):89–123

Rinkes ZL, Weintraub MN, DeForest JL, Moorhead DL (2011) Microbial substrate preference and community dynamics during decomposition of *Acer saccharum*. Fungal Ecol 4(6):396–407

Robak K, Balcerek M (2018) Review of second generation bioethanol production from residual biomass. Food Technol Biotech 56(2):174–187

Rogers SW, Ong SK, Kjartanson BH, Golchin J, Stenback GA (2002) Natural attenuation of polycyclic aromatic hydrocarbon-contaminated sites. Pract Period Hazard Toxic Radioact Waste Manage 6(3):141–155

Saraf M, Pandya U, Thakkar A (2014) Role of allelochemicals in plant growth promoting rhizobacteria for biocontrol of phytopathogens. Microbiol Res 169(1):18–29

Sawulski P, Clipson N, Doyle E (2014) Effects of polycyclic aromatic hydrocarbons on microbial community structure and PAH ring hydroxylating dioxygenase gene abundance in soil. Biodegradation 25(6):835–847

Schimel J, Schaeffer S (2012) Microbial control over carbon cycling in soil. Front Microbiol 3(348)

Schuster A, Schmoll M (2010) Biology and biotechnology of *Trichoderma*. Appl Microbiol Biotechnol 87(3):787–799

Sharma PK, Gothalwal R (2017) Trichoderma: a potent fungus as biological control agent. In: Singh JS, Seneviratne G (eds) Agro-environmental sustainability: volume 1: managing crop health. Springer, Cham, pp 113–125

Shipman JA, Berleman JE, Salyers AA (2000) Characterization of four outer membrane proteins involved in binding starch to the cell surface of *Bacteroides thetaiotaomicron*. J Bacteriol 182(19):5365–5372

Silvia Cretoiu M, Korthals GW, Visser JHM, van Elsas JD (2013) Chitin amendment increases soil suppressiveness toward plant pathogens and modulates the Actinobacterial and Oxalobacteraceal communities in an experimental agricultural field. Appl Environ Microbiol 79(17):5291–5301

Smil V (2000) PHOSPHORUS IN THE ENVIRONMENT: natural flows and human interferences. Annu Rev Energ Environ 25(1):53–88

Smith P (2012) Soils and climate change. Curr Opin Env Sust 4(5):539–544

Srivastava V, McKee LS, Bulone V (2017) Plant cell walls, eLS: essentials for life science. Chichester, Wiley

Strakowska J, Blaszczyk L, Chelkowski J (2014) The significance of cellulolytic enzymes produced by *Trichoderma* in opportunistic lifestyle of this fungus. J Basic Microbiol 54(Suppl 1):S2–S13

Taha M, Foda M, Shahsavari E, Aburto-Medina A, Adetutu E, Ball A (2016) Commercial feasibility of lignocellulose biodegradation: possibilities and challenges. Curr Opin Biotechnol 38:190–197

Taillefer M, Arntzen MØ, Henrissat B, Pope PB, Larsbrink J (2018) Proteomic dissection of the cellulolytic machineries used by soil-dwelling bacteroidetes. mSystems 3(6)

Tandrup T, Frandsen KEH, Johansen KS, Berrin JG, Lo Leggio L (2018) Recent insights into lytic polysaccharide monooxygenases (LPMOs). Biochem Soc Trans 46(6):1431–1447

Tilman D, Cassman KG, Matson PA, Naylor R, Polasky S (2002) Agricultural sustainability and intensive production practices. Nature 418:671

Toymentseva AA, Schrecke K, Sharipova MR, Mascher T (2012) The LIKE system, a novel protein expression toolbox for *Bacillus subtilis* based on the liaI promoter. Microb Cell Factories 11(1):143

Traving SJ, Thygesen UH, Riemann L, Stedmon CA (2015) A model of extracellular enzymes in free-living microbes: which strategy pays off? Appl Environ Microbiol 81(21):7385

van Dijl J, Hecker M (2013) *Bacillus subtilis*: from soil bacterium to super-secreting cell factory. Microb Cell Factories 12(1):3

Verhertbruggen Y, Marcus SE, Haeger A, Verhoef R, Schols HA, McCleary BV, McKee L, Gilbert HJ, Knox JP (2009) Developmental complexity of arabinan polysaccharides and their processing in plant cell walls. Plant J 59(3):413–425

von Blottnitz H, Curran MA (2007) A review of assessments conducted on bio-ethanol as a transportation fuel from a net energy, greenhouse gas, and environmental life cycle perspective. J Clean Prod 15(7):607–619

von Freiesleben P, Spodsberg N, Stenbaek A, Stalbrand H, Krogh K, Meyer AS (2018) Boosting of enzymatic softwood saccharification by fungal GH5 and GH26 endomannanases. Biotechnol Biofuels 11:194

Waldrop MP, Balser TC, Firestone MK (2000) Linking microbial community composition to function in a tropical soil. Soil Biol Biochem 32(13):1837–1846

Wallenstein MD, Burns RG (2011) Ecology of extracellular enzyme activities and organic matter degradation in soil: a complex community-driven process. In: Dick RP (ed) Methods of soil enzymology. Soil Science Society of Americ, Inc., Madison

Wang W, Zhong Z, Wang Q, Wang H, Fu Y, He X (2017) Glomalin contributed more to carbon, nutrients in deeper soils, and differently associated with climates and soil properties in vertical profiles. Sci Rep 7(1):13003

Wang X, Yao B, Su X (2018) Linking enzymatic oxidative degradation of lignin to organics detoxification. Int J Mol Sci 19(11)

Westereng B, Cannella D, Wittrup Agger J, Jorgensen H, Larsen Andersen M, Eijsink VG, Felby C (2015) Enzymatic cellulose oxidation is linked to lignin by long-range electron transfer. Sci Rep 5:18561

Westereng B, Loose JSM, Vaaje-Kolstad G, Aachmann FL, Sorlie M, Eijsink VGH (2018) Analytical tools for characterizing cellulose-active lytic polysaccharide monooxygenases (LPMOs). Methods Mol Biol 1796:219–246

Whiteley M, Diggle SP, Greenberg EP (2017) Progress in and promise of bacterial quorum sensing research. Nature 551:313–320

Yan J, Wang L, Fu PP, Yu H (2009) Photomutagenicity of 16 polycyclic aromatic hydrocarbons from the US EPA priority pollutant list. Mutat Res 557(1):99–108

Ye J, Yin H, Peng H, Bai J, Li Y (2014) Pyrene removal and transformation by joint application of alfalfa and exogenous microorganisms and their influence on soil microbial community. Ecotoxicol Environ Saf 110:129–135

You Y, Wang J, Huang X, Tang Z, Liu S, Sun OJ (2014) Relating microbial community structure to functioning in forest soil organic carbon transformation and turnover. Ecol Evol 4(5):633–647

Zhang YHP (2011) What is vital (and not vital) to advance economically-competitive biofuels production. Process Biochem 46(11):2091–2110

Zhou H, Gu W, Sun W, Hay AG (2018) A microbial community snapshot of windrows from a commercial composting facility. Appl Microbiol Biotechnol 102(18):8069–8077

Zhu Y, McBride MJ (2014) Deletion of the *Cytophaga hutchinsonii* type IX secretion system gene sprP results in defects in gliding motility and cellulose utilization. Appl Microbiol Biotechnol 98(2):763–775

Zhu Y, McBride MJ (2017) The unusual cellulose utilization system of the aerobic soil bacterium *Cytophaga hutchinsonii*. Appl Microbiol Biotechnol 101(19):7113–7127

Chapter 11
Role of Microbes in Degradation of Chemical Pesticides

Supriyanka Rana, Priyatharishini Mardarveran, Reena Gupta, Lakhveer Singh, and Zularisam ab Wahid

Abstract Global population explosion and food security concerns have resulted in an increased use of toxic pesticides to prevent the cash food crops from pest-infestation or to minimize yield loss. These xenobiotic compounds are known to cause hazardous effect on human health and its inhabiting environment. The current chapter aims to summarize the innate ability of the soil microbial communities to metabolize the toxic pesticide compounds. Microbial-mediated pesticide degradation is a sustainable approach to restore the pesticide-infested environments back to its previous ecologically clean and balanced state. Researches based on the steering effect of various factors on the growth of pesticide biograders (viz. bacteria, fungi, cynobacteria) are only few, and change in the microbial dynamics and associated mechanistics of biodegradation, with changing pesticide type, are yet to be fully understood). However, advent of advanced tools such as genomics, proteomics, transcriptomics, and metabolomics has tremendously helped researchers to gain the basic mechanistic understanding of microbial community dynamics and associated metabolic pathways involved in pesticide biodegradation, in order to make knowledge-based decisions to design better strategies to enhance pesticide degradation potential of microbes by manipulating its metabolic networks using genetic engineering approaches. This chapter will address the current state of the art of researches taking place in the area of microbe-assisted pesticide (xenobiotic compounds) degradation along with the integrative role of omics approaches in microbial-mediated bioremediation.

Keywords Microorganism · Biodegradation · Bioremediation · Pesticide

---

S. Rana · P. Mardarveran · L. Singh · Z. a. Wahid (✉)
Faculty of Engineering Technology, University Malaysia Pahang, Lebuhraya Tun Razak Gambang, Kuantan, Malaysia
e-mail: zularisam@ump.edu.my

R. Gupta
Department of Biotechnology, Himachal Pradesh University, Shimla, Himachal Pradesh, India

A. Kumar, S. Sharma (eds.), *Microbes and Enzymes in Soil Health and Bioremediation*, Microorganisms for Sustainability 16,
https://doi.org/10.1007/978-981-13-9117-0_11

11.1 Introduction

Pesticides have been a part of global expenditure in attaining and retaining the food security. Great reliance on the application of pesticides to gain a higher yield of crops suggests its significance in the modern world. Many pesticide and fertilizers are released into the surrounding environment each year to prevent pest infestations or to prevent crop yield losses (Isman 2015). Unfortunately, frequent applications of chemical pesticides have found to cause many problems to humans, beneficial insects, animals, and birds (Naqqash et al. 2016). For instance, health implications has been observed in occupationally involved workers or inhabitants cultivating agricultural crops in continuous exposure with the pesticides. Bioaccumulation of the pesticides in food chains is one of the reported cause of severe health implications in human beings (Chen et al. 2015a) along with the occurrence of breeding / or larval developmental issues in exposed nontarget flora and fauna (Goulson 2013; Woodcock et al. 2016). Also, excessive usage of pesticides significantly contaminates our natural and groundwater resources. Furthermore, regular application of one or more pesticides on agricultural fields leads to the development of multipesticide-resistant strains of pathogens or insect populations is another serious issue which yet remained to be fully understood and resolved (McCaffery and Nauen 2006). Native microbes present in pesticide-exposed habitat also get adversely affected by its applications due to the unfamiliarity of the applied chemicals, which compel them to drastically change their population dynamics accordingly. Continuous exposure to pesticides forces them to evolve (and adapt) by devising mechanisms to utilize or reduce the chemicals to evade the associated toxicity of pesticides; thereby, genetical evolution of pesticide-degrading genes and mobile DNA elements in native microbes arm them with the high pesticide-metabolizing efficiency for its adaptability and survival (Khan and Rao 2019). A solution to this issue is to use biopesticides instead of chemical pesticides which are environmentally sustainable and nonhazardous (Pavela and Benelli 2016). So, bioremediating the pesticide-infested soil is a major challenge that calls for more sustainable solutions (like using biopesticides) to nullify or minimize the long-term hazards associated with conventional agricultural practices.

The microorganisms have emerged as potential natural "biodegraders" capable to biodegrade various pesticides and to bioremediate the toxic chemical-infested land (Joutey et al. 2013). Presence of high levels of obstinate pesticides such as dichlorodiphenyltrichloroethane, Hexachlorocyclohexanes, (Saadati et al. 2012; Zhu et al. 2019), atrazine (Jablonowski et al. 2011), and dieldrin (Beyer and Gale 2013), still haunts the researchers with its recalcitrance. The major polluting compounds associated with majority of pesticides are, polyaromatics, chlorinates nitrobenzenes, dioxins, polychlorinated biphenyls, chlorophenols, polycyclic aromatic hydrocarbons (Mitra et al. 2019; Omenn 2013; Silva et al. 2019). Previously, many researchers have successfully reclaimed the pesticide infested soil by applying various biological methods; however, most of the methods could bioremediate either one or two pesticides / or compounds only (Acosta-Cortés et al. 2019; Barba et al. 2019; Rong et al. 2019; Shabbir et al. 2018). Few recent studies have addressed this issue by finding the microbial solution (such as, by applying the multienriched cultures / or microbial biofilms instead of pure cultures) to biodegrade the multipesticides in one go (Krishna and Philip 2008; Sniegowski and Springael 2015; Zhang et al. 2019). Over the past century, many conventional approaches have been deduced to clean the pesticides from infested soils, which includes, physical degradation and chemical degradation. However, toxic intermediates formed during these processes had been a contradiction to the world's long-term sustainability goals (Coats 1991; Sun and Pignatello 1992). So, microbe-assisted biodegradation of pesticide is a suitable alternative to bioremediate the contaminated soils with no or minimal toxic intermediates. Microbial degradation of the chemical pesticides is a slow, but a steady process, which makes use of the versatile enzymatic potential of the effective microbial communities in reclaiming the pesticide infested soil (Ortiz-Hernández et al. 2013b). Microbes being a chief driving force in pesticide biodegradation, understanding of metabolic pathways which help it to biodegrade the pesticides becomes essential for researchers to make more informed decisions to find novel biological solutions for ever-increasing pesticide problem (Ortiz-Hernández et al. 2018). A variety of microbial strains have successfully helped in biodegrading the different pesticide types (Table 11.1). This table is the depiction of the all successful attempts, which have been made till date to treat various recalcitrant pesticide types by using different microbial species (Chaussonnerie et al. 2016). This chapter will address the current state of the art of researches taking place in the area of microbe-assisted pesticide (xenobiotic compounds) degradation and integrative role of omics approaches in microbial-mediated bioremediation to innovate better solution for pesticide problems.

11.2 Microbe-Assisted Pesticide Biodegradation: Principles and Mechanism

Increased level of pesticide in agricultural soil and potable water pose long-term environmental and human health risks (Hatzikioseyian 2010). Biodegradation is a process of a complete breakdown of the organic compound into its inorganic constituents. Such microbial transformation helps in pesticide detoxification assisted by the co-metabolism of multimicrobial communities found in soil (Ortiz-Hernández et al. 2013b). In situ biodegradation of the pesticides using bio-beds is an economic and effective tool for the bioaugmentation of polluted sites (Ortiz-Hernández et al. 2013a). Excellent biocatalytic potential of few microbial strains is a result of the versatility, being bestowed by the microbial enzymes helps in the biodegradation of the complex organic compounds of almost any kind. Among microbial communities, bacteria, fungi, and actinomycetes are the main transformers and pesticide degraders (Ortiz-Hernández et al. 2011). Fungi and bacteria are the chief extracellular enzyme-producing microorganisms which are extensively used to biodegrade the harmful chemical pesticides, for instance, fungi are known to form extended mycelial networks coupled with highly specific catabolic enzymes, which makes it

Table 11.1 Classification of pesticides and their implications

Pesticide	Implications	Example	Effective microbial species against pesticide	References
λ-Cyhalothrin	Causes serious environmental issues Enters in water sources from either residential or agricultural runoff Enters humans via ingestion of food or drinking water or inhalation or dermal contact Cause significant toxicity and health effects, such as neurotoxicity, cytotoxicity, and endocrine disruption which can damage mammalian reproduction Elevated risk of mutagenicity, carcinogenicity, as well as childhood leukemia Highly toxic to aquatic invertebrates and fish		*Cunninghamella elegans*, *Bacillus thuringiensis*, *Serratia* sp., *Streptomyces parvulus*, *Brevibacterium aureum*, *Klebsiella* sp., *Sphingobium* sp., *Ochrobactrum anthropi*	Chen et al. (2015b), Chen et al. (2013), Lin et al. (2011), Palmer-Brown et al. (2019), Wang et al. (2009), Wu et al. (2006), Zhai et al. (2012), and Zhang et al. (2010)

Organochlorine	Leads to environmental and human health issues Accumulate in the fatty tissue of animals, are transferred through the food chain Have endocrine-disrupting potential and carcinogenic action Associated with neurodevelopmental effects in children, endocrine disorders, effects on embryonic development, lipid metabolism and hematological and hepatic alterations	Dichlorodiphenyltrichloroethane (DDT), dichlorodiphenyldichloroethylene (DDE), aldrin, lindane, chlordane, mirex, dieldrin, endosulfan, heptachlor, dicofol, methoxychlor, and α-, β-, and γ-hexachlorocyclohexane	*Stenotrophomonas* sp., *Alcaligenes eutrophus*, *Alcaligenes faecalis*, *Aspergillus niger*, *Streptomyces* sp., *Fusarium ventricosum*, *Pandoraea* sp., *Aspergillus sydoni*, *Achromobacter xylosoxidans*, *Pseudomonas aeruginosa*, *Anabaena* sp., *Trametes hirsute*, *Sphingobium japonicum*, *Arthrobacter fluorescens*, *Arthrobacter giacomelloi*, *Phlebia*, *Trichoderma harzianum*, *Klebsiella oxytoca*, *Trametes hirsutus*	Bhalerao and Puranik (2007), Cao et al. (2013), De Paolis et al. (2013), Erdem and Cutright (2016), Fuentes et al. (2011), Goswami et al. (2009), Hussain et al. (2009a), Kamei et al. (2011), Katayama and Matsumura (1993), Kong et al. (2018), Kong et al. (2013), Kwon et al. (2005), Lee et al. (2003), Nagata et al. (2016), Okeke et al. (2002), Pan et al. (2016), Siddique et al. (2003), Singh and Kuhad (1999), Singh and Singh (2011), and Xiao et al. (2011)

(continued)

Table 11.1 (continued)

Pesticide	Implications	Example	Effective microbial species against pesticide	References
Organophosphorus	Have endocrine-disrupting potential Adversely effects the function of cholinesterase enzymes Reduce the insulin secretion Disrupts proteins/carbohydrates/fat metabolism inside cell Have genotoxic effects Interrupts mitochondrial function. Causes problems in nervous and endocrine systems Leads to health issues like cardiovascular diseases, male reproductive system-related issues, neurological disorders, dementia, non-Hodgkin's lymphoma, decreased gestational duration Glyphosate leads to hypertension, diabetes, strokes, autism, kidney failure, Parkinson's and Alzheimer's diseases, and cancer	Methyl-parathion, malathion, methyl-diazinon, parathion, dimethoate, glyphosate, fenitrothion, chlorpyrifos, terbufos	*Rhodobacter sphaeroides*, *Bacillus amyloliquefaciens*, *Pseudomonas putida*, *Flavobacterium* sp., *Burkholderia* sp., *Enterobacter* sp., *Pseudomonas aeruginosa*, *Serratia marcescens*, *Chlorella vulgaris*, *Sphingobium* sp., *Lactobacillus, Plantarum*, *Radopholus similis*	Abo-Amer (2011), Ahn et al. (2018), and Cabrera et al. (2010), Cycoń et al. (2009), Deshpande et al. (2001), Hayatsu et al. (2000), Kurade et al. (2016), Li et al. (2018), Meng et al. (2019), Rani and Lalithakumari (1994), Sethunathan and Yoshida (1973), Singh (2009), Singh et al. (2004), and Wu et al. (2019)

Carbamates	Associated with endocrine-disrupting activity Leads to reproductive disorders, cause cellular metabolic mechanisms and mitochondrial dysfunction In vitro found to possess cytotoxic and genotoxic effects on hamster ovarian cells Capable to induce apoptosis and necrosis in human immune cells (95), natural killer cells, T lymphocytes Leads to neurobehavioral effects, dementia, and non-Hodgkin's lymphoma	Aldicarb, carbofuran, ziram, sevin, carbaryl, phenmedipham, propoxur	*Ascochyta* sp., *Xylaria* sp., *Ochrobactrum anthropi*, *Paracoccus* sp., *Pseudomonas* species, *Micrococcus* sp., *Pseudomonas aeruginosa*, *Pichia anomala*, *Enterobacter cloacae*, *Rhodotorula mucilaginosa*, *Pseudaminobacter* sp., *Exophiala jeanselmei*, *Klebsiella pneumoniae*	Balomajumder (2019), Doddamani and Ninnekar (2001), Fareed et al. (2017), Kadakol et al. (2011), Kamanavalli and Ninnekar (2000), Kim et al. (2017), Li et al. (2019), Naqvi et al. (2013), Owen et al. (1996), Peng et al. (2008), Pujar et al. (2019), Wu et al. (2013), and Yang et al. (2011)
Triazines	Causes endocrine-disrupting effects and reproductive toxicity Leads to breast cancer incidence, oxidative stress (cytotoxicity), and dopaminergic effects Associated with reproductive toxicity and delays in sexual maturation (in vitro studies)	Atrazine, simazine, ametrine, metamitron	*Rhodococcus* sp., *Klebsiella pneumoniae*, *Acetobacterium malicum*, *Phanerochaete chrysosporium*, *Arthrobacter* sp., *Arthrobacter aurescens*, *Citricoccus* sp., *Daldinia concentrica*	Adrian and Arnett (2004), Bankole et al. (2018), Fang et al. (2016), Fernando and Aust (1991), Fournier et al. (2002), Jiang et al. (2019), Viegas et al. (2019), Yang et al. (2018), and Zhao et al. (2002)

(continued)

Table 11.1 (continued)

Pesticide	Implications	Example	Effective microbial species against pesticide	References
Triazinone			*Rhodococcus* sp.	Fang et al. (2016)
Synthetic pyrethroids	Displays endocrine-disrupting activity Adversely affect the reproductive system and reproductive behavior (in vitro) Found to cause DNA damages in human sperm and adverse effect on human reproductive health Can cause developmental neurotoxicity	Fenvalerate, cypermethrin, sumithrin, fenpropathrin, deltamethrin, permethrin, cyfluthrin	*Bacillus subtilis*, *Micrococcus* sp., *Stenotrophomonas maltophilia*, *Aspergillus niger*, *Clostridium* sp., *Bacillus licheniformis*, *Sphingomonas* sp., *Azoarcusindigens*, *Streptomyces aureus*, *Sphingobium faniae*, *Pseudomonas aeruginosa*, *Acinetobacter* sp., *Bacillus cereus*, *Acinetobacter baumannii*, *Photobacterium*, *Ganghwense*	Chen et al. (2011b), Deng et al. (2015), Guo et al. (2010), GÜR et al. (2014), Jin et al. (2014), Liu et al. (2014), Ma et al. (2013), Sharma et al. (2016), Song et al. (2015), Tallur et al. (2008), Wang et al. (2019), Zhan et al. (2018), and Zhang et al. (2016)
Neonicotinoid	Have adverse effect on bees Negative effect on the endocrine and reproductive systems of animals Induces enzyme aromatase that leads to breast cancer	Imidacloprid, thiacloprid, guadipyr	*Ensifer adhaerens*, *Pigmentiphaga* sp., *Ensifer meliloti*, *Bacillus* sp., *Variovorax boronicumulan*, *Phanerochaete sordida*	Ge et al. (2014), Mori et al. (2017), Sun et al. (2017), Wang et al. (2013), and Zhou et al. (2013)

most suitable for the bioremediation processes. Microbes biotransform pesticides by introducing minor structural changes to its molecule, ultimately rendering it non-toxic. Such biotransformed particles are more susceptible to bacterial biodegradation (Burns and Wallenstein 2010; Singh 2016). In this way, microbes contribute to in improving the bioaugmentation scenarios. Presence of other indigenous microbial communities along with assimilable organic carbon has found to influence the microbe's ability to biodegrade the pesticide in an aquatic habitat and make it difficult to predict bioaugmentation potential of a specific microbe; however, recently a biokinetic model has been proposed which can be used to predict the carbofuran biotransformation potential of bacteria *Novosphingobium* sp. (Liu et al. 2019). Also, knowledge of such biodegradation data might be of significant help to the fore coming researcher to design safer and nonhazardous chemicals.

11.3 Bacteria-Mediated Pesticide Degradation

Despite positive role of pesticides in reducing the vector-borne diseases and enhancing the crop/food production, its unregulated applications have led to serious consequences in terms of environmental pollution and health hazards (Umadevi et al. 2017). Over the past few decades, few bacteria have been reported as a bioremediating agent and have widely been applied to detoxify harmful pesticides. These biologically active soil-bacterial species are capable to use pesticides as its sole carbon and energy at ambient atmospheric conditions. Several bacterial genera are known to biodegrade various pesticides, which includes *Bacillus*, *Pseudomonas*, *Flavobacterium*, *Moraxella*, *Acinetobacter*, *Arthrobacter*, *Paracoccus*, *Aerobacter*, *Alcaligenes*, *Burkholderia* and *Sphingomonas*, *Rhodococcus*, *Gliocladium*, *Trichoderma*, and *Penicillium* (Table 11.1). For instance, bacteria *Bacillus thuringiensis* can biodegrade cyhalothrin (Chen et al. 2015b). Bacteria secrete various extracellular enzymes (such as dehydrogenase, ligninase, oxygenase, peroxidases, phosphor-tri-esterase, hydrolases, dehalogenase, laccase, and organophosphorus acid anhydrolase) to detoxify itself from intracellular toxicity arose out of excessive pesticides uptake from contaminated soil (Aislabie and Lloyd-Jones 1995; Karigar and Rao 2011; Tang et al. 2015).

Microbial mechanistic of pesticide biodegradation can be explained by taking *Bacillus thuringiensis*-mediated cyhalothrin degradation as an example to understand how this microbe transforms pesticide into simpler compounds (Chen et al. 2015b) (Fig. 11.1). A novel pathway of cyhalothrin degradation has been studied in *Bacillus thuringiensis* which can easily metabolize the 3-phenoxy-benzoic acid, a common hazardous metabolite of pyrethroids. Insights into the promising role of microbial metabolism in cyhalothrin degradation which help them to bioremediate the pyrethroid-contaminated environment acquaint us with the fundamental mechanism behind pesticide degradation. The bacteria *Bacillus thuringiensis* cleave the ester linkage and diaryl bond of pesticides and release 3-phenoxyphenyl acetonitrile and N-(2-isopropoxy-phenyl)-4-phenoxy-benzamide as the metabolites during degradation of cyhalothrin. The primary step of pyrethroid degradation is ester hydro-

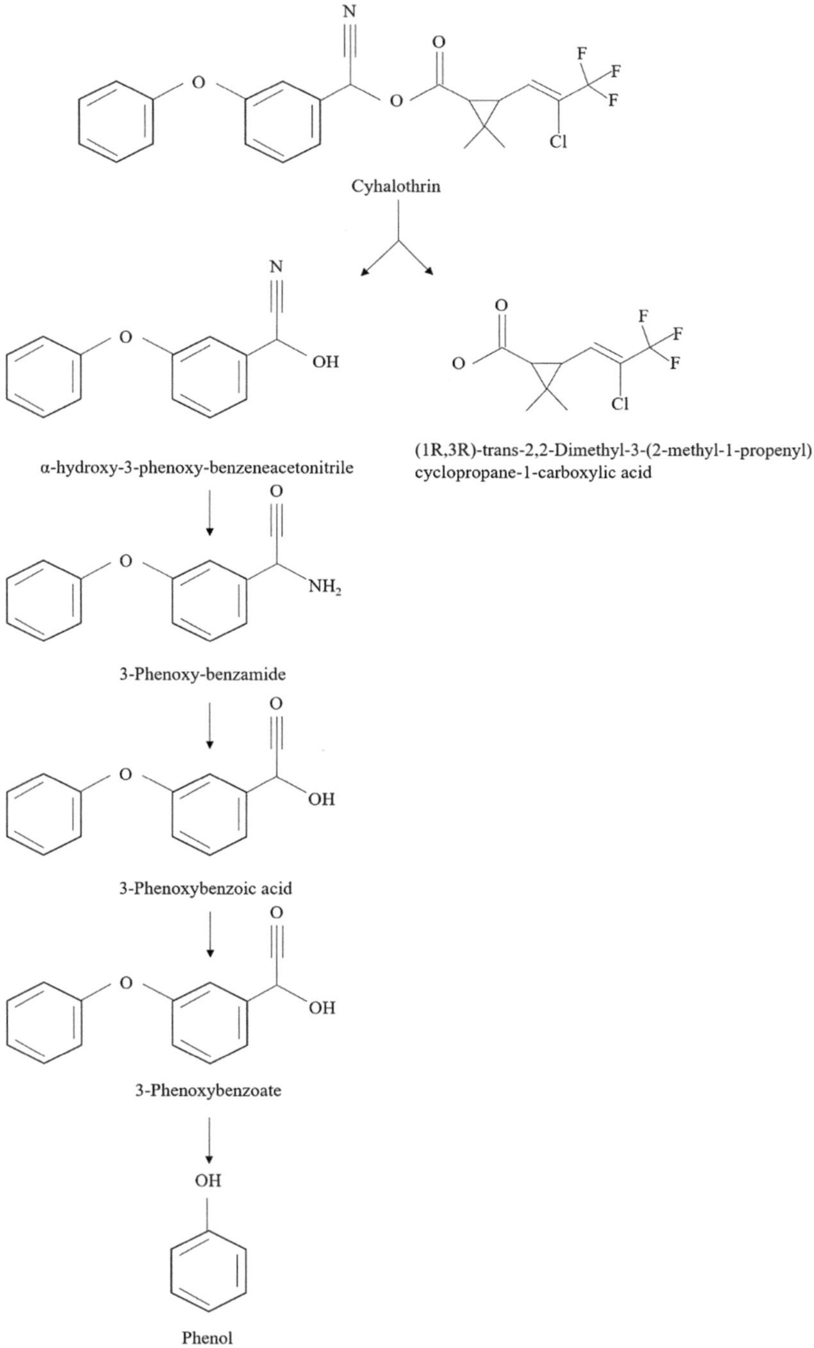

Fig. 11.1 Mechanism of cyhalothrin biodegradation in *Bacillus thuringiensis*. (Adapted from Chen et al. 2015b)

lysis via carboxylesterases which is a common detoxification step to deal with pyrethroids in various microbial species (Chen et al. 2015b). The pyrethroid biodegradation pathways have also been studies for few other microbes as well (Chen et al. 2013; Chen et al. 2011a, 2012; Tallur et al. 2008).

11.4 Fungal-Mediated Pesticide Degradation

Fungi can be applied for biodegradation and bioremediation of pesticides, including lindane, methamidophos, endosulfan, chlorpyrifos, atrazine, cypermethrin, dieldrin, methyl parathion, heptachlor, and so on. Fungal strains involve various processes, including hydroxylation, demethylation, dechlorination, deoxygenation, esterification, dehydrochlorination, and oxidation, which assist it to biodegrade complex xenobiotic compounds with various functional groups (Maqbool et al. 2016). Pesticide degradation potential of fungi depends on various factors like soil moisture content, nutrient availability, pH, temperature, and oxygen level. Involvement of various fungal enzymes like laccase, hydrolase, peroxidase, esterase, dehydrogenase, manganese peroxidase, and lignin peroxidase in pesticide degradation has been reported several times (Ali et al. 2019; Maqbool et al. 2016). Recent advancements in fungal biodegradation of pesticides targeting the processes, pathways, genes/enzymes, and factors affecting its biodegradation have helped in making better strategies for pesticide removal. The chief fungal species, which has sucessufully been validated for its excellent pesticide degradation potential include, *Fusarium oxysporum*, *Aspergillus niger*, *Penicillium*, *Lentinula edodes*, *Lecanicillium*, and so on. These fungi can degrade pesticides through a series of enzymatic reactions (extracellular enzyme and lignin-degrading enzyme, i.e., peroxidase and laccase) (Ali et al. 2019). For instance, planktonic and biofilm cultures of fungi *Cunninghamella elegans* can biotransform cyhalothrin after its initial biosorption, thereby hydrolyzing the complex ester bonds catalyzed by trifluoromethyl group and cytochrome P450. However, biodegradation ability of fungi in planktonic form has reported much faster than in biofilm form (Palmer-Brown et al. 2019). For example, fungi *Aspergillus niger* can metabolize the cypermethrin and its intermediates which are depicted in Fig. 11.2. Potential of fungi as a pesticide bioremediating agent is yet to be fully harnessed to exploit it as a tool for in situ degradation and remediation. Sustainable and clean applications of using naturally available bioresources (like fungi) are the only way forward to create a pesticide-free land to ensure the growth of chemical-free crops and fruits in the coming future.

11.5 Strategies to Improve Microbe-Mediated Remediation Using Omics Approaches

Application of the microorganisms for xenobiotic degradation requires an overall understanding of various (physiological, microbiological, ecological, biochemical, and molecular) aspects which participates in xenobiotic transformation (Singh and

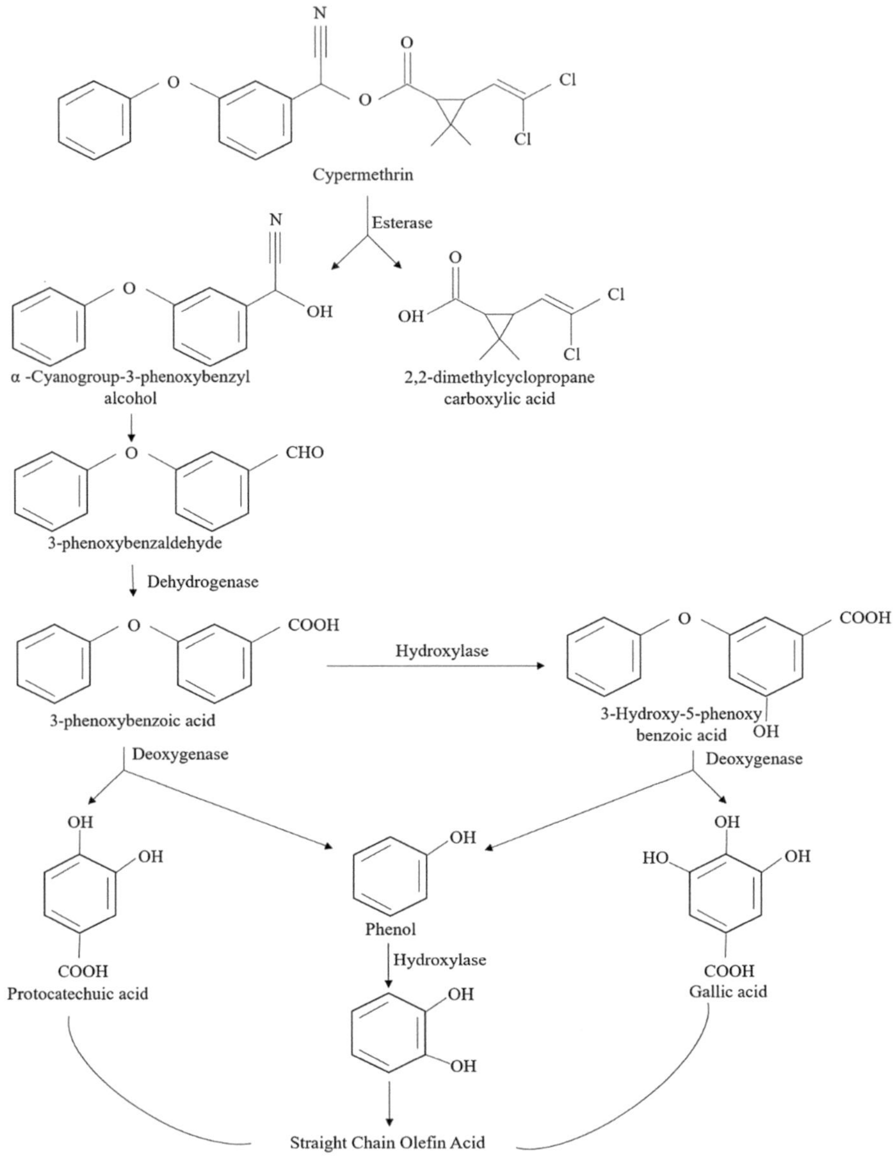

Fig. 11.2 Mechanism of cypermethrin biodegradation by fungi *Aspergillus niger*. (Adapted from Deng et al. 2015)

Walker 2006). In situ bioremediation potential of microbes has been hindered due to the limited system-wide understanding of interactions occurring between degrading genes of multimicrobial pesticide degraders, interconnectedness of its metabolic network, and susceptibility to environmental variability in any specific environment. The advent of high-throughput technology has helped to gain that understanding and to find the relationships between pesticides, microbial communities involved in its degradation and to reveal novel pesticide biodegradation processes. A holistic metagenomic/transcriptomic approach is applied to predict microbial degradability in the context of the ecology of contaminated habitats (Malla et al. 2018). Metagenomics and metatranscriptomics approaches have allowed exploration of unknown xenobiotic-degrading microbial communities, thereby facilitating the overall examination metabolic pathways involved in pesticide degradation (Bharagava et al. 2019; Dangi et al. 2019). For instance, catabolic genes associated with xenobiotics degradation can be characteristically annotated and linked to the identified taxonomic group/genus of the microbial communities. Metagenomics database can be used to mine the transcripts/genes connected with direct or indirect degradation of pesticides, which can further be used to engineer the microbial strain with advanced traits to bioremediate the contaminated soil (Jaiswal et al. 2019). Dawn of the eco-genomic tools has allowed the researchers to profile these microbes directly in its inhabiting environment to catch its metabolic process in action (via next-generation DNA sequencing coupled with network analysis) and has presented researchers with a way forward to unreveal the existing correlation between microbial taxonomy, function, and environmental variables at molecular level.

11.6 Future Outlook

The major hindrance in the commercial adoption of microbe-driven bioremediation as model bioremediation technology is the difficulty of reproducing or replicating its lab-scale remediation efficacy at field scale. Limited novel steadfast techniques make it difficult to validate the effectiveness of microbes in bioremediating the contaminated site. Advanced tools and techniques are needed for the prediction of different pesticide types and real-time prediction of its concentration in a contaminated environment, and this will possibly help the agriculturist and researchers to make informed decisions about pesticide-infested soil. A site contaminated with multipesticides with uneven distribution of target substances in soil along with low solubilities and high binding capacities poses another challenge which can prevent microbial activity, so better strategies are required which can address such situations, for example, mixed cultures or biofilm-based studies could lead us to microbial consortium which collectively feasts and co-metabolizes the multipesticides simultaneously. Applying a combination of genetically modified indigenous and nonindigenous strains with enhanced degradative can be effective for simultaneous co-metabolization of pesticides. Applying conventional means like plowing and biological means like earthworms can co-assist microbes by improving the bioremediation conditions for them. Moreover, application of the whole cell microbes to bioremediate the pesticides-infested soil may pose few challenges that hinders its effectiveness against pesticides due to inaccessibility of its cell wall entrapped cellular enzymes/or metabolites which however can be resolved by applying cell-free microbial extracts (containing only cell metabolites and pesticide-degrading enzymes) to achieve best results, though more researches are needed to devise better strategies to increase the microbial bioavailability/bioaccessibility to toxic pesticides. The selection of appropriate strain (aerobic or anaerobic) should be made only after considering various factors such as the geographical position of the infested site and pesticide concentration at the different subsurface level of soil to obtain maximum benefits. Further advancement in biotechnological innovations has the potential to revolutionize the bioremediation sector as biological and clean bioremediation solutions can only ensure the sustainability of our planet in the coming future.

11.7 Conclusion

Presently, microbes-based pesticide degradation approach is considered as an ideal and most sustainable option. Since the last past decade, increased application of microbe-mediated bioremediation approaches has validated its effectiveness in terms of its excellent bioremediation potential to bioremediate the pesticide-infested areas. Microbes works synergistically to accomplish the pesticide biodegradation by using various processes like co-metabolism using various biocatalytic-enzymes to break the complex xenobiotic compounds into simpler harmless compounds. So, improving the microbe's pesticide degradation potential by understanding its metabolic behaviors using high-throughput technologies will surely pave a way to take the microbe-mediated bioremediation based processes or technologies to the next level.

References

Abo-Amer A (2011) Biodegradation of diazinon by Serratia marcescens DI101 and its use in bioremediation of contaminated environment. J Microbiol Biotechnol 21(1):71–80

Acosta-Cortés AG, Martinez-Ledezma C, López-Chuken UJ, Kaushik G, Nimesh S, Villarreal-Chiu JF (2019) Polyphosphate recovery by a native Bacillus cereus strain as a direct effect of glyphosate uptake. ISME J 13:1497

Adrian NR, Arnett CM (2004) Anaerobic biodegradation of hexahydro-1, 3, 5-trinitro-1, 3, 5-triazine (RDX) by Acetobacterium malicum strain HAAP-1 isolated from a methanogenic mixed culture. Curr Microbiol 48(5):332–340

Ahn J-H, Lee S-A, Kim S-J, You J, Han B-H, Weon H-Y, Lee S-W (2018) Biodegradation of organophosphorus insecticides with P S bonds by two Sphingobium sp. strains. Int Biodeterior Biodegradation 132:59–65

Aislabie J, Lloyd-Jones G (1995) A review of bacterial-degradation of pesticides. Soil Res 33(6):925–942

Ali M, Husain Q, Ishqi HM (2019) Chapter-2 fungal peroxidases mediated bioremediation of industrial pollutants. Fungal bioremediation: fundamentals and applications

Balomajumder C (2019) Simultaneous biodegradation of mixture of carbamates by newly isolated Ascochyta sp. CBS 237.37. Ecotoxicol Environ Saf 169:590–599
Bankole PO, Adekunle AA, Govindwar SP (2018) Biodegradation of a monochlorotriazine dye, cibacron brilliant red 3B-A in solid state fermentation by wood-rot fungal consortium, Daldinia concentrica and Xylaria polymorpha: co-biomass decolorization of cibacron brilliant red 3B-A dye. Int J Biol Macromol 120:19–27
Barba S, Villaseñor J, Cañizares P, Rodrigo MA (2019) Strategies for the electrobioremediation of oxyfluorfen polluted soils. Electrochim Acta 297:137–144
Beyer WN, Gale RW (2013) Persistence and changes in bioavailability of dieldrin, DDE, and heptachlor epoxide in earthworms over 45 years. Ambio 42(1):83–89
Bhalerao TS, Puranik PR (2007) Biodegradation of organochlorine pesticide, endosulfan, by a fungal soil isolate, *Aspergillus niger*. Int Biodeterior Biodegradation 59(4):315–321
Bharagava RN, Purchase D, Saxena G, Mulla SI (2019) Applications of metagenomics in microbial bioremediation of pollutants: from genomics to environmental cleanup, microbial diversity in the genomic era. Elsevier, pp 459–477
Burns RG, Wallenstein MD (2010) Microbial extracellular enzymes and natural and synthetic polymer degradation in soil: current research and future prospects. Gilkes R & N Prakongkep
Cabrera J, Pocasangre L, Pattison A, Sikora R (2010) Terbufos biodegradability and efficacy against Radopholus similis in soils from banana cultivation having different histories of nematicide use, and the effect of terbufos on plant growth of in vitro-propagated Musa AAA cv. Grande Naine Intl J Pest Manage 56(1):61–67
Cao X, Yang C, Liu R, Li Q, Zhang W, Liu J, Song C, Qiao C, Mulchandani A (2013) Simultaneous degradation of organophosphate and organochlorine pesticides by Sphingobium japonicum UT26 with surface-displayed organophosphorus hydrolase. Biodegradation 24(2):295–303
Chaussonnerie S, Saaidi P-L, Ugarte E, Barbance A, Fossey A, Barbe V, Gyapay G, Brüls T, Chevallier M, Couturat L (2016) Microbial degradation of a recalcitrant pesticide: Chlordecone. Front Microbiol 7:2025
Chen S, Dong YH, Chang C, Deng Y, Zhang XF, Zhong G, Song H, Hu M, Zhang L-H (2013) Characterization of a novel cyfluthrin-degrading bacterial strain Brevibacterium aureum and its biochemical degradation pathway. Bioresour Technol 132:16–23
Chen S, Hu Q, Hu M, Luo J, Weng Q, Lai K (2011a) Isolation and characterization of a fungus able to degrade pyrethroids and 3-phenoxybenzaldehyde. Bioresour Technol 102(17):8110–8116
Chen S, Lai K, Li Y, Hu M, Zhang Y, Zeng Y (2011b) Biodegradation of deltamethrin and its hydrolysis product 3-phenoxybenzaldehyde by a newly isolated Streptomyces aureus strain HP-S-01. Appl Microbiol Biotechnol 90(4):1471–1483
Chen S, Luo J, Hu M, Geng P, Zhang Y (2012) Microbial detoxification of bifenthrin by a novel yeast and its potential for contaminated soils treatment. PLoS One 7(2):e30862
Chen M, Chang C-H, Tao L, Lu C (2015a) Residential exposure to pesticide during childhood and childhood cancers: a meta-analysis. Pediatrics 136(4):719–729
Chen S, Deng Y, Chang C, Lee J, Cheng Y, Cui Z, Zhou J, He F, Hu M, Zhang L-H (2015b) Pathway and kinetics of cyhalothrin biodegradation by *Bacillus thuringiensis* strain ZS-19. Sci Rep 5:8784
Coats JR (1991) Pesticide degradation mechanisms and environmental activation
Cycoń M, Wójcik M, Piotrowska-Seget Z (2009) Biodegradation of the organophosphorus insecticide diazinon by Serratia sp. and pseudomonas sp. and their use in bioremediation of contaminated soil. Chemosphere 76(4):494–501
Dangi AK, Sharma B, Hill RT, Shukla P (2019) Bioremediation through microbes: systems biology and metabolic engineering approach. Crit Rev Biotechnol 39(1):79–98
De Paolis M, Lippi D, Guerriero E, Polcaro C, Donati E (2013) Biodegradation of α-, β-, and γ-hexachlorocyclohexane by Arthrobacter fluorescens and Arthrobacter giacomelloi. Appl Biochem Biotechnol 170(3):514–524

Deng W, Lin D, Yao K, Yuan H, Wang Z, Li J, Zou L, Han X, Zhou K, He L (2015) Characterization of a novel β-cypermethrin-degrading *Aspergillus niger* YAT strain and the biochemical degradation pathway of β-cypermethrin. Appl Microbiol Biotechnol 99(19):8187–8198

Deshpande N, Dhakephalkar P, Kanekar P (2001) Plasmid-mediated dimethoate degradation in Pseudomonas aeruginosa MCMB-427. Lett Appl Microbiol 33(4):275–279

Doddamani HP, Ninnekar HZ (2001) Biodegradation of carbaryl by a Micrococcus species. Curr Microbiol 43(1):69–73

Erdem Z, Cutright TJ (2016) Biodegradation potential of 1, 1, 1-trichloro-2, 2-bis (p-chlorophenyl) ethane (4, 4′-DDT) on a sandy-loam soil using aerobic bacterium Alcaligenes eutrophus A5. Environ Eng Sci 33(3):149–159

Fang H, Xu T, Cao D, Cheng L, Yu Y (2016) Characterization and genome functional analysis of a novel metamitron-degrading strain Rhodococcus sp MET via both triazinone and phenyl rings cleavage. Sci Reports 6:32339

Fareed A, Zaffar H, Rashid A, Maroof Shah M, Naqvi TA (2017) Biodegradation of N-methylated carbamates by free and immobilized cells of newly isolated strain Enterobacter cloacae strain TA7. Biorem J 21(3–4):119–127

Fernando T, Aust S (1991) Biodegradation of munition waste. In: TNT (2, 4, 6-trinitrotoluene), and RDX (hexahydro-1, 3, 5-trinitro-1, 3, 5-triazine) by Phanerochaete chrysosporium, ACS symposium series-American. Chemical Society (USA), College Station

Fournier D, Halasz A, Spain J, Fiurasek P, Hawari J (2002) Determination of key metabolites during biodegradation of hexahydro-1, 3, 5-trinitro-1, 3, 5-triazine with Rhodococcus sp. strain DN22. Appl Environ Microbiol 68(1):166–172

Fuentes MS, Sáez JM, Benimeli CS, Amoroso MJ (2011) Lindane biodegradation by defined consortia of indigenous Streptomyces strains. Water Air Soil Pollut 222(1–4):217–231

Ge F, Zhou L-Y, Wang Y, Ma Y, Zhai S, Liu Z-H, Dai Y-J, Yuan S (2014) Hydrolysis of the neonicotinoid insecticide thiacloprid by the N2-fixing bacterium Ensifer meliloti CGMCC 7333. Int Biodeterior Biodegradation 93:10–17

Goswami S, Vig K, Singh DK (2009) Biodegradation of α and β endosulfan by Aspergillus sydoni. Chemosphere 75(7):883–888

Goulson D (2013) An overview of the environmental risks posed by neonicotinoid insecticides. J Appl Ecol 50(4):977–987

Guo P, Wang B-Z, Hang B-J, Li L, Li S-P, He J (2010) Sphingobium faniae sp. nov., a pyrethroid-degrading bacterium isolated from activated sludge treating wastewater from pyrethroid manufacture. Int J Syst Evol Microbiol 60(2):408–412

Gür Ö, Özdal M, Algur ÖF (2014) Biodegradation of the synthetic pyrethroid insecticide α-cypermethrin by Stenotrophomonas maltophilia OG2. Turk J Biol 38(5):684–689

Hatzikioseyian A (2010) Principles of bioremediation processes. In: Trends in Bioremediation and Phytoremediation. Research Signpost, Trivandrum, pp 23–54

Hayatsu M, Hirano M, Tokuda S (2000) Involvement of two plasmids in fenitrothion degradation by Burkholderia sp. strain NF100. Appl. Environ. Microbiology 66(4):1737–1740

Hussain S, Arshad M, Shaharoona B, Saleem M, Khalid A (2009a) Concentration dependent growth/non-growth linked kinetics of endosulfan biodegradation by Pseudomonas aeruginosa. World J Microbiol Biotechnol 25(5):853–858

Isman MB (2015) A renaissance for botanical insecticides? Pest Manag Sci 71(12):1587–1590

Jablonowski ND, Schäffer A, Burauel P (2011) Still present after all these years: persistence plus potential toxicity raise questions about the use of atrazine. Environ Sci Pollut Res 18(2):328–331

Jaiswal S, Singh DK, Shukla P, (2019) Gene editing and systems biology tools for pesticide bioremediation: a review. Front Microbiol:10

Jiang Z, Zhang X, Wang Z, Cao B, Deng S, Bi M, Zhang Y (2019) Enhanced biodegradation of atrazine by Arthrobacter sp. DNS10 during co-culture with a phosphorus solubilizing bacteria: Enterobacter sp. P1. Ecotoxicol Environ Saf 172:159–166

Jin Z, Guo Q, Zhang Z, Yan T (2014) Biodegradation of type II pyrethroids and major degraded products by a newly isolated Acinetobacter sp. strain JN8. Can J Microbiol 60(8):541–545

Joutey NT, Bahafid W, Sayel H, El Ghachtouli N (2013) Biodegradation: involved microorganisms and genetically engineered microorganisms. In: Biodegradation-life of science. InTechOpen, Croatia

Kadakol JC, Kamanavalli CM, Shouche Y (2011) Biodegradation of Carbofuran phenol by free and immobilized cells of Klebsiella pneumoniae ATCC13883T. World J Microbiol Biotechnol 27(1):25–29

Kamanavalli C, Ninnekar H (2000) Biodegradation of propoxur by Pseudomonas species. World J Microbiol Biotechnol 16(4):329–331

Kamei I, Takagi K, Kondo R (2011) Degradation of endosulfan and endosulfan sulfate by white-rot fungus Trametes hirsuta. J Wood Sci 57(4):317–322

Karigar CS, Rao SS (2011) Role of microbial enzymes in the bioremediation of pollutants: a review. Enzyme Res 2011:1

Katayama A, Matsumura F (1993) Degradation of organochlorine pesticides, particularly endosulfan, by Trichoderma harzianum. Environ Toxicol Chem Intl J 12(6):1059–1065

Khan A, Rao TS (2019) Molecular evolution of xenobiotic degrading genes and mobile DNA elements in soil bacteria, microbial diversity in the genomic era. Elsevier pp 657–678

Kim H, Kim D-U, Lee H, Yun J, Ka J-O (2017) Syntrophic biodegradation of propoxur by Pseudaminobacter sp. SP1a and Nocardioides sp. SP1b isolated from agricultural soil. Int Biodeterior Biodegradation 118:1–9

Kong L, Zhu S, Zhu L, Xie H, Su K, Yan T, Wang J, Wang J, Wang F, Sun F (2013) Biodegradation of organochlorine pesticide endosulfan by bacterial strain *Alcaligenes faecalis* JBW4. J Environ Sci 25(11):2257–2264

Kong L, Zhang Y, Zhu L, Wang J, Wang J, Du Z, Zhang C (2018) Influence of isolated bacterial strains on the in situ biodegradation of endosulfan and the reduction of endosulfan-contaminated soil toxicity. Ecotoxicol Environ Saf 160:75–83

Krishna KR, Philip L (2008) Biodegradation of lindane, methyl parathion and carbofuran by various enriched bacterial isolates. J Environ Sci Health B 43(2):157–171

Kurade MB, Kim JR, Govindwar SP, Jeon B-H (2016) Insights into microalgae mediated biodegradation of diazinon by Chlorella vulgaris: microalgal tolerance to xenobiotic pollutants and metabolism. Algal Res 20:126–134

Kwon G-S, Sohn H-Y, Shin K-S, Kim E, Seo B-I (2005) Biodegradation of the organochlorine insecticide, endosulfan, and the toxic metabolite, endosulfan sulfate, by Klebsiella oxytoca KE-8. Appl Microbiol Biotechnol 67(6):845–850

Lee S-E, Kim J-S, Kennedy IR, Park J-W, Kwon G-S, Koh S-C, Kim J-E (2003) Biotransformation of an organochlorine insecticide, endosulfan, by Anabaena species. J Agric Food Chem 51(5):1336–1340

Li C, Ma Y, Mi Z, Huo R, Zhou T, Hai H, Kwok L-Y, Sun Z, Chen Y, Zhang H (2018) Screening for *Lactobacillus plantarum* strains that possess organophosphorus pesticide-degrading activity and Metabolomic analysis of Phorate degradation. Front Microbiol 9

Li F, Di L, Liu Y, Xiao Q, Zhang X, Ma F, Yu H (2019) Carbaryl biodegradation by Xylaria sp. BNL1 and its metabolic pathway. Ecotoxicol Environ Saf 167:331–337

Lin Q, Chen S, Hu M, Haq MU, Yang L, Li H (2011) Biodegradation of cypermethrin by a newly isolated actinomycetes HU-S-01 from wastewater sludge. Intl J Environ Sci Technol 8(1):45–56

Liu F, Chi Y, Wu S, Jia D, Yao K (2014) Simultaneous degradation of cypermethrin and its metabolite, 3-phenoxybenzoic acid, by the cooperation of bacillus licheniformis B-1 and Sphingomonas sp. SC-1. J Agric Food Chem 62(33):8256–8262

Liu L, Helbling DE, Kohler H-PE, Smets BF (2019) Modelling carbofuran biotransformation by Novosphingobium sp. KN65. 2 in the presence of coincidental carbon and indigenous microbes. Environ Sci Water Res Technol 5:798–807

Ma Y, Chen L, Qiu J (2013) Biodegradation of beta-cypermethrin by a novel Azoarcus indigens strain HZ5. J Environ Sci Health B 48(10):851–859

Malla MA, Dubey A, Yadav S, Kumar A, Hashem A, Abd Allah EF (2018) Understanding and designing the strategies for the microbe-mediated remediation of environmental contaminants using omics approaches. Front Microbiol 9:1132

Maqbool Z, Hussain S, Imran M, Mahmood F, Shahzad T, Ahmed Z, Azeem F, Muzammil S (2016) Perspectives of using fungi as bioresource for bioremediation of pesticides in the environment: a critical review. Environ Sci Pollut Res 23(17):16904–16925

McCaffery A, Nauen R (2006) The insecticide resistance action committee (IRAC): public responsibility and enlightened industrial self-interest. Outlooks Pest Manage 17(1):11–14

Meng D, Jiang W, Li J, Huang L, Zhai L, Zhang L, Guan Z, Cai Y, Liao X (2019) An alkaline phosphatase from *Bacillus amyloliquefaciens* YP6 of new application in biodegradation of five broad-spectrum organophosphorus pesticides. J Environ Sci Health B:1–8

Mitra S, Corsolini S, Pozo K, Audy O, Sarkar SK, Biswas JK (2019) Characterization, source identification and risk associated with polyaromatic and chlorinated organic contaminants (PAHs, PCBs, PCBzs and OCPs) in the surface sediments of Hooghly estuary, India. Chemosphere 221:154–165

Mori T, Wang J, Tanaka Y, Nagai K, Kawagishi H, Hirai H (2017) Bioremediation of the neonicotinoid insecticide clothianidin by the white-rot fungus Phanerochaete sordida. J Hazard Mater 321:586–590

Nagata Y, Tabata M, Ohtsubo Y, Tsuda M (2016) Biodegradation of organochlorine pesticides. In: Manual of environmental microbiology, 4th edn. American Society of Microbiology, Washington, DC, pp 5.1. 2-1-5.1. 2-30

Naqqash MN, Gökçe A, Bakhsh A, Salim M (2016) Insecticide resistance and its molecular basis in urban insect pests. Parasitol Res 115(4):1363–1373

Naqvi T, Armughan A, Ahmed N, Ahmed S (2013) Biodegradation of carbamates by Pseudomonas aeruginosa. Minerva Biotecnologica 25(4):207–211

Okeke BC, Siddique T, Arbestain MC, Frankenberger WT (2002) Biodegradation of γ-hexachlorocyclohexane (lindane) and α-hexachlorocyclohexane in water and a soil slurry by a Pandoraea species. J Agric Food Chem 50(9):2548–2555

Omenn GS (2013) Environmental biotechnology: reducing risks from environmental chemicals through biotechnology. Springer, New York

Ortiz-Hernández ML, Sánchez-Salinas E, Olvera-Velona A, Folch-Mallol JL (2011) Pesticides in the environment: impacts and its biodegradation as a strategy for residues treatment, pesticides-formulations, effects, fate. INTECH, Croatia

Ortiz-Hernández ML, Sanchez-Salinas E, Castrejon Godinez ML, Dantan Gonzalez E, Popoca Ursino EC (2013a) Mechanisms and strategies for pesticide biodegradation: opportunity for waste, soils and water cleaning. Revista Internacional de Contaminación Ambiental 29

Ortiz-Hernández ML, Sánchez-Salinas E, Dantán-González E, Castrejón-Godínez ML (2013b) Pesticide biodegradation: mechanisms, genetics and strategies to enhance the process. In: Biodegradation-life of science. IntechOpen, Croatia

Ortiz-Hernández ML, Castrejón-Godínez ML, Popoca-Ursino EC, Cervantes-Dacasac FR, Fernández-Lópezd M (2018) Strategies for biodegradation and bioremediation of pesticides in the environment. In: Strategies for bioremediation of organic and inorganic pollutants. CRC Press, Boca Raton, pp 95–115

Owen S, Otani T, Masaoka S, Ohe T (1996) The biodegradation of low-molecular-weight urethane compounds by a strain of *Exophiala jeanselmei*. Biosci Biotechnol Biochem 60(2):244–248

Palmer-Brown W, de Melo Souza PL, Murphy CD (2019) Cyhalothrin biodegradation in *Cunninghamella elegans*. Environ Sci Pollut Res 26(2):1414–1421

Pan X, Lin D, Zheng Y, Zhang Q, Yin Y, Cai L, Fang H, Yu Y (2016) Biodegradation of DDT by Stenotrophomonas sp. DDT-1: characterization and genome functional analysis. Scientific Rep 6:21332

Pavela R, Benelli G (2016) Essential oils as ecofriendly biopesticides? Challenges and constraints. Trends Plant Sci 21(12):1000–1007

Peng X, Zhang JS, Li YY, Li W, Xu GM, Yan YC (2008) Biodegradation of insecticide carbofuran by Paracoccus sp. YM3. J Environ Sci Health B 43(7):588–594

Pujar, N.K., Laad, S., Premakshi, H., Pattar, S.V., Mirjankar, M., Kamanavalli, C.M., 2019. Biodegradation of phenmedipham by novel Ochrobactrum anthropi NC-1. 3 Biotech 9(2):52

Rani NL, Lalithakumari D (1994) Degradation of methyl parathion by pseudomonas putida. Can J Microbiol 40(12):1000–1006

Rong X, Zhao G, Fein JB, Yu Q, Huang Q (2019) Role of interfacial reactions in biodegradation: a case study in a montmorillonite, Pseudomonas sp. Z1 and methyl parathion ternary system. J Hazard Mater 365:245–251

Saadati N, Abdullah MP, Zakaria Z, Rezayi M, Hosseinizare N (2012) Distribution and fate of HCH isomers and DDT metabolites in a tropical environment – case study Cameron Highlands–Malaysia. Chem Cent J 6(1):130

Sethunathan N, Yoshida T (1973) A Flavobacterium sp. that degrades diazinon and parathion. Can J Microbiol 19(7):873–875

Shabbir M, Singh M, Maiti S, Kumar S, Saha SK (2018) Removal enactment of organo-phosphorous pesticide using bacteria isolated from domestic sewage. Bioresour Technol 263:280–288

Sharma, A., Gangola, S., Khati, P., Kumar, G., Srivastava, A., 2016. Novel pathway of cypermethrin biodegradation in a bacillus sp. strain SG2 isolated from cypermethrin-contaminated agriculture field. 3 biotech 6(1), 45

Siddique T, Okeke BC, Arshad M, Frankenberger WT (2003) Biodegradation kinetics of endosulfan by Fusarium ventricosum and a Pandoraea species. J Agric Food Chem 51(27):8015–8019

Silva V, Mol HG, Zomer P, Tienstra M, Ritsema CJ, Geissen V (2019) Pesticide residues in European agricultural soils–a hidden reality unfolded. Sci Total Environ 653:1532–1545

Singh BK (2009) Organophosphorus-degrading bacteria: ecology and industrial applications. Nat Rev Microbiol 7(2):156

Singh SN (2016) Microbe-induced degradation of pesticides. Springer

Singh B, Kuhad R (1999) Biodegradation of lindane (γ-hexachlorocyclohexane) by the white-rot fungus Trametes hirsutus. Lett Appl Microbiol 28(3):238–241

Singh NS, Singh DK (2011) Biodegradation of endosulfan and endosulfan sulfate by Achromobacter xylosoxidans strain C8B in broth medium. Biodegradation 22(5):845–857

Singh BK, Walker A (2006) Microbial degradation of organophosphorus compounds. FEMS Microbiol Rev 30(3):428–471

Singh BK, Walker A, Morgan JAW, Wright DJ (2004) Biodegradation of chlorpyrifos by Enterobacter strain B-14 and its use in bioremediation of contaminated soils. Appl Environ Microbiol 70(8):4855–4863

Sniegowski K, Springael D (2015) Establishment of multiple pesticide biodegradation capacities from pesticide-primed materials in on-farm biopurification system microcosms treating complex pesticide-contaminated wastewater. Pest Manag Sci 71(7):986–995

Song H, Zhou Z, Liu Y, Deng S, Xu H (2015) Kinetics and mechanism of fenpropathrin biodegradation by a newly isolated *Pseudomonas aeruginosa* sp. strain JQ-41. Curr Microbiol 71(3):326–332

Sun Y, Pignatello JJ (1992) Chemical treatment of pesticide wastes. Evaluation of iron (III) chelates for catalytic hydrogen peroxide oxidation of 2, 4-D at circumneutral pH. J Agric Food Chem 40(2):322–327

Sun S-L, Yang W-L, Guo J-J, Zhou Y-N, Rui X, Chen C, Ge F, Dai Y-J (2017) Biodegradation of the neonicotinoid insecticide acetamiprid in surface water by the bacterium Variovorax boronicumulans CGMCC 4969 and its enzymatic mechanism. RSC Adv 7(41):25387–25397

Tallur PN, Megadi VB, Ninnekar HZ (2008) Biodegradation of cypermethrin by Micrococcus sp. strain CPN 1. Biodegradation 19(1):77–82

Tang A, Wang B, Liu Y, Li Q, Tong Z, Wei Y (2015) Biodegradation and extracellular enzymatic activities of Pseudomonas aeruginosa strain GF31 on β-cypermethrin. Environ Sci Pollut Res 22(17):13049–13057

Umadevi S, Ayyasamy P, Rajakumar S (2017) Biological perspective and role of Bacteria in pesticide degradation, bioremediation and sustainable technologies for cleaner environment. Springer, Cham, pp 3–12

Verma JP, Jaiswal DK, Sagar R (2014) Pesticide relevance and their microbial degradation: a-state-of-art. Rev Environ Sci Biotechnol 13(4):429–466

Viegas CA, Silva VP, Varela VM, Correia V, Ribeiro R, Moreira-Santos M (2019) Evaluating formulation and storage of Arthrobacter aurescens strain TC1 as a bioremediation tool for terbuthylazine contaminated soils: efficacy on abatement of aquatic ecotoxicity. Sci Total Environ 668:714

Wang B-Z, Guo P, Hang B-J, Li L, He J, Li S-P (2009) Cloning of a novel pyrethroid-hydrolyzing carboxylesterase gene from Sphingobium sp. strain JZ-1 and characterization of the gene product. Appl Environ Microbiol 75(17):5496–5500

Wang G, Yue W, Liu Y, Li F, Xiong M, Zhang H (2013) Biodegradation of the neonicotinoid insecticide Acetamiprid by bacterium Pigmentiphaga sp. strain AAP-1 isolated from soil. Bioresour Technol 138:359–368

Wang T, Hu C, Zhang R, Sun A, Li D, Shi X (2019) Mechanism study of cyfluthrin biodegradation by Photobacterium ganghwense with comparative metabolomics. Appl Microbiol Biotechnol 103(1):473–488

Woodcock BA, Isaac NJ, Bullock JM, Roy DB, Garthwaite DG, Crowe A, Pywell RF (2016) Impacts of neonicotinoid use on long-term population changes in wild bees in England. Nat Commun 7:12459

Wu PC, Liu YH, Wang ZY, Zhang XY, Li H, Liang WQ, Luo N, Hu JM, Lu JQ, Luan TG (2006) Molecular cloning, purification, and biochemical characterization of a novel pyrethroid-hydrolyzing esterase from Klebsiella sp. strain ZD112. J Agric Food Chem 54(3):836–842

Wu Q, Zhao Y, Wang D, Xu Y (2013) Immobilized Rhodotorula mucilaginosa: a novel urethanase-producing strain for degrading ethyl carbamate. Appl Biochem Biotechnol 171(8):2220–2232

Wu P, Zhang Y, Chen Z, Wang Y, Zhu F, Cao B, Wu Y, Li N (2019) The organophosphorus pesticides in soil was degraded by Rhodobacter sphaeroides after wastewater treatment. Biochem Eng J 141:247–251

Xiao P, Mori T, Kamei I, Kondo R (2011) Metabolism of organochlorine pesticide heptachlor and its metabolite heptachlor epoxide by white rot fungi, belonging to genus Phlebia. FEMS Microbiol Lett 314(2):140–146

Yale R, Sapp M, Sinclair C, Moir J (2017) Microbial changes linked to the accelerated degradation of the herbicide atrazine in a range of temperate soils. Environ Sci Pollut Res 24(8):7359–7374

Yang L, Chen S, Hu M, Hao W, Geng P, Zhang Y (2011) Biodegradation of carbofuran by Pichia anomala strain HQ-C-01 and its application for bioremediation of contaminated soils. Biol Fertil Soils 47(8):917

Yang X, Wei H, Zhu C, Geng B (2018) Biodegradation of atrazine by the novel Citricoccus sp. strain TT3. Ecotoxicol Environ Saf 147:144–150

Zhai Y, Li K, Song J, Shi Y, Yan Y (2012) Molecular cloning, purification and biochemical characterization of a novel pyrethroid-hydrolyzing carboxylesterase gene from Ochrobactrum anthropi YZ-1. J Hazard Mater 221:206–212

Zhan H, Wang H, Liao L, Feng Y, Fan X, Zhang L, Chen S (2018) Kinetics and novel degradation pathway of permethrin in *Acinetobacter baumannii* ZH-14. Front Microbiol 9:98

Zhang C, Jia L, Wang S, Qu J, Li K, Xu L, Shi Y, Yan Y (2010) Biodegradation of beta-cypermethrin by two Serratia spp. with different cell surface hydrophobicity. Bioresour Technol 101(10):3423–3429

Zhang H, Zhang Y, Hou Z, Wang X, Wang J, Lu Z, Zhao X, Sun F, Pan H (2016) Biodegradation potential of deltamethrin by the Bacillus cereus strain Y1 in both culture and contaminated soil. Int Biodeterior Biodegradation 106:53–59

Zhang X, Gao Y, Zang P, Zhao Y, He Z, Zhu H, Song S, Zhang L (2019) Study on the simultaneous degradation of five pesticides by *Paenibacillus polymyxa* from Panax ginseng and the characteristics of their products. Ecotoxicol Environ Saf 168:415–422

Zhao J-S, Halasz A, Paquet L, Beaulieu C, Hawari J (2002) Biodegradation of hexahydro-1, 3, 5-trinitro-1, 3, 5-triazine and its mononitroso derivative hexahydro-1-nitroso-3, 5-dinitro-1, 3, 5-triazine by Klebsiella pneumoniae strain SCZ-1 isolated from an anaerobic sludge. Appl Environ Microbiol 68(11):5336–5341

Zhou G-C, Wang Y, Zhai S, Ge F, Liu Z-H, Dai Y-J, Yuan S, Hou J-Y (2013) Biodegradation of the neonicotinoid insecticide thiamethoxam by the nitrogen-fixing and plant-growth-promoting rhizobacterium Ensifer adhaerens strain TMX-23. Appl Microbiol Biotechnol 97(9):4065–4074

Zhu X, Dsikowitzky L, Kucher S, Ricking M, Schwarzbauer J (2019) Formation and fate of point-source non-extractable DDT-related compounds on their environmental aquatic-terrestrial pathway. Environ Sci Technol 53(3):1305–1314

Chapter 12
Biodegradation of Pesticides in Brazil and Other Tropical Countries: Experimental and In Silico Studies

Maricy Raquel Lindenbah Bonfá, Rodrigo Matheus Pereira, Rômulo Penna Scorza Júnior, and Caio César Achiles do Prado

Abstract Biodegradation is the most effective tool to attenuate pesticide contamination in soil; as a process that may also improve soil health and prevent contamination of water resources, it consequently averts threats to human and animal health. However, many factors influence the process, mainly the presence of microorganisms and certain enzymes that catalyze the reduction of pesticide molecules. Climate also impacts biological degradation due to its effects on microbiological activity and the solubility of compounds. Given that Brazil is one of the largest consumers of pesticides in the world, the study of pesticide biodegradation in such tropical zones is very important. Current research on this subject involves both laboratory experiments (i.e., isolation of biodegrading microorganisms) and in silico investigations (i.e., taxonomic and functional metagenomics and bioinformatics approaches with applications in metabolic pathways). This chapter provides an overview of how such research is conducted in the field of biodegradation, including a review of research taking place in Brazil and other tropical zones on the biodegradation of three widely used pesticides in Brazil: atrazine, fipronil, and glyphosate.

Keywords Atrazine · Fipronil · Glyphosate · Bioremediation · Metabolic pathway · Metagenomic

M. R. L. Bonfá (✉) · R. M. Pereira
Faculty of Biological and Environmental Sciences, Universidade Federal da Grande Dourados, Dourados, MS, Brazil
e-mail: maricybonfa@ufgd.edu.br

R. P. Scorza Júnior
Embrapa Agropecuária Oeste, Dourados, MS, Brazil

C. C. A. do Prado
University of São Paulo, Engineering School of Lorena, Lorena, SP, Brazil

A. Kumar, S. Sharma (eds.), *Microbes and Enzymes in Soil Health and Bioremediation*, Microorganisms for Sustainability 16,
https://doi.org/10.1007/978-981-13-9117-0_12

12.1 Introduction

Pesticides constitute a diverse group of inorganic and organic chemicals including herbicides, insecticides, nematicides, fungicides, and soil fumigants (Verma et al. 2014). Humans have benefited from the use of pesticides in various fields such as the production of higher yields of safe agricultural products, repelling home pests, and controlling infectious diseases—among which malaria eradication programs represent a highly successful application of insecticide use (Mostafalou and Abdollahi 2017). However, many research studies report the health problems that pesticides may cause in the short-term (alterations in complete blood count, hepatic and renal function, and nerve conduction velocities and amplitudes) and over long-term exposure (associated with increased abnormality in nerve conduction, especially in sensory nerves) (Hu et al. 2015). In addition to the health problems caused by pesticides, they constitute xenobiotic compounds that persist for long periods of time in the environment, the detoxification of which is of the utmost importance (Narwal and Gupta 2017).

Biodegradation is a significant natural tool for the decontamination of xenobiotics such as pesticides. In this process, microorganisms, i.e., bacteria, archaea, and fungi, utilize these compounds as their sole source of carbon and energy. Such microorganisms are highly adaptable with versatile catabolic pathways to degrade these persistent compounds through the evolution of new genes (Suenaga et al. 2010), giving them high metabolic plasticity that together with high reproductive rates and the horizontal transfer of genes ensure their capability to biodegrade xenobiotics (Kumar et al. 2016).

The bioremediation process of contaminated environments is extremely dependent on biodegradation; therefore, it is important to understand the metabolic and genetic factors involved, as well as the microbiota diversity capable of biodegrading such contaminants. Bioremediation occurs when the natural biodegradation process is monitored to verify the decontamination level and stimulated by biostimulation, aeration, and bioaugmentation methods.

The biodegradation and bioremediation processes depend on several biotic factors such as the richness and abundance of species harboring genes that encode for xenobiotic degradation enzymes, as well as abiotic factors such as temperature, humidity, presence of other carbon sources and macro and micronutrients, and the physical characteristics of the place where the contamination occurred. In this way, the soil and its composition can be a factor that either assists or limits the biodegradation process. Though agricultural soils in tropical regions require the use of high quantities of pesticides due to intense agricultural activity, studies on the behavior and degradation of pesticides in these environments are scarce (Laabs et al. 2002).

Brazil is one of the largest consumers of pesticides in the world, with growth in product sales up 200% from 2000 to 2013 (Gerage et al. 2017). This increase in consumption is due to high agricultural productivity. The transformation of natural ecosystems into agriculture can trigger disturbances in the carbon cycle equilibrium, particularly in tropical regions (Sul et al. 2013).

Consequently, studies involving the diversity of microbial groups, metabolic diversity, bioremediation tests, and assessments of the persistence of pesticides in tropical soils are essential. Bioremediation represents a novel tool for such processes, providing an eco-friendly, economical, and efficient method for the detoxification of pesticides, given that physical and chemical methods remain incomplete and costly (Uqab et al. 2016).

Prior to applying the bioremediation process to any newly released pollutant in the environment, an in silico study is required to predict possible degradation pathways using various computational tools. There are many databases and computer programs available to perform bioinformatics analysis, thereby assisting the development and implementation of microbial bioremediation (Khan 2013).

Other biodegradation/bioremediation techniques involve the evaluation of the diversity and function of the microbiota present in soils contaminated with pesticides through metagenomics, metatranscriptomics, and metaproteomics.

12.2 Pesticides and Agriculture in the Tropics

Pesticides have been widely used in agriculture to protect crops against attack by weeds, insects, and fungi. The potential for crop losses due to the attack of weeds, for example, is reported to be 23, 37, 40, 30, 37, and 36% for wheat, rice, maize, potatoes, soybean, and cotton, respectively (Oerke 2005). At the same time, farmers must boost crop productivity to handle the responsibility of feeding an ever-increasing population. Given that the availability of additional agricultural land is limited in many countries, including Brazil, the only option is to increase crop productivity on existing land using sustainable agriculture.

One way to overcome the challenges of sustainability, profitability, and crop productivity is to use the best available technology and innovation. In this context, pesticides are considered an important technology for agriculture. Schreinemachers and Tipraqsa (2012) state that a 1% increase in crop production per hectare is associated with an increase of 1.8% in pesticide use per hectare. The benefits of pesticide use to maintain and improve living standards remain significant enough to warrant their continued use. However, there are well-known environmental problems associated with this approach, including the contamination of water resources in the tropics (Albuquerque et al. 2016; Gama et al. 2017).

In the tropics, pest dynamics can be quite different from temperate regions. The absence of a severe winter can lead to the presence of pests throughout the year as long as they are able to locate hosts. Moreover, the favorable climatic conditions in the tropics exacerbate the damaging function of pests more so than in temperate regions. Thus, pesticide consumption in certain tropical countries with significant area devoted to agricultural is high, as for example in Brazil. However, when expressed in units per area, pesticide use in Brazil is about 7 kg ha^{-1} (IBGE, 2015), which is below that of Japan and some European countries such as Germany, France, and Italy.

After the application of pesticides on crops, a significant portion is deposited in soil. Thus, the soil compartment plays an important role in the fate of pesticides in the environment. Once in the soil compartment, different processes occur related to the transport, retention, and degradation of pesticides. It is important to keep in mind that soil is a dynamic ecosystem; therefore, pesticides can be transported to other compartments (i.e., air and water), degraded by microorganisms, and retained by soil particles (Navarro et al. 2007). Transport or the movement of pesticides is the main cause of water resources contamination (i.e., surface and groundwater). Carter (2000) states that pesticide losses from soil due to volatilization, leaching, and runoff amount to ~2–90%, 1–5%, and 0.001–0.25% of the amount applied, respectively.

The degradation of pesticides in soil is important to decrease the amount available for transportation, thereby diminishing the risk of water resources contamination.

Degradation is the main process responsible for the removal of pesticides from soil by attenuating the residues; it is classified into two types: biological (or biodegradation) and chemical degradation. Although it would appear to be easy to separate the chemical and biological degradation of pesticides, in practice they are very closely linked and are therefore treated as biochemical degradation (Navarro et al. 2007). Chemical degradation of pesticides in soil is considered a multi-step process involving hydrolysis, oxidation, reduction, and conjugation reactions. Many factors influence chemical degradation, including those related to environmental conditions (e.g., temperature and pH) and the chemical structure of pesticides. Biodegradation of pesticides in soil is broadly defined as the biologically catalyzed reduction in molecular complexity. In practice, we expect that the biodegradation of pesticides in soil will lead to the generation of smaller molecules or degradation products by soil microorganisms (i.e., bacteria and fungi) and their enzymes. During biodegradation, some microorganisms are able to use pesticide molecules as carbon and nitrogen sources (Fenner et al. 2013).

In the tropics, the degradation of pesticides in soil is facilitated by high temperature and precipitation, unlike that in temperate regions. Moreover, tropical soils exhibit different adsorption properties compared to temperate ones, which influence the bioavailability of pesticides for degradation. Thus, data extrapolation of pesticide degradation studies from temperate to tropical regions may not be valid. Unfortunately, only a few pesticide degradation studies have been conducted on tropical soils, including those involving Brazilian soil conditions (Brum et al. 2013; Bonfleur et al. 2015; Portilho et al. 2015; Scorza Júnior and Franco 2013).

12.3 Pesticide Biodegradation: Tools

The first studies on pesticide biodegradation were published in the second half of the last century (Alexander 1966). Since then, considerable technological progress has been made with regard to the techniques used. Studies of the biodegradation of xenobiotics involve several steps, as shown in Fig. 12.1 and discussed below.

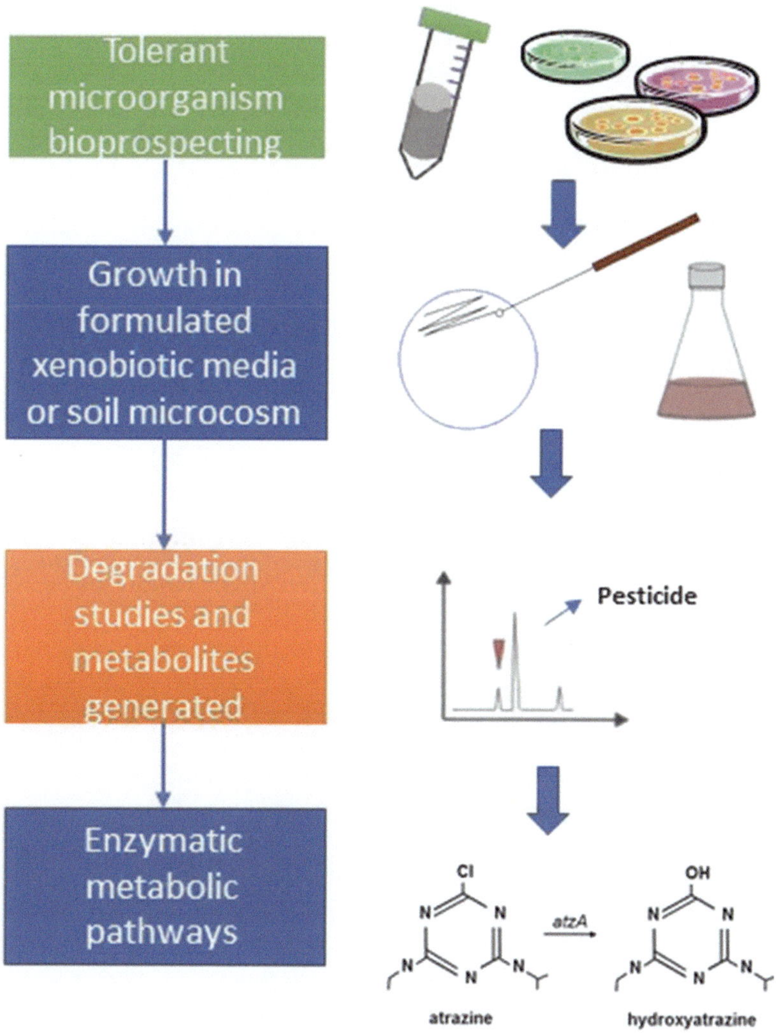

Fig. 12.1 Flowchart illustrating the steps generally involved in biodegradation studies

12.3.1 Bioprospecting of Microorganisms Able to Metabolize Pesticides

Although the term "bioprospecting" refers mainly to commercial research, in practice it is difficult to distinguish the term when the objective is both basic discovery and commercial applications (Antunes et al. 2016).

Bioprospecting of microorganisms in environments contaminated with pesticides aims to isolate microorganisms able to tolerate and grow in pesticides, using these chemicals as their sole source of carbon and/or nitrogen. Given that the isolated microbiota would be adapted to the target compound, it is also possible to use the microorganisms from such collections to carry out xenobiotic biodegradation studies. However, naturally occurring pesticide-degrading microorganisms may be relatively rare in unspoiled environments and non-exposed agricultural soils (Bartha 1990). A typical example of microorganism isolation and a model for atrazine biodegradation involved a strain known as *Pseudomonas* sp. ADP, isolated from an herbicide spill site (Mandelbaum et al. 1995).

Microbiological bioprospecting may be either culture-dependent or -independent (Fig. 12.2). In the former, there is the advantage of isolating the microorganisms capable of growing in culture media prepared in the laboratory, allowing them to be maintained and studied in the lab.

However, only 1% of microbiological diversity can be accessed by this method, resulting in studies restricted to only a small portion of the microorganisms that inhabit the soil. Bioprospecting using culture-independent methods (metagenomics) will be discussed later in this chapter.

12.3.2 Metabolization and Biodegradation (Analytical Methods)

Subsequent to the isolation of microorganisms with the capacity to metabolize pesticides, studies require experiments to evaluate the degradation of compounds. This stage can not only be carried out using the isolated microorganisms to verify degradation in the contaminated environment itself (bioremediation in situ) but also with methods that try to simulate the environment (micro- or mesocosms).

However, analytical methods that are routinely used to assess pesticide degradation [gas chromatography-mass spectrometry (GC-MS) or liquid chromatography-tandem mass spectrometry (LC-MS/MS)] do not distinguish transformations from other processes such as dilution or sorption unless combined with stringent mass balance modeling of the environmental system in question (Fenner et al. 2013). An alternative would be to perform mass balance with ^{14}C, but investigations with radioactively labeled substrates cannot be conducted in situ.

Another crucial issue in the study of biodegradation/bioremediation processes is ensuring careful experimental design that seeks to understand the different types of

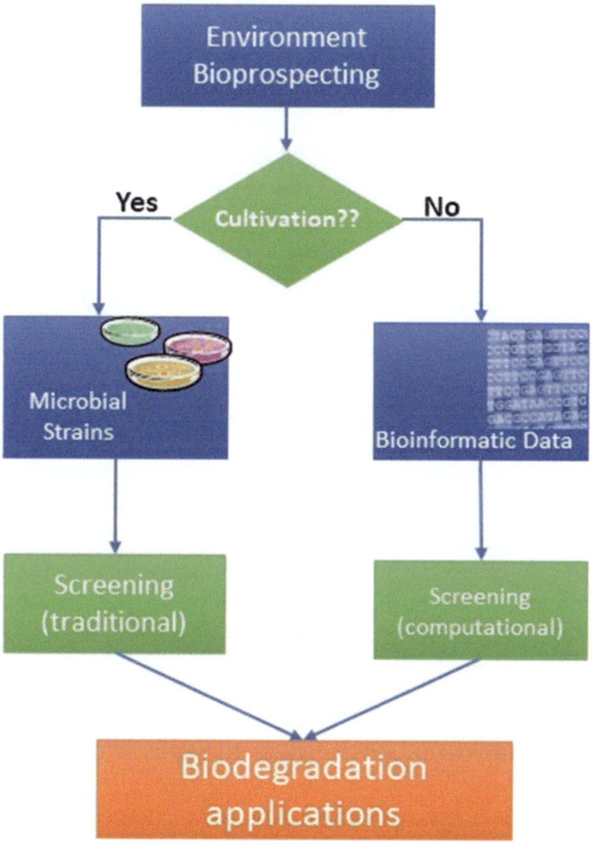

Fig. 12.2 Microbiological bioprospecting flowchart, culture-dependent or -independent

degradation to which pesticides are subjected, as photodegradation, chemical degradation, sorption, may occur in addition to biodegradation. These non-biological processes should be assessed when calculating the actual biodegradation of the compound.

Ideally, the compound under evaluation should be mineralized to CO_2 and water; however, in addition to mineralization, cometabolism also occurs, thereby generating metabolites that may be even more toxic than the original compound. Such metabolites must be detected by the methods mentioned above.

12.3.3 Enzymatic Metabolic Pathways

By observing the metabolites generated as a consequence of the biodegradation of pesticides, it is possible to infer enzymatic metabolic routes for each pesticide; these degradation routes may vary depending on the microorganism studied. The degradation routes of pesticides such as atrazine have been extensively investigated for over 20 years; they have been well established based on the analysis of genomic libraries, the screening of clones for the phenotypic expression of genes, gene sequencing, protein coding, and comparison with bioinformatics databases (Souza et al. 1995; De Souza et al. 1996).

Recently, the omics sciences have sought more robust tools in the study of metabolic pathways. Newly developed techniques such as transcriptomics, proteomics, and combined omics analysis offer remarkable promise as tools to address long-standing questions regarding the molecular mechanisms involved in the control of mineralization pathways (Singh and Nagaraj 2006). Transcriptomic and proteomic analyses enable researchers to predict and identify unknown catabolic pathways using mRNA and protein data that provide precise information on gene sequences and/or protein characteristics, while metabolomic analysis can be a powerful tool to validate the pathways (Cao et al. 2009). These techniques will be dealt with in greater detail later in this chapter.

12.4 Biodegradation of Pesticides: Studies in Tropical Areas

Despite the lack of information available for tropical regions, a review of recent studies on the biodegradation of three pesticides used frequently on Brazilian crops—atrazine, fipronil, and glyphosate—was undertaken to better understand the fate of pesticides in the country's soil and water resources.

12.4.1 Atrazine

Atrazine, an s-triazine herbicide, is used intensively on sugarcane, corn, and sorghum crops. Despite its low solubility of 34 mg L^{-1} in neutral aqueous medium, it has been detected in surface, rain, and ground waters at concentrations exceeding the permissible limit of contamination in water (Oturan et al. 2012). Researchers have shown that atrazine has toxic effects on algae, aquatic plants, aquatic insects, fish, and mammals (Sene et al. 2010) and its chlorinated metabolites are considered endocrine disruptors (Mamián et al. 2009; Sene et al. 2010). Consequently, atrazine was banned in several countries (Chan and Chu 2003), though its presence and that of its metabolites will persist in natural waters for several years (Oturan et al. 2012).

Due to the persistence and toxicity of this pesticide, it is important to identify its biodegradation pathways. Upon atrazine's initial use in the 1950s, only the non-specific pathway of the P450 monooxygenases, which do not provide energy for microbial growth, had been determined (Udikoviç-Koliç et al. 2012).

After their introduction, s-triazine compounds were found to be poor targets for biodegradation; more recently, however, it has been demonstrated that the environmental half-lives of these compounds have decreased substantially. This decrease indicates that microbes have developed new mechanisms to facilitate the degradation of s-triazine compounds. Early investigations of the breakdown of atrazine by microbes found that degradation proceeded mostly via dealkylation of the N-alkyl substituents on the s-triazine ring, usually with no subsequent ring cleavage (Fig. 12.3). More recently, most atrazine-degrading bacteria have been found to utilize a pathway that does involve ring cleavage. At the first stage, acyl substituents are removed from the compounds, resulting in the formation of cyanuric acid. At the second stage, the ring is cleaved and complete degradation to carbon dioxide and ammonia follows. A set of nearly identical s-triazine-catabolic genes has been found

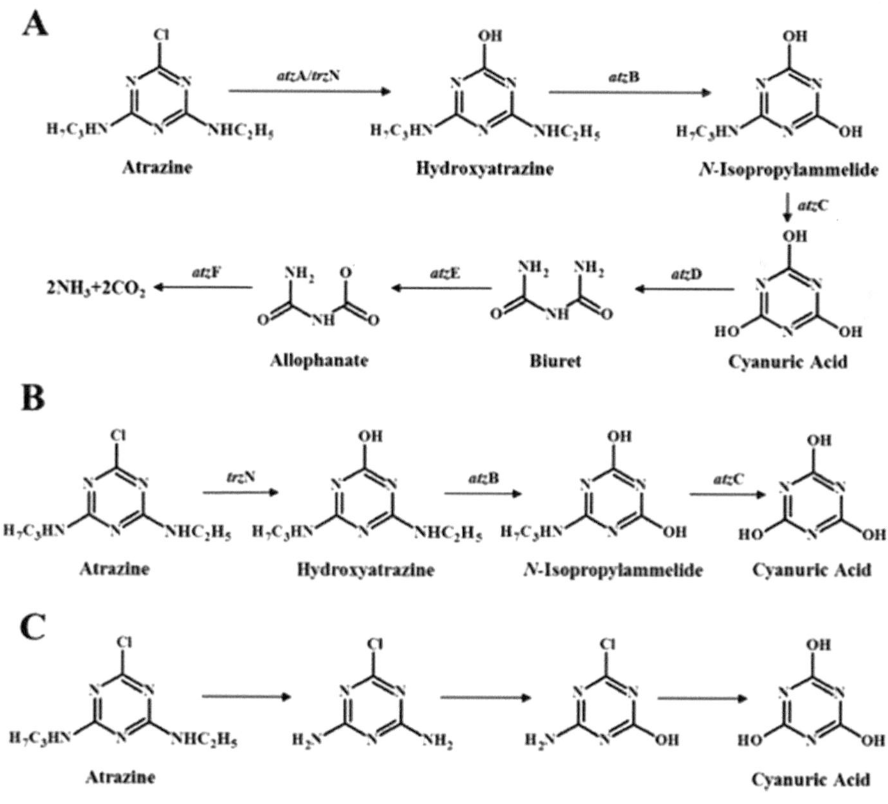

Fig. 12.3 The catabolic pathways of atrazine in *Pseudomonas* sp. ADP (**a**) and *Arthrobacter aurescens* TC1 (**b**) and a mechanism unclear pathway (**c**) (Huang et al. 2017)

worldwide in diverse bacterial genera; in almost all cases, the genes are present on plasmids, packaged via flanking insertion sequence elements. The genes encoding the pathway shown are named either atz or trz, depending on the organism. MONOMER-13542, encoded by the G-10313 gene, is found in gram-positive bacteria and has a very wide substrate specificity. The corresponding protein in gram-negative bacteria is CPLX-2201, encoded by the G-124 gene, which has a much narrower specificity; it catalyzes the hydrolytic displacement of chloride and fluoride substituents exclusively from the s-triazine ring (Geer et al. 2009).

Table 12.1 shows recent studies on atrazine biodegradation in Brazil and other tropical regions. Among the articles found, two stand out because they present possible metabolic routes that are different from those commonly found. Mesquini et al. (2015) isolated the actinomycete *Streptomyces* sp. atz2 from sugarcane leaves that degrade atrazine in a short half-life (4.1 days) yielding a still unknown metabolite of MS/MS with a confirmed mass of *m/z* 96.4. In addition, ecotoxicological assays performed with *Daphnia similis* have confirmed that this metabolite is non-toxic.

In a study on the basidiomycete *Pleurotus ostreatus* INCQS 40310, the authors suggested that, unlike other fungi in the same group that use extracellular oxidative enzymes or the P450 complex system, this fungus degrades atrazine using a different metabolic route, forming desethylatrazine (DEA) and deisopropylatrazine (DIA) when cultivated under agitation and DEA in static culture. Furthermore, they found a rapid degradation of atrazine (90% in 10 days of cultivation under agitation) when compared to other fungi of the same group (Balesteros et al. 2014).

12.4.2 Fipronil

Fipronil, a phenylpyrazole insecticide registered in the USA since 1996 (Gunasekara et al. 2007), is labeled for use with a large number of crops and is effective against a wide range of insect pests (Mandal et al. 2014). The half-life of fipronil in soil varies greatly, ranging from 3 days to 7 months (Bobe et al. 1998; Tingle et al. 2003); however, in studies performed on sugarcane soils in Brazil, half-life values were 15–105 days (Silva et al. 2016).

Fipronil is considered a "new generation" insecticide because its mode of action, involving interference with the normal function of g-aminobutyric acid (GABA) gated channels, differs from classical insecticides such as organophosphates and carbamates, to which some insects have developed resistance (Gunasekara et al. 2007).

Several studies on toxicity report that fipronil and its metabolites can be toxic to many organisms, including non-target insects such as bees (El Hassani et al. 2005; Bernadou et al. 2009), aquatic vertebrates (Stehr et al. 2006), and mammals (Tomlin 1997; Tingle et al. 2003; Das et al. 2006; Leghait et al. 2009; Romero et al. 2016).

The degradation of fipronil forms fipronil-sulfone, fipronil-sulfide (biotic and abiotic oxidation), and fipronil-desulfinyl (photolysis). Biotic degradation is the

Table 12.1 Atrazine degradation studies in Brazil and other tropical regions

Microorganism	Degradation pathway	Sampling	Country/ region	References
Agrobacterium tumefaciens	atzA, atzB, atzC, atzD, atzE, atzF, trzD and trzN	Agricultural soil (15-year history of atrazine application)	University of California, Riverside, USA	Smith et al. (2005)
Caulobacter crescentus				
Pseudomonas putida				
Sphingomonas yanoikuyae				
Nocardia sp.				
Rhizobium sp.				
Flavobacterium oryzihabitans				
Variovorax paradoxus				
Consortium (SMC)	Hydrolytic dehalogenation of ATZ to hydroxyatrazine	Contaminated soil (100-year-old mix load site) Reading, PA	Newdak, DE, EUA	Chirnside et al. (2009)
Pseudomonas aeruginosa	atzA, atzB, atzC, atzD, atzE and atzF	Amazon Forest	Amazon, Brazil	Fernandes et al. (2014)
Streptomyces sp. atz2	Possible new pathway	Leaves of sugarcane	Campinas, SP, Brazil	Mesquini et al. (2015)
Pseudomonas fluorescence	No data	Microbe bank of soil biology research department, Karaj, Iran	Tehran, Iran	Chegini et al. (2015)
Pseudomonas putida				
Ochrobactrum oryzae	No data	–	Shiraz, Iran	Shiri (2016)
Bacillus spp.	No data	The samples were obtained from the soil alluvial	New Delhi, India	Dutta et al. (2016)
Pseudomonas spp.				
Burkholderia spp.				
Pseudomonas sp. *Achromobacter* sp.	atzA, atzB, atzC, atzD, atzE and atzF atzA, atzB and atzC	Corn fields with historical use of atrazine	Ribeirão Preto, SP, Brazil	Tonelli Fernandes et al. (2018)
Pseudomonas sp. strain ACB	No data	Bacteria was isolated from the rhizoplanes of the roots	Delhi, India	James et al. (2018)
Arthrobacter sp. strain PKB				
Pseudomonas sp. strain PCB				
Pleurotus ostreatus INCQS 40310	Different metabolites formed in static cultivation and agitation	Provided by the Federal University of Lavras	Lavras, MG, Brazil	Balesteros et al. (2014)

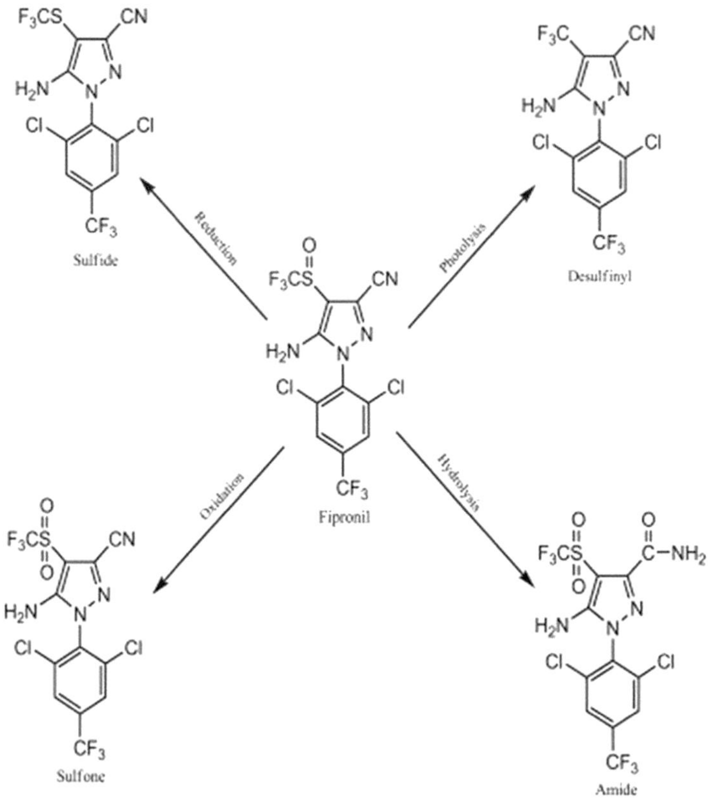

Fig. 12.4 The degradation of fipronil to fipronil-sulfide, fipronil-sulfone, fipronil-desulfinyl, and fipronil-amide via reductive, oxidative, photolytic, hydrolytic, and biotic reaction mechanisms (Mandal et al. 2013)

primary route for the degradation of fipronil in the soil. Fipronil-amides formed in smaller proportions by hydrolysis (Uniyal et al. 2016a) (Fig. 12.4).

Regarding the following translation: Research on the biodegradation of fipronil is relatively recent, with the first article on the topic appearing in 2002, published by Ying and Kookana. By virtue of these studies, the metabolites generated from biodegradation have now been identified via chromatography and detection by mass spectrometry. However, these publications do not associate biotic degradation with specific enzymes.

Table 12.2 compiles the main research on the biodegradation of fipronil in soil using isolated microorganisms in tropical regions. As described previously, the formation of fipronil-sulfide by reduction and fipronil-sulfone by oxidation generally occurs whenever the degradation process is biotic. Among the works cited, only the bacterium *Stenotrophomonas acidaminiphila* was also able to form fipronil-amide by hydrolysis (Uniyal et al. 2016a). Among the eukaryotes, *Aspergillus glaucus* AJAG1 forms the non-specific extracellular ligninolytic enzymes manganese

Table 12.2 Fipronil degradation studies in Brazil and other tropical regions

Microorganism	Degradation pathway	Sampling	Country/ region	References
Paracoccus sp.	No data	Soil under cotton fields (Punjab Agricultural University)	Punjab, India	Kumar et al. (2012)
Bacillus thuringiensis	No data	Soil from sugarcane fields, with history of extensive pesticide practice, in the Gurdaspur district	Punjab, Índia	Mandal et al. (2013)
Bacillus firmus	Fipronil degrades to its major metabolites by reduction to sulfide and oxidation to sulfone	Soil from sugarcane fields, with history of extensive pesticide practice, in the Gurdaspur district	Punjab, Índia	Mandal et al. (2014)
Acinetobacter calcoaceticus Acinetobacter oleivorans	No data	Soil from the rhizospheric zone situated at crop research Centre of the G.B.P.U.A.T. Pantnagar.	Srinagar Garhwal, Uttarakhand, India	Uniyal et al. (2016b)
Stenotrophomonas acidaminiphila	Fipronil degrades to its major metabolites by reduction to sulfide, oxidation to sulfone and hydrolysis to amide	Soil from the rhizospheric zone situated at crop research Centre of the G.B.P.U.A.T. Pantnagar	Srinagar Garhwal, Uttarakhand, India	Uniyal et al. (2016a)
Burkholderia thailandensis	Fipronil degrades to its major metabolites by reduction to sulfide and oxidation to sulfone	Soil from sugarcane fields, which had a long-term track-record of use of fipronil, in the city of São Carlos	São Carlos-SP, Brazil	Cappelini et al. (2018)
Aspergillus glaucus AJAG1	MnP and LiP	Soil from *Abelmoschus esculentus* field (commonly known as lady's finger)	Vellore, Tamil Nadu, India	Gajendiran and Abraham (2017)

peroxidase (MnP) and lignin peroxidase (LiP) during the degradation of fipronil (Gajendiran and Abraham 2017).

As evidenced by the literature review provided above, gaps exist in the understanding of fipronil biodegradation, especially regarding metabolic pathways, enzymes, and the characterization of the relevant genes. Thus, research involving systems biology should be undertaken.

12.4.3 Glyphosate

In the time since the herbicide glyphosate [N-(phosphonomethyl) glycine] was commercialized in 1974, it has become the most widely used herbicide in the world, due largely to the wide-scale adoption of transgenic, glyphosate-resistant (GR) crops after their introduction in 1996 (Duke et al. 2013).

The health impacts of glyphosate remain highly controversial. It was at first widely assumed that this pesticide would not cause health or environmental impacts because the target glyphosate enzyme, belonging to the shikimate pathway, does not exist in animal cells; it was therefore assumed that it would not cause harm to organisms lacking this pathway. Animal tests involving exposure to glyphosate generally yielded few observed effects, with results compared to those exhibited upon exposure to sodium chloride. Furthermore, studies on the environmental fate of glyphosate were also very promising: it was commonly understood that the compound was susceptible to rapid and complete decomposition via photolysis and microbial degradation to aminomethylphosphonic acid (AMPA), which was considered to be physiologically neutral. Glyphosate accumulation and soil mobility were considered insignificant, in that the propensity of the pesticide to be absorbed by plant roots and its impact on the soil microbiota appeared to be minor (Sviridov et al. 2014).

However, after more extensive independent studies were conducted, the interruption of embryonic development caused by glyphosate was verified. This was based on evidence of the adverse effects of herbicides containing glyphosate in people living in areas where they are heavily used. Agricultural workers handling these compounds reported pregnancy problems; women exposed during pregnancy showed an increase in the percentage of offspring with congenital malformations, including microcephaly, anencephaly, and cranial malformations (Savitz et al. 1997; Benítez-Leite et al. 2009).

These reports have led to more in-depth studies; currently, glyphosate is no longer considered innocuous by the scientific community. To eliminate glyphosate-related health and environmental risks, the development of an effective and eco-friendly bioremediation strategy is critical (Zhan et al. 2018).

Glyphosate can be used as the sole source of phosphorus, carbon, and/or nitrogen. Three main intermediate metabolites of glyphosate metabolism—AMPA, sarcosine, and acetylglyphosate (Fig. 12.5)—have been found which are further metabolized through different metabolic pathways. The most frequently detected metabolite of glyphosate degradation is AMPA, the intracellular degradation of which is impossible, causing it to be released to the environment resulting in secondary contamination (Zhan et al. 2018).

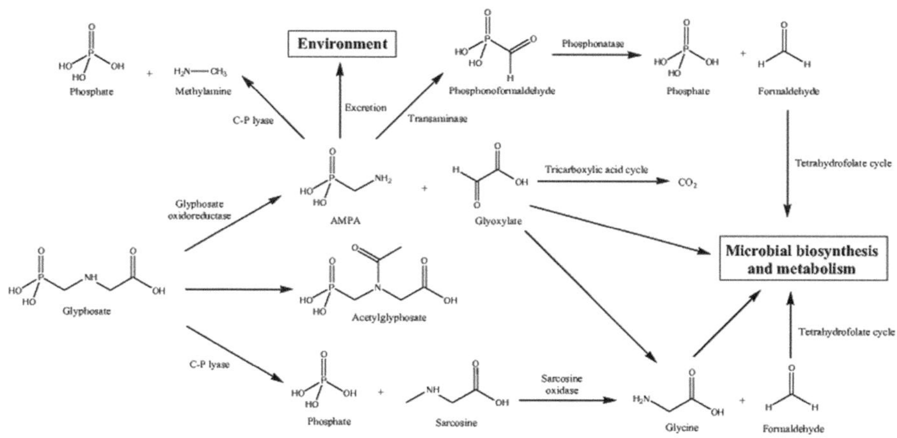

Fig. 12.5 Microbial mechanisms of glyphosate degradation (Zhan et al. 2018)

Zhan et al. 2018 published the most recent research on the metabolic routes of glyphosate degradation, describing three possible pathways that are dependent on microorganisms, with each pathway associated with different genes, hence different enzymes. (a) Cleavage of carboxymethylene-nitrogen (C-N) bond, catalyzed by glyphosate oxidoreductase (GOX) yielding AMPA and glyoxylate; (b) direct cleavage of carbon-phosphorus (C-P) bond, catalyzed by C-P lyase yielding sarcosine. Both of these degradation pathways may involve C-P lyase to break C-P bond in AMPA molecule; and (c) *Ochrobactrum anthropi* GPK3 has another totally different AMPA degradation pathway, where it was metabolized to phosphonoformaldehyde by transaminase and then catabolized to formaldehyde by phosphonatase (Sviridov et al. 2014).

Another monooxygenase enzyme reported was glycine oxidase (GO) can target glyphosate as it is a derivative of glycine. Both GOX/GO enzymes catalyze the breakdown of glyphosate to AMPA and glyoxylate at a low rate of activity (Iyer et al. 2018).

Research has reported the possibility of including enzymes capable of biodegrading herbicide residues in glyphosate-resistant plants. One possibility would be the insertion of the enzyme gene (BliGO) from *Bacillus licheniformis*. Transgenic crops with these characteristics are important from the environmental point of view (Zhang et al. 2016).

A review of recent articles on glyphosate biodegradation in tropical areas was carried out, revealing that in the last 10 years, few studies have been conducted in these areas, as can be seen in Table 12.3. These studies report that certain bacteria [*Bacillus cereus* CB4 (Fan et al. 2012) and *Comamonas odontotermitis* P2] had more than one degradation pathway (Firdous et al. 2017). Among fungi, the metabolic pathways for only *Fusarium oxysporum* strains 91,148 and 55.1 have been studied; they apparently use a different degradation pathway. Interestingly, according

Table 12.3 Glyphosate degradation studies in Brazil and other tropical regions

Microorganism	Degradation pathway	Country/region	Sampling	References
Bacillus cereus CB4	Glyphosate oxidoreductase and C-P lyase	Glyphosate-polluted soil in the herbicide plant located in Chengdu, Sichuan Province	Chengdu, China	Fan et al. (2012)
Bacillus subtilis Bs-15	No data	Was isolated from the rhizospheric soil of a pepper plant	Shandong, China	Yu et al. (2015)
Ochrobactrum sp.	The release of AMPA strongly suggests that herbicide degradation relies upon glyphosate oxidoreductase activity	Soils from with 5-year history of glyphosate treatment (Ghazvin, sari, Khoramabad, and Garmsar)	Tehran, Iran	Hadi et al. (2013)
*Comamonasodontotermitis*P2	Glyphosate oxidoreductase and C-P lyase	Soil in Sydney with glyphosate application history	Sydney, Australia	Firdous et al. (2017)
Fusarium Oxysporum strain (91,148 and 55.1)	AMPA was not observed	Isolated from sugarcane soil	Porto Alegre, Brazil	Castro et al. (2007)
Fusarium sp. FRP1	No data	The soil samples from the tropical forest area at Malang, East Java, Indonesia	Malang, Indonesia	Arfarita et al. (2011)
Scopulariopsis sp. FRP2				
Trichoderma sp. FRP3				
Trichoderma viride FRP3	No data	The soil samples from the tropical forest area at Malang, East Java, Indonesia	Malang, Indonesia	Arfarita et al. (2013)

to a recent review, all fungi studied showed the AMPA degradation pathway (Zhan et al. 2018).

As stated previously, pesticides function differently depending on climate; it is therefore of the utmost importance that more studies be done on the degradation and metabolization of glyphosate by microorganisms in tropical climates.

12.5 In Silico Studies of Pesticides Biodegradation

Bioinformatics, when applied to bioremediation, makes it possible to study the toxicity of pesticides and other chemical reagents, as well as microorganisms and their biodegradation metabolic pathways, in silico. Through the different tools of bioinformatics, it is possible to undertake preliminary analyses before going to the field to apply bioremediation treatments, thus allowing for the optimization of results and greater efficiency in remediating the environment. Data banks, metabolic pathway prediction systems, programs, web servers, and biological systems, all of which integrate various information from the omics sciences, can collaborate for in silico bioremediation analyses by studying the biodegradation of toxic chemicals present in the environment (Khan et al. 2013; Arora and Bae 2014; Shukla 2017; Chibwe et al. 2017; Malla et al. 2018).

12.5.1 Toxicity

In order for bioremediation to achieve maximum effectiveness, it is necessary to discover the level of toxicity of the compounds that are contaminating the environment. The survival of the microorganisms used in the biodegradation process will depend on the understanding of the toxicity of the compounds to be degraded (Khan et al. 2013).

The development of computational methods to assess chemical toxicity is occurring very rapidly. Many models for predicting chemical toxicity have been developed recently (Arora and Bae 2014).

The resolution of paths and different terms, as well as the different results obtained through the omics, should be used to construct the most relevant SAR (Q) models and thus assist in the development of the area of environmental toxicology (Cronin 2017).

A simple use tool can estimate the toxicity of chemicals using the Quantitative Structure Activity Relationship (QSAR) methodologies. The Toxicity Estimating Software Tool (TEST) can be accessed here: https://www.epa.gov/chemical-research/toxicity-estimation-software-tool-test

Another useful online tool to determine chemical toxicity is the ACTOR software that can be accessed at: https://actor.epa.gov/actor/home.xhtml.

There is a large set of databases and other tools for studying the toxicity of chemicals that can be accessed in the site area: compound toxicity databases from the OMIC tools site (https://omictools.com/) (Henry et al. 2014).

12.5.2 Databases and Tools

Obtaining information about the function of the metabolic pathways of different organisms can be very useful for studying possible alternatives in silico before starting a bioremediation treatment. Table 12.4 provides some interesting biological databases in this regard: Metacyc, Biocyc, and the most famous of all, KEGG.

Among the three biological databases, it can be said that KEGG was the pioneer in the field of analysis of in silico metabolic pathways; although it is not new, it is still the most frequently used as the basis for several other bioinformatics applications in the analysis of metabolic pathways.

Microorganisms are an essential part of bioremediation procedures; as important are the enzymes involved in the decomposition of toxic chemical compounds present in the environment. Below is a description of two databases related to enzymes: Brenda and Enzyme Portal in Table 12.4.

Other useful databases for biodegradation and bioremediation studies found in Table 12.4 are String, BioGRID, BioCatNet, XMetDB, BioSurfDB, Aromadeg, Bionemo, and OXDbase.

In Table 12.4, it is possible to find some databases on toxicity, metabolic pathways, enzymes, and several related to biodegradation.

Table 12.5 shows the webserver, pathways prediction system, and systems biology that can be applied to biodegradation/bioremediation. Three prediction systems used to predict the biodegradation of contaminants are listed in Table 12.5: EnviPath, Eawag-Soil package, and PathPhred. Another very interesting tool can also be observed in Table 12.5.

12.5.3 Omics Sciences Applied to Pesticide Degradation

Metagenomics is a technique that involves the study of the microbial diversity present in different environments. This technique is independent of the traditional methods of isolation of microorganisms, which allowed the discovery of many previously unknown microorganisms. The concept became more widely known from the work of Jo Handelsman et al. (1998) and grew almost exponentially as new generation sequencing technology evolved, becoming widely adopted around the world. Several research projects on the soil metagenome were initiated around the world, one of the largest being Terragenome (Thompson et al. 2017). In Brazil, the largest joint research initiative using metagenome is the Brazilian Microbiome Project (BMP) (Pylro et al. 2014).

An important observation about the metagenomic technique is that its function can be explained by a photograph at a given time, failing to identify the entire dynamics of the microbiota in an environment with the many variations that occur over time. To achieve this goal, many researchers collect samples at different times or situations to try to understand the dynamics of the microbiota. Although it is only

Table 12.4 Databases on toxicity, metabolic pathways, enzymes, and several related to biodegradation

Database	URL	Description
Metacyc	https://metacyc.org/	A curated database of experimentally elucidated metabolic pathways from all domains of life (Caspi et al. 2018)
STRING	https://string-db.org/	STRING is a database of known and predicted protein-protein interactions (Szklarczyk et al. 2017)
BioCyc	https://biocyc.org/	BioCyc.org is a microbial genome Web portal that combines thousands of genomes with additional information inferred by computer programs, imported from other databases and curated from the biomedical literature by biologist curators (Karp et al. 2017)
BioGRID	https://thebiogrid.org/	The Biological General Repository for Interaction Datasets is a database dedicated to the annotation and archival of protein, genetic and chemical interactions for all major model organism species and humans (Chatr-Aryamontri et al. 2017)
BioCatNet	https://biocatnet.de/	The BioCatNet is a site that contains links to a collection of family-specific enzyme databases, with a focus on enzymes of interest for biocatalysis (Buchholz et al. 2016)
XMetDB	http://www.xmetdb.org/xmetdb	*XMetDB* is an open access and open source database and web interface for the submission and retrieval of experimental metabolite data for drugs and other xenobiotics (Spjuth et al. 2016)
BioSurfDB	http://www.biosurfdb.org/	The main goal of this repository is to gather information on the characterization of biological compounds and mechanisms involved in biosurfactant production and/or biodegradation and make it available in a curated way and associated with a number of computational tools to support studies of genomic and metagenomic data (Oliveira et al. 2015)
Aromadeg	http://aromadeg.siona.helmholtz-hzi.de/	Database for phylogenomics of aerobic bacterial degradation of aromatics (Duarte et al. 2014)
KEGG	https://www.genome.jp/kegg/pathway.html	KEGG metabolic maps are part of the annotation platform for nearly every bacterial and archaeal genome sequencing project (Kanehisa et al. 2012)
Bionemo	http://bionemo.bioinfo.cnio.es/whatsin.html	Stores manually curated information about proteins and genes directly implicated in the Biodegradation metabolism (Carbajosa et al. 2009)
OxDbase	http://crdd.osdd.net/raghava/oxdbase/	The database contains information on over 240 oxygenases including both dioxygenases and monooxygenases involved in the biodegradation on xenobiotic compounds (Arora et al. 2009)
BRENDA	https://www.brenda-enzymes.org/	BRENDA is a comprehensive protein function database, containing enzymatic and metabolic information extracted from the primary literature (Schomburg et al. 2017)

(continued)

Table 12.4 (continued)

Database	URL	Description
Enzyme portal	https://www.ebi.ac.uk/enzymeportal/	Enzyme Portal provides an interface to all European Bioinformatics Institute (EMBL-EBI) data about enzymes (Pundir et al. 2010)
CompTox chemistry dashboard	https://comptox.epa.gov	A community data resource for environmental chemistry (Williams et al. 2017)

Table 12.5 Webserver, pathways prediction system, and systems biology that can be applied to biodegradation/bioremediation

System biology	URL	Description
MetNetX	https://www.metanetx.org/	MetaNetX.org is an online platform for accessing, analyzing and manipulating genome-scale metabolic networks (GSM) as well as biochemical pathways (Moretti et al. 2016)
BioSystems	https://www.ncbi.nlm.nih.gov/biosystems/	Provides integrated access to biological systems and their component genes, proteins, and small molecules, as well as literature describing those biosystems and other related data throughout Entrez (Geer et al. 2009)
Prediction system		
enviPath	https://envipath.org/	A platform created as an improvement over EAWAG-PPS by addressing combinatorial explosion concerns that arise from the use of EAWAG-PPS (Wicker et al. 2016)
Eawag-soil	https://envipath.org/package	EAWAG-SOIL package contains pathway information from soil degradation studies, stored in a biotransformation reaction scheme in the object pathway (Latino et al. 2017)
PathPhred	http://www.genome.jp/tools/pathpred/	Predicts pathways for microbial biodegradation of environmental compounds and biosynthesis of plant secondary metabolites (Moriya et al. 2010)
novoStoic	http://www.maranasgroup.com/metrxn	The novoStoic allows us to exploit enzyme plasticity by suggesting homologs to perform the hypothesized conversion when natural options are not available (Kumar et al. 2018)
Webserver		
iPath 3.0 (webserver)	http://pathways.embl.de/	Interactive Pathways Explorer (iPath) is a web-based tool for the visualization, analysis, and customization of various pathway maps (Darzi et al. 2018)

a moment, the metagenomic technique is extremely efficient in detecting all microorganisms present in the environment under analysis, many of them still unknown.

Although several metagenomics studies are conducted in Brazil, few studies involve development of metagenomics in areas contaminated with pesticides and applications in bioremediation. One study of metagenomics and pesticides by Souza et al. (2018), involving evaluation of the effects of the planting system on the soil microbiota, found that agricultural soils under conservative managements may represent a hotspot for bioprospection of hydrolases. Hydrolases are widely used for bioremediation of pesticide-contaminated environments (Ortiz-Hernández et al. 2013). Other published articles use simple techniques like DGGE to evaluate the effect of pesticides such as glyphosate, endolsulfan, and imazapyr on the soil microbiota (Botelho and Prata 2008), (Souza et al. 2013), (Babujia et al. 2016).

The metagenomic is very useful and accurate to reveal the microbiota present in the analyzed environment accurately. However, but for functional evaluation metatranscriptomic and metaproteomic techniques are the most indicated.

Environmental transcriptomics or metatranscriptomics is the study of the changes in the gene expression profiles of microbes and their regulation in natural environments at a specific place and time by sequencing the mRNA transcripts, which are randomly extracted from microbial communities (Biswas and Sarkar 2018). Singh et al. (2018) analyzed the metatranscriptome data of 20 wheat rhizosphere samples to decipher the taxonomic microbial communities and their multifunctionalities linked with the degradation of organic soil contaminants. The analysis revealed a total of 21 different metabolic pathways for the degradation of aromatic compounds and 6 for xenobiotic degradation. Abundance of transcripts related to the degradation of aromatic amine compounds, carbazoles, benzoates, naphthalene, ketoadipate pathway, phenols, biphenyls, and xenobiotics indicated abundant degradation capabilities in the soils. The results highlighted a potentially dominant role of crop rhizosphere-associated microbial communities in the remediation of contaminant aromatic compounds.

In metaproteomics, the complete proteome of a microbial community is analyzed (Maron et al. 2007). The goals of metaproteomics analysis are to identify bio-indicators of environmental health, to track new functional genes and complex metabolic pathways, and to review microbial ecology concepts with a functional point of view. The metaproteomics workflow can be separated into four steps. The first step is biological sample preparation, the second one is mass spectrometry analysis, the third one is the conversion of results to an exchangeable format, and the last one is identification (Zoun 2018). A recent study in Brazil characterized the bacterial communities in the phyllospheres of four tree species of the Atlantic Forest (*Mollinediaschottiana*, *Ocotea dispersa*, *Ocotea teleiandra*, and *Tabebuia serratifolia*) and their metaproteomes to examine the basic protein functional groups expressed in the phyllosphere (Lambais et al. 2017).

12.6 Conclusion

In conclusion, there remains a huge gap in information on pesticide-degrading microorganisms present in crops and tropical environments. The issue is relevant both in the terms of the understanding of metabolic pathways of degradation and in the bioremediation of contaminated areas. There are many methods involving both experimental and in silico studies that can be applied to pesticide biodegradation. Given that Brazil has some of the greatest biodiversity on the planet, many promising organisms with potential uses in the biodegradation of toxic compounds may be present in the diverse environments in the country.

Acknowledgments I would like to acknowledge Federal University of Grande Dourados, Embrapa Agropecuária Oeste, and University of São Paulo for the financial support.

5 References

Albuquerque AF, Ribeiro JS, Kummrow F et al (2016) Pesticides in Brazilian freshwaters: a critical review. Environ Sci Process Impacts 18:779–787. https://doi.org/10.1039/C6EM00268D
Alexander M (1966) Biodegradation of pesticides 1. In: Pesticides and their effects on soils and water, n. pesticidesandth, pp 78–84.
Antunes A, Simões MF, Grötzinger SW, et al (2016) Bioprospecting archaea: focus on extreme halophiles. In: Bioprospecting Success, Potential and Constraints. pp 81–112
Arfarita N, Imai T, Kanno A et al (2011) Screening of soil-born Fungi from Forest soil using glyphosate herbicide as the sole source of phosphorus. J Water Environ Technol 9:391–400. https://doi.org/10.2965/jwet.2011.391
Arfarita N, Imai T, Kanno A et al (2013) The potential use of trichoderma viride strain FRP3 in biodegradation of the herbicide glyphosate. Biotechnol Biotechnol Equip 27:3518–3521. https://doi.org/10.5504/bbeq.2012.0118
Arora PK, Bae H (2014) Integration of bioinformatics to biodegradation. Biol Proced Online 16:1–10. https://doi.org/10.1186/1480-9222-16-8
Arora PK, Kumar M, Chauhan A et al (2009) OxDBase: a database of oxygenases involved in biodegradation. BMC Res Notes 2:1–8. https://doi.org/10.1186/1756-0500-2-67
Babujia LC, Silva AP, Nakatani AS et al (2016) Impact of long-term cropping of glyphosate-resistant transgenic soybean [Glycine max (L.) Merr.] on soil microbiome. Transgenic Res 25:425–440. https://doi.org/10.1007/s11248-016-9938-4
Balesteros MR, De Sá LRV, Pereira PM et al (2014) Monitoring of atrazine biodegradation by Pleurotus ostreatus INCQS 40310 through the simultaneous analysis of atrazine and its derivatives by HPLC. Biocatal Biotransformation 32:23–33. https://doi.org/10.3109/10242422.2013.870557
Bartha R (1990) Isolation of microorganisms that metabolize xenobiotic compounds. In: Isolation of Biotechnological Organisms from Nature. pp 283–307
Benítez-Leite S, Macchi ML, Acosta M (2009) Malformaciones congénitas asociadas a agrotóxicos. Arch Pediatr Uruguay 80:237–247. https://doi.org/10.4067/S0370-41062009000400010
Bernadou A, Démares F, Couret-Fauvel T et al (2009) Effect of fipronil on side-specific antennal tactile learning in the honeybee. J Insect Physiol 55:1099–1106. https://doi.org/10.1016/j.jinsphys.2009.08.019
Biswas R, Sarkar A (2018) 'Omics' tools in soil microbiology: the state of the art. In: Advances in soil microbiology: recent trends and future prospects, 3rd edn. Springer, Singapore, pp 35–64

Bobe A, Cooper JF, Coste CM, Muller MA (1998) Behaviour of fipronil in soil under Sahelian plain field conditions. Pestic Sci 52:275–281. https://doi.org/10.1002/(SICI)1096-9063(199803)52:3<275::AID-PS720>3.0.CO;2-S
Bonfleur EJ, Tornisielo VL, Regitano JB, Lavorenti A (2015) The effects of glyphosate and atrazine mixture on soil microbial population and subsequent impacts on their fate in a tropical soil. Water Air Soil Pollut 226. https://doi.org/10.1007/s11270-014-2190-8
Botelho GR, Prata MC (2008) Efeito de glyphosate e imazaquin na comunidade bacteriana do rizoplano de soja (. 633–642
Brum CS, Franco AA, Scorza Júnior RP (2013) Degradação do herbicida sulfentrazone em dois solos de Mato Grosso do Sul. Rev Bras Eng Agrícola e Ambient 17:558–564
Buchholz PCF, Vogel C, Reusch W et al (2016) BioCatNet: a database system for the integration of enzyme sequences and biocatalytic experiments. Chembiochem 17:2093–2098. https://doi.org/10.1002/cbic.201600462
Cao B, Nagarajan K, Loh KC (2009) Biodegradation of aromatic compounds: current status and opportunities for biomolecular approaches. Appl Microbiol Biotechnol 85:207–228. https://doi.org/10.1007/s00253-009-2192-4
Cappelini LTD, Alberice JA, Eugênio PFM, et al (2018) Burkholderia thailandensis 29:1934–1943
Carbajosa G, Trigo A, Valencia A, Cases I (2009) Bionemo: molecular information on biodegradation metabolism. Nucleic Acids Res 37:598–602. https://doi.org/10.1093/nar/gkn864
Carter ADD (2000) Herbicide movement in soils: principles, pathways and processes. Weed Res 40:113–122. https://doi.org/10.1046/j.1365-3180.2000.00157.x
Caspi R, Billington R, Fulcher CA et al (2018) The MetaCyc database of metabolic pathways and enzymes. Nucleic Acids Res 46:D633–D639. https://doi.org/10.1093/nar/gkx935
Castro JV, Peralba MCR, Ayub MAZ (2007) Biodegradation of the herbicide glyphosate by filamentous fungi in platform shaker and batch bioreactor. J Environ Sci Heal - Part B Pestic Food Contam Agric Wastes 42:883–886. https://doi.org/10.1080/03601230701623290
Chan KH, Chu W (2003) Modeling the reaction kinetics of Fenton's process on the removal of atrazine. Chemosphere 51:305–311
Chatr-Aryamontri A, Oughtred R, Boucher L et al (2017) The BioGRID interaction database: 2017 update. Nucleic Acids Res 45:D369–D379. https://doi.org/10.1093/nar/gkw1102
Chegini S, Sani B, Darvishi HH (2015) The Effect of Pseudomonas Bacteria on Maize Growth and Atrazine Biodegradation in Soil 7:951–956
Chibwe L, Titaley IA, Hoh E, Simonich SLM (2017) Integrated framework for identifying toxic transformation products in complex environmental mixtures. Environ Sci Technol Lett 4:32–43. https://doi.org/10.1021/acs.estlett.6b00455
Chirnside AEM, Ritter WF, Radosevich M (2009) Biodegradation of aged residues of atrazine and alachlor in a mix-load site soil. Soil Biol Biochem 41:2484–2492. https://doi.org/10.1016/j.soilbio.2009.09.005
Cronin MTD (2017) (Q)SARs to predict environmental toxicities: current status and future needs. Environ Sci Process Impacts 19:213–220. https://doi.org/10.1039/c6em00687f
Darzi Y, Letunic I, Bork P, Yamada T (2018) IPath3.0: interactive pathways explorer v3. Nucleic Acids Res 46:W510–W513. https://doi.org/10.1093/nar/gky299
Das PC, Cao Y, Cherrington N et al (2006) Fipronil induces CYP isoforms and cytotoxicity in human hepatocytes. Chem Biol Interact 164:200–214. https://doi.org/10.1016/j.cbi.2006.09.013
De Souza ML, Wackett LP, Mandelbaum RT (1995) Cloning , characterization, and expression of a gene region from Pseudomonas sp . strain ADP involved in the dechlorination of atrazine. Cloning, Characterization, and Expression of a Gene Region from Pseudomonas sp. Strain ADP Involved in the Dechlo
De Souza ML, Sadowsky MJ, Wackett LP (1996) Atrazine chlorohydrolase from Pseudomonas sp. strain ADP: gene sequence, enzyme purification, and protein characterization. J Bacteriol 178:4894–4900. https://doi.org/10.1128/jb.178.16.4894-4900.1996

Duarte M, Jauregui R, Vilchez-Vargas R, et al (2014) AromaDeg, a novel database for phylogenomics of aerobic bacterial degradation of aromatics. Database (Oxford) 2014:bau118. https://doi.org/10.1093/database/bau118

Duke SO, Lydon J, Koskinen WC et al (2013) Erratum: glyphosate effects on plant mineral nutrition, crop rhizophere microbiota, and plant disease in glyphosate-resistant crops. J Agric Food Chem (2012) 60:42 (10375–10397) https://doi.org/10.1021/jf302436u. J Agric Food Chem 61:12745. https://doi.org/10.1021/jf405497g

Dutta A, Vasudevan V, Nain L, Singh N (2016) Characterization of bacterial diversity in an atrazine degrading enrichment culture and degradation of atrazine, cyanuric acid and biuret in industrial wastewater. J Environ Sci Heal - Part B Pestic Food Contam Agric Wastes 51:24–34. https://doi.org/10.1080/03601234.2015.1080487

El Hassani AK, Dacher M, Gauthier M, Armengaud C (2005) Effects of sublethal doses of fipronil on the behavior of the honeybee (Apis mellifera). Pharmacol Biochem Behav 82:30–39. https://doi.org/10.1016/j.pbb.2005.07.008

Fan J, Yang G, Zhao H et al (2012) Isolation, identification and characterization of a glyphosate-degrading bacterium, Bacillus cereus CB4, from soil. J Gen Appl Microbiol 58:263–271. https://doi.org/10.2323/jgam.58.263

Fenner K, Canonica S, Wackett LP, Elsner M (2013) Evaluating pesticide degradation in the environment: blind spots and emerging opportunities. Science 341(80):752–758. https://doi.org/10.1126/science.1236281

Fernandes AFT, da Silva MBP, Martins VV et al (2014) Isolation and characterization of a Pseudomonas aeruginosa from a virgin Brazilian Amazon region with potential to degrade atrazine. Environ Sci Pollut Res 21:13974–13978. https://doi.org/10.1007/s11356-014-3316-7

Firdous S, Iqbal S, Anwar S (2017) Optimization and modeling of glyphosate biodegradation by a novel Comamonas odontotermitis P2 through response surface methodology. Pedosphere 0160. https://doi.org/10.1016/S1002-0160(17)60381-3

Gajendiran A, Abraham J (2017) Biomineralisation of fipronil and its major metabolite, fipronil sulfone, by Aspergillus glaucus strain AJAG1 with enzymes studies and bioformulation. 3 Biotech 7:1–15. https://doi.org/10.1007/s13205-017-0820-8

Gama AF, Cavalcante RM, Duaví WC et al (2017) Occurrence, distribution, and fate of pesticides in an intensive farming region in the Brazilian semi-arid tropics (Jaguaribe River, Ceará). J Soils Sediments 17:1160–1169. https://doi.org/10.1007/s11368-016-1597-9

Geer LY, Marchler-Bauer A, Geer RC et al (2009) The NCBI BioSystems database. Nucleic Acids Res 38:492–496. https://doi.org/10.1093/nar/gkp858

Gerage JM, Meira APG, da Silva MV (2017) Food and nutrition security: pesticide residues in food. Forum Nutr 42:3. https://doi.org/10.1186/s41110-016-0028-4

Gunasekara AS, Truong T, Goh KS et al (2007) Environmental fate and toxicology of fipronil. J Pestic Sci 32:198–199. https://doi.org/10.1584/jpestics.R07-02

Hadi F, Mousavi A, Noghabi KA et al (2013) New bacterial strain of the genus Ochrobactrum with glyphosate-degrading activity. J Environ Sci Heal – Part B Pestic Food Contam Agric Wastes 48:208–213. https://doi.org/10.1080/03601234.2013.730319

Handelsman J, Rondon M, Brady SF, Jon Clardy RMG (1998) Molecular biological access to the chemistry of unknown soil microbes: a new frontier for natural products. Chem Biol 5:245–249. https://doi.org/10.1016/S1074-5521(98)90108-9

Henry VJ, Bandrowski AE, Pepin AS et al (2014) OMICtools: an informative directory for multi-omic data analysis. Database (Oxford) 2014:1–5. https://doi.org/10.1093/database/bau069

Hu R, Huang X, Huang J et al (2015) Long- and short-term health effects of pesticide exposure: a cohort study from China. PLoS One 10:1–13. https://doi.org/10.1371/journal.pone.0128766

Huang X, He J, Yan X et al (2017) Microbial catabolism of chemical herbicides: microbial resources, metabolic pathways and catabolic genes. Pestic Biochem Physiol 143:272–297. https://doi.org/10.1016/j.pestbp.2016.11.010

Iyer R, Iken B, Damania A, Krieger J (2018) Whole genome analysis of six organophosphate-degrading rhizobacteria reveals putative agrochemical degradation enzymes with broad

substrate specificity. Environ Sci Pollut Res 25:13660–13675. https://doi.org/10.1007/s11356-018-1435-2
James A, Singh DK, Khankhane PJ (2018) Enhanced atrazine removal by hydrophyte-bacterium associations and in vitro screening of the isolates for their plant growth-promoting potential. Int J Phytoremediation 20:89–97. https://doi.org/10.1080/15226514.2017.1337068
Kanehisa M, Goto S, Sato Y et al (2012) KEGG for integration and interpretation of large-scale molecular data sets. Nucleic Acids Res 40:109–114. https://doi.org/10.1093/nar/gkr988
Karp PD, Billington R, Caspi R et al (2017) The BioCyc collection of microbial genomes and metabolic pathways. Brief Bioinform:1–9. https://doi.org/10.1093/bib/bbx085
Khan F (2013) In silico approach for the bioremediation of toxic pollutants. J Pet Environ Biotechnol 04. https://doi.org/10.4172/2157-7463.1000161
Khan F, Sajid M, Cameotra SS (2013) In silico approach for the bioremediation of toxic pollutants. J Pet Environ Biotechnol 04. https://doi.org/10.4172/2157-7463.1000161
Kumar R, Singh B, Gupta VK (2012) Biodegradation of fipronil by Paracoccus sp. in different types of soil. Bull Environ Contam Toxicol 88:781–787. https://doi.org/10.1007/s00128-012-0578-y
Kumar M, Varma A, Kumar V (2016) Ecogenomics based microbial enzyme for biofuel industry. Sci Int 4:1–11. https://doi.org/10.17311/sciintl.2016.1.11
Kumar A, Wang L, Ng CY, Maranas CD (2018) Pathway design using de novo steps through uncharted biochemical spaces. Nat Commun 9:9. https://doi.org/10.1038/s41467-017-02362-x
Laabs V, Amelung W, Pinto AA et al (2002) Pesticides in surface water, sediment, and rainfall of the northeastern Pantanal Basin, Brazil. J Environ Qual 31:1636. https://doi.org/10.2134/jeq2002.1636
Lambais MR, Barrera SE, Santos EC et al (2017) Phyllosphere Metaproteomes of trees from the Brazilian Atlantic Forest show high levels of functional redundancy. Microb Ecol 73:123–134. https://doi.org/10.1007/s00248-016-0878-6
Latino DARS, Wicker J, Gütlein M et al (2017) Eawag-soil in enviPath: a new resource for exploring regulatory pesticide soil biodegradation pathways and half-life data. Environ Sci Process Impacts 19:449–464. https://doi.org/10.1039/c6em00697c
Leghait J, Gayrard V, Picard-Hagen N et al (2009) Fipronil-induced disruption of thyroid function in rats is mediated by increased total and free thyroxine clearances concomitantly to increased activity of hepatic enzymes. Toxicology 255:38–44. https://doi.org/10.1016/j.tox.2008.09.026
Malla MA, Dubey A, Yadav S et al (2018) Understanding and designing the strategies for the microbe-mediated remediation of environmental contaminants using omics approaches. Front Microbiol 9:9. https://doi.org/10.3389/fmicb.2018.01132
Mamián M, Torres W, Larmat FE (2009) Electrochemical degradation of atrazine in aqueous solution at a platinum electrode. Port Electrochim Acta 27:371–379. https://doi.org/10.4152/pea.200903371
Mandal K, Singh B, Jariyal M, Gupta VK (2013) Microbial degradation of fipronil by Bacillus thuringiensis. Ecotoxicol Environ Saf 93:87–92. https://doi.org/10.1016/j.ecoenv.2013.04.001
Mandal K, Singh B, Jariyal M, Gupta VK (2014) Bioremediation of fipronil by a Bacillus firmus isolate from soil. Chemosphere 101:55–60. https://doi.org/10.1016/j.chemosphere.2013.11.043
Mandelbaum RT, Allan DL, Wackett LP (1995) Isolation and characterization of a Pseudomonas sp. that mineralizes the s-Triazine herbicide atrazine. Appl Environ Microbiol 61:1451–1457
Maron P-A, Ranjard L, Mougel C, Lemanceau P (2007) Metaproteomics: a new approach for studying functional microbial ecology. Microb Ecol 53:486–493. https://doi.org/10.1007/s00248-006-9196-8
Mesquini JA, Sawaya ACHF, López BGC et al (2015) Detoxification of atrazine by endophytic Streptomyces sp. isolated from sugarcane and detection of nontoxic metabolite. Bull Environ Contam Toxicol 95:803–809. https://doi.org/10.1007/s00128-015-1673-7
Moretti S, Martin O, Van Du Tran T et al (2016) MetaNetX/MNXref - reconciliation of metabolites and biochemical reactions to bring together genome-scale metabolic networks. Nucleic Acids Res 44:D523–D526. https://doi.org/10.1093/nar/gkv1117

Moriya Y, Shigemizu D, Hattori M et al (2010) PathPred: an enzyme-catalyzed metabolic pathway prediction server. Nucleic Acids Res 38:138–143. https://doi.org/10.1093/nar/gkq318

Mostafalou S, Abdollahi M (2017) Pesticides: an update of human exposure and toxicity. Arch Toxicol 91:549–599. https://doi.org/10.1007/s00204-016-1849-x

Narwal SK, Gupta R (2017) Biodegradation of xenobiotic compounds. In: Handbook of Research on Inventive Bioremediation Techniques. IGI Global, pp 186–212

Navarro S, Vela N, Navarro G (2007) An overview on the environmental behaviour of pesticide residues in soils. Spanish J Agric Res 5:357–375. https://doi.org/10.5424/sjar/2007053-5344

Oerke EC (2005) Crop losses to pests. J Agric Sci 144:31–43. https://doi.org/10.1017/S0021859605005708

Oliveira JS, Araújo W, Sales AIL et al (2015) BioSurfDB: knowledge and algorithms to support biosurfactants and biodegradation studies. Database 2015:1–8. https://doi.org/10.1093/database/bav033

Ortiz-Hernández ML, Sanchez-Salinas E, Dantn Gonzlez E, Luisa M (2013) Pesticide biodegradation: mechanisms, genetics and strategies to enhance the process. In: biodegradation – life of science. InTech, p 64

Oturan N, Brillas E, Oturan MA (2012) Unprecedented total mineralization of atrazine and cyanuric acid by anodic oxidation and electro-Fenton with a boron-doped diamond anode. Environ Chem Lett 10:165–170. https://doi.org/10.1007/s10311-011-0337-z

Portilho IIR, Scorza Júnior RP, Salton JC et al (2015) Persistência de inseticidas e parâmetros microbiológicos em solo sob sistemas de manejo. Ciência Rural 45:22–28

Pundir S, Onwubiko J, Zaru R et al (2010) An update on the enzyme portal: an integrative approach for exploring enzyme knowledge. Aust J Pharm 91:26. https://doi.org/10.1093/protein/gzx008

Pylro VS, Roesch LFW, Ortega JM et al (2014) Brazilian microbiome project: revealing the unexplored microbial diversity-challenges and prospects. Microb Ecol 67:237–241. https://doi.org/10.1007/s00248-013-0302-4

Romero A, Ramos E, Ares I et al (2016) Fipronil sulfone induced higher cytotoxicity than fipronil in SH-SY5Y cells: protection by antioxidants. Toxicol Lett 252:42–49. https://doi.org/10.1016/j.toxlet.2016.04.005

Savitz DA, Arbuckle T, Kaczor D, Curtis KM (1997) Male pesticide exposure and pregnancy outcome. Am J Epidemiol 146:1025–1036. https://doi.org/10.1093/oxfordjournals.aje.a009231

Schomburg I, Jeske L, Ulbrich M et al (2017) The BRENDA enzyme information system–from a database to an expert system. J Biotechnol 261:194–206. https://doi.org/10.1016/j.jbiotec.2017.04.020

Schreinemachers P, Tipraqsa P (2012) Agricultural pesticides and land use intensification in high, middle and low income countries. Food Policy 37:616–626. https://doi.org/10.1016/j.foodpol.2012.06.003

Scorza Júnior RP, Franco AA (2013) A temperatura e umidade na degradação de fipronil em dois solos de Mato Grosso do Sul. Ciência Rural 43:1203–1209

Sene L, Converti A, Secchi GAR, de Simão RCG (2010) New aspects on atrazine biodegradation. Brazilian Arch Biol Technol 53:487–496. doi: https://doi.org/10.1590/S1516-89132010000200030

Shiri MA (2016) Isolation of Atrazine Degrading Bacteria in Semi-Salinity Medium. 4:

Shukla N (2017) Bioinformatics in environmental bioremediation – a review. Int J os Sci Res Sci Technol 3:195–205

Silva RO, Júnior RPS, Bonfá MRL et al (2016) Degradation and sorption of fpronil and atrazine in Latossols with organic residues from sugarcane crop. Ciência Rural 46:1172–1177. https://doi.org/10.1590/0103-8478cr20150696

Singh OV, Nagaraj NS (2006) Transcriptomics, proteomics and interactomics: unique approaches to track the insights of bioremediation. Briefings Funct Genomics Proteomics 4:355–362. https://doi.org/10.1093/bfgp/eli006

Singh DP, Prabha R, Gupta VK, Verma MK (2018) Metatranscriptome analysis deciphers multifunctional genes and enzymes linked with the degradation of aromatic compounds and pesticides in the wheat rhizosphere. Front Microbiol 9:1–15. https://doi.org/10.3389/fmicb.2018.01331

Smith D, Alvey S, Crowley DE (2005) Cooperative catabolic pathways within an atrazine-degrading enrichment culture isolated from soil. FEMS Microbiol Ecol 53:265–273. https://doi.org/10.1016/j.femsec.2004.12.011

Souza RA, Babujia LC, Silva AP et al (2013) Impact of the ahas transgene and of herbicides associated with the soybean crop on soil microbial communities. Transgenic Res 22:877–892. https://doi.org/10.1007/s11248-013-9691-x

Souza RC, Cantão ME, Nogueira MA et al (2018) Outstanding impact of soil tillage on the abundance of soil hydrolases revealed by a metagenomic approach. Brazilian J Microbiol 49:723–730. https://doi.org/10.1016/j.bjm.2018.03.001

Spjuth O, Rydberg P, Willighagen EL et al (2016) XMetDB: an open access database for xenobiotic metabolism. J Cheminform 8:1–8. https://doi.org/10.1186/s13321-016-0161-3

Stehr CM, Linbo TL, Incardona JP, Scholz NL (2006) The developmental neurotoxicity of fipronil: notochord degeneration and locomotor defects in zebrafish embryos and larvae. Toxicol Sci 92:270–278. https://doi.org/10.1093/toxsci/kfj185

Suenaga H, Mitsuoka M, Ura Y, et al (2010) Anna Esteve Garcia, Narcís Ferrer Ledo i Marc Guasch Rodríguez GRUP4 MATERIALS I MÈTODES. 183:5444. https://doi.org/10.1128/JB.183.18.5441

Sul WJ, Asuming-Brempong S, Wang Q et al (2013) Tropical agricultural land management influences on soil microbial communities through its effect on soil organic carbon. Soil Biol Biochem 65:33–38. https://doi.org/10.1016/j.soilbio.2013.05.007

Sviridov AV, Shushkova TV, Ermakova IT, et al (2014) Glyphosate: safety risks, biodegradation, and bioremediation. In: Current Environmental Issues and Challenges. Springer, Dordrecht/Heidelberg/New York/London, pp 183–196

Szklarczyk D, Morris JH, Cook H et al (2017) The STRING database in 2017: quality-controlled protein-protein association networks, made broadly accessible. Nucleic Acids Res 45:D362–D368. https://doi.org/10.1093/nar/gkw937

Thompson LR, Sanders JG, McDonald D et al (2017) A communal catalogue reveals Earth's multiscale microbial diversity. Nature 551:457–463. https://doi.org/10.1038/nature24621

Tingle CCD, Rother JA, Dewhurst CF, et al (2003) Fipronil: environmental fate, ecotoxicology, and human health concerns. In: Reviews of Environmental Contamination and Toxicology. pp 1–66

Tomlin C (1997) The pesticide manual: a world compendium, 11th edn. British Crop Protection Council and the Royal Society of Chemistry, Alton

Tonelli Fernandes AF, Braz VS, Bauermeister A et al (2018) Degradation of atrazine by Pseudomonas sp. and Achromobacter sp. isolated from Brazilian agricultural soil. Int Biodeterior Biodegrad 130:17–22. https://doi.org/10.1016/j.ibiod.2018.03.011

Udikoviç-Koliç N, Scott C, Martin-Laurent F (2012) Evolution of atrazine-degrading capabilities in the environment. Appl Microbiol Biotechnol 96:1175–1189. https://doi.org/10.1007/s00253-012-4495-0

Uniyal S, Paliwal R, Sharma RK, Rai JPN (2016a) Degradation of fipronil by Stenotrophomonas acidaminiphila isolated from rhizospheric soil of Zea mays. 3 Biotech 6:1–10. https://doi.org/10.1007/s13205-015-0354-x

Uniyal S, Paliwal R, Verma M et al (2016b) Isolation and characterization of Fipronil degrading Acinetobacter calcoaceticus and Acinetobacter oleivorans from Rhizospheric zone of Zea mays. Bull Environ Contam Toxicol 96:833–838. https://doi.org/10.1007/s00128-016-1795-6

Uqab B, Mudasir S, Nazir R (2016) Review on bioremediation of pesticides. J Bioremed Biodegr. https://doi.org/10.4172/2155-6199.1000343

Verma JP, Jaiswal DK, Sagar R (2014) Pesticide relevance and their microbial degradation: a-state-of-art. Rev Environ Sci Biotechnol 13:429–466. https://doi.org/10.1007/s11157-014-9341-7

Wicker J, Lorsbach T, Gütlein M et al (2016) enviPath - the environmental contaminant biotransformation pathway resource. Nucleic Acids Res 44:D502–D508. https://doi.org/10.1093/nar/gkv1229

Williams AJ, Grulke CM, Edwards J et al (2017) The CompTox chemistry dashboard: a community data resource for environmental chemistry. J Cheminform 9:1–27. https://doi.org/10.1186/s13321-017-0247-6

Ying GG, Kookana R (2002) Laboratory and field studies on the degradation of fipronil in a soil. Aust J Soil Res 40:1095–1102. https://doi.org/10.1071/SR02018

Yu XM, Yu T, Yin GH et al (2015) Glyphosate biodegradation and potential soil bioremediation by Bacillus subtilis strain Bs-15. Genet Mol Res 14:14717–14730. https://doi.org/10.4238/2015.November.18.37

Zhan H, Feng Y, Fan X, Chen S (2018) Recent advances in glyphosate biodegradation. Appl Microbiol Biotechnol 102:5033–5043. https://doi.org/10.1007/s00253-018-9035-0

Zhang K et al (2016) Characterization and directed evolution of BliGO, a novel glycine oxidase from Bacillus licheniformis. Enzym Microb Technol 85:12–18. https://doi.org/10.1016/j.enzmictec.2015.12.012

Zoun R (2018) Internet of Metaproteomics. 2018 IEEE 34th Int Conf Data Eng 1714–1718. https://doi.org/10.1109/ICDE.2018.00221

Chapter 13
Microbial Degradation of Phenolic Compounds

Sikandar I. Mulla, Ram Naresh Bharagava, Dalel Belhaj, Ganesh Dattatraya Saratale, Ashok Kumar, Aruliah Rajasekar, Preeti N. Tallur, Vijayalakshmi A. Edalli, Anyi Hu, and Chang-Ping Yu

Abstract Human beings use various synthetic products in day-to-day life. The vigorous manufacturing process needs various chemical compounds with different functional groups. Hence, due to human activities several of such molecules (including phenol

S. I. Mulla (✉)
Department of Biochemistry, School of Applied Sciences, Reva University, Bangalore, Karnataka, India

Key Laboratory of Urban Pollutant Conversion, Institute of Urban Environment, Chinese Academy of Sciences, Xiamen, China

R. N. Bharagava
Laboratory of Bioremediation and Metagenomics Research (LBMR), Department of Microbiology (DM), Babasaheb Bhimrao Ambedkar University (A Central University), Lucknow, Uttar Pradesh, India

D. Belhaj
Department of Life Sciences, Laboratory of Biodiversity and Aquatic Ecosystems Ecology and Planktonology, University of Sfax-Tunisia, FSS, CP, Sfax, Tunisia

G. D. Saratale
Department of Food Science and Biotechnology, Dongguk University-Seoul, Goyang-si, Gyeonggi-do, Republic of Korea

A. Kumar
Department of Biotechnology and Bioinformatics, Jaypee University of Information Technology, Waknaghat, Solan, Himachal Pradesh, India

A. Rajasekar
Environmental Molecular Microbiology Research Laboratory, Department of Biotechnology, Thiruvalluvar University, Vellore, Tamilnadu, India

P. N. Tallur
Government of Arts & Science College, Karwar, Karnataka State, India

V. A. Edalli
Department of Biochemistry, Karnatak University, Dharwad, Karnataka, India

A. Hu · C.-P. Yu
Key Laboratory of Urban Pollutant Conversion, Institute of Urban Environment, Chinese Academy of Sciences, Xiamen, China

A. Kumar, S. Sharma (eds.), *Microbes and Enzymes in Soil Health and Bioremediation*, Microorganisms for Sustainability 16,
https://doi.org/10.1007/978-981-13-9117-0_13

and its derivatives) were continuously present in the environmental surroundings which have been detected by advanced analytical tools. On the other hand, several reports revealed that most of these contaminants were toxic/hazardous in nature and some of them have consist carcinogenic and mutagenic properties. Hence, many of such contaminants including phenolic compounds were listed in United States Environmental Protection Agency list. For this reason, several researchers took a major step with the aim of detoxification/degradation of such contaminants by various treatment techniques (including biological methods) around the world. Considering this, here, we discuss some of these chemical contaminations and toxic effects and also their degradation/removal by microorganisms.

Keywords Biodegradation · Cresol · Chlorophenol · Phenol · Biological mediators · Carcinogenic · Mutagenic

13.1 Introduction

Several types of consumable products have been produced every year using different functional group containing substrates to improve the quality of lifestyle of human beings. During manufacturing process, knowingly or unknowingly, these chemicals are released into the environment. In addition, natural products are also released from a variety of foodstuff and human systems (Rather et al. 2017). Most of these environmental contaminants have shown their toxic effects towards living organisms (Edalli et al. 2016, 2018; Hoskeri et al. 2014; Mulla et al. 2017; Tallur et al. 2015). Additionally, different kinds of phenolic contaminants have shown carcinogenic and mutagenic effects on living organisms. Furthermore, advancement in analytical tools enhanced the detection level of several unidentified molecules, even at less than nanogram per litre. Hence, researchers recommended that the detected level of trace contaminants might still possess their adverse effect on living organisms and human systems (Templeton et al. 2009). Therefore, the presence of such contaminants in the ecosystem is of great worry because they may lead to severe impairment towards useful living organisms (Mulla et al. 2016a). On the other hand, different treatment methods were applied for the removal of such chemicals. Detoxification of chemical contaminants including phenolic compounds by biological (microorganisms) treatment was found to be a cost-effective and eco-friendly process.

Generally, microorganisms are freely available in the environment (soil and water), and they play a crucial role in the maintenance of biogeochemical cycle. Most of microorganisms possess the ability to degrade a variety of natural and synthetic chemicals (Alexander 1981; Arora and Bae 2014; Hoskeri et al. 2011; Li et al. 2018; Megadi et al. 2010; Mulla et al. 2012, 2013, 2017, 2018; Yu et al. 2013). Among the microorganisms, bacteria such as *Acinetobacter*, *Alcaligenes*, *Azotobacter*, *Bacillus*, *Flavobacterium*, *Micrococcus*, *Mycobacterium*, *Nocardia*, *Pseudomonas*, *Rhodococcus* and *Sphingomonas* species and fungal culture like *Aspergillus*, *Cunninghamella*, *Neurospora*, *Phanerochaete* and *Trichosporon* species have shown maximum versatility towards the degradation of various contaminants (Arora and Bae 2014; Fewson 1981; Mulla et al. 2016b, c). In this chapter, we have mainly focused on pollution and toxic effects of few phenolic compounds and also their degradative pathways in microorganisms.

13.2 Phenolic Chemicals as Environmental Contaminants

Different groups (nitro, chlorinated, sulfonated, hydroxyl, etc.) containing chemicals are the common contaminants, and they are found in large quantities in the environmental surroundings including industrial wastewater (Hoskeri et al. 2011; Mulla et al. 2011a, b, 2014, 2017, 2018; Talwar et al. 2014). Phenolic compounds are one of the important constituents used in various products due to their unique properties like they are easily soluble in aqueous solutions (especially water) and greatly mobile. For example, chlorophenols are used as biocides and antiseptics for home and hospital, and also few of them are used as chemical intermediates especially in the manufacture of the herbicides. Hence, chlorophenol production has increased in recent times due to industrial production and agriculture activities (Igbinosa et al. 2013). On the other hand, even at very minute concentration, these contaminants (for example, 0.005 mg/L of phenol) can cause odour and taste problems (Jogdand 2003). Hence, several phenolic derivatives are considered to be priority pollutants (Arora and Bae 2014; Nedal et al. 2007). For example, literature data reveals that the most of phenolic compounds are found to be very toxic and health hazardous which can cause cardiac arrhythmias, renal diseases, skin cancer and even death (Anku et al. 2017; Atlow et al. 1984; Igbinosa et al. 2013). Additionally, because of their lipophilic nature, these contaminants are increasingly accumulated within the food chain. Hence, several chlorinated phenolic compounds including 2,4-dichlorophenol, 2,4,5-trichlorophenol and pentachlorophenol were listed in the United States Environmental Protection Agency (Olaniran and Igbinosa 2011) and also in Water Quality Standard Database list as priority contaminants (Sharma et al. 1997). Therefore, the elimination of such contaminants is of great interest.

On the other hand, nitro group containing phenols are common components and they were extensively used in the production of consumer products (pesticides, pharmaceuticals, dyes and other useful chemicals) (Mulla et al. 2011a, b, 2012, 2013). In addition, the unique property of the nitro group leads to the usage of nitroaromatic compounds in the production of explosives (Mulla et al. 2014). On the other hand, due to the large quanties of such molecules utilization, they continue to persist in the environment (Mulla et al. 2014).

The primary source of cresols is creosote which is vastly utilized for the treatment of woods (Ehrlich et al. 1983; Goerlitz et al. 1985). Hence, cresols are often detected in leachate and also contaminate groundwater by coal gasification activities (Stuermer et al. 1982). Generally, these molecules are moderately soluble in water. Because of this, they often migrate with the groundwater and thereby cause widespread contamination of the aquatic system(s). Hence, these chemicals are of eco-geological worry due to their toxicity and mobility in the surface environments

(Londry et al. 1999). For example, *p*-cresol is among the isomers which is the main component of several consumer products (disinfectants, fumigants, preservatives, antioxidants, fragrance and dye industries, etc.). The primary sources of *p*-cresol are coal gasification plants, fractionation of coal tar and diverse synthetic processes (Muller et al. 2001). On the other hand, *p*-cresol has highly toxic and corrosive property. It also causes nervous system depression (Tallur et al. 2006). *m*-Cresol is another isomer and was vastly used in the preparation of various consumer products such as resin, herbicides, pharmaceuticals and surfactants (Shivaraman and Pandey 2000). Conversely, it was detected at high concentrations in industrial wastewater. And its solubility in water was found to be more than 24 g/L at room temperature (25 °C). Hence, it is a major risk to water body including groundwater and surface water (which are the main sources of drinking for living organisms) (Kavitha and Palanivelu 2005; Yan et al. 2006).

o-Cresol is commonly used in the synthesis of pesticides, epoxy resins, dyes, industrial disinfectants and cleaning agents and it is also used as a solvent. Inspite of its usefullness, o-cresol possess hazardous/toxic property. For example, interaction with *o*-cresol might cause severe skin burns and eye damage and has an adverse effect on the nervous system, cardiovascular system, kidney, lungs and liver. In addition, it also has carcinogenic property. Hence, cresols were included as primary contaminants in the United States Environmental Protection Agency list (Kavitha and Palanivelu 2005; Yan et al. 2006).

13.3 Microbial Degradation of Phenol and Its Derivatives

Various treatment processes like physicochemical methods and biological methods have been implicated to degrade phenol and its derivatives from industrial wastewater. Physical method, for example, absorption technique consists of activated sludge, carbon blacks, powdered activated carbon and pyrolysed rice husk among the materials that have been used and reported in the literature (Dominguez-Vargas et al. 2009; Loo et al. 2010; Zhao et al. 2008). On the other hand, chemical methods including photo-Fenton reaction were used for the treatment of natural or synthetic aromatic compounds (Gernjak et al. 2003). Solvent-impregnated resin system was used for the treatment of phenol and thiophenol(s) from water (Cuypers et al. 2010). Electrochemical detoxification was also used for the treatment of contaminated wastewater (Heyl and Jorissen 2006). There are reports on the degradation of phenols by phytoremediation and plant was used as a biological mediator during the process (Saiyood et al. 2010). In addition, algae such as *Ankistrodesmus braunii* and *Scenedesmus quadricauda* were shown to have the ability to degrade phenolic compound(s) (Pinto et al. 2003).

Use of microorganism(s) (bacterial and fungal cultures) is an alternative biological (other than plants and algae) approach towards bioremediation of phenolic contaminants in industrial wastewater and effluent. Because, these bological mediators were easily available in the environment. The microorganisms like *Alcaligenes* sp., *Bacillus subtilis*, *Nocardioides* sp., *Ralstonia eutropha*, *Sphingobium amiense* and *Sphingomonas* sp. were shown to have the ability to degrade phenol and its derivatives (El-Sayed et al. 2009; Mannisto et al. 2001; Padilla et al. 2000; Ushiba et al. 2003). Hence, in the following section, we discuss the mechanism of degradation pathways of phenol, chlorophenols and cresols in microorganisms.

13.3.1 Phenol

Microorganisms have shown the ability to utilize phenol as a growth substrate (Iqbal et al. 2018; Liu et al. 2016). In some cases, phenol was metabolized with the production of catechol and subsequently transformed via either *ortho*-ring cleavage pathway (catechol-1,2-dioxygenase enzyme) or *meta*-ring cleavage pathway (using catechol-2,3-dioxygenase). For example, bacterial cultures like *Alcaligenes* sp. A7–2, *Pseudomonas cepacia* strain CMA1, *Pseudomonas stutzeri* strain SPC2 and *Streptomyces setonii* (Aneez Ahamed and Kunhi 1996; Antai and Crawford 1983; Menke and Rehm 1992; Stockinger et al. 1992) degraded phenol through *ortho*-ring cleavage pathway, while bacterial species like *Bacillus stearothermophilus*, *Pseudomonas aeruginosa*, *Pseudomonas fluorescens* PU1, *Pseudomonas putida* and thermophilic bacteria *Bacillus* sp. A2 (Gurujeyalakshmi and Oriel 1989; Mahiudddin et al. 2012; Morsen and Rehm 1990; Mutzel et al. 1996; Ribbons 1970) were able to degrade phenol via *meta*-ring cleavage pathway. Recently, porous carbonaceous gels-immobilized cells of *Bacillus* sp. SAS19 were used for the removal of phenol and it was found that immobilized cells even at third cycle (reused) showed complete (100%) degradation of phenol at 1600 mg/L within 24 h (Ke et al. 2018). Furthermore, under anaerobic condition, phenol was degraded via 4-hydroxybenzoate in iron-reducing bacterial culture (GS-15) (Lovley and Lonergan 1990).

On the other hand, in fungal culture *(Aspergillus fumigatus)*, metabolism of phenol proceeded through two different routes. In the first route, it was converted to catechol by hydroxylation and subsequently transformed to 3-oxoadipate by intradiol mechanism. In the other pathway, phenol was converted to 1,2,4-trihydroxybenzene via hydroquinone and subsequently converted to malylacetate through *ortho*-ring fission (Jones et al. 1995). In addition, various research groups have demonstrated the use of fungal cultures like *Aspergillus niger, Aspergillus wentii, Phanerochaete chrysosporium, Pleurotus ostreatus, Pleurotus sajor caju, Rhizopus oryzae* and *Trametes versicolor* in the mycological degradation of phenol(s) (Fountoulakis et al. 2002; Justino et al. 2011; Mantzavinos and Kalogerakis 2005; Prabu and Udayasoorian 2005; Rocha-Santos et al. 2010). In addition, few yeast cultures (*Cryptococcus elinovii* H1, *Candida tropical*, *Candida tropicalis strain K1*, *Candida tropicalis strain* K11, *Meyerozyma guilliermondii strain K7*, *Pichia guilliermondii strain K2 and Trichosporon cutaneum*) were also shown to have the ability to degrade phenol (Karimi and Hassanshahian 2016; Morsen and Rehm 1990; Neujahr and Gaal 1973; Neujahr et al. 1974; Neujahr and Varga 1970; Yan et al. 2005). Microbial degradation of phenol is illustrated in Fig. 13.1.

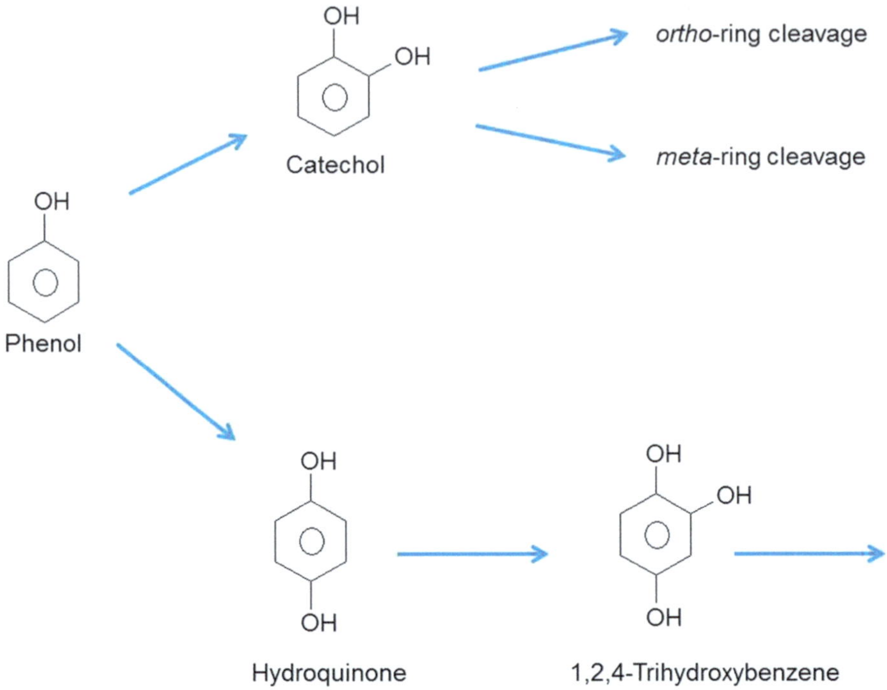

Fig. 13.1 Biodegradation of phenol by microorganisms

13.3.2 Chlorophenols

The chlorophenols are among the major class of environmental contaminants because of their toxic in nature and they continuously appear in the environmental surroundings (Haggblom 1990). Therefore, such kinds of chemicals have been the major objectives around the world and their possible bioremediation using biological mediators were focused (Madsen and Aamand 1991; Mohn and Kennedy 1992). Different types of microorganisms have shown the ability to metabolize mono-, di-, tri-, tetra- and pentachlorophenol (Arora and Bae 2014). For example, 4-chlorophenol is one of the simplest chlorinated compounds containing a single chlorine atom on its benzene ring. Bacteria like *Alcaligenes* sp. A7–2, *Alcaligenes xylosoxidans* JH1, *Arthrobacter chlorophenolicus* A6, *Arthrobacter ureafaciens* CPR706, *Herbaspirillum chlorophenolicum* CPW301, *Pseudomonas knackmussii* B-13, *Rhodococcus opacus* 1G and *Ralstonia pickettii* LD1 (Bae et al. 1996a, b; Fava et al. 1995; Finkel'shtein et al. 2000; Hollender et al. 2000; Im et al. 2004; Knackmuss and Hellwig 1978; Menke and Rehm 1992; Stolz et al. 2007; Westerberg et al. 2000) were shown to have the ability to utilize 4-chlorophenol as a growth substrate. Generally, 4-chlorophenol degradation proceeds in a different direction (either a chlorocatechol pathway or hydroquinone pathway) by bacteria. For example, it is initially transformed to 4-chlorocatechol by a dioxygenase enzyme and subsequently enters tricarboxylic acid (TCA) cycle via either *ortho*-ring cleavage (catechol-1,2-dioxygenase) with the formation of 3-chloromuconate or *meta*-ring cleavage (catechol-2,3-dioxygenase) with the generation of 5-chloro-2-hydoxymuconic semi-aldehyde (Arora and Bae 2014). In addition, 4-chlorocatechol pathway also proceeds via 1,2,4-trihydroxy benzene with the release of chlorine ions. In the other pathway (hydroquinone pathway), 4-chlorophenol is initially degraded to hydroquinone by monooxygenase with the release of chlorine atoms and subsequently converted to 1,2,4-trihydroxybenzene. Microbial degradation of 4-chlorophenol is shown in Fig. 13.2.

Recently, a research group demonstrated that stabilization and formulation will improve the stability of an organism *(Arthrobacter chlorophenolicus* A6) up to 60% around 3 months at 4 °C. The stabilized cells of *Arthrobacter chlorophenolicus* A6 after storing was shown to have the capability to degrade 4-chlorophenol in the same way as newly grown cells in two dissimilar set-ups by hygienized and non-treated soils (Bjerketorp et al. 2018). On the other hand, multiple enzymes (monooxygenase, CphC-I, and dioxygenase, CphA-I) were expressed by cloning a responsible gene *(cphC-I* and *cphA-I)* in *Arthrobacter chlorophenolicus* A6. The over expressed enzyme (CphC-I) was immobilized onto montmorillonite and was found to degrade 4-chlorophenol into hydroquinone (Kwean et al. 2018).

In addition, various organisms like *Alcaligenes xylosoxidans* JH1, *Alcaligenes* sp. A7–2, *Ralstonia pickettii* LD1, *Rhodococcus opacus* 1G and *Streptomyces rochei* 303 (Fava et al. 1995; Finkel'shtein et al. 2000; Golovleva et al. 1991; Hollender et al. 2000; Menke and Rehm 1992) were also shown to have the ability to utilize 2-chlorophenol as a growth substrate. In bacteria, 2-chlorophenol was initially transformed to 3-chlorocatechol and subsequently proceeded through either

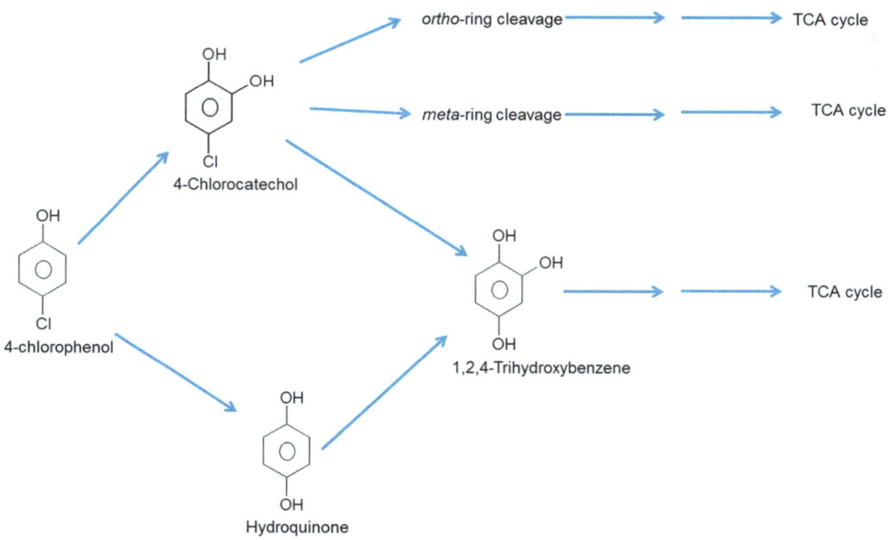

Fig. 13.2 Biodegradation of 4-chlorophenol by microorganisms

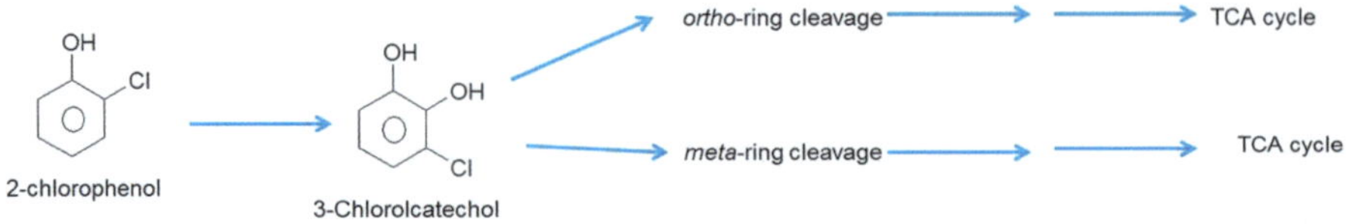

Fig. 13.3 Biodegradation of 2-chlorophenol by microorganisms

ortho-ring cleavage [there are two different *ortho*-ring cleavage pathways; the first one proceeds with the formation of 2-chloro-cis, cis-muconate (Solyanikova and Golovleva 2004) and the second pathway proceeds with the formation of chloromuconate cycloisomerase (Moiseeva et al. 2001; Moiseeva et al. 2002)] or *meta-ring* cleavage pathway. Catabolism of 2-chlorophenol in microorganism(s) is illustrated in Fig. 13.3.

On the other hand, microbial degradation of 3-chlorophenol occurs with the formation of either 3-chlorocatechol or 4-chlorocatethol and subsequently enters the TCS cycle through either *ortho-ring* cleavage or *meta*-ring cleavage (Arora and Bae 2014).

Furthermore, in bacteria, pentachlorophenol (a polychlorinated phenol) is transformed to 1,2,4-trihydroxybenzene via dichlorotrihydroxybenzene and tetrachlorohydroquinone and thereby enters the TCS cycle through the formation of a ring-cleavage product, 2-chloromaleylacetate by 2,6-dichloro-1,4-hydroquinone −1,2-dioxygenase (Fig. 13.4) (Arora and Bae 2014).

13.3.3 Cresols

Various microorganisms have shown the ability to degrade isomers of cresol. *p*-Cresol was among the isomers which proceeds in different pathways in bacteria (Fig. 13.5).

In the first pathway, *p*-cresol was initially converted to 4-methyl catechol and finally enters into TCS cycle via *meta*-ring cleavage pathway (Bayly et al. 1966; Hopper 1978). However, in other pathway, *p*-cresol was transformed to protocatechuic acid via 4-hydroxybenzoic acid and subsequently enters TCS cycle via *ortho-ring* fission (Dagley and Patel 1957; Hopper and Taylor 1975; Jones et al. 1993). On the other hand, Tallur and co-group demonstrated that during *p*-cresol degradation, various by-products like 4-hydroxybenzyl alcohol, 4-hydroxy benzaldehyde, 4-hydroxybenzoic acid and gentisic acid were generated and subsequently a ring fission by-product, maleyl pyruvate, was formed by gentisate-1, 2-dioxygenase in *Bacillus* sp. strain PHN 1 (Tallur et al. 2006). Additionally, immobilized cells of *Bacillus* sp. strain PHN 1 in different entrapments like polyurethane foam (PUF), polyacrylamide, alginate and agar were used for the degradation of *p*-creosol and higher concentrations (20 mM and 40 mM) of *p*-cresol degradation in polyurethane foam-immobilized cells were found compared to *Bacillus* cells immobilized in

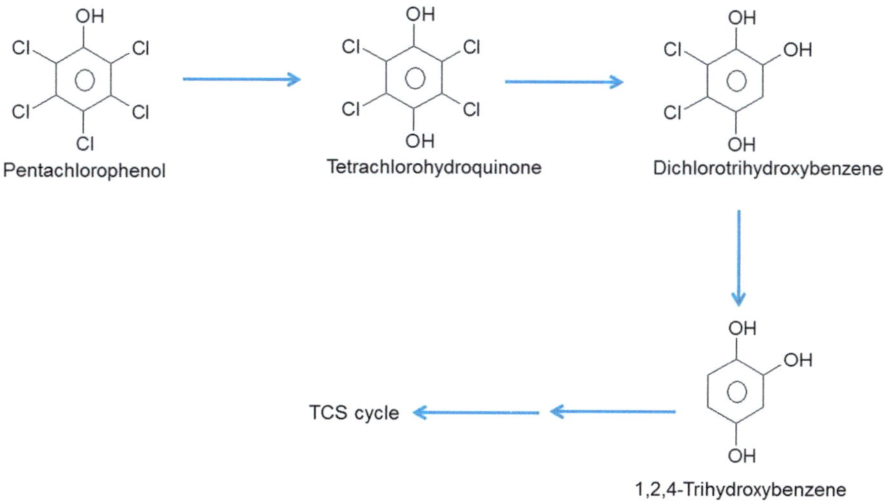

Fig. 13.4 Biodegradation of pentachlorophenol by microorganisms

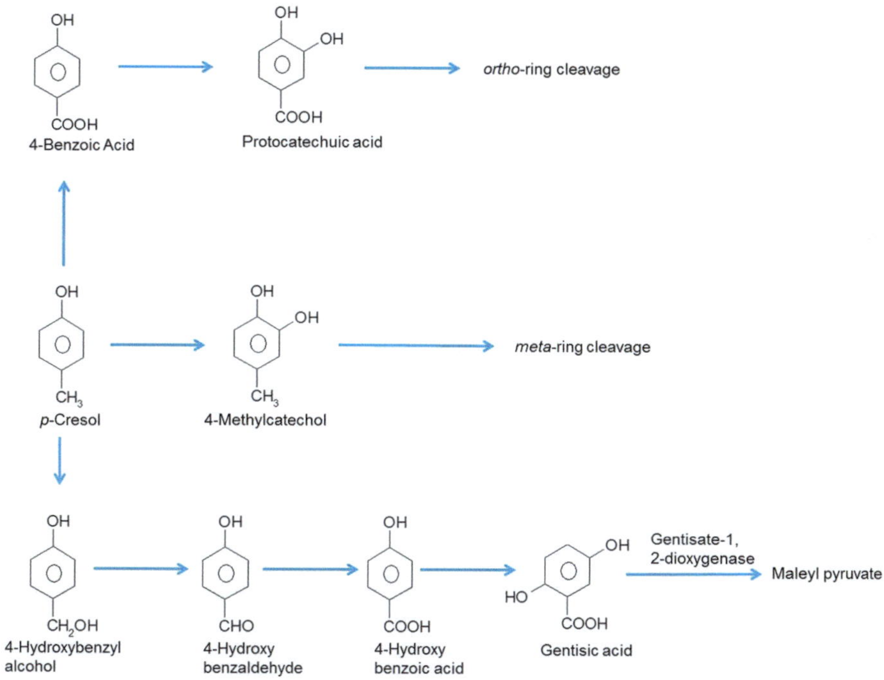

Fig. 13.5 Biodegradation of *p*-cresol by microorganisms

other matrices and also freely suspended cells of the same organism (Tallur et al. 2009). There are also few reports on the degradation of *p*-cresol under anaerobic conditions (Franchi et al. 2018b; Rudolphi et al. 1991; Suflita et al. 1989). However, during the degradation of phenol and *p*-cresol, a dissimilar role of the active microbial community (consisting a dominant *Syntrophorhabdus* and *Bacillus* species) compared to the total microbial community was observed and it was also found that *bamA* played a role in the degradation process (Franchi et al. 2018a).

Furthermore, in *Pseudomonas putida*, *m*-cresol was degraded to 3-hydroxybenzoic acid by oxidation process and subsequently transformed to gentisic acid (Hopper and Taylor 1975). In another pathway, *m*-cresol was transformed to 3-methylcatechol and was further transformed through *meta*-ring cleavage ring pathway (Hopper and Taylor 1975). Microbial degradation of *m*-cresol is shown in Fig. 13.6.

In addition, there are also numerous reports on the degradation of *m*-cresol under anaerobic conditions (Londry et al. 1997; Ramanand and Suflita 1991; Roberts et al. 1990).

On the other hand, *o*-cresol was highly resistant to biodegradation compared to *para*- and *meta*-isomers (Flyvbjerg et al. 1993). In the bacterial culture, *o*-cresol was initially transformed to 3-methylcatechol and subsequently enters the TCS cycle via 4-hydroxyl-2-oxo-valeric acid (Fig. 13.7) (Aneez Ahamed et al. 2001). On the other hand, in *Penicillium frequentans* Bi 712, co-metabolism of *o*-cresol proceeds with the generation of various by-products like methylhydroquinone, methyl-p-benzo-

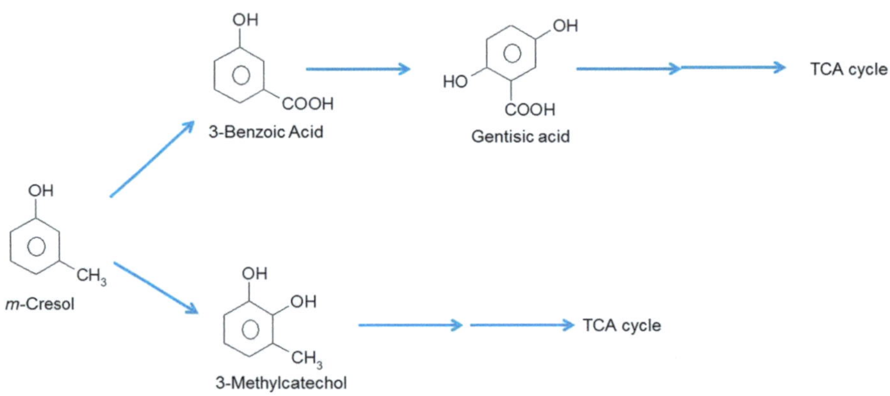

Fig. 13.6 Biodegradation of *m*-cresol by microorganisms

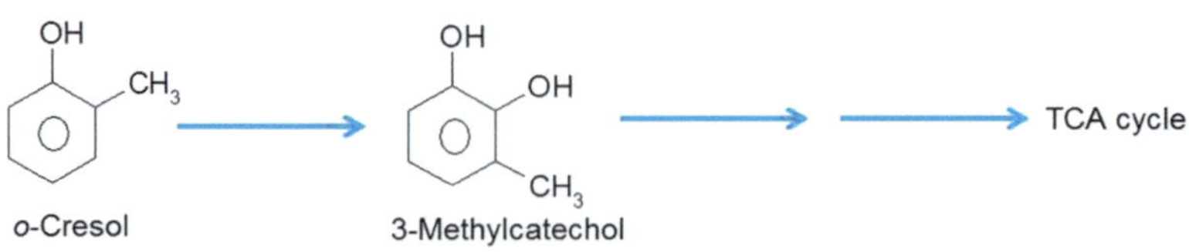

Fig. 13.7 Biodegradation of *o*-cresol by microorganisms

quinone, 2-methyl-5-hydroxyhydroquinone and 2-methyl-5-hydroxy-p-benzoquinone (Hofrichter et al. 1995). Recently, a research group studied the impact of *o*-cresol on various vegetables and found that mustard germination was reduced to around 64% and 12% at the concentration 25 and 50 mg *o*-cresol/kg soil, whereas the germination was restored when a bacterium, *Pseudomonas monteilii* SHY, was used to degrade *o*-cresol (Krishnan et al. 2018).

13.4 Conclusions

Overall, research studies around the universe demonstrated that after the usage of consumer products, part of phenolic compounds with their metabolites are present in the environmental surrounding. In recent times, some of the undetectable contaminants can be detected even at trace level (microgram to nanogram/L) mainly due to the availability of improved analytical tools. In fact, most of these contaminants also possess different kinds of toxicity which has been proven by various researchers. Hence, several research groups studied phenolic contaminants degradation using various methods. However, the biological method (especially using microbial cultures as a mediator) was found cost-effective compared to other methods (including physical and chemical). Furthermore, genomic and proteomic studies open up a new trend to study such contaminants (phenolic compounds) degradation mechanisms in microorganisms. However, it can further be improved by using an integrated process (a combination of chemical and biological methods) and also modification in organisms' genome.

Acknowledgement The authors are very thankful to Dr. H Z. Ninnekar for his valuable suggestions during the preparation of manuscript.

References

Alexander M (1981) Biodegradation of chemicals of environmental concern. Science 211:132–138

Aneez Ahamed PY, Kunhi AAM (1996) Degradation of phenol through *ortho*-cleavage pathway by *Pseudomonas stutzeri* strain SPC2. Lett Appl Microbiol 22:26–29

Aneez Ahamed PY, Kunhi AAM, Divakar S (2001) New metabolic pathway for *o*-cresol degradation by *Pseudomonas* sp. CP4 as evidenced by ^{1}H NMR spectroscopic studies. World J Microbiol Biotechnol 17:371–377

Anku WW, Mamo MA, Govender PP (2017) Phenolic compounds in water: sources, reactivity, toxicity and treatment methods. In: Soto-Hernandez M, Palma-Tenango M, Garcia-Mateos MR (eds) Phenolic compounds-natural sources importance and applications, 1st edn. InTech, Zagrep, pp 419–443

Antai SP, Crawford DL (1983) Degradation of phenol by *Streptomyces setonii*. Can J Microbiol 29:142–143

Arora PK, Bae HS (2014) Bacterial degradation of chlorophenols and their derivatives. Microb Cell Factories 13:31

Atlow SC, Bonadonna-Aparo L, Klibanov AM (1984) Dephenolization of industrial wastewaters catalyzed by polyphenol oxidase. Biotechnol Bioeng 26:599–603

Bae HS, Lee JM, Kim YB, Lee ST (1996a) Biodegradation of the mixtures of 4-chlorophenol and phenol by *Comamonas testosteroni* CPW301. Biodegradation 7(6):463–469

Bae HS, Lee JM, Lee ST (1996b) Biodegradation of 4-chlorophenol via a hydroquinone pathway by *Arthrobacter ureafaciens* CPR706. FEMS Microbiol Lett 145:125–129

Bayly RC, Dagley S, Gibson DT (1966) The metabolism of cresols by species of *Pseudomonas*. Biochem J 101:293–301

Bjerketorp J, Roling WFM, Feng XM, Garcia AH, Heipieper HJ, Hakansson S (2018) Formulation and stabilization of an *Arthrobacter* strain with good storage stability and 4-chlorophenol-degradation activity for bioremediation. Appl Microbiol Biotechnol 102:2031–2040

Cuypers R, Sudholter EJ, Zuilhof H (2010) Hydrogen bonding in phosphine oxide/phosphate-phenol complexes. ChemPhysChem 11:2230–2240

Dagley S, Patel MD (1957) Oxidation of p-cresol and related compounds by a *Pseudomonas*. Biochem J 66:227–233

Dominguez-Vargas JR, Navarro-Rodriguez JA, de Heredia JB, Cuerda-Correa EM (2009) Removal of chlorophenols in aqueous solution by carbon black low-cost adsorbents. Equilibrium study and influence of operation conditions. J hazard mater 169:302–308

Edalli VA, Mulla SI, Eqani SAMAS, Mahadevan GD, Sharma R, Shouche Y, Kamanavalli CM (2016) Evaluation of *p*-cresol degradation with polyphenoloxidase (PPO) immobilized in various matrices. 3 Biotech 6:229

Edalli VA, Patil KS, Le VV, Mulla SI (2018) An overview of aniline and chloroaniline compounds as environmental significances. Bioeng Biosci 1: SBB.000519.2018

Ehrlich GG, Godsy EM, Goerlitz DF, Hultz MF (1983) Microbial ecology of a creosote-contaminated aquifer at St. Louis park, Minnesota. Dev Ind Microbiol 24:235–245

El-Sayed WS, Ismaeil M, El-Beih F (2009) Isolation of 4-chlorophenol-degrading bacteria, *Bacillus Subtilis* OS1 and *Alcaligeges* sp. OS2 from petroleum oil-contaminated soil and characterization of its catabolic pathway. Aust J Basic Appl Sci 3:776–783

Fava F, Armenante PM, Kafkewitz D (1995) Aerobic degradation and dechlorination of 2-chlorophenol, 3-chlorophenol and 4-chlorophenol by a *Pseudomonas pickettii* strain. Lett Appl Microbiol 21:307–312

Fewson CA (1981) Biodegradation of aromatics with industrial relevance. In: Leisinger T, Cook AM, Hutter R, Nuesch J (eds) Microbial degradation of xenobiotics and recalcitrant compounds. Academic Press London, pp 141–179

Finkel'shtein ZI, Baskunov BP, Golovlev EL, Moiseeva OV, Vervoort J, Rietjens I, Golovleva LA (2000) Dependence of transformation of chlorophenols by *Rhodococci* on position and number of chlorine atoms in the aromatic ring. Mikrobiologiia 69:49–57

Flyvbjerg J, Jorgensen C, Arvin E, Jensen BK, Olsen SK (1993) Biodegradation of ortho-cresol by a mixed culture of nitrate-reducing bacteria growing on toluene. Appl Environ Microbiol 59:2286–2292

Fountoulakis MS, Dokianakis SN, Kornaros ME, Aggelis GG, Lyberatos G (2002) Removal of phenolics in olive mill wastewaters using the white-rot fungus *Pleurotus ostreatus*. Water Res 36:4735–4744

Franchi O, Bovio P, Ortega-Martinez E, Rosenkranz F, Chamy R (2018a) Active and total microbial community dynamics and the role of functional genes *bamA* and *mcrA* during anaerobic digestion of phenol and *p*-cresol. Bioresour Technol 264:290–297

Franchi O, Rosenkranz F, Chamy R (2018b) Key microbial populations involved in anaerobic degradation of phenol and p-cresol using different inocula. Electron J Biotechnol 35:33–38

Gernjak W, Krutzler T, Glaser A, Malato S, Caceres J, Bauer R, Fernandez-Alba AR (2003) Photo-Fenton treatment of water containing natural phenolic pollutants. Chemosphere 50:71–78

Goerlitz DF, Troutman DE, Godsy EM, Franks BJ (1985) Migration of wood-preserving chemicals in contaminated groundwater in a sand aquifer at Pensacola, Florida. Environ Sci Technol 19:955–961

Golovleva LA, Zaborina O, Pertsova R, Baskunov B, Schurukhin Y, Kuzmin S (1991) Degradation of polychlorinated phenols by *Streptomyces rochei* 303. Biodegradation 2:201–208
Gurujeyalakshmi G, Oriel P (1989) Isolation of phenol-degrading *Bacillus stearothermophilus* and partial characterization of the phenol hydroxylase. Appl Environ Microbiol 55:500–502
Haggblom M (1990) Mechanisms of bacterial degradation and transformation of chlorinated monoaromatic compounds. J Basic Microbiol 30(2):115–141
Heyl A, Jörissen J (2006) Electrochemical detoxification of waste water without additives using solid polymer electrolyte (SPE) technology. J Appl Electrochem 36:1281–1290
Hofrichter M, Bublitz F, Fritsche W (1995) Cometabolic degradation of *o*-cresol and 2,6-dimethylphenol by *Penicillium frequentans* Bi 712. J Basic Microbiol 5:303–313
Hollender J, Hopp J, Dott W (2000) Cooxidation of chloro- and methylphenols by *Alcaligenes xylosoxidans* JH1. World J Microbiol Biotechnol 16:445–450
Hopper DJ (1978) Incorporation of [^{18}O]water in the formation of p-hydroxybenzyl alcohol by the p-cresol methylhydroxylase from *Pseudomonas putida*. Biochem J 175:345–347
Hopper DJ, Taylor DG (1975) Pathways for the degradation of *m*-cresol and *p*-cresol by *Pseudomonas putida*. J Bacteriol 122:1–6
Hoskeri RS, Mulla SI, Shouche YS, Ninnekar HZ (2011) Biodegradation of 4-chlorobenzoic acid by *Pseudomonas aeruginosa* PA01 NC. Biodegradation 22:509–516
Hoskeri RS, Mulla SI, Ninnekar HZ (2014) Biodegradation of chloroaromatic pollutants by bacterial consortium immobilized in polyurethene foam and other matrices. Biocatal Agric Biotechnol 3:390–396
Igbinosa EO, Odjadjare EE, Chigor VN, Igbinosa IH, Emoghene AO, Ekhaise FO, Igiehon NO, Idemudia OG (2013) Toxicological profile of chlorophenols and their derivatives in the environment: the public health perspective. Sci World J 2013:460215
Im WT, Bae HS, Yokota A, Lee ST (2004) *Herbaspirillum chlorophenolicum* sp. nov., a 4-chlorophenol-degrading bacterium. Int J Syst Evol Microbiol 54:851–855
Iqbal A, Arshad M, Hashmi I, Karthikeyan R, Gentry TJ, Schwab AP (2018) Biodegradation of phenol and benzene by endophytic bacterial strains isolated from refinery wastewater-fed *Cannabis sativa*. Environ Technol 39:1705–1714
Jogdand SN (2003) Environmental biotechnology (industrial pollution management). Himalaya Publishing House, Mumbai
Jones KH, Trudgill PW, Hopper DJ (1993) Metabolism of *p*-cresol by the fungus *Aspergillus fumigatus*. Appl Environ Microbiol 59:1125–1130
Jones KH, Trudgill PW, Hopper DJ (1995) Evidence of two pathways for the metabolism of phenol by *Aspergillus fumigatus*. Arch Microbiol 163:176–181
Justino C, Marques AG, Rodrigues D, Silva L, Duarte AC, Rocha-Santos T, Freitas AC (2011) Evaluation of tertiary treatment by fungi, enzymatic and photo-Fenton oxidation on the removal of phenols from a kraft pulp mill effluent: a comparative study. Biodegradation 22:267–274
Karimi M, Hassanshahian M (2016) Isolation and characterization of phenol degrading yeasts from wastewater in the coking plant of Zarand, Kerman. Braz J Microbiol 47:18–24
Kavitha V, Palanivelu K (2005) Destruction of cresols by Fenton oxidation process. Water Res 39:3062–3072
Ke Q, Zhang Y, Wu X, Su X, Wang Y, Lin H, Mei R, Hashmi MZ, Chen C, Chen J (2018) Sustainable biodegradation of phenol by immobilized *Bacillus* sp. SAS19 with porous carbonaceous gels as carriers. J Environ Manag 222:185–189
Knackmuss HJ, Hellwig M (1978) Utilization and cooxidation of chlorinated phenols by Pseudomonas sp. B 13. Arch Microbiol 117:1–7
Krishnan SN, Nayarisseri A, Rajamanickam U (2018) Biodegradation effects of *o*-cresol by *Pseudomonas monteilii* SHY on mustard seed germination. Bioinformation 14(6):271–278
Kwean OS, Cho SY, Yang JW, Cho W, Park S, Lim Y, Shin MC, Kim HS, Park J (2018) 4-Chlorophenol biodegradation facilitator composed of recombinant multi-biocatalysts immobilized onto montmorillonite. Bioresour Technol 259:268–275

Li M, Zhao X, Zhang X, Wu D, Leng S (2018) Biodegradation of 17beta-estradiol by bacterial co-culture isolated from manure. Sci Rep 8:3787

Liu Z, Xie W, Li D, Peng Y, Li Z, Liu S (2016) Biodegradation of phenol by bacteria strain *Acinetobacter calcoaceticus* PA isolated from phenolic wastewater. Int J Environ Res Public Health 13:300

Londry KL, Fedorak PM, Suflita JM (1997) Anaerobic degradation of *m*-cresol by a sulfate-reducing bacterium. Appl Environ Microbiol 63:3170–3175

Londry KL, Suflita JM, Tanner RS (1999) Cresol metabolism by the sulfate-reducing bacterium *Desulfotomaculum* sp. strain Groll. Can J Microbiol 45:458–463

Loo YM, Lim PE, Seng CE (2010) Treatment of *p*-nitrophenol in an adsorbent-supplemented sequencing batch reactor. Environ Technol 31:479–487

Lovley DR, Lonergan DJ (1990) Anaerobic oxidation of toluene, phenol, and *p*-cresol by the dissimilatory iron-reducing organism, GS-15. Appl Environ Microbiol 56:1858–1864

Madsen T, Aamand J (1991) Effects of sulfuroxy anions on degradation of pentachlorophenol by a methanogenic enrichment culture. Appl Environ Microbiol 57:2453–2458

Mahiudddin M, Fakhruddin AN, Abdullah Al M (2012) Degradation of phenol via *meta* cleavage pathway by *Pseudomonas fluorescens* PU1. ISRN Microbiol 2012:741820

Männistö MK, Tiirola MA, Puhakka JA (2001) Degradation of 2,3,4,6-tetrachlorophenol at low temperature and low dioxygen concentrations by phylogenetically different groundwater and bioreactor bacteria. Biodegradation 12:291–301

Mantzavinos D, Kalogerakis N (2005) Treatment of olive mill effluents Part I. Organic matter degradation by chemical and biological processes--an overview. Environ Int 31:289–295

Megadi VB, Tallur PN, Mulla SI, Ninnekar HZ (2010) Bacterial degradation of fungicide captan. J Agric Food Chem 58:12863–12868

Menke B, Rehm HJ (1992) Degradation of mixtures of monochlorophenols and phenol as substrates for free and immobilized cells of *Alcaligenes* sp. A7-2. Appl Microbiol Biotechnol 37:655–661

Mohn WW, Kennedy KJ (1992) Reductive dehalogenation of chlorophenols by *Desulfomonile tiedjei* DCB-1. Appl Environ Microbiol 58:1367–1370

Moiseeva OV, Belova OV, Solyanikova IP, Schlomann M, Golovleva LA (2001) Enzymes of a new modified ortho-pathway utilizing 2-chlorophenol in *Rhodococcus opacus* 1CP. Biochem Biokhimiia 66:548–555

Moiseeva OV, Solyanikova IP, Kaschabek SR, Groning J, Thiel M, Golovleva LA, Schlomann M (2002) A new modified *ortho* cleavage pathway of 3-chlorocatechol degradation by Rhodococcus opacus 1CP: genetic and biochemical evidence. J Bacteriol 184(19):5282–5292

Morsen A, Rehm HJ (1990) Degradation of phenol by a defined mixed culture immobilized by adsorption on activated carbon and sintered glass. Appl Microbiol Biotechnol 33:206–212

Mulla SI, Hoskeri RS, Shouche YS, Ninnekar HZ (2011a) Biodegradation of 2-nitrotoluene by *Micrococcus* sp. strain SMN-1. Biodegradation 22:95–102

Mulla SI, Manjunatha TP, Hoskeri RS, Tallur PN, Ninnekar HZ (2011b) Biodegradation of 3-nitrobenzoate by *Bacillus flexus* strain XJU-4. World J Microbiol Biotechnol 27:1587–1592

Mulla SI, Talwar MP, Hoskeri RS, Ninnekar HZ (2012) Enhanced degradation of 3-nitrobenzoate by immobilized cells of *Bacillus flexus* strain XJU-4. Biotechnol Bioproc E 17:1294–1299

Mulla SI, Talwar MP, Bagewadi ZK, Hoskeri RS, Ninnekar HZ (2013) Enhanced degradation of 2-nitrotoluene by immobilized cells of *Micrococcus* sp. strain SMN-1. Chemosphere 90:1920–1924

Mulla SI, Talwar MP, Ninnekar HZ (2014) Bioremediation of 2,4,6-Trinitrotoluene explosive residues. In: Singh SN (ed) Biological remediation of explosive residues. Environmental Science and Engineering. Springer, Cham, pp 201–233

Mulla SI, Hu A, Wang Y, Sun Q, Huang SL, Wang H, Yu CP (2016a) Degradation of triclocarban by a triclosan-degrading *Sphingomonas* sp. strain YL-JM2C. Chemosphere 144:292–296

Mulla SI, Sun Q, Hu A, Wang Y, Ashfaq M, Eqani SA, Yu CP (2016b) Evaluation of sulfadiazine degradation in three newly isolated pure bacterial cultures. PLoS One 11:e0165013

Mulla SI, Wang H, Sun Q, Hu A, Yu CP (2016c) Characterization of triclosan metabolism in *Sphingomonas* sp. strain YL-JM2C. Sci Rep 6:21965

Mulla SI, Ameen F, Tallur PN, Bharagava RN, Bangeppagari M, Eqani S, Bagewadi ZK, Mahadevan GD, Yu CP, Ninnekar HZ (2017) Aerobic degradation of fenvalerate by a Gram-positive bacterium, *Bacillus flexus* strain XJU-4. 3 Biotech 7:320

Mulla SI, Hu A, Sun Q, Li J, Suanon F, Ashfaq M, Yu CP (2018) Biodegradation of sulfamethoxazole in bacteria from three different origins. J Environ Manag 206:93–102

Muller JA, Galushko AS, Kappler A, Schink B (2001) Initiation of anaerobic degradation of p-cresol by formation of 4-hydroxybenzylsuccinate in desulfobacterium cetonicum. J Bacteriol 183:752–757

Mutzel A, Reinsecheid UM, Antranikian G, Muller R (1996) Isolation and characterization of a thermophilic bacillus strain, that degrades phenol and cresols as sole carbon source at 70 °C. Appl Microbiol Biotechnol 46:593–596

Nedal M, Sobhi B, Isam S (2007) Effect of adsorption and bead size of immobilized biomass on the rate of biodegradation of phenol at high concentration levels. Ind Eng Chem Res 46:6820–6824

Neujahr HY, Gaal A (1973) Phenol hydroxylase from yeast. Purification and properties of the enzyme from *Trichosporon cutaneum*. Eur J Biochem 35:386–400

Neujahr HY, Varga JM (1970) Degradation of phenols by intact cells and cell-free preparations of *Trichosporon cutaneum*. Eur J Biochem 13:37–44

Neujahr HY, Lindsjo S, Varga JM (1974) Oxidation of phenols by cells and cell-free enzymes from *Candida tropicalis*. Antonie Van Leeuwenhoek 40:209–216

Olaniran AO, Igbinosa EO (2011) Chlorophenols and other related derivatives of environmental concern: properties, distribution and microbial degradation processes. Chemosphere 83:1297–1306

Padilla L, Matus V, Zenteno P, Gonzalez B (2000) Degradation of 2,4,6-trichlorophenol via chlorohydroxyquinol in *Ralstonia eutropha* JMP134 and JMP222. J Basic Microbiol 40:243–249

Pinto G, Pollio A, Previtera L, Stanzione M, Temussi F (2003) Removal of low molecular weight phenols from olive oil mill wastewater using microalgae. Biotechnol Lett 25:1657–1659

Prabu PC, Udayasoorian C (2005) Biodecolorization of phenolic paper mill effluent by ligninolytic fungus *Trametes versicolor*. J Biol Sci 5:558–561

Ramanand K, Suflita JM (1991) Anaerobic degradation of m-cresol in anoxic aquifer slurries: carboxylation reactions in a sulfate-reducing bacterial enrichment. Appl Environ Microbiol 57:1689–1695

Rather IA, Koh WY, Paek WK, Lim J (2017) The sources of chemical contaminants in food and their health implications. Front Pharmacol 8:830

Ribbons DW (1970) Specificity of monohydric phenol oxidations by meta cleavage pathways in Pseudomonas aeruginosa strain Tl. Arch Microbiol 74:103–115

Roberts DJ, Fedorak PM, Hrudey SE (1990) CO2 Incorporation and 4-hydroxy-2-methylbenzoic acid formation during anaerobic metabolism of *m*-cresol by a methanogenic consortium. Appl Environ Microbiol 56:472–478

Rocha-Santos T, Ferreira F, Silva L, Freitas AC, Pereira R, Diniz M, Castro L, Peres I, Duarte AC (2010) Effects of tertiary treatment by fungi on organic compounds in a kraft pulp mill effluent. Environ Sci Pollut Res Int 17:866–874

Rudolphi A, Tschech A, Fuchs G (1991) Anaerobic degradation of cresols by denitrifying bacteria. Arch Microbiol 155:238–248

Saiyood S, Vangnai AS, Thiravetyan P, Inthorn D (2010) Bisphenol A removal by the Dracaena plant and the role of plant-associating bacteria. J Hazard Mater 178:777–785

Sharma HA, Barber JT, Ensley HE, Polito MA (1997) A comparison of the toxicity and metabolism of phenol and chlorinated phenols by *Lemna gibba*, with special reference to 2,4,5-trichlorophenol. Environ Toxicol Chem 16:346–350

Shivaraman N, Pandey R (2000) Characterization and biodegradation of phenolic wastewater. J Indian Assoc Environ Manag 27:12–15

Solyanikova IP, Golovleva LA (2004) Bacterial degradation of chlorophenols: pathways, biochemica, and genetic aspects. J Environ Sci Health B 39:333–351

Stockinger J, Hinteregger C, Loidl M, Ferschl A, Streichsbier F (1992) Mineralization of 3-chloro-4-methylaniline via an *ortho*-cleavage pathway by *Pseudomonas cepacia* strain CMA1. Appl Microbiol Biotechnol 38:421–428

Stolz A, Busse HJ, Kampfer P (2007) *Pseudomonas knackmussii* sp. nov. Int J Syst Evol Microbiol 57:572–576

Stuermer DH, Ng DJ, Orris CJ (1982) Organic contaminants in groundwater near an underground coal gasification site in North-Eastern Wyoming. Environ Sci Technol 16:582–587

Suflita JM, Liang L, Saxena A (1989) The anaerobic biodegradation of *o*-, *m*- and *p*-cresol by sulfate-reducing bacterial enrichment cultures obtained from a shallow anoxic aquifer. J Ind Microbiol 4:255–266

Tallur PN, Megadi VB, Kamanavalli CM, Ninnekar HZ (2006) Biodegradation of *p*-cresol by *Bacillus* sp. strain PHN 1. Curr Microbiol 53:529–533

Tallur PN, Megadi VB, Ninnekar HZ (2009) Biodegradation of *p*-cresol by immobilized cells of *Bacillus* sp. strain PHN 1. Biodegradation 20:79–83

Tallur PN, Mulla SI, Megadi VB, Talwar MP, Ninnekar HZ (2015) Biodegradation of cypermethrin by immobilized cells of *Micrococcus* sp. strain CPN 1. Braz J Microbiol 46:667–672

Talwar MP, Mulla SI, Ninnekar HZ (2014) Biodegradation of organophosphate pesticide quinalphos by *Ochrobactrum* sp. strain HZM. J Appl Microbiol 117:1283–1292

Templeton MR, Graham N, Voulvoulis N (2009) Emerging chemical contaminants in water and wastewater. Philos Trans A, Math Phy Eng Sci 367:3873–3875

Ushiba Y, Takahara Y, Ohta H (2003) *Sphingobium amiense* sp. nov., a novel nonylphenol-degrading bacterium isolated from a river sediment. Int J Syst Evol Microbiol 53:2045–2048

Westerberg K, Elvang AM, Stackebrandt E, Jansson JK (2000) *Arthrobacter chlorophenolicus* sp. nov., a new species capable of degrading high concentrations of 4-chlorophenol. Int J Syst Evol Microbiol 50:2083–2092

Yan J, Jianping W, Hongmei L, Suliang Y, Zongding H (2005) The biodegradation of phenol at high initial concentration by the yeast *Candida tropicalis*. Bioch Eng J 24:243–247

Yan J, Jianping W, Jing B, Daoquan W, Zongding H (2006) Phenol biodegradation by the yeast *Candida tropicalis* in the presence of *m*-cresol. Bioch Eng J 29:227–234

Yu CP, Deeb RA, Chu KH (2013) Microbial degradation of steroidal estrogens. Chemosphere 91:1225–1235

Zhao J, Li Y, Zhang C, Zeng Q, Zhou Q (2008) Sorption and degradation of bisphenol A by aerobic activated sludge. J Hazard Mater 155:305–311

Chapter 14
Bioremediation of Soil Contaminated with Arsenic

María del Carmen Molina, Luis Fernando Bautista, Ignacio Belda, Manuel Carmona, Eduardo Díaz, Gonzalo Durante-Rodríguez, Sara García-Salgado, Jaime López-Asensio, Pilar Martínez-Hidalgo, María Ángeles Quijano, James F. White, and Natalia González-Benítez

Abstract Human-industrial activity causes a remarkable increase in the arsenic (As) environmental concentrations, with a potential impact in plant and animal health, and may cause severe losses in biodiversity. This metalloid is bioaccumulative through the food chain and highly associated with different types of cancers. To overcome the inherent drawbacks of physicochemical removal techniques, biological treatments arose as adequate and cost-effective remediation alternatives for As pollution. An interest arises from the endophytes, which live inside the host plant and have been studied for their plant growth-promoting properties, production of bioactive molecules, biocontrol processes, and As detoxification. The integration of bioremediation with multiple omic technologies provides, moreover, innovative

M. d. C. Molina (✉)
Department of Biology and Geology, Physic and Inorganic Chemistry. ESCET, Madrid, Spain

Department of Plant Biology and Plant Pathology, Rutgers University, School of Environmental and Biological Sciences (SEBS), New Brunswick, NJ, USA
e-mail: carmen.molina@urjc.es

L. F. Bautista
Department of Chemical and Environmental Technology, ESCET, Universidad Rey Juan Carlos, Madrid, Spain

I. Belda · Jaime López-Asensio · P. Martínez-Hidalgo · N. González-Benítez
Department of Biology and Geology, Physic and Inorganic Chemistry. ESCET, Madrid, Spain

M. Carmona · E. Díaz · G. Durante-Rodríguez
Department of Microbial and Plant Biotechnology, Centro de Investigaciones Biológicas, Madrid, Spain

S. García-Salgado · M. Á. Quijano
Department of Civil Engineering: Hydraulic and Land Planning, ETSIC, Universidad Politécnica de Madrid, Madrid, Spain

J. F. White
Department of Plant Biology and Plant Pathology, Rutgers University, School of Environmental and Biological Sciences (SEBS), New Brunswick, NJ, USA

A. Kumar, S. Sharma (eds.), *Microbes and Enzymes in Soil Health and Bioremediation*, Microorganisms for Sustainability 16,
https://doi.org/10.1007/978-981-13-9117-0_14

approaches to handle As remediation. The aim of this review is to show the latest knowledge, advances, and applications in arsenic bioremoval. We will focus on the following items: (1) human and environmental health, (2) biological tools for remediation with an emphasis in plants-microbiome interactions and omic technologies, (3) advances in As speciation analysis, and (4) As biosensors.

Keywords Arsenic · Bioremediation. · Bioreactors · Analytical methods · Omics · Biosensor.

14.1 Introduction

Living soils house the largest deposit of genes from fungi, bacteria, protozoa, invertebrates, algae, etc. Therefore, the soil is considered the most dynamic, complex, and biodiverse habitat that exists providing many benefits for humans (Wall et al. 2015). However, they are subjected to important human disturbance being the main global change driver (Smith et al. 2016). Degraded soils cover 24% of the global land area (35 Mkm^2; Bai et al. 2008) and one third are polluted. The intense anthropogenic activities and the expansion of the industry have led to a large-scale increase in the release of toxic metals (As, Cr, Pb, Hg, Cd, U, etc.) into the environment (Horta et al. 2015). Toxic metals have affected the dynamics of the complex ecosystems present in the pedosphere, due to its toxicity, nonbiodegradable nature, and bioaccumulation capacity throughout the food chain (Gall et al. 2015). Arsenic (As) is a metalloid widely distributed occurring both in organic and inorganic forms and in natural and anthropogenic environments (soil and water). As are present in soils under different chemical forms or types of binding, which affect its bioavailability, mobility, and toxicity, due to its transfer to aquatic media and uptake by plants, with the subsequent introduction into the food chain (Zhao et al. 2010). The forms of As present in soils depend on the type and amounts of sorbing components of the soil, the pH, and the redox potential (Anawar et al. 2018). Thus, As(V) is the main As species in aerobic soils. It has a strong affinity for iron oxides/hydroxides in soil; therefore, the concentrations of arsenate in soil solutions are usually low (Zhao et al. 2010). However, in reducing environments such as flooded paddy soils, As(III) is the dominant As species. In fact, flooding of paddy soils leads to mobilization of arsenite into the soil solution and enhanced As bioavailability (Kumarathilaka et al. 2018). Regarding organic species of As (DMA, MMA, and TMAO), they also can be found in soils although their concentrations usually account for less than 5% of As total (Huang et al. 2011).

Since the beginning of the twentieth century, As was known as a causal factor of different types of cancers (O'Donovan 1924). However, it was not until the 1970s when scientific interest in the presence of As in the soil began as a potential source of this carcinogen (Fig. 14.1). Hot spots in the As distribution are South and North America, Asia, and Central Africa (Amini et al. 2008). Among the main anthropo-

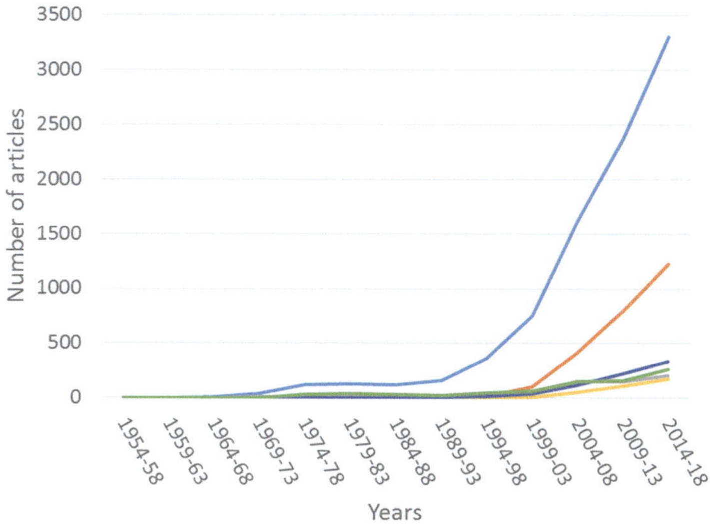

Fig. 14.1 Scientific production in terms of number of published papers whose subject was: As contaminated soils (*sensu lato*) (ligh blue), As and phytoremediation (orange), As and mycoremediation (gray), As and rizhosphere (yellow), As contaminated soils and prokaryotes (dark blue), As and microbiome from plant metaorganism (green)

genic sources of As in the environment, we can highlight the smelting of metals (specially copper), pharmaceutics and medical waste incineration, manufacturing, pesticides, cattle care, dyeing activity, fossil fuel utilization, wood burning, and semiconductor production, among others (Wang et al. 2017a; Gupta et al. 2019; Murcott 2012; Government of New South Wales 2017; Kant 2012; Shankar et al. 2014). The environmental impact of As is mainly displayed in two ways: (i) the in situ impact, as a contaminant in soil, air, and water – not only affecting biodiversity in animals and plants but also modifying or limiting microbial populations – and (ii) its presence in food chain, as a potent toxic and carcinogen, affecting human health. Both aspects are intimately related since As arrives at the food chain via plant uptake and vegetable accumulation that, at the same time, affects the feeding of farmed animals (Santra et al. 2013).

There are several physicochemical methods capable of removing As from contaminated water such as membranes, coagulation, anion exchange, disposable iron media, and softening adsorption (Bibi et al. 2017; Nidheesh and Singh 2017; Wang et al. 2018). However, the elimination or stabilization of As in contaminated soils is not feasible, in most cases, using this type of treatment. The use of indigenous organisms (mainly plants, fungi, and prokaryotes) to eliminate or stabilize the As of soils, through their metabolism, started in the 1990s (Fig. 14.1), and it has proved to be a successful eco-friendly option. Different terms have been used to describe the process to clean up contaminated environments based on the major microorganism responsible for recovery. As a general rule, when the biological agent is used, the term utilized is "bioremediation" (Kumar et al. 2011); but this term is also used when sensu stricto microorganisms are employed (Sing 2014). The utilization of plants to remove the pollutants is known as "phytoremediation" (Wang et al. 2011), and the use of fungi is named "mycoremediation" (Barrech et al. 2018). The contribution of these techniques to the contaminated soils' recovery is shown in Fig. 14.1. The uptake and accumulation capacity of As in plants varies widely, from plants known as "excluders" that have limited capacity of As translocation from roots to leaves to "hyperaccumulator" species that are able to uptake and translocate large amount of As to different plant tissues. The presence of As in plants was first described by Hengl et al. (1930), but has not been considered as an approach to remove pollutants from the environment until the end of the twentieth century (Fig. 14.1). Phytoremediation can also be divided into diverse techniques (Ma et al. 2016) depending if the pollutant is converted into less toxic forms (phytodegradation) and volatile species (phytovolatilization), accumulated in the aerial part (phytoextraction), accumulated in the root (phytostabilization), or metabolized by the rhizosphere microorganisms (rhizodegradation; Tangahu et al. 2011). The different strategies (bio-, phyto-, and mycoremediation) are frequently addressed in isolation; however, an implementation in the recovery systems requires the assembly of all elements of the system. Interactions between plants and microorganisms show complex interactions playing a pivotal role in the removal of toxic metals (Basu et al. 2018).

As-tolerant microbes have been already described more than a century ago (Green 1918; Green and Kestell 1920; Thom and Raper 1932). Current efforts have been focused in the identification of genes involved in As metabolism (Dowdle et al. 1996), the conversion to volatile species (Qin et al. 2006), and the genetic modification of microorganisms to improve their As tolerance (Kostal et al. 2004). Although the scientific studies are still scarce (Fig. 14.1), there is clear evidence that it may be possible to optimize bioremediation technologies. Emerging integrative approaches, such as (meta-)genomics, (meta-)transcriptomics, (meta-)bolomics, and (meta-)proteomics studies, are powerful tools to sequence partially or completely the As-metabolizing bacteria genome (Maizel et al. 2015) and to study the metagenome in As-contaminated soil (Luo et al. 2014) and the proteomic response to As stress (Belfiore et al. 2013). In summary, the eruption of omic and high-throughput technologies in bioremediation represents a pool of innovative methods that allows us to handle deep analysis and large amounts of data in each experiment (Fig. 14.2).

Chemical and geological analysis (Rinklebe et al. 2016) in combination with genomic and metagenomic techniques will provide insights into the specific roles of the complex biochemical pathways in the global As biogeochemical cycle. In addition, transcriptomic and proteomic techniques enable the scrutiny of the expression of those marker genes as indicators of enzymatic activity in response to the presence of As species, and metabolomic technologies inform about the As-derivative synthetized during the metabolic network established (Zhu et al. 2017; see Fig. 14.2). Other innovative technologies are underway in this subject, such as modeling of attenuation and environmental fate (Wallis et al. 2010), the use of nanoparticles in

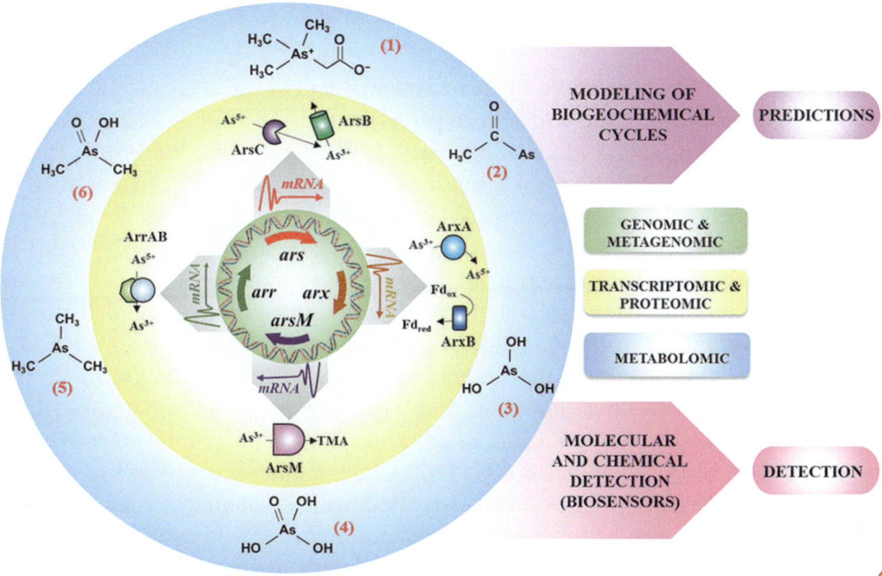

Fig. 14.2 General scheme of analytical technologies useful on arsenic bioremediation. Genomic techniques are represented in the green circle, and some examples of marker genes are presented: *ars* (arsenic resistance), *arr* (respiration of arsenate), *arx* (oxidation of arsenite) or *arsM* (methylation of arsenic species). Transcriptomic and proteomic are represented in the yellow circle: the clusters of genes are transcribed (zig-zag line) and the functions of the transcribed genes are cartooned. ArsC reduces arsenate (As^{5+}) to arsenite (As^{3+}) that is exported out of the cell by ArsB. ArxA oxidizes arsenite to arsenate with the collaboration of ArxB assisted by an oxidized ferredoxin (Fd_{ox}) that is then transformed into reduced ferredoxin (Fd_{red}). ArsM methylates arsenite to trimethylarsine (TMA) and the ArrAB proteins reduces arsenate into arsenite in a respiratory event. The blue circle represents some of the arsenic derivative metabolites produced as a consequence of the metabolism of arsenic compounds. Some examples are presented: arsenobetaine (1), acetylarsenic (2), arsenite (3), arsenate (4), trimethylarsine (5) and cacodylic acid (6). All the information obtained from the *omic* technologies can be used as support to develop molecular and chemical detection system (biosensor) and to perform predictions of environmental dynamics based on biochemical cycles modeling

controlling As mobilization (Gil-Díaz et al. 2014; Huang et al. 2018), process improvement through the use of organic amendments (Beesley et al. 2014; Onireti et al. 2017), bioaugmentation and biostimulation techniques (Chen et al. 2017a), or the use of dual-sensing bioreporters (Yoon et al. 2016).

There are many perspectives of analysis to approach the problem of the As contamination in soil environments. In the present chapter, we will focus on the following items: human and environmental health, biological tools for remediation, and advances in analytical and detection methods.

14.2 Human and Environmental Health

There is a major concern caused by environmental and health risks associated with the natural or anthropogenic widespread presence of As in soils and further migration to underground and surface waters worldwide. Therefore, the World Health Organization (WHO 2016) set up a safe limit of 10 μg/L for As concentration in drinking water. Dietary exposure to As, especially of inorganic As (iAs) forms, which are the most toxic forms, is a major concern in human health (EFSA 2014). Long-term exposures to As from drinking water and food can cause minor skin lesions, but it has also been associated with cardiovascular disease and diabetes. In addition, it is a known carcinogen able to cause skin, lung, bladder, liver, or kidney tumors, being lung cancer the most common cause of As-related mortality (WHO 2018). The greatest As threat to public health is related to groundwater contamination. As is naturally present at hazardous concentrations in the groundwater of many countries, including Argentina, Bangladesh, Chile, China, India, Mexico, and the United States. Drinking water, crops irrigated with contaminated water and/or growing in contaminated soil, and food prepared with contaminated water are the main sources of exposure. Figure 14.3 shows ranges and boundaries in total As concentrations detected in different water (Fig. 14.3a) and terrestrial (Fig. 14.3b)

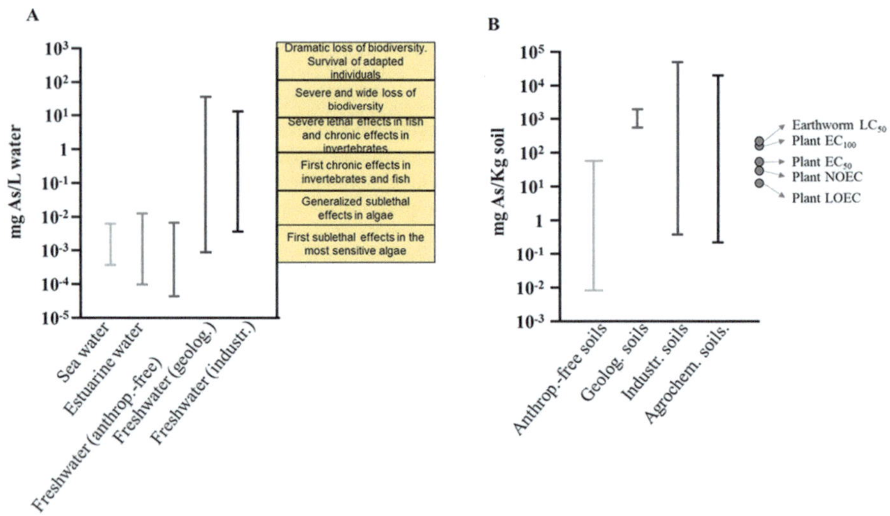

Fig. 14.3 Concentrations of arsenic in water (A) and soils (B), and its general biological effects. Data obtained from WHO (2001) report. Abbreviations: (i) 'anthrop.-free' means: anthropogenic input unlikely; (ii) 'geolog.' means: volcanic/geothermal origin; (iii) 'industry.' means: mining/chemical manufacture; (iv) 'agrochem.' means: treated with pesticides, sheep dips; (v) LC/EC mean: lethal/effective concentration; (vi) NOEC/LOEC mean: No observed/Lowest observed effect concentration; (vii) EC50/EC100 mean: concentration of a substance (toxic) at which 50%/100% of the population are affected; (viii) LC50 means: concentration of a substance causing dead in a 50% of the population

environments, indicating some reference values related to its general biological effect. As expected, the human-industrial activity causes a remarkable increase in the environmental concentrations of As, enhancing its potential impact in animal and plant health, even promoting severe losses in biodiversity (WHO 2018). Unfortunately, the majority of the data available from public surveys is still reported as total As, without information of the different As species present in the samples. Consequently, a risk assessment not considering the different species would lead to an overestimation of the health risk related to dietary As exposure. However, as reported by Yang et al. (2018) in a study performed in soils from China, the carcinogenic risk of As was found as relatively unacceptable in both industrial and agricultural regions.

Ingestion of As derivatives has been established as the main exposure pathway followed by dermal absorption. The general hazardous risk of noncarcinogenic As effects in human populations is in the following order: children, adult females, and adult males. However, adult females have the highest As-associated carcinogenic risk followed by adult males and children. For all the age classes except infants and toddlers, the main contributors to dietary exposure to iAs are foods belonging to "grain-based processed products" (in particular, wheat bread and rolls, rice, and rice-derived). Other food groups that contribute to iAs exposure are milk and dairy products (especially in infants and toddlers), vegetables, shellfish and seaweeds, and drinking water (Fig. 14.4). It is estimated that, in the United States and especially among the Native American communities, there are more than two million people who are exposed to concentrations higher than the maximum contaminant level allowed (>10 μg/L, according to the Environmental Protection Agency)

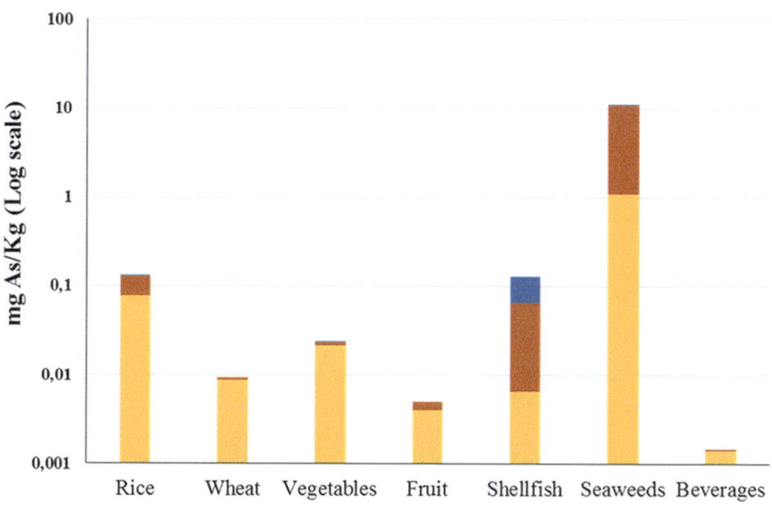

Fig. 14.4 Concentration and species distribution of As in food defined as major contributors of inorganic As (iAs), highly toxic (yellow), organic As (usually methylated) less toxic (brown) and non-toxic organic As (blue). Data obtained from Cubadda et al. (2017) and Lynch et al. (2014)

(Powers et al. 2019). Millions more are exposed to As below this concentration (Amini et al. 2008), which is of concern since the evidence suggests that there is no safe threshold (Schmidt 2014). The consumption of drinking water with moderate concentration of As, which is estimated to affect about 100 million people globally, may lead to a broad range of diseases from skin lesions to circulatory, respiratory, reproductive, and neurological complications, diabetes, hepatic, and renal dysfunction, and most of them may lead to the development of malignant tumors (Chen et al. 2009; reviewed in Abdul et al. 2015). Thus, it is possible to distinguish the effects of As on human health depending on the organ system affected. Different symptoms may appear in different parts of the integumentary system, where the skin is known to be particularly susceptible, showing the initial manifestations of As poisoning. With higher frequency in men than in women, and usually appearing 5–10 years after the exposure, the most common skin injuries are pigmentation, melanosis, and keratosis (Lindberg et al. 2008; Rahman et al. 2009). The brain appears to be a key target of As toxicity since its permeability through the blood-brain barrier. Both acute and chronic exposures to As may lead to central and peripheral neuropathies, but it typically affects peripheral nerves causing symptoms such as paresthesia, pain, and numbness in the limbs (Vahidnia et al. 2007; Mathew et al. 2010). The main mechanisms related to As-induced neurotoxicity are oxidative stress, disorganization of cytoskeletal structure, and neuronal apoptosis (via p38 and JNK kinases expression; Mundey et al. 2013; Namgung and Xia 2001).

Inhalation of As is not as common as its ingestion; however, some reports link mineral mining with a respiratory illness such as chronic cough, laryngitis, bronchitis, and rhinitis as a consequence of As exposure (Parvez et al. 2010). Moreover, long-term inhalation and ingestion of iAs could have deleterious effects on cardiovascular system functioning (Lewtas 2007) demonstrating a strong correlation between As exposure and atherosclerosis (via atherogenesis) and, although still debated, hypertension (Simeonova and Luster 2004). Since its metabolism/detoxification in the human body takes place in the liver, hepatic lesions may appear as a result of As acute and chronic exposure. Several injuries may occur depending on the doses of exposure. Hepatic diseases range from liver enlargement to more severe complications such as hepatic fibrosis, noncirrhotic portal fibrosis, cirrhosis, and liver cancer and sometimes lead to liver failure (Liu et al. 2002; Kapaj et al. 2006). Direct induction of apoptosis and oxidative stress are, again, among others, the main mechanisms involved in As-related hepatic toxicity and might also affect the renal system during the process of As elimination.

Finally, As can also affect the reproductive system causing infertility problems. In males, gonad dysfunction appears through a reduced synthesis of testosterone and cell apoptosis/necrosis (Davila-Esqueda et al. 2012; Shen et al. 2013). In females, As exposure through drinking water during pregnancy causes complications from premature delivery to fetal loss (Chakraborti et al. 2003). As a teratogen, As can also affect fetus development, producing growth retardation or fetal death, but in most cases, birth defects are accumulated leading to an increase of infant mortality (Wu et al. 2011).

14.3 Biological Tools for Remediation

14.3.1 Microorganisms in As-Contaminated Soil

The heavy metal and metalloid toxicity is a consequence of their affinity for different cellular components by forming metal-biomolecule complexes that might cause diverse adverse effects. At high concentrations, heavy metals and metalloids can inhibit essential metabolic functions and cause cell death (Hobman and Crossman 2014; Silver and Hobman 2007). To survive in environments contaminated with heavy metals, microorganisms have developed resistance or tolerance to high levels of these metals (Ahmed 2012), and many specific genes have been detected for resistance to toxic ions of heavy metals. It is possible to ascribe the microorganism resistance mechanisms to two classes: (i) the first depends on cellular metabolic activity, processes of oxidation, reduction, methylation, secretion, or intracellular accumulation, and (ii) the second mechanism does not depend on this metabolic activity; it is a passive process of uptake mediated by cell wall components, exopolysaccharides, proteins, or siderophores (Rajendran et al. 2003).

Genes responsible for As resistance have been described in many isolated microorganisms (Zhu et al. 2014) and also in environmental metagenomic samples (Zhu et al. 2017). Arsenate (AsV) and arsenite (AsIII) enter into the cell most probably through phosphate (Pi) transporters and aquaglyceroporins, respectively. The more widely spread genes in bacteria are organized in the *ars* cluster (Fig. 14.2), mainly arranged as *arsRCDAB* (Stolz et al. 2006; Ben Fekih et al. 2018). The *arsR* gene encodes a transcriptional repressor that controls the whole cluster (Busenlehner et al. 2003) and responds to the arsenite as inducer (Wu and Rosen 1993); *arsC* gene encodes the arsenate reductase responsible for the reduction of arsenate to arsenite (Mukhopadhyay et al. 2002); *arsAB* genes encode the energy-dependent arsenite translocator (Rosen 1999; 2002); and gene *arsD* encodes a metallochaperone that increases affinity of the transporter ArsAB for the arsenite (Lin et al. 2007). In addition to the *ars* genes of As resistance, some bacteria are able to use arsenate as an electron acceptor or arsenite as an electron donor. The *arr* genes are responsible for the anaerobic respiratory reduction of arsenate to arsenite (Silver and Phung 2005), and the arsenotrophic oxidation of arsenite is a transformation that can occur in oxic or anoxic conditions catalyzed by arsenite oxidases encoded by either the *aio* cluster (aerobic environments) or the *arx* cluster (anaerobic environments) (van Lis et al. 2013; Zargar et al. 2010). There are other genes with strong relevance in As resistance but less represented in microorganisms. Some bacteria, for example, are able to methylate As oxyanions with the participation of enzymes coded by *arsM* or *arsH* (Bentley and Chasteen 2002; Yuan et al. 2008; Ye et al. 2007). The presence of these genes can be detected by genetic analysis after isolation and cultivation of bacteria or by screening through metagenomic technologies that can analyze total DNA present in a given amount of soil. However, the As-resistant genes are widespread in nature, and their presence is not a conclusive probe to determine a record of As contamination in a given environment. Nevertheless, most of As-resistant genes are organized in clusters tightly controlled by regulators that ensure their expression only when As compounds are present in the medium (Andres and Bertin 2016). Thus, the environmental transcriptomic analysis can be used as a powerful tool to monitor bacterial activity in As-contaminated environments (Sun et al. 2004; Evans 2015). Besides the identification of the expression of genes related to the As resistance, environmental metabolomic is a comprehensive method able to detect metabolites released by microorganisms into the environment (VerBerkmoes et al. 2009). Thus, metabolomic analysis is a powerful tool to detect marker analytes in soils or water that unequivocally can be correlated with bacterial As metabolism such as methylated compounds [mono-(MMA), di-(DMA), tri-methylarsenic acid, trimethylarsine oxide (TMAO)] or volatile compounds like trimethylarsine (TMA) (Bentley and Chasteen 2002; Qin et al. 2006). Moreover, the recent understanding of the role of some As-derivative metabolites synthetized by bacteria such as arsenobetaine (Hoffmann et al. 2018), arsenosugars (Xue et al. 2018), or many other organoarsenic compounds (Chen and Rosen 2016) might also increase the number of molecules that can be used as markers of enzymatic transformation of As species in environmental samples.

From the above, integrating all the multiple omic technologies become crucial to elucidate the dynamic and complex interactions between microbial communities and the As biogeochemical cycle in the environment (Zhu et al. 2017). Interestingly, modeling approaches linking all omic data analyses will also predict the dynamics of As species in soil and waters providing capable tools to improve remediation technologies (Dunivin et al. 2018).

14.3.2 Plant Growth-Promoting Microorganisms (PGPMOs) to Improve Phytoremediation Approaches

Recent studies have shown that plant microbiomes (archaea, bacteria, protists, fungi, and viruses) and their symbiotic interactions play important roles in plant growth and response to abiotic and biotic stresses, helping to adapt the plant to the niche occupied (Mueller and Sachs 2015; Sim et al. 2019). In particular, plant growth-promoting microorganisms (PGPMOs) are a variety of microbes such as bacteria, cyanobacteria, and fungi including arbuscular mycorrhizal fungi (Mishra et al. 2017), representing 80% of the plants. PGPMOs are actively involved in plants growth and yield buffering the biotic and abiotic stress through diverse mechanisms, such as pathogen protection, phytohormone production, and nutrient acquisition (Vacheron et al. 2013, Ma et al. 2016, Martínez-Hidalgo and Hirsch 2017). Pathogen defense can be carried out directly through the production of antibiotics or enzymes that affect the growth of the pathogen such as β-glucanase chitinases (Martínez-Hidalgo et al. 2014, Martínez-Hidalgo et al. 2017) or indirectly by inducing the defensive systems of plants (Martínez-Hidalgo et al. 2015). PGPMOs are important producers of phytohormones such as auxin, gibberellin, and cytokinin that directly affect the growth of plants (Olanrewaju et al. 2017). The production of siderophores by the PGPMOs occurs under Fe-limiting conditions improving the uptake of Fe in the form of ferric ions (Fe^{3+}) and the increase in bioavailability of other essential nitrates through mineralization of organic matter that improves the nutrition and growth (Martínez-Hidalgo et al. 2014; Johnstone and Nolan 2015; Etesami 2018). Different studies conducted using various bacteria have shown that PGPMOs improve both plant growth and tolerance to As. The As stabilization and elimination mechanisms in these helper microorganisms seem similar to those described in non-symbiotic fungi and bacteria (Molina et al., in press). The number of publications on the successful application of endophytic microorganism inoculants to plants for bioremediation is extensive and increasing (Fig. 14.1). A plethora of bacteria such as *Kocuria* sp. and *Bacillus* sp. (Mallick et al. 2018), *Variovorax* sp. and *Phyllobacterium* sp. (Mesa et al. 2017), *Agrobacterium radiobacter* (Wang et al. 2011), *Rhizoglomus intraradices* and *Glomus etunicatum* (Wang et al. 2011; Wu et al. 2015; Spagnoletti and Lavado 2015), *Enterobacter* sp. (Nie et al. 2002), or *Bacillus thuringiensis* (Babu et al. 2013) have shown to be PGPMOs and offer resistance to As. In addition, fungi associated with plants such as *Trichoderma* (Tripathi et al. 2017) or *Piriformospora indica* (Mohd et al. 2017) and arbuscular mycorrhizal (AM) fungi (Chen et al. 2017b) have shown to be good candidates as PGPM reducing the As stress to the host plants. Despite this fact, the problems associated with heavy metal and metalloid contamination, particularly with As, are numerous, and its investigation should not be neglected. Recently, the posttranscriptional regulation of gene expression using RNA-induced silencing complexes (RISCs) mediated by siRNAs (noncoding RNA molecules involved gene expression regulation) has been considered as a potential tool to improve the plant-PGPMO interaction and bioremediation in heavy metal-contaminated soils. Other tools recently discovered are the riboswitches (RNA elements) that regulate mRNA expression and the ribozymes (catalytic RNAs) able to initiate or inhibit gene expression. These new tools are becoming powerful for bioremediation studies providing clear mechanisms of gene regulations (Du Toit 2015; Furukawa et al. 2015; Topp and Gallivan 2010).

14.3.3 Metaorganisms

Plants must be considered as a complex plurigenomic organism (metaorganism) formed by the plant itself, its microbiome, and the set of interspecific interactions that are established (Thijs et al. 2016). The microbiome is complex and is part of the rhizosphere, endosphere, or phyllosphere. The potential microbiome-host interactions can be favorable or competitive (Novotná and Suárez 2018). Previous studies have shown how certain bacteria favor the formation of mycorrhizae (Duponnois and Garbaye 1991; Vivas et al. 2003), while others inhibit the growth of fungal pathogens (Berg et al. 2005; Fikri et al. 2018). However, microbiome interactions are not static and change with their host at different life cycle stages or in response to changing environmental conditions. Microbiome interactions can evolve between trophic states of pathogenesis, symbiosis, mutualism, and parasitism (Newton et al. 2010). Despite lack of data, it is reasonable to think that an equilibrium will be established between favoring and competitive interactions within the complex host-microbiome in response to abiotic factors, such as environmental stress.

To further the knowledge about microbe-host interactions in response to abiotic stress, our study research group studied the relationships between bacterial and fungal endophytes isolated from *Jasione montana* L., collected from soils highly contaminated with As (García-Salgado et al. 2012; Gutiérrez-Ginés et al. 2015). Prokaryotes and fungi were identified by the molecular markers 16S rDNA and ITS rDNA, respectively. Five fungal (*Curvularia* sp. MC-L1, *Fusarium* sp. MC-A, *Fusarium* sp. MC-D, *Fusarium* sp. MC-J, and fungus MC-H) and eight bacteria (*Kocuria* sp. MC-K2, *Arthrobacter* sp. MC-D3a, *Kocuria* sp. MC-D3b, *Pantoea* sp. MC-J, *Kocuria rosae* MC-D2, *Pantoea conspicua* MC-K1, *Arthrobacter* sp. MC-D3a, and *Rhodococcus rhodochrous* MC-D1) were finally used, and a mixture of all endobacteria was also prepared. All fungal endophytes were tolerant to arsenate (Table 14.1) although the As minimum lethal concentrations (AsV-MLC) were lower than those for bacteria (> 300 mM). *Arthrobacter* sp. MC-D3a did not survive at arsenate concentrations higher than 7 mM (Table 14.1). The dual cultures of the selected fungi with single or a mixture of endophytes bacteria caused fungal phenotypic changes, such as growth inhibition percentages depending on the culture medium used (LB, Luria-Bertani agar, frequently used to bacteria and PDA, Potato Dextrose Agar, more suitable for fungi) (Table 14.1). Some endobacteria can decrease fungal development with values even above the 50% of the inhibition, whereas the mixture appears to increase (e.g., *Curvularia* sp. MC-L1 vs endobacteria mixture) or reduce (e.g., *Fusarium* sp. MC-D vs bacteria mixture) the growth inhibition percentage if we compare with the effect of the single bacteria (Table 14.1). This ability of endophytic bacteria to modulate the growth of potentially pathogenic fungi has been previously described (Fikri et al. 2018). Other physiological and phenotypic changes like the suppression in the formation of sporangia (Fig. 14.5e) or the production of excreted compounds of unknown nature have been observed (Fig. 14.6). Previous reports have also shown how *Enterobacter cloacae* prevented the germination of a pathogenic fungus (van Dijk and Nelson 2000) and how *Acinetobacter* sp. reduced the endophytic fungus colony diameter and spore germination rate (Wang et al. 2013). Moreover, we observed how fungus MC-H (Fig. 14.5f) produced chlamydospores (thick-walled resting spores) in the border with *P. conspicua*, as a mechanism of defense against bacteria (Li et al. 2012). When metaorganisms are subject to abiotic stresses, interactions are established and modulated and may change in response to the environmental stress (Fig. 14.6). Our results showed, during dual culture experiments, different responses of growth inhibition under As conditions (Fig. 14.7). Fungus MC-H, growing with *R. rhodochrous* MC-D1 or *Kocuria* sp. MC-K2 under As conditions, showed how it increased growth (30% and 60%, respectively) but controlled the reproductive machinery, inhibiting the sporangia development. These patterns were opposite under favorable

Table 14.1 Percentage of growth inhibition of endophytes fungi therefore to growth with several endobacteria or endobacteria mixture isolated from *J. montana*. In parentheses the AsV minimum lethal concentration. n.a.= not available

		Fungus MC-H	*Curvularia* sp. MC-L1 (220 mM)	*Fusarium* sp. MC-A (220 mM)	*Fusarium* sp. MC-D (220 mM)	*Fusarium* sp. MC-J (70 mM)
LB	*Kocuria* sp. MC-K2 (450 mM)	60	53.8	0	33	n.a.
	P. conspicua MC-K1 (450 mM)	67	42.3	17	76	n.a.
	K. rosae MC-D2 (450 mM)	62	42.3	0	19	n.a.
	R. rhodochorus MC-D1 (450 mM)	60	53.8	0	0	n.a.
	Anthrobacter sp. MC-D3a (7 mM)	52	40	23	0	n.a.
	Kocuria sp. MC-D3b (300 mM)	0	40	0	0	n.a.
	Pantoea sp. MC-J (300 mM)	60	54	5	14	n.a.
	Endobacteria Mixture	69	73	13	33	0
PDA	*Kocuria* sp. MC-K2 (450 mM)	0	0	0	0	0
	P. conspicua MC-K1 (450 mM)	58	48	4	0	0
	K. rosae MC-D2 (450 mM)	0	0	0	0	0
	R. rhodochorus MC-D1 (450 mM)	0	0	0	0	0
	Anthrobacter spo. MC-D3a (7 mM)	0	0	12	0	0
	Kocuria sp. MC-D3b (300 mM)	0	24	0	0	0
	Pantoea sp. MC-J (200 mM)	24	40	0	25	15
	Endobacteria Mixture	61	52	35	27	80

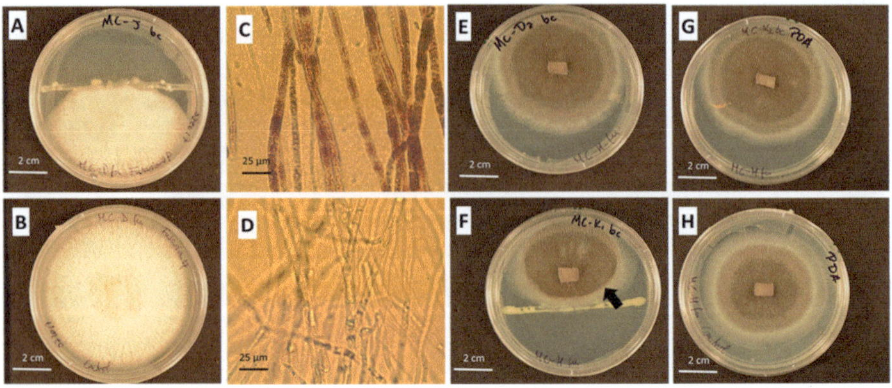

Fig. 14.5 Dual culture test in PDA at room temperature after 18 days. Inhibition of *Fusarium* sp. MC-D by *Pantoea* sp. MC-J (**A**). *Fusarium* sp. Control (**B**). *Fusarium* sp. MC-D hyphae invaded by *Pantoea* sp. on the border, Stained with 3, 30-diaminobenzidine tetrachloride (White et al. 2014) (**C**). Detail of *Fusarium* sp MC-D control (**D**). Fungus MC-H vs K. *rosae* MC-D2 with suppression in the production of sporophytes (**E**). MC-H vs P. *conspicua* with growth inhibition and chlamydospores production (arrow) on the border (**F**). MC-H vs *Kocuria* sp. without phenotypic changes apparent (**F**). MC-H axenic culture (**H**)

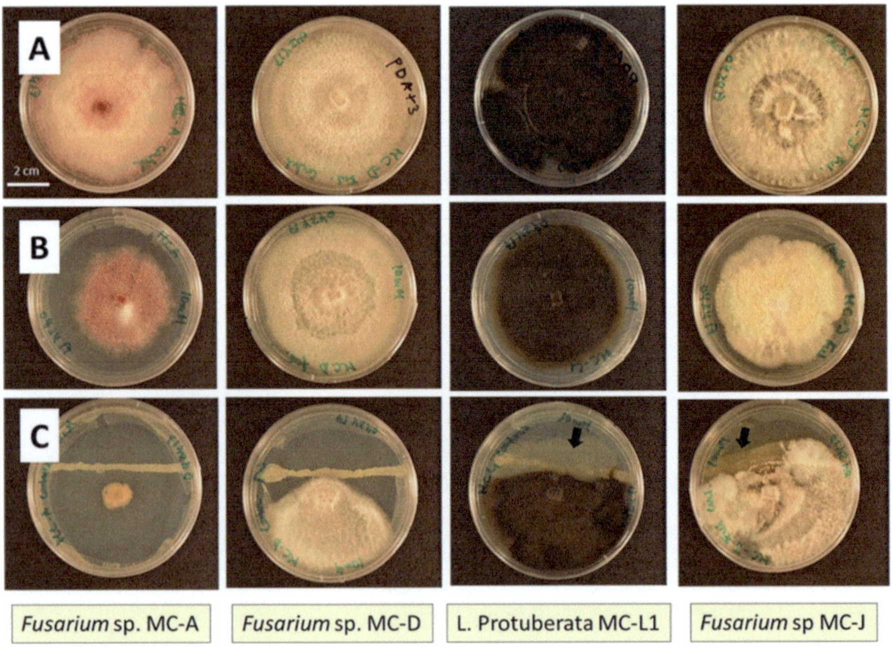

Fig. 14.6 Fungus growing on PDA control, at room temperature, after 18 days (**A**), on 10 mM arsenate PDA (**B**) and dual culture test between single endophyte fungus and mixture endophyte bacteria (**C**). Arrows show unknown exolites production

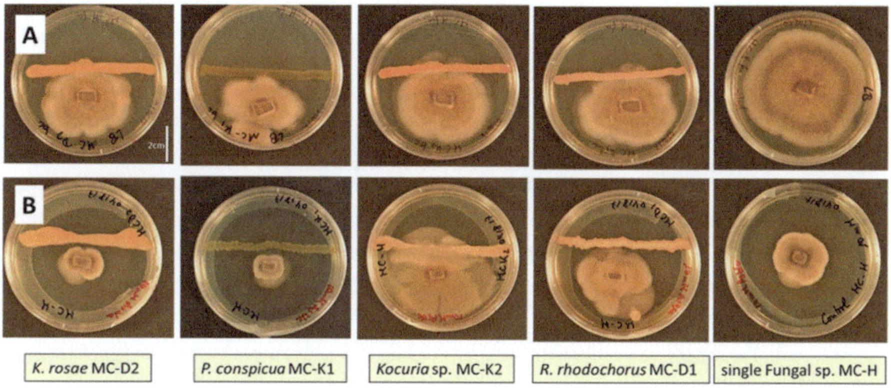

Fig. 14.7 Dual culture test between fungus MC-H and several endophytic bacteria isolated from J. *montana* on PDA (**A**) and on 10 mM arsenate PDA (**B**)

conditions, where *R. rhodochrous* MC-D1 or *Kocuria* sp. MC-K2 inhibited the growth of the fungus MC-H. These results suggest that under stress conditions, positive interactions in detriment of the competitive ones are favored (Liancourt et al. 2017).

A plant bacterial endophyte can also penetrate the hyphal wall of the fungus and settle inside the hyphae (Fig. 14.5a, b, c, and d) suggesting a fungal growth control by symbiotic bacteria (Fig. 14.5 a and b). Endobacteria have been isolated from AM cytoplasm (Bianciotto and Bonfante 2002; Bonfante and Anca 2009; Naumann et al. 2010) that are able to modify gene expression and physiology of the fungus (Salvioli et al. 2010). These bacteria can enhance the growth of AM fungi (Adams et al. 2009: Bonfante and Anca 2009) and be transmitted horizontally (Moebius et al. 2014) and vertically (Sharma et al. 2008; Bonfante and Anca 2009). In the association of AM fungi-bacteria, and Ghignone (2016) demonstrated that fungal infection with the endobacterium increased the fungal sporulation events, raised the fungal bioenergetic capacity, and elicited mechanisms to detoxify reactive oxygen species. Moreover, Chen et al. (2016) established a relationship between diversity of endobacteria and virulence of the fungus. In relation to pathogenic fungi, some endobacteria are responsible for fungal pathogenicity (Partida-Martinez and Hertweck 2005), while others modulate their antagonistic effects (Minerdi et al. 2008). These results indicate that bacteria living in the cytoplasm of fungi still represent an unexplored area of biology.

Despite the lack of studies on microbiomes, interactions (pathogenesis, mutualisms, or parasitism) depend on the specificity of the response, the type of stresses, and the scale of the interactions. Therefore, the idea of a metaorganism (host-microbiome interactions), linked with the omics strategies, will provide a successful tool for heavy metal decontamination process.

14.3.4 Enzymes and Bioreactors

To overcome the inherent drawbacks of physicochemical techniques, biological treatments arose as adequate and cost-effective remediation alternatives for As pollution. Bioremediation systems exploit microbial metabolic machinery ability, as whole cells or their isolated enzymes, to catalyze precipitation-dissolution processes, sequestration reactions, or biotransformations of As and As compounds (Plewniak et al. 2018). Unlike physicochemical technologies, biological technologies are much more effective at very low concentration ranges, even at the picomolar level (Sevcenco et al. 2015).

Many prokaryotic species are known to be able to include As within their metabolism. In addition, many bacterial genes involved in As metabolic pathways and resistance have been identified (Fig. 14.2). However, despite many microbial species and genes encoding As-related enzymes, only some of them have been described in pilot or industrial scale bioremediation processes developed in bioreactors.

The use of single enzymes immobilized on solid supports increases their stability and permits their repeated use in consecutive cycles of treatment, improving the economic viability of the whole process since the cost of enzymes at industrial scale is usually large. Arsenate reductase from *Pseudomonas alcaligenes* cross-linked immobilized on alginate beads has been used for the remediation of water containing arsenate at trace levels (< 1 ppm), yielding a biosorption capacity of 96.2 μg/g (Banerjee et al. 2017). Large enzymes such as ferritin, from the hyperthermophilic archeon *Pyrococcus furiosus*, showed a remarkable capacity to bind arsenate by interacting with the iron oxyhydroxide encapsulated inside ferritin nanocages (Sevcenco et al. 2015). This biosorption process is attractive for scaling up due to the developed heterologous overexpression of the gene that encodes ferritin from *P. furiosus* in *Escherichia coli*. This protein showed high thermostability and the ability to reuse the biosorbent.

Besides immobilized isolated enzymes, whole-cell biomass can be used as effective biosorbent for As sequestration from water. Biosorption presents, as a benefit, high elimination performance, low cost and minimum use of chemical and biological sludge. This technology can be applied either as living or as dead cells without clear evidence of which of the two alternatives is more effective since the results are sometimes contradictory and the biosorption mechanisms are complex and not clearly defined (AsadiHaris et al. 2018; Hlihor et al. 2017; de Bashan and Bashan 2010). However, the use of dead cells has a series of advantages, such as that the biomass can be reused, the system can be operated under extreme pH conditions (favorable for sorption but not compatible with living cells), and it is not necessary to use any growth media (AsadiHaris et al. 2018).

The ex situ bioremediation of As-polluted water, sludge, and soil can be carried out in bioreactors using a wide range of microorganisms harnessing their metabolism to perform a variety of transformations. For example, sulfate-reducing bacteria (SRB) are known to use sulfate as the terminal electron acceptor for their metabolism and, thus, produce insoluble metal or metalloid sulfides. For As, the removal efficiency by the action of SRB depends not only on the specific microbial strain, but also on the presence of different carbon sources and other metals within the medium. A SRB consortium isolated from an antimony mine slurry achieved up to 96% As (III) and As (V) removal when Fe (II) was present and ethanol as carbon source was added in the anaerobic pilot bioreactor (Liu et al. 2018). Higher As removal efficiency, up to 99.8%, can be reached in a continuous attached growth reactor in the absence of oxygen with simultaneous nitrate depletion using a bacterial consortium obtained from a sewage treatment plant (Shakya and Ghosh 2018). In summary, the use of controlled bioreactors is an efficient approach to remove As contamination reducing time consumption although is more expensive than biosorption techniques.

14.4 Advances in Analytical Methods

14.4.1 Sample Treatment Methods for Speciation Analysis

The making of adequate decisions for the recovery of systems contaminated with arsenic involves the use of appropriate techniques and protocols that allow us to make a precise approximation of the concentration and As species present. The As speciation analysis in soils requires the application of single and sequential extraction methods. Single extraction methods are generally preferred due to their simplicity and efficiency for mobility studies of toxic elements, which is related to the environmentally accessible metal fraction when soil conditions change, and their potential bioavailability, related to the easily accessible metal fraction to plants and soil microorganisms. For this purpose, weak neutral salt solutions ($CaCl_2$ or $NaNO_3$) are used for the leaching of heavy metals present in exchangeable fractions in soils (Alvarenga et al. 2013), whereas ethylenediaminetetraacetic acid (EDTA) and acetic acid solutions are used to estimate the possible bioavailability of heavy metals from environmental samples to living organisms (García-Salgado and Quijano 2016). Ultrasonic and microwave energy have been applied to reduce the extraction time and the sample-extractant consumption (Arain et al. 2008; De la Calle et al. 2013; García-Salgado and Quijano 2016; Li et al. 2014; Relić et al. 2013; Wang et al. 2015). García-Casillas et al. (2014) obtained quantitative recoveries for BCR (Community Bureau of Reference) 486 and 700, reducing extraction times from hours to a few minutes.

For As extraction, the use of EDTA can be insufficient to remove both cationic and anionic metal species in contaminated soils. It has been proposed the combination of this solvent with organic reducing agents, such as oxalic, ascorbic, citric, or malic acids, or their salts, which can also be used by their own (Nguyen Van et al. 2017; Wei et al. 2018), or with dithionite (Wang et al. 2017b). Martínez-Sánchez et al. (2011) have proposed dithionite-citrate buffered with sodium bicarbonate as the most effective solvent for As extraction from soils affected by old mining activities. Fleming et al. (2013) used ammonium acetate to study the extractability and bioavailability of As in historically contaminated orchard soil. The hydroxylamine hydrochloride, which is a solvent commonly used in one of the steps of the sequential extraction methods, can also be used for single As extraction in soils (Palumbo-Roe et al. 2015). Another solvent applied for As extraction is 1 M ammonium nitrate, according to the German DIN 19730:1997, which describes a method for the extraction of readily available trace elements from soils by shaking (Antoniadis et al. 2017). Finally, phosphoric acid and phosphate mixtures have been also used for As extraction from soils, to evaluate the As exchangeable fraction (García-Salgado et al. 2012; Sadee et al. 2016), as well as ammonium sulfate for weakly retained As (Moreno-Jiménez et al. 2010).

Regarding sequential extraction methods, they are used for the partitioning of heavy metals into different soil fractions: the water soluble and exchangeable, bound to carbonates, to Fe/Mn oxides, to organic matter, and the residual fraction (Tessier et al. 1979). This procedure was simplified by the BCR and later modified by Rauret et al. (1999). The main shortcomings from these conventional methods are high extraction time and reagent consumption, lack of selectivity, and poor reproducibility. Improvements on them are focused on (a) acceleration of batch leaching by sonication or microwave treatment (Rusnák et al. 2010), (b) reduction of sample handling by the application of continuous flow techniques (Savonina et al. 2012), (c) reduction of matrix effect by matrix separation or matrix matched calibration, and (d) application of internal standardization (Heltai et al. 2015).

Alternative sequential extraction methods have been developed for As fractionation in soils, because of the anionic nature of As ions unlike the heavy metals (Javed et al. 2013; Kreidie et al. 2011; Larios et al. 2012; Shiowatana et al. 2001, Tan et al. 2018; Wenzel et al. 2001). For example, several of these schemes have been proposed to replace the acetic acid solution by alkaline medium, for releasing As from the exchangeable fraction (Javed et al. 2013; Larios et al. 2012; Shiowatana et al. 2001; Tan et al. 2018). Also, alkaline solutions are used for dissolving the As associated with Fe/Al oxides/hydroxides (Larios et al. 2012; Shiwatana et al. 2001; Wang et al. 2017c), reporting higher percentages than those obtained with hydroxylamine solution. Several of these procedures increase the number of fractions (to 8 or 10), in order to differentiate between the As bound to amorphous or crystalline Fe, Al, and Mn oxyhydroxides, and therefore reduce the As bound to the residual fraction. In this way, authors reported the use of oxalate, citrate, or ascorbic acid solutions (Javed et al. 2013; Kreidie et al. 2011; Larios et al. 2012; Wenzel et al. 2001).

Conventional and As-specific sequential extraction methods have been applied to highly polluted soils (Kalyvas et al. 2018; Kim et al. 2014; Larios et al. 2013; Moreno-Jiménez et al. 2010; Wang et al. 2017c). The authors reported As contents lower than 10% in bioavailable fractions (soluble + exchangeable), while As was predominantly bound to amorphous and crystalline Fe oxyhydroxides (up to about 50%). Nevertheless, the absence of commercially available reference materials certified in As concentrations bound to the different soil fractions makes the validation of this kind of methods difficult, so recovery studies must be performed (Larios et al. 2013).

Apart from chemical methods, other extraction procedures such as diffusive gradients in thin-film technique (DGT) have proved to be effective for the determination of the bioavailability of trace elements in flooded soils (Zhang et al. 2018). Also, the effect of nanomaterials on As volatilization and extraction from this kind of soils has been studied (Huang et al. 2018).

14.4.2 Biosensors in as Analytical Methods

A biosensor is a device that presents a combination of biotechnology and microelectronics (Gronow 1984). It comprises (i) a biological component such as an enzyme, an antibody, a DNA, or a whole cell; (ii) a transducer, e.g., electrochemical, optical, or thermal; and (iii) a signal amplifier. Biosensors can be designed to detect a multitude of molecules, e.g., xenobiotics, pesticides, heavy metals, and many other pollutants (Saleem 2013). Various types of biosensors of As species (more commonly, arsenite) have been developed, and they can be grouped into whole-cell-based biosensors and cell-free-based biosensors (Kaur et al. 2015; Pothier et al. 2018).

The design of whole-cell As biosensors is mainly based on the ArsR transcriptional regulator that control the expression of the *Pars* promoter controlling the *ars* cluster. This protein is able to recognize arsenite or arsenate (Busenlehner et al. 2003; Wu and Rosen 1993) allowing the expression of a gene fusion of the *Pars* promoter and some reporter genes encoding β-galactosidase (*lacZ*) (Date et al. 2010; Cortés-Salazar et al. 2013; Huang et al. 2015), luciferase (Bakhrat et al. 2011; Sharma et al. 2013; Hou et al. 2014), green fluorescent protein (Chen et al. 2012; Truffer et al. 2014; Li et al. 2015; Ravikumar et al. 2017; Aye et al. 2018), or carotenoids (Fujimoto et al. 2006; Yoshida et al. 2008). However, some As biosensors are based on proteins encoded by the *ars* cluster like the ArsA-ArsD protein pair able to recognize As(III) (Liu et al. 2012).

The cell-free biosensors of As are primarily based on the ability of different biomolecules (DNA, proteins, aptamers, or nanomaterials) to interact with some As species. DNA can interact with As by electrostatic forces through the grooves of the double helix or by intercalation between the stacked base pairs of native DNA (Arora et al. 2007). Although these biosensors are able to detect the very low amount of As, their specificity is low (Liu and Wei 2008; Solanki et al. 2009). Some proteins have also shown their ability to sense As through a mechanism based on the affinity of some As oxyanions to bind and oxidize the sulfur groups of the proteins (Sarkar et al. 2010; Sanllorente-Méndez et al. 2012; Irvine et al. 2017). Aptamers are oligonucleotide or peptides modified to bind specifically a selected number of analytes. Some As aptamers are ultrasensitive to arsenite in aqueous detection, and they base their detection on gold nanoparticle aggregation (Wu et al. 2012; Wu et al. 2013; Pan et al. 2018). Nanomaterial-modified electrode interfaces for electrochemical sensing of As are based on unique chemical, physical, and electronic properties of the nanoparticles, enhancing the sensitivity, selectivity, field portability, and multiplexed detection capability of these kind of biosensors (Song et al. 2016; Vaishanav et al. 2017; An and Jang 2017; Kempahanumakkagari et al. 2017).

A high number of As biosensors have been developed in the last few years to detect As species in diverse environments. However, some limitations such as stability, sensibility, or specificity are still pending for solutions. New technologies such as synthetic biology or surface plasmon resonance are called to bypass some of the limitations of the current As biosensors (Fig. 14.7) (Kaur et al. 2015).

14.5 Conclusion

As speciation analysis requires the application of extraction procedures. In soils or sediments, this is carried out through sequential extraction methods, which permit discrimination between different As solid-phase associations. These analytical approaches allow us to inquire into soil composition and determine which remediation technique is more appropriate. Today, it is known that the interactions of plants with their microbiome and particularly with the PGPMOs will improve the effectiveness of plant-metaorganism. Therefore, through the resources that nature offers, plant endophytes and PGPMOs from As-tolerant plants can be used to improve bioremediation approaches. Microbiome interactions depend on the specificity of the response, the type of stress, and the scale of the interactions. Recent tools discovered, such as riboswitches and posttranscriptional regulation of gene expression, have been considered as potential tools to improve plant-PGPMO interactions. Other technologies such as biosensors, synthetic biology, or surface plasmon resonance have been developed to detect efficiently As species in diverse environments. Chemical and geological analysis and the idea of metaorganisms (host-microbiome interactions) linked with *omics* strategies will provide successful eco-friendly tools to remove As from contaminated environments.

Acknowledgments The authors thank Kathryn L. Kingsley and Dr. I. Irizarry (Rutgers University), Dr. S.K. Verma (Banaras Hindu University), and Dr. R. Simarro (Rey Juan Carlos University) for their invaluable help and collaboration. Molina M.C. greatly thanks Rutgers University for her time there as a Visiting Scientist.

References

Abdul KSM, Jayasinghe SS, Chandana EPS et al (2015) Arsenic and human health effects: a review. Environ Toxicol Phar 40:828–846

Adams AS, Currie CR, Cardoza Y et al (2009) Effects of symbiotic bacteria and tree chemistry on the growth and reproduction of bark beetle fungal symbionts. Can J For Res 39:1133–1147. https://doi.org/10.1139/X09-034

Ahmed M (2012) Implications of bacterial resistance against heavy metals in bioremediation: a review. IIOAB J 3:39–46

Alvarenga P, Laneiro C, Palma P et al C (2013) A study on As, Cu, Pb and Zn (bio)availability in an abandoned mine area (São Domingos, Portugal) using chemical and ecotoxicological tools. Environ Sci Pollut Res 20:6539–6550. doi:https://doi.org/10.1007/s11356-013-1649-2

Amini M, Abbaspour KC, Berg M et al (2008) Statistical modeling of global geogenic arsenic contamination in groundwater. Environ Sci Technol 42:3669–3675. https://doi.org/10.1021/es702859e

An JH, Jang J (2017) A highly sensitive FET-type aptasensor using flower-like MoS2 nanospheres for real-time detection of arsenic(III). Nanoscale 9:7483–7492. https://doi.org/10.1039/c7nr01661a

Anawar HM, Rengel Z, Damon P et al (2018) Arsenic-phosphorus interactions in the soil-plant-microbe system: dynamics of uptake, suppression and toxicity to plants. Environ Pollut 233:1003–1012. https://doi.org/10.1016/j.envpol.2017.09.098

Andres J, Bertin PN (2016) The microbial genomics of arsenic. FEMS Microbiol Rev 40:299–322. https://doi.org/10.1093/femsre/fuv050

Antoniadis V, Shaheen SM, Boersch J et al (2017) Bioavailability and risk assessment of potentially toxic elements in garden edible vegetables and soils around a highly contaminated former mining area in Germany. J Environ Manag 186:192–200. https://doi.org/10.1016/j.jenvman.2016.04.036

Arain MB, Kazi TG, Jamali MK et al (2008) Speciation of heavy metals in sediment by conventional, ultrasound and microwave assisted single extraction methods: a comparison with modified sequential extraction procedure. J Hazard Mater 154:998–1006. https://doi.org/10.1016/j.jhazmat.2007.11.004

Arora K, Prabhakar N, Chand S et al (2007) Immobilization of single stranded DNA probe onto polypyrrole-polyvinyl sulfonate for application to DNA hybridization biosensor. Sensors Actuators B Chem 126:655–663. https://doi.org/10.1016/J.SNB.2007.04.029

AsadiHaris S, Altowayti WAH, Ibrahim Z et al (2018) Arsenic biosorption using pretreated biomass of psychrotolerant Yersinia sp. strain SOM-12D3 isolated from Svalbard, Arctic. Environ Sci Pollut Res 25(28):27959–27970. https://doi.org/10.1007/s11356-018-2799-z

Aye SL, Fujiwara K, Doi N (2018) A dual system using compartmentalized partnered replication for selection of arsenic-responsive transcriptional regulator. J Biochem 164:341–348. https://doi.org/10.1093/jb/mvy055

Babu AG, Kim JD, Oh BT (2013) Enhancement of heavy metal phytoremediation by Alnus firma with endophytic Bacillus thuringiensis GDB-1. J Hazard Mater 250:477–483. https://doi.org/10.1016/j.jhazmat.2013.02.014

Bai ZG, Dent DL, Olsson L et al (2008) Global assessment of land degradation and improvement. 1. Identification by remote sensing. Report 2008/01, ISRIC – World Soil Information, Wageningen

Bakhrat A, Eltzov E, Finkelstein Y et al (2011) UV and arsenate toxicity: a specific and sensitive yeast bioluminescence assay. Cell Biol Toxicol 27:227–236. https://doi.org/10.1007/s10565-011-9184-8

Banerjee S, Banerjee A, Sarkar P (2017) Statistical optimization of arsenic biosorption by microbial enzyme via Ca-alginate beads. J Environ Sci Health 53(5):436–442. https://doi.org/10.1080/10934529.2017.1409009

Barrech D, Ali I, Tareen M (2018) 1. A review on Mycoremediation—the fungal bioremediation. Pure Appl Biol (PAB) 7(1):343–348. https://doi.org/10.19045/bspab.2018.70042

Basu S, Rabara RC et al (2018) Engineering PGPMOs through gene editing and systems biology: a solution for phytoremediation? Trends Biotechnol 36(5):499–510. https://doi.org/10.1016/j.tibtech.2018.01.011

Beesley L, Inneh OS, Norton GJ et al (2014) Assessing the influence of compost and biochar amendments on the mobility and toxicity of metals and arsenic in a naturally contaminated mine soil. Environ Pollut 186:195–202. https://doi.org/10.1016/j.envpol.2013.11.026

Belfiore C, OF O, Farías ME (2013) Proteomic approach of adaptive response to arsenic stress in Exiguobacterium sp. S17, an extremophile strain isolated from a high-altitude Andean Lake stromatolite. Extremophiles 17:421–431. https://doi.org/10.1007/s00792-013-0523-y

Ben Fekih I, Zhang C, Li YP et al (2018) Distribution of arsenic resistance genes in prokaryotes. Front Microbiol 9:2473. https://doi.org/10.3389/fmicb.2018.02473
Bentley R, Chasteen TG (2002) Microbial methylation of metalloids: arsenic, antimony, and bismuth. Microbiol Mol Biol Rev 66:250–271. https://doi.org/10.1128/MMBR.66.2.250-271.2002
Berg G, Krechel A, Ditz M et al (2005) Endophytic and ectophytic potato-associated bacterial communities differ in structure and antagonistic function against plant pathogenic fungi. FEMS Microbiol Ecol 51:215–229. https://doi.org/10.1016/j.femsec.2004.08.006
Bianciotto V, Bonfante P (2002) Arbuscular mycorrhizal fungi: a specialised niche for rhizospheric and endocellular bacteria. Antonie Leeuwenhoek 81:365–371
Bibi S, Kamran MA, Sultana J et al (2017) Occurrence and methods to remove arsenic and fluoride contamination in water. Environ Chem Lett 15:125–149
Bonfante P, Anca IA (2009) Plants, mycorrhizal fungi, and bacteria: a network of interactions. Annu Rev Microbiol 63:363–383. https://doi.org/10.1146/annurev.micro.091208.073504
Busenlehner LS, Pennella MA, Giedroc DP (2003) The SmtB/ArsR family of metalloregulatory transcriptional repressors: structural insights into prokaryotic metal resistance. FEMS Microbiol Rev 27:131–143. https://doi.org/10.1016/S0168-6445(03)00054-8
Chakraborti D, Mukherjee SC, Pati S et al (2003) Arsenic groundwater contamination in Middle Ganga Plain, Bihar, India: a future danger? Environ Health Persp 111:1194–1201. https://doi.org/10.1289/ehp.5966
Chen C, Chen X, Xie T et al (2016) Diverse bacterial symbionts of insect-pathogentic fungi and possible impact on the maintenance of virulence during infection. Symbiosis 69:47–58. https://doi.org/10.1007/s13199-015-0371-x
Chen J, Rosen B (2016) Organoarsenicalbiotransformations by Shewanella putrefaciens. Environ Sci Technol 50:7956–7963. https://doi.org/10.1021/acs.est.6b00235
Chen J, Zhu YG, Rosen BP (2012) A novel biosensor selective for organoarsenicals. Appl Environ Microbiol 78:7145–7147. https://doi.org/10.1128/AEM.01721-12
Chen P, Li J, Wang H et al (2017a) Evaluation of bioaugmentation and biostimulation on arsenic remediation in soil through biovolatilization. Environ Sci Pollut Res Int 24:21739–21749. https://doi.org/10.1007/s11356-017-9816-5
Chen XW, Wu FY, Li H et al (2017b) Mycorrhizal colonization status of lowland rice (Oryza sativa L.) in the southeastern region of China. Environ Sci Pollut Res 24:5268–5276. https://doi.org/10.1007/s11356-016-8287-4
Chen Y, Parvez F, Gamble M et al (2009) Arsenic exposure at low-to-moderate levels and skin lesions, arsenic metabolism, neurological functions, and biomarkers for respiratory and cardiovascular diseases: review of recent findings from the Health Effects of Arsenic Longitudinal Study (HEALS) in Bangladesh. Toxicol Appl Pharm 239:184–192. https://doi.org/10.1016/j.taap.2009.01.010
Cubadda F, Jackson B, Kuzurius-Spencer M et al (2017) Human exposure to dietary inorganic arsenic and other arsenic species: state of knowledge, gaps and uncertainties. Sci Total Environ 579:1228–1239. https://doi.org/10.1016/j.scitotenv.2016.11.108
Date A, Pasini P, Daunert S (2010) Integration of spore-based genetically engineered whole-cell sensing systems into portable centrifugal microfluidic platforms. Anal Bioanal Chem 398:349–356. https://doi.org/10.1007/s00216-010-3930-2
Davila-Esqueda ME, Jimenez-Capdeville ME, Delgado JM et al (2012) Effects of arsenic exposure during the pre- and postnatal development on the puberty of female offspring. Exp Toxicol Pathol 64:25–30. https://doi.org/10.1016/j.etp.2010.06.001
De la Calle I, Cabaleiro N, Lavilla I et al (2013) Ultrasound-assisted single extraction tests for rapid assessment of metal extractability from soils by total reflection X-ray fluorescence. J Hazard Mater 260:202–209. https://doi.org/10.1016/j.jhazmat.2013.05.021
de Bashan LE, Bashan Y (2010) Immobilized microalgae for removing pollutants: review of practical aspects. Biores Technol 101:1611–1627. https://doi.org/10.1016/j.biortech.2009.09.043
Dowdle PR, Laverman AM, Oremland RS (1996) Bacterial dissimilatory reduction of arsenic (V) to arsenic (III) in anoxic sediments. Appl Environ Microbiol 62:1664–1669

Du Toit A (2015) Bacterial genetics: Metalloregulatory riboswitches. Nat Rev Microbiol 13:249

Dunivin TK, Yeh, SY Shade A (2018) Targeting microbial arsenic resistance genes: a new bioinformatic toolkit informs arsenic ecology and evolution in soil genome and metagenomes bioRxiv 445502

Duponnois R, Garbaye J (1991) Effect of dual inoculation of Douglas fir with the ectomycorrhizal fungus Laccaria laccata and mycorrhization helper bacteria (MHB) in two bare-root forest nurseries. Plant Soil 138:169–176. https://doi.org/10.1007/BF00012243

ESFA (European Food Safety Authority) (2014) Dietary exposure to inorganic arsenic in the European population. EFSA J 12:3597

Etesami H (2018) Bacterial mediated alleviation of heavy metal stress and decreased accumulation of metals in plant tissues: mechanisms and future prospects. Ecotoxicol Environ Saf 147:175–191. https://doi.org/10.1016/j.ecoenv.2017.08.032

Evans (2015) Considerations for the use of transcriptomics in identifying the "genes that matter" for environmental adaptation. J Exp Biol 218:1925–1935

Fikri ASI, Rahman IA, Nor NSM et al (2018) Isolation and identification of local bacteria endophyte and screening of its antimicrobial property against pathogenic bacteria and fungi. AIP Conf Proc 1:020072. https://doi.org/10.1063/1.5027987

Fleming M, Tai Y, Zhuang P et al (2013) Extractability and bioavailability of Pb and As in historically contaminated orchard soil: effects of compost amendments. Environ Pollut 177:90–97. https://doi.org/10.1016/j.envpol.2013.02.013

Fujimoto H, Wakabayashi M, Yamashiro H (2006) Whole-cell arsenite biosensor using photosynthetic bacterium Rhodovulum sulfidophilum. Appl Microbiol Biotechnol 73:332–338. https://doi.org/10.1007/s00253-006-0483-6

Furukawa K, Ramesh A, Zhou Z et al (2015) Bacterial riboswitches cooperatively bind Ni2+ or Co2+ ions and control expression of heavy metal transporters. Mol Cell 57(6):1088–1098

Gall JE, Boyd RS, Rajakaruna N (2015) Transfer of heavy metals through terrestrial food webs: a review. Environ Monit Assess 187:201. https://doi.org/10.1007/s10661-015-44363

García-Casillas D, García-Salgado S, Quijano MA (2014) Accuracy evaluation of ultrasound probe sonication and microwave-assisted extraction systems for rapid single extraction of metals in soils. Anal Methods 6:8403–8412. https://doi.org/10.1039/c4ay01788a

García-Salgado S, García-Casillas D, Quijano-Nieto MA et al (2012) Arsenic and heavy metal uptake and accumulation in native plant species from soils polluted by mining activities. Water Air Soil Pollut 223:559–572. https://doi.org/10.1007/s11270-011-0882-x

García-Salgado S, Quijano MA (2016) Rapid metal extractability test from polluted mining soils by ultrasound probe sonication and microwave-assisted extraction systems. Environ Sci Pollut Res 23:24567–24577. https://doi.org/10.1007/s11356-016-7999-9

Gil-Díaz M, Alonso J, Rodríguez-Valdés E et al (2014) Reducing the mobility of arsenic in brownfield soil using stabilised zero-valent iron nanoparticles. J Environ Sci Health A 49:1361–1369. https://doi.org/10.1080/10934529.2014.928248

Government of New South Wales-NSW (2017) Arsenic and DDT residues at cattle dip yards. Available via NSW Webpage https://wwwdpinswgovau/__data/assets/pdf_file/0009/532458/Arsenic-and-ddt-residues-at-cattle-dip-yardspdf Accessed 5 Jan 2019

Green HH, Kestell NH (1920) Behaviour of bacteria towards arsenic. In: Union of South Africa, of Agriculture, 5th and 6th Reports of the Director of Veterinary Research, pp 701–706

Green, HH (1918) Description of a bacterium isolated from a cattle dipping tank, which reduces arsenate to arsenite. In: Union of South Africa, of Agriculture, 5th and 6th Reports of the Director of Veterinary Research, pp 613–624

Gronow M (1984) Biosensors. Trends Biochem Sci 9:336–340. https://doi.org/10.1016/0968-0004(84)90055-0

Gupta R, Sati B, Gupta A (2019) Treatment and recycling of wastewater from pharmaceutical industry. In: Singh RL, Singh RP (eds) Advances in biological treatment of industrial waste water and their recycling for a sustainable future. Springer, Singapore, pp 267–302

Gutiérrez-Ginés MJ, Pastor J, Hernández AJ (2015) Heavy metals in native mediterranean grassland species growing at abandoned mine sites: Ecotoxicological assessment and phytoremediation of polluted soils. In: Sherameti I, Varma A (eds) Heavy metal contamination of soils, Soil Biology, vol 44. Springer, Cham, pp 159–178

Heltai G, Fekete I, Halasz G et al (2015) Multi-elemental inductively coupled plasma-optical emission spectroscopic calibration problems for sequential extraction procedure of fractionation of heavy metal content from aquatic sediments. Hung J Ind Chem 43:7–12. https://doi.org/10.1515/hjic-2015-0002

Hengl F, Reckendorfee P, Beran F (1930) The arsenic and Lead on grapes, in grape must and wine as a result of control measures. Gartenbauwissenschaft 4:1

Hlihor RM, Figueiredo H, Tavares T et al (2017) Biosorption potential of dead and living *Arthrobacter viscosus* biomass in the removal of Cr(VI): batch and column studies, proc. Saf Environ Protect 108:44–56. https://doi.org/10.1016/j.psep.2016.06.016

Hobman JL, Crossman LC (2014) Bacterial antimicrobial metal ion resistance. J Medical Microbiol 64:471–497. https://doi.org/10.1099/jmm.0.023036-0

Hoffmann T, Warmbold B, Smits SHJ et al (2018) Arsenobetaine: an ecophysiologically important organoarsenical confers cytoprotection against osmotic stress and growth temperatures extremes. Environ Microbiol 20:305–323. https://doi.org/10.1111/1462-2920.13999

Horta A, Malone B, Stockmann U et al (2015) Potential of integrated field spectroscopy and spatial analysis for enhanced assessment of soil contamination: a prospective review. Geoderma 241:180–209. https://doi.org/10.1016/j.geoderma.2014.11.024

Hou QH, Ma AZ, Lv D et al (2014) The impacts of different long-term fertilization regimes on the bioavailability of arsenic in soil: integrating chemical approach with Escherichia coli arsRp::luc-based biosensor. Appl Microbiol Biotechnol 98:6137–6146. https://doi.org/10.1007/s00253-014-5656-0

Huang CW, Wei CC, Liao VHC (2015) A low cost color-based bacterial biosensor for measuring arsenic in groundwater. Chemosphere 141:44–49. https://doi.org/10.1016/j.chemosphere.2015.06.011

Huang JH, Hu KN, Decker B (2011) Organic arsenic in the soil environment: speciation, occurrence, transformation, and adsorption behavior. Water Air Soil Pollut 219:401–415. https://doi.org/10.1007/s11270-010-0716-2

Huang Q, Zhou S, Lin L et al (2018) Effect of nanomaterials on arsenic volatilization and extraction from flooded soils. Environ Pollut 239:118–128. https://doi.org/10.1016/j.envpol.2018.03.091

Irvine G, Tan S, Stillman M (2017) A simple metallothionein-based biosensor for enhanced detection of arsenic and mercury. Biosensors 7:14. https://doi.org/10.3390/bios7010014

Javed MB, Kachanoski G, Siddique T (2013) A modified sequential extraction method for arsenic fractionation in sediments. Anal Chim Acta 787:102–110. https://doi.org/10.1016/j.aca.2013.05.050

Johnstone TC, Nolan EM (2015) Beyond iron: non-classical biological functions of bacterial siderophores. Dalton Trans 44:6320–6329. https://doi.org/10.1039/C4DT03559C

Kalyvas G, Gasparatos D, Massas I (2018) A critical assessment on arsenic partitioning in mine-affected soils by using two sequential extraction protocols. Arch Agron Soil Sci 64:1549–1563. https://doi.org/10.1080/03650340.2018.1443443

Kant R (2012) Textile dyeing industry an environmental hazard. Nat Sci 4:22–26. https://doi.org/10.4236/ns.2012.41004

Kapaj S, Peterson H, Liber K et al (2006) Human health effects from chronic arsenic poisoning–a review. J Environ Sci Heal A 41:2399–2428. https://doi.org/10.1080/10934520600873571

Kaur H, Kumar R, Babu JN et al (2015) Advances in arsenic biosensor development – a comprehensive review. Biosens Bioelectron 63:533–545. https://doi.org/10.1016/j.bios.2014.08.003

Kempahanumakkagari S, Deep A, Kim KH et al (2017) Nanomaterial-based electrochemical sensors for arsenic – a review. Biosens Bioelectron 95:106–116. https://doi.org/10.1016/j.bios.2017.04.013

Kim EJ, Yoo J-C, Baek K (2014) Arsenic speciation and bioaccessibility in arsenic-contaminated soils: sequential extraction and mineralogical investigation. EnviromPollut 186:29–35. https://doi.org/10.1016/j.envpol.2013.11.032

Kostal J, Yang R, Wu CH et al (2004) Enhanced arsenic accumulation in engineered bacterial cells expressing ArsR. Appl Environ Microbiol 70:4582–4587. https://doi.org/10.1128/AEM.70.8.4582-4587.2004

Kreidie N, Armiento G, Cibin G et al (2011) An integrated geochemical and mineralogical approach for the evaluation of arsenic mobility in mining soils. J Soils Sediments 11:37–52. https://doi.org/10.1007/s11368-010-0274-7

Kumar A, Bisht BS, Joshi VD et al (2011) Review on bioremediation of polluted environment: a management tool. Int J Environ Sci Technol 1:1079

Kumarathilaka P, Seneweera S, Meharg A et al (2018) Arsenic speciation dynamics in paddy rice soil-water environment: sources, physico-chemical, and biological factors – a review. Water Res 140:403–414. https://doi.org/10.1016/j.watres.2018.04.034

Larios R, Fernández-Martínez R, Álvarez R et al (2012) Comparison of three sequential extraction procedures for fractionation of arsenic from highly polluted mining sediments. Anal Bioanal Chem 402:2909–2921. https://doi.org/10.1007/s00216-012-5730-3

Larios R, Fernández-Martínez R, Rucandio I (2013) Assessment of a sequential extraction procedure for arsenic partitioning and application to samples from different pollution sources. Anal Methods 5:4096–4104. https://doi.org/10.1039/c3ay40102b

Lewtas J (2007) Air pollution combustion emissions: characterization of causative agents and mechanisms associated with cancer, reproductive, and cardiovascular effects. Mutat Res 636:95–133. https://doi.org/10.1016/j.mrrev.2007.08.003

Li L, Liang J, Hong W et al (2015) Evolved bacterial biosensor for arsenite detection in environmental water. Environ Sci Technol 49:6149–6155. https://doi.org/10.1021/acs.est.5b00832

Li L, Ma M, Huang R et al (2012) Induction of chlamydospore formation in Fusarium by cyclic lipopeptide antibiotics from Bacillus subtilis C2. J Chem Ecol 38:966–974. https://doi.org/10.1007/s10886-012-0171-110.1007/s10886-012-0171-1

Li Q, Zhou JL, Chen B et al (2014) Toxic metal contamination and distribution in soils and plants of a typical metallurgical industrial area in southwest of China. Environ Earth Sci 72:2101–2109. https://doi.org/10.1007/s12665-014-3118-8

Liancourt P, Le Bagousse-Pinguet Y, Rixen C et al (2017) SGH: stress or strain gradient hypothesis? Insights from an elevation gradient on the roof of the world. Ann Bot 120:29–38. https://doi.org/10.1093/aob/mcx037

Lin Y-F, Yang J, Rosen BP (2007) ArsD: an As(III) metallochaperone for the ArsB As(III)-translocating ATPase. J Bioenerg Biomembr 39:453–458. https://doi.org/10.1007/s10863-007-9113-y

Lindberg AL, Rahman M, Persson LÅ et al (2008) The risk of arsenic induced skin lesions in Bangladeshi men and women is affected by arsenic metabolism and the age at first exposure. Toxicol Appl Pharm 230:9–16. https://doi.org/10.1016/j.taap.2008.02.001

Liu C, Balsamo V, Sun D et al (2012) A 3D localized surface plasmon resonance biosensor for the study of trivalent arsenic binding to the ArsA ATPase. Biosens Bioelectron 38:19–26. https://doi.org/10.1016/J.BIOS.2012.04.026

Liu F, Zhang G, Liu S, Fu Z et al (2018) Bioremoval of arsenic and antimony from wastewater by a mixed culture of sulfate-reducing bacteria using lactate and ethanol as carbon sources. Int Biodet Biodegr 126:152–159. https://doi.org/10.1016/j.ibiod.2017.10.011

Liu J, Zheng BS, Aposhian HV et al (2002) Chronic arsenic poisoning from burning high-arsenic-containing coal in Guizhou, China. Environ Health Persp 110:119–122. https://doi.org/10.1289/ehp.02110119

Liu Y, Wei W (2008) Layer-by-layer assembled DNA functionalized single-walled carbon nanotube hybrids for arsenic(III) detection. Electrochemcommun 10:872–875. https://doi.org/10.1016/J.ELECOM.2008.03.013

Luo J, Bai Y, Liang J et al (2014) Metagenomic approach reveals variation of microbes with arsenic and antimony metabolism genes from highly contaminated soil. PLoS One 9:e108185. https://doi.org/10.1371/journal.pone.0108185

Lynch HN, Greenberg GI, Pollock MC et al (2014) A comprehensive evaluation of inorganic arsenic in food and considerations for dietary intake analyses. Sci Total Environ 496:299–313. https://doi.org/10.1016/j.scitotenv.2014.07.032

Ma Y, Oliveira RS, Freitas H et al (2016) Biochemical and molecular mechanisms of plant-microbe-metal interactions: relevance for phytoremediation. Front Plant Sci 7:918. https://doi.org/10.3389/fpls.2016.00918

Maizel D, Utturkar SM, Brown SD et al (2015) Draft genome sequence of Brevibacterium linens AE038-8, an extremely arsenic-resistant bacterium. Genome Announc 3:e00316–e00315. https://doi.org/10.1128/genomeA.00316-15

Mallick I, Bhattacharyya C, Mukherji S et al (2018) Effective rhizoinoculation and biofilm formation by arsenic immobilizing halophilic plant growth promoting bacteria (PGPB) isolated from mangrove rhizosphere: a step towards arsenic rhizoremediation. Sci Total Environ 610:1239–1250. https://doi.org/10.1016/j.scitotenv.2017.07.234

Martínez-Hidalgo P, Hirsch AM (2017) The nodule microbiome: N2-fixing rhizobia do not live alone. Phytobiomes J 1:70–82. https://doi.org/10.1094/PBIOMES-12-16-0019-RVW

Martínez-Hidalgo P, Galindo-Villardón P, Trujillo ME et al (2014) Micromonospora from nitrogen fixing nodules of alfalfa (Medicago sativa L.). a new promising plant probiotic Bacteria. Sci Rep 4:6389. https://doi.org/10.1038/srep06389

Martínez-Hidalgo P, García JM, Pozo MJ (2015) Induced systemic resistance against Botrytis cinerea by Micromonospora strains isolated from root nodules. Front Microbiol 6:922. https://doi.org/10.3389/fmicb.2015.00922

Martínez-Sánchez MJ, Martínez-López S, García-Lorenzo ML et al (2011) Evaluation of arsenic in soils and plant uptake using various chemical extraction methods in soils affected by old mining activities. Geoderma 160:535–541. https://doi.org/10.1016/j.geoderma.2010.11.001

Mathew L, Vale A, Adcock JE (2010) Arsenical peripheral neuropathy. Pract Neurol 10:34–38. https://doi.org/10.1136/jnnp.2009.201830

Mesa V, Navazas A, González-Gil R et al (2017) Use of endophytic and rhizosphere bacteria to improve phytoremediation of As-contaminated industrial soils by autochthonous Betula celtiberica. Appl Environ Microbiol 83:03411–03416. https://doi.org/10.1128/AEM.03411-16

Minerdi D, Moretti M, Gilardi G et al (2008) Bacterial ectosymbionts and virulence silencing in a Fusarium oxysporum strain. Environ Microbiol 10:1725–1741. https://doi.org/10.1111/j.1462-2920.2008.01594.x

Mishra J, Singh R, Arora NK (2017) Alleviation of heavy metal stress in plants and remediation of soil by rhizosphere microorganisms. Front Microbiol 8:1706. https://doi.org/10.3389/fmicb.2017.01706

Moebius N, Üzüm Z, Dijksterhuis J et al (2014) Active invasion of bacteria into living fungal cells. Elife 3:e03007. https://doi.org/10.7554/eLife.03007.001

Mohd S, Shukla J, Kushwaha AS et al (2017) Endophytic fungi Piriformospora indica mediated protection of host from arsenic toxicity. Front Microbiol 8:754. https://doi.org/10.3389/fmicb.2017.00754

Molina MC, White JF, Kingsley KL et al (in press) Seed endophytes of *Jasionemontana*: arsenic detoxification workers in an eco-friendly factory. In: Verma SK, White JF (eds) Seed endophytes. Biology and biotechnology. Springer, Switzerland. https://doi.org/10.1007/978-3-030-10504-4

Moreno-Jiménez E, Manzano R, Esteban E et al (2010) The fate of arsenic in soils adjacent to an old mine site (Bustarviejo, Spain): mobility and transfer to native flora. J Soils Sediments 10:301–312. https://doi.org/10.1007/s11368-009-0099-4

Mueller UG, Sachs J (2015) Engineering microbiomes to improve plant and animal health. Trends Microbiol 23:606–617

Mukhopadhyay R, Rosen BP, Phung LT et al (2002) Microbial arsenic: from geocycles to genes and enzymes. FEMS Microbiol Rev 26:311–325. https://doi.org/10.1111/j.1574-6976.2002.tb00617.x

Mundey MK, Roy M, Roy S et al (2013) Antioxidant potential of Ocimum sanctum in arsenic induced nervous tissue damage. BJVP 6:95–101

Murcott S (ed) (2012) Arsenic contamination in the world: an international source- book. IWA Publishing, London

Namgung U, Xia Z (2001) Arsenic induces apoptosis in rat cerebellar neurons via activation of JNK3 and p38 MAP kinases. Toxicol Appl Pharm 174:130–138. https://doi.org/10.1006/taap.2001.9200

Naumann M, Schüßler A, Bonfante P (2010) The obligate endobacteria of arbuscular mycorrhizal fungi are ancient heritable components related to the Mollicutes. ISME J 4:862. https://doi.org/10.1038/ismej.2010.21

Newton AC, Fitt BD, Atkins SD et al (2010) Pathogenesis, parasitism and mutualism in the trophic space of microbe–plant interactions. Trends Microbiol 18:365–373. https://doi.org/10.1016/j.tim.2010.06.002

Nguyen Van T, Osanai Y, Do Nguyen H et al (2017) Arsenic speciation and extraction and the significance of biodegradable acid on arsenic removal—An approach for remediation of arsenic-contaminated soil. Int J Environ Res Public Health 14:990–1006. https://doi.org/10.3390/ijerph14090990

Nidheesh PV, Singh TA (2017) Arsenic removal by electrocoagulation process: recent trends and removal mechanism. Chemosphere 181:418–432

Nie L, Shah S, Rashid A et al (2002) Phytoremediation of arsenate contaminated soil by transgenic canola and the plant growth-promoting bacterium Enterobacter cloacae CAL2. Plant Physiol Biochem 40:355–361. https://doi.org/10.1016/S0981-9428(02)01375-X

Novotna A, Suárez JP (2018) Molecular detection of bacteria associated with Serendipita sp., a mycorrhizal fungus from the orchid Stanhopea connata Klotzsch in southern Ecuador. Botany Letters 165(2):307–313. https://doi.org/10.1016/j.myc.2017.07.007

O'Donovan WJ (1924) Arsenic cancer of occupational origin. Br J Dermatol 36:477

Palumbo-Roe B, Wragg J, Cave M (2015) Linking selective chemical extraction of iron oxyhydroxides to arsenic bioaccessibility in soil. Environ Pollut 207:256–265. https://doi.org/10.1016/j.envpol.2015.09.026

Pan J, Li Q, Zhou D, Chen J (2018) Ultrasensitive aptamer biosensor for arsenic (III) detection based on label-free triple-helix molecular switch and fluorescence sensing platform. Talanta 189:370–376. https://doi.org/10.1016/j.talanta.2018.07.024

Partida-Martinez LP, Hertweck C (2005) Pathogenic fungus harbours endosymbiotic bacteria for toxin production. Nature 437:884–888. https://doi.org/10.1038/nature03997

Parvez F, Chen Y, Brandt-Rauf PW (2010) A prospective study of respiratory symptoms associated with chronic arsenic exposure in Bangladesh: findings from the health effects of arsenic longitudinal study (HEALS). Thorax 65:528–533. https://doi.org/10.1136/thx.2009.119347

Plewniak F, Crognale S, Rossetti S et al (2018) A genomic outlook on bioremediation: the case of arsenic removal. Front Microbiol 9:820. https://doi.org/10.3389/fmicb.2018.00820

Pothier MP, Hinz AJ, Poulain AJ (2018) Insights into arsenite and arsenate uptake pathways using a whole cell biosensor. Front Microbiol 9:2310. https://doi.org/10.3389/fmicb.2018.02310

Powers M, Yracheta J, Harvey D et al (2019) Arsenic in groundwater in private wells in rural North Dakota and South Dakota: water quality assessment for an intervention trial. Environ Res 168:41–47. https://doi.org/10.1016/j.envres.2018.09.016

Qin J, Rosen BP, Zhang Y et al (2006) Arsenic detoxification and evolution of trimethylarsine gas by a microbial arsenite S-adenosylmethionine methyltransferase. Proc Natl Acad Sci U S A 1037:2075–2080. https://doi.org/10.1073/pnas.0506836103

Rahman MM, Ng JC, Naidu R (2009) Chronic exposure of arsenic via drinking water and its adverse health impacts on humans. Environ Geochem Hlth 31:189–200. https://doi.org/10.1007/s10653-008-9235-0

Rajendran P, Muthukrishnan J, Gunasekaran P (2003) Microbes in heavy metal remediation. Indian J Exp Biol 41:935–944

Rauret G, Lopez-Sanchez JF, Sahuquillo A et al (1999) Improvement of the BCR three step sequential extraction procedure prior to the certification of new sediment and soil reference materials. J Environ Monit 1:57–61. https://doi.org/10.1039/A807854H

Ravikumar Y, Nadarajan SP, Lee CS et al (2017) Engineering an FMN-based iLOV protein for the detection of arsenic ions. Anal Biochem 525:38–43. https://doi.org/10.1016/j.ab.2017.02.012

Relić D, Đorđević D, Sakan S et al (2013) Conventional, microwave, and ultrasound sequential extractions for the fractionation of metals in sediments within the Petrochemical Industry, Serbia. Environ Monit Assess 185:7627–7645. https://doi.org/10.1007/s10661-013-3124-4

Rinklebe J, Shaheen SM, Schröter F et al (2016) Exploiting biogeochemical and spectroscopic techniques to assess the geochemical distribution and release dynamics of chromium and lead in a contaminated floodplain soil. Chemosphere 150:390–397. https://doi.org/10.1016/j.chemosphere.2016.02.021

Rosen BP (1999) Families of arsenic transporters. Trends Microbiol 7:207–212

Rosen BP (2002) Biochemistry of arsenic detoxification. FEBS Lett 529:86–92

Rusnák R, Halasz G, Horvath M et al (2010) Preliminary results on the intensification of the BCR sequential extraction with sonication for sediments, soils, and gravitation dust sediment samples. Toxicol Environ Chem 92:443–452. https://doi.org/10.1080/02772240903300139

Sadee BA, Foulkes ME, Hill SJ (2016) A study of arsenic speciation in soil, irrigation water and plant tissue: a case study of the broad bean plant, Vicia faba. Food Chem 210:362–370. https://doi.org/10.1016/j.foodchem.2016.04.066

Saleem M (2013) IOP Conference series: material science and engineering 51. Bristol, UK, p 012012

Salvioli A, Chiapello M, Fontaine et al (2010) Endobacteria affect the metabolic profile of their host Gigasporamargasrita, an arbuscular mycorrhizal fungus. Environ Microbiol Rep 12:2083–2095. https://doi.org/10.1111/j.1462-2920.2010.02246.x

Salvioli A, Ghignone S, Novero et al (2016) Symbiosis with an endobacterium increases the fitness of a mycorrhizal fungus, raising its bioenergetic potential. ISME J 10:130. https://doi.org/10.1038/ismej.2015.91

Sanllorente-Méndez S, Domínguez-Renedo O, Arcos-Martínez MJ (2012) Development of acid phosphatase based amperometric biosensors for the inhibitive determination of As(V). Talanta 93:301–306. https://doi.org/10.1016/J.talanta.2012.02.037

Santra SC, Samal AC, Bhattacharya P (2013) Arsenic in food chain and community health risk: a study in gangetic West Bengal. Procedia Environ Sci 18:2–13

Sarkar P, Banerjee S, Bhattacharyay D et al (2010) Electrochemical sensing systems for arsenate estimation by oxidation of L-cysteine. Ecotoxicol Environ Saf 73:1495–1501. https://doi.org/10.1016/J.ECOENV.2010.07.004

Savonina EZ, Fedotov PS, Wennrich R (2012) Fractionation of Sb and As in soil and sludge samples using different continuous-flow extraction techniques. Anal Bioanal Chem 403:1441–1449. https://doi.org/10.1007/s00216-012-5927-5

Schmidt CW (2014) Low-dose arsenic: in search of a risk threshold. Environ Health Perspect 122:A130–A134. https://doi.org/10.1289/ehp.122-A130

Sevcenco AM, Paravidino M, Vrouwenvelder JS (2015) Phosphate and arsenate removal efficiency by thermostable ferritin enzyme from Pyrococcus furiosus using radioisotopes. Water Res 76:181–186. https://doi.org/10.1016/j.watres.2015.02.054

Shakya AK, Ghosh PK (2018) Simultaneous removal of arsenic and nitrate in absence of iron in an attached growth bioreactor to meet drinking water standards: importance of sulphate and empty bed contact time. J Clean Prod 186:304–312. https://doi.org/10.1016/j.jclepro.2018.03.139

Shankar S, Shanker U, Shikha (2014) Arsenic contamination of groundwater: a review of sources, prevalence, health risks, and strategies for mitigation. Sci World J 304524. https://doi.org/10.1155/2014/304524

Sharma M, Schmid M, Rothballer M et al (2008) Detection and identification of bacteria intimately associated with fungi of the order Sebacinales. Cell Immunol 10:2235–2246. https://doi.org/10.1111/j.1462-5822.2008.01202.x

Sharma P, Asad S, Ali A (2013) Bioluminescent bioreporter for assessment of arsenic contamination in water samples of India. J Biosci 38:251–258. https://doi.org/10.1007/s12038-013-9305-z

Shen H, Xu W, Zhang J (2013) Urinary metabolic biomarkers link oxidative stress indicators associated with general arsenic exposure to male infertility in a han chinese population. Envir Sci Tech 47:8843–8851. https://doi.org/10.1021/es402025n

Shiowatana J, McLaren RF, Chanmekha N et al (2001) Fractionation of arsenic in soil by a continuous-flow sequential extraction method. J Environ Qual 30:1940–1949. https://doi.org/10.2134/jeq2001.1940

Silver S, Hobman JL (2007) Mercury microbiology: resistance systems, environmental aspects, methylation and human health. In: Nies DH, Silver S (eds) Molecular microbiology heavy metals. Springer, Berlin, pp 357–370

Silver S, Phung LT (2005) Genes and enzymes involved in bacterial oxidation and reduction of inorganic arsenic. Appl Environ Microb 71:599–608. https://doi.org/10.1128/AEM.71.2.599-608.2005

Sim CSF, Chen SH, Ting ASY (2019) Endophytes: emerging tools for the bioremediation of pollutants. In: Emerging and eco-friendly approaches for waste management. Springer, Singapore, pp 189–217. https://doi.org/10.1007/978-981-10-8669-4_10

Simeonova PP, Luster MI (2004) Arsenic and atherosclerosis. Toxicol Appl Pharm 198:444–449. https://doi.org/10.1016/j.taap.2003.10.018

Singh R (2014) Microorganism as a tool of bioremediation technology for cleaning environment: a review. Proc Int Acad Ecol Environ Sci 4:1

Smith P, House JI, Bustamante M et al (2016) Global change pressures on soils from land use and management. Glob Chang Biol 22:1008–1028

Solanki PR, Prabhakar N, Pandey MK et al (2009) Surface plasmon resonance-based DNA biosensor for arsenic trioxide detection. Int J Environ Anal Chem 89:49–57. https://doi.org/10.1080/03067310802398872

Song L, Mao K, Zhou X et al (2016) A novel biosensor based on Au@Ag core–shell nanoparticles for SERS detection of arsenic(III). Talanta 146:285–290. https://doi.org/10.1016/j.talanta.2015.08.052

Spagnoletti FN, Lavado RS (2015) The arbuscular mycorrhiza Rhizophagus intraradices reduces the negative effects of arsenic on soybean plant. J Agron 5:188–199. https://doi.org/10.3390/agronomy5020188

Stolz JF, Basu P, Santini JM et al (2006) Arsenic and selenium in microbial metabolism. Annu Rev Microbiol 60:107–130. https://doi.org/10.1146/annurev.micro.60.080805.142053

Sun Y, Polishchuk EA, Radoja U et al (2004) Identification and quantification of arsC genes in environmental samples by using real-time PCR. J Microbiol Meth 58:335–349. https://doi.org/10.1016/j.mimet.2004.04.015

Tan D, Long JM, Li BY et al (2018) Fraction and mobility of antimony and arsenic in three polluted soils: a comparison of single extraction and sequential extraction. Chemosphere 213:533–540. https://doi.org/10.1016/j.chemosphere.2018.09.089

Tangahu BV, Sheikh Abdullah SR, Basri H et al (2011) A review on heavy metals (As, Pb, and Hg) uptake by plants through phytoremediation. Int J Chem Eng 2011:1–32

Tessier A, Campbell PGC, Bisson M (1979) Sequential extraction procedure for the speciation of particulate trace metals. Anal Chem 51:844–851. https://doi.org/10.1021/ac50043a017

Thijs S, Sillen W, Rineau F et al (2016) Towards an enhanced understanding of plant–microbiome interactions to improve phytoremediation: engineering the metaorganism. Front Microbiol 107:1–15. https://doi.org/10.3389/fmicb.2016.00341

Thom C, Raper KB (1932) The arsenic fungi of Gosio. Science 76:548–550

Topp S, Gallivan JP (2010) Emerging applications of riboswitches in chemical biology. ACS Chem Biol 5:139–148

Tripathi P, Singh PC, Mishra A et al (2017) Arsenic tolerant Trichoderma sp. reduces arsenic induced stress in chickpea (Cicer arietinum). Environ Pollut 223:137–145. https://doi.org/10.1016/j.envpol.2016.12.073

Truffer F, Buffi N, Merulla D et al (2014) Compact portable biosensor for arsenic detection in aqueous samples with Escherichia coli bioreporter cells. Rev Sci Instrum 85:015120. https://doi.org/10.1063/1.4863333

Vacheron J, Desbrosses G, Bouffaud ML et al (2013) Plant growth-promoting rhizobacteria and root system functioning. Front Plant Sci 4:356. https://doi.org/10.3389/fpls.2013.00356

Vahidnia A, Van der Voet G, De Wolff F (2007) Arsenic neurotoxicity—a review. Hum Exp Toxicol 26:823–832

Vaishanav SK, Korram J, Pradhan P et al (2017) Green luminescent CdTe quantum dot based fluorescence nano-sensor for sensitive detection of arsenic (III). J Fluoresc 27:781–789. https://doi.org/10.1007/s10895-016-2011-0

van Dijk K, Nelson EB (2000) Fatty acid competition as a mechanism by which Enterobacter cloacae suppresses Pythium ultimum sporangium germination and damping-off. J Appl Environ Microbiol 66:5340–5347. https://doi.org/10.1128/AEM.66.12.5340-5347.2000

Van Lis R, Nitschke W, Duval S et al (2013) Arsenic as bioenergetic substrates. BiochBiophysic Acta 1827:176–188. https://doi.org/10.1016/j.bbabio.2012.08.007

VerBerkmoes NC, Denef VJ, Hettich RL et al (2009) Functional analysis of natural microbial consortia using community proteomics. Nat Rev Microbiol 7:196–205. https://doi.org/10.1038/nrmicro2080

Vivas A, Marulanda A, Ruiz-Lozano JM et al (2003) Influence of a Bacillus sp. on physiological activities of two arbuscular mycorrhizal fungi and on plant responses to PEG-induced drought stress. Mycorrhiza 13:249–256. https://doi.org/10.1007/s00572-003-0223-z

Wall DH, Nielsen UN, Six J (2015) Soil biodiversity and human health. Nature 528:69. https://doi.org/10.1038/nature15744

Wallis I, Prommer H, Simmons CT et al (2010) Evaluation of conceptual and numerical models for arsenic mobilization and attenuation during managed aquifer recharge. Environ Sci Technol 44:5035–5041. https://doi.org/10.1021/es100463q

Wang C, Liu H, Zhang Y et al (2018) Review of arsenic behavior during coal combustion: volatilization, transformation, emission and removal technologies. Prog Energy Combust Sci 68:1–28. https://doi.org/10.1016/j.pecs.2018.04.001

Wang JM, Jiang JG, Li D et al (2015) Removal of Pb and Zn from contaminated soil by different washing methods: the influence of reagents and ultrasound. Environ Sci Pollut R 22:20084–20091. https://doi.org/10.1007/s11356-015-5219-7

Wang Q, Xiong D, Zhao P et al (2011) Effect of applying an arsenic-resistant and plant growth-promoting rhizobacterium to enhance soil arsenic phytoremediation by Populus deltoides LH05-17. J Appl Microbiol 111:1065–1074. https://doi.org/10.1111/j.1365-2672.2011.05142.x

Wang QM, Guo XY, Tian QH (2017a) Reaction mechanism and distribution behavior of arsenic in the bottom blown copper smelting process. Metals 7:302. https://doi.org/10.3390/met7080302

Wang X, Dong HC, Feng L et al (2017c) Comparison of three sequential extraction procedures for arsenic fractionation in highly polluted sites. Chemosphere 178:402–410. https://doi.org/10.1016/j.chemosphere.2017.03.078

Wang X, Ma FJ, Zhang Q et al (2017b) An evaluation of different soil washing solutions for remediating arsenic-contaminated soils. Chemosphere 173:368–372. https://doi.org/10.1016/j.chemosphere.2017.01.068

Wang S, Wang W, Jin Z, Du B, Ding Y, Ni T, Jiao F (2013) Screening and diversity of plant growth promoting endophytic bacteria from peanut. Afr J Microbiol Res 7:875–884. https://doi.org/10.5897/AJMR12.1500

Wei M, Chen J, Wang Q (2018) Remediation of sandy soil contaminated by heavy metals with Na(2)EDTA washing enhanced with organic reducing agents: element distribution and spectroscopic analysis. Eur J Soil Sci 69:719–731. https://doi.org/10.1111/ejss.12560

Wenzel W, Kirchbaumer N, Prohaska T et al (2001) Arsenic fractionation in soils using an improved sequential extraction procedure. Anal Chim Acta 436:309–323. https://doi.org/10.1016/S0003-2670(01)00924-2

White JF Jr, Torres MS, Somu MP et al (2014) Hydrogen peroxide staining to visualize intracellular bacterial infections of seedling root cells. Microsc Res Tech 77:566–573. https://doi.org/10.1002/jemt.22375

WHO (World Health Organization) (2016) The public health impact of chemicals: knowns and unknowns. Geneva

WHO (World Health Organization) (2018) Arsenic. Available via WHO Webpage. https://www.who.int/news-room/fact-sheets/detail/arsenic. Accessed 5 Jan 2019

Wu F, Hu J, Wu S et al (2015) Grain yield and arsenic uptake of upland rice inoculated with arbuscular mycorrhizal fungi in As- spiked soils. Environ Sci Pollut Res 22:8919–8926. https://doi.org/10.1007/s11356-012-1440-9

Wu J, Chen G, Liao Y (2011) Arsenic levels in the soil and risk of birth defects: a population-based case-control study using GIS technology. J Environ Health 74:20–25

Wu J, Rosen B (1993) Metalloregulated expression of the ars operon. J Biol Chem 268:52–58

Wu Y, Wang F, Zhan S et al (2013) Regulation of hemin peroxidase catalytic activity by arsenic-binding aptamers for the colorimetric detection of arsenic(III). RSC Adv 3:25614. https://doi.org/10.1039/c3ra44346a

Wu Y, Zhan S, Xing H et al (2012) Nanoparticles assembled by aptamers and crystal violet for arsenic(III) detection in aqueous solution based on a resonance Rayleigh scattering spectral assay. Nanoscale 4:6841. https://doi.org/10.1039/c2nr31418e

Xue XM, Ye J, Raber G et al (2018) Identification of steps in the pathway of arsenosugars biosynthesis. Environ Sci Technol. https://doi.org/10.1021/acs.est.8b04389

Yang Q, Li Z, Lu X (2018) A review of soil heavy metal pollution from industrial and agricultural regions in China: pollution and risk assessment. Sci Total Environ 642:690–700. https://doi.org/10.1016/j.scitotenv.2018.06.068

Ye J, Yang HC, Rosen B et al (2007) Crystal structure of the flavoprotein ArsH from Sinorhizobium meliloti. FEBS Lett 581:3996–4000. https://doi.org/10.1016/j.febslet.2007.07.039

Yoon Y, Kim S, Chae Y et al (2016) Simultaneous detection of bioavailable arsenic and cadmium in contaminated soils using dual-sensing bioreporters. Appl Microbiol Biotechnol 100:3713–3722. https://doi.org/10.1007/s00253-016-7338-6

Yoshida K, Inoue K, Takahashi Y et al (2008) Novel carotenoid-based biosensor for simple visual detection of arsenite: characterization and preliminary evaluation for environmental application. Appl Environ Microbiol 74:6730–6738. https://doi.org/10.1128/AEM.00498-08

Yuan C, Lu X, Quin J et al (2008) Volatile arsenic species released from Escherichia coli expressing the AsIII S-adenosylmethionine methyltransferase gene. Environ Technol 42:3201–3206

Zargar K, Hoeft S, Oremlanl et al (2010) Identification of a novel arsenite oxidase gene, arxA, in the haloalkaliphilic, arsenite-oxidating bacterium Alkalilimnicola ehrlichii strain MLHE-1. J Bacteriol 192:3755–3762. https://doi.org/10.1016/j.febslet.2007.07.039

Zhang SH, Wang Y, Pervaiz A et al (2018) Comparison of diffusive gradients in thin-films (DGT) and chemical extraction methods for predicting bioavailability of antimony and arsenic to maize. Geoderma 332:1–9. https://doi.org/10.1016/j.geoderma.2018.06.023

Zhao FJ, McGrath SP, Meharg AA (2010) Arsenic as a food chain contaminant: mechanisms of plant uptake and metabolism and mitigation strategies. Annu Rev Plant Biol 61:535–559. https://doi.org/10.1146/annurev-arplant-042809-112152

Zhu YG, Xue XM, Kappler A et al (2017) Linking genes to microbial biogeochemical cyclic: lesson from arsenic. Environ Sci Technol 51:7326–7339. https://doi.org/10.1021/acs.est.7b00689

Zhu YG, Yoshinaga M, Zhao FJ et al (2014) Earth abides arsenic biotransformations. Annu Rev Earth Planet Sci 42:443–467. https://doi.org/10.1146/annurev-earth-060313-054942

Chapter 15
Biosurfactants in Bioremediation and Soil Health

Kuttuvan Valappil Sajna and Lalitha Devi Gottumukkala

Abstract Owing to their versatile properties, many biosurfactants are implicated in the cleanup of oil spills, heavy metals, and organopollutants. A number of biosurfactants have the potential to be used in the household detergent formulation as they are good stain remover and are quite compatible with enzymes and other additives used in detergents. This chapter presents an in-depth evaluation of the use of biosurfactants in bioremediation and in approaches for maintaining soil quality. However, many studies on the exogenous supplementation of biosurfactants in bioremediation showed the contradictory effect on biodegradation of pollutants. Hence, a thorough investigation of the efficacy and toxicity of biosurfactants is to be performed before implementing the biosurfactants in bioremediation. Biosurfactants can be a potential replacement to chemical surfactants only if they meet the large-scale production and cheap prices of synthetic surfactants. Use of inexpensive substrates, employing high yield strain, and developing cost-effective downstream processing are some of the approaches to reduce the cost of biosurfactants.

Keywords Biosurfactants · Bioremediation · Soil health · Oil spill cleanup · Rhamnolipids · Sophorolipids · Polyaromatic hydrocarbon · Corexit 9500A · Soil washing

K. V. Sajna (✉)
Department of Biochemistry, Indian Institute of Science, Bengaluru, India
e-mail: sajnak@iisc.ac.in

L. D. Gottumukkala
Celignis Biomass Analysis Laboratory, University of Limerick, Limerick, Ireland

A. Kumar, S. Sharma (eds.), *Microbes and Enzymes in Soil Health and Bioremediation*, Microorganisms for Sustainability 16,
https://doi.org/10.1007/978-981-13-9117-0_15

15.1 Introduction

Surfactants are indispensable compounds in the modern world as cleansers, emulsifying agents, wetting agents, foaming agents, or dispersing agents. Because of their amphiphilic nature, surfactants have a tendency to adsorb at water-air, solid-water, and liquid-liquid interfaces and reduce the surface tension or interfacial tension. When surfactants cover the surface more closely, surface tension is reduced significantly. Another fundamental feature of surfactants is micellization or self-assembly. Surfactants alter the free energy, enthalpy, and entropy of the system (Mehta et al. 2010). The tendency of surfactants to adsorb at the water-air phase is determined by Gibbs free energy of adsorption, and Gibbs free energy of micellization represents the tendency of surfactants to form micelles at an appropriate concentration. Surface tension of the surfactant tail, the water-tail interface tension, as well the surfactant tail area contactable with the water molecules influence the Gibbs free energy of adsorption. During micellization, the partial molar volume of surfactants is changed depending on the average distance between the surfactants and the water molecules as well as between the surfactants molecules (Zdziennicka et al. 2018).

Biosurfactants are natural amphiphilic compounds produced by microbes. The hydrophobic moiety is usually fatty acid or fatty alcohol, and the hydrophilic moieties are sugars/carbohydrates or amino acids/peptides. Biosurfactants are classified into low molecular weight and high molecular weight biosurfactants. Low molecular weight biosurfactants effectively reduce the surface tension or interfacial tension and consist of glycolipids and lipopeptides. High molecular weight biosurfactants are usually extracellular amphiphilic polysaccharide or proteins such as lipopolysaccharides or lipopeptides and are good emulsifying agents (Banat et al. 2010, Roz and Rosenberg 2001). Glycolipids are produced by microbes such as *Pseudomonas aeruginosa* (rhamnolipids), *Candida bombicola* (sophorolipids), *Rhodococcus erythropolis* (trehalose lipids), and *Pseudozyma antarctica* (mannosylerythritol lipids). Some lipopeptide biosurfactant producers are *Bacillus licheniformis* (lichenysin), *Pseudomonas fluorescens* (viscosin), *Serratia marcescens* (serrawettin), and *B. subtilis* (surfactin). Some popular polymeric biosurfactants are emulsan (*Acinetobacter calcoaceticus*), alasan (*Acinetobacter radioresistens*), biodispersan (*A. calcoaceticus*), and liposan (*C. lipolytica*) (Sarubbo et al. 2015).

Biosurfactants have gained attention as an alternative to chemical surfactants due to the increasing awareness of environmental protection. Various attributes of biosurfactants are structural diversity, low toxicity, high biodegradability, and performance at extreme conditions. Compared to chemical surfactants, they are structurally diverse and are relatively high molecular weight compounds with more number of functional groups, which result in improved functionality and performances. Many biosurfactants outperform conventional surfactants in terms of surface activity and detergency. Apart from all these, biosurfactants are mild to humans, have relative low aquatic toxicity and derived completely from renewable sources. Hence, biosurfactants have huge applicability in many industries, such as personal care, domestic and industrial cleaning, agriculture, and enhanced oil recovery.

15.2 Physicochemical Attributes of Biosurfactants

Many biosurfactants have higher surface activity with low critical micelle concentration (CMC). Rhamnolipids exhibit better surface activity than sodium lauryl sulfate (SDS) due to their larger molecular area at the air-liquid interface. The CMC of rhamnolipid produced by *Pseudomonas aeruginosa PA1* is 25.7 mg/L, while CMC of SDS is 2.6 g/L. In addition to this, rhamnolipids form a stable emulsion and were found to have potential in nano-/microsphere formulation of thermoplastic polymethylmethacrylate (Mendes et al. 2015). The high surface activity of rhamnolipids is due to their high molecular weight and multiple oxygenated structures. They exhibit good frothability with more viscous and elastic froth phase and have the potential to be used as a replacement for conventional chemical frothers in the mineral processing industry (Khoshdast et al. 2012).

Biosurfactants are capable of emulsifying two immiscible liquid by reducing the interfacial tension. Biosurfactants such as rhamnolipids and surfactin are excellent emulsifier for a range of hydrocarbons such as aromatic compounds and vegetable oils, and their activity is comparable to SDS. However, the emulsifying activity of biosurfactant is pH dependent and greater emulsifying activity was observed at basic pH. This attribute is particularly advantageous for demulsification of strong emulsions formed by surfactin in the area of enhanced oil recovery (EOR). Surfactins demulsified by decreasing the pH are readily reusable and thus improving the economic viability of the process (Long et al. 2017, Lovaglio et al. 2011).

Biosurfactants effectively solubilize hydrophobic compounds in the aqueous system. Polyaromatic hydrocarbon (PAH) solubilization was found to follow a linear trend with the concentration of glycolipids biosurfactants above critical micelle concentration (CMC). This happens through micellar solubilization, where PAH disperse into the hydrophobic core of the micelles. Temperature, pH, ionic strength, and structural complexity of hydrophobic compounds are the various parameters affecting the solubility of hydrophobic compounds. Furthermore, the solubility of PAH can be greatly enhanced by mixing the biosurfactants such as rhamnolipids and sophorolipids (Li et al. 2015, Song et al. 2016).

Structural complexity and mosaic distribution of charge and polarity of biosurfactants molecules contribute to their superior performance and biological activity such as membrane binding. Biosurfactants bind weakly to protein and are less denaturing than chemical surfactants. Hence, biosurfactants are compatible with industrial enzymes and other additives in the different industrial formulations (Otzen 2017, Madsen et al. 2015).

Structural variation in the hydrophilic/hydrophobic moiety or degree of acetylation can alter the physicochemical properties of biosurfactants such as self-assembly and adsorption properties. Based on the hydrophobic moiety, sophorolipids are of two types- lactone and free acid forms (Fig. 15.1) (Penfold et al. 2011). In the presence of anionic surfactant sodium dodecyl benzene sulfonate [LAS], acidic [AS] and lactonic sophorolipids [LS] exhibit a complex and unusual phase behavior, depending on the concentration and composition (Table 15.1) (Penfold et al. 2011, 2012).

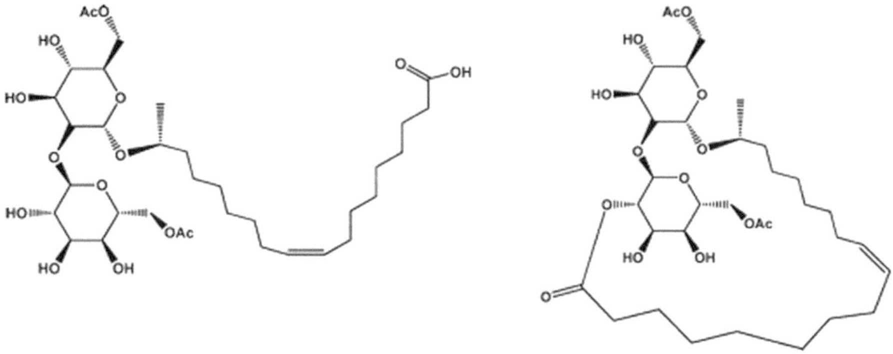

Fig. 15.1 Structure of acidic sophorolipids (AS) and lactonic sophorolipids (LS). (Penfold et al. 2011)

Table 15.1 Phase behavior of acidic (AS) and lactonic (LS) variants sophorolipids in addition of anionic surfactant sodium dodecyl benzene sulfonate (LAS). (Penfold et al. 2011)

Sophorolipid combination	Concentration	Phase behavior
LS	0.2–3 mM	Small unilamellar vesicles
LS	7 mM	Larger unilamellar vesicles
LS	10–20 mM	Disordered dilute phase of tubules
AS	0.5–50 mM	Small globular micelles
AS/LS	5–30 mM	Micellar structure of AS dominated
AS/LAS	5–30 mM AS, 60:40 AS/LAS	Globular micellar structure dominated
AS/LS/LAS	5–30 mM LS:AS (1:1)/LAS mixtures at fixed composition (60:40)	Predominantly globular micelle structure
LS/LAS	10 mM LS, 90:10–10:90 LS/LAS composition	Micellar/lamellar coexistence evolved into pure micellar phase

15.3 Biosurfactants in Bioremediation

Bioremediation is the process to restore the contaminated site by biological means. This can be achieved by the addition of living organisms (bioaugmentation) or by addition of nutrients or microbial metabolites such as biosurfactants to stimulate the growth of indigenous population, which degrade the pollutants (biostimulation) (Das and Chandran 2011). Bioremediation can be used for the treatment of oil spills, metal contamination, and organic pollutants such as nitroaromatic compounds and halogenated biphenyls. Bioremediation has various advantages such as relatively low cost, less energy requirement, and efficiency of treatment. There are two methods of treatments – in situ and ex situ. Criteria for choosing the treatment methods are the degree of contamination, geographical area, environmental conditions, and feasibility (Barnes et al. 2002, Saxena et al. 2012). Biosorption is an emerging cost-effective cleanup technology for the removal of metals and organic

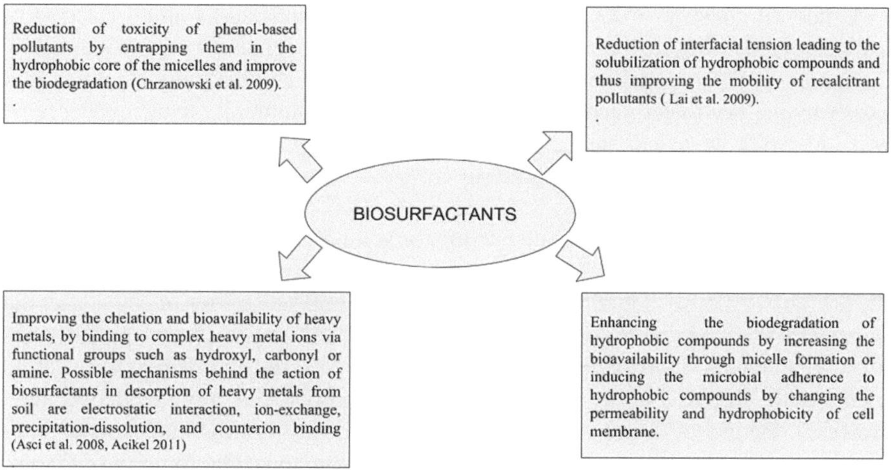

Fig. 15.2 Mechanisms of action of biosurfactants in natural and induced bioremediation of different pollutants. (Lawniczak et al. 2013)

pollutants, by using live or dead organisms, or their components. This involves various physicochemical phenomena like adsorption, absorption, ion exchange, surface complexation, and precipitation (Fomina and Gadd 2014).

Figure 15.2 shows the mechanisms of action of biosurfactants in natural and induced bioremediation of different pollutants (Lawniczak et al. 2013, Santos et al. 2016).

15.3.1 Biosurfactants for Oil Spill Cleanup

Oil spills are a major cause of environmental pollution and have a drastic effect on marine and terrestrial ecosystem. Oil spills cause a havoc on local fauna and flora, ravage the farmland cultivation, and make the affected area unfit for habitation and human activities such as fishing and swimming. Aftermath effect of oil contamination can be seen by carcinogenic heavy metal and PAH accumulation in the food chain and decreased photosynthesis in the affected area. More than 2400 animals had been killed and 1000 plant species had been destroyed by an oil spill in Colombia in 2018 (Zachos 2018). People exposed to an oil spill can have acute and chronic health effects. Cough, headache, vomiting, diarrhea, and shortness of breath are some of the immediate health issues, and hematological, hepatic, pulmonary, and cardiac functions of people exposed to spill, especially cleanup workers, were aberrated years after the incidents (D'Andrea and Reddy 2018). According to US Clean Water Act and Oil Pollution Act of 1990, bioremediation agents or chemical agents such as dispersants, sinking agents, miscellaneous oil spill control agent, and burning agents can be added to combat oil spills (National Oil and Hazardous Substances Pollution Contingency Plan).

Crude oil consists of alkanes, cycloalkanes, aromatics, and a small fraction of asphaltenes. During degradation, alkanes are readily degraded by microbes while aromatics remain recalcitrant. After an oil spill, microbial community acts synergistically to degrade the crude oil. Analysis of microbial community in deepwater horizon oil spill revealed the presence of alkane-degrading *Marinobacter* and polyaromatic hydrocarbon-degrading *Alpha-* and *Gammaproteobacteria* (Dombrowski et al. 2016).

Dispersants are emulsifying, dispersing, or solubilizing agents and contain three constituents – surfactants, solvents, and additives. Addition of dispersants is among the primary response, added to mitigate the surface and subsurface oil slick after the oil spill. Surfactants in dispersants reduce the interfacial tension between water and oil, thus enhance the dissolution of oil into the water. Ideally, dispersants should increase the bioavailability of crude oil and improve the biodegradation rate. However, the use of dispersants for treatment of oil spill is still controversial, considering their detrimental effects on the marine ecosystem. The toxicity of dispersant along with crude oil is more pronounced than crude alone and can slow down the biodegradation by altering the ingenious microbial community. Furthermore, simulation study showed that dispersant can alter the microbial community and negatively affect biodegradation rate (Kleindienst et al. 2015). Dispersants are highly toxic to marine life, can bioaccumulate and increase the PAH uptake by fish during oil spill (Ramachandran et al. 2004).

Synthetic dispersants have been used to combat the oil spills. Approximately two million gallons of Corexit 9500A was used to disperse the deepwater horizon oil spill. Formulation of a safer dispersant can address the environmental concern raised by synthetic surfactants (Athas et al. 2014).

Corexit 9500A is highly toxic to marine organisms such as zooplankton and octocorals. Studies showed that Corexit 9500A can have a devastating effect on the coral reef as it increases the mortality of coral larvae and could change the marine biodiversity and dynamics of the marine food chain (Almeda et al. 2014, Frometa et al. 2017, Goodbody-Gringley et al. 2013). One of the active ingredients in Corexit 9500A is surfactant dioctyl sodium sulfosuccinate (DOSS), which cause the pulmonary and dermatological adverse effect in the oil spill cleanup workers (Anderson et al. 2011). Environmental samples collected from deepwater horizon oil spill still contain DOSS 6 years after the spillage (White et al. 2014). Hence, continuous monitoring of dispersant and damage assessment posttreatment is needed to evaluate the use of dispersant as a measure to mitigate the future oil spills (Passow et al. 2017). Many of the components used in dispersant formulation are also used in the household detergent formulation and can make some household detergent more toxic than dispersant such as Corexit 9500A (Word et al. 2015).

In this scenario, biosurfactants are a potential alternative to synthetic dispersants as they are quite efficacious in action and totally environmentally safe. Biosurfactant from *Candida bombicola* was found to be promising as a dispersant due to its excellent dispersant activity and stability at different temperatures and pH and in presence of salt (Freitas et al. 2016). It was shown that the efficiency of rhamnolipid to disperse the crude oil was decreased after settling. Hence, the addition of additives is necessary to increase the stability of the emulsion. An environmentally benign silica nanoparticle modified with rhamnolipid resulted in a stable oil-in-water emulsion and worked well as a dispersant for crude oil in seawater system (Holakoo 2011, Pi et al. 2015). BioSURF, a rhamnolipid-based commercial dispersant formulation from Bionetix® International, is designed to combat oil slick and oil spills on rocks, beaches, and soil surfaces. Hydrocarbon degradation rate is also enhanced, as the BioSURF is fortified with micronutrients (Amanda 2018).

Lipopolysaccharide produced by *Acinetobacter calcoaceticus* disrupted the oil slick on the water surface, improved the dissolution of hydrocarbon to form a stable emulsion, and enhanced the natural biodegradation than the chemical dispersants. Microbial adhesion to hydrocarbon was also improved in the presence of lipopolysaccharide (Crescenzi et al. 2002). A biological oil spill dispersing agent containing biosurfactants such as sophorolipids, rhamnolipids, trehalose lipids, lipoprotein, and other auxiliary agents was found promising when applied onsite during oil leakage or oil pollution (Zheng 2012). Even a mixture of biosurfactants-chemical dispersant can exhibit higher efficiency in oil removal and can reduce the impact of secondary pollution caused by chemical surfactants. A lipopeptide-sodium dihexyl sulfosuccinate formulation based on hydrophilic-lipophilic deviation concept showed better oil dispersion and improved solubilization of crude oil in the water column (Rongsayamanont et al. 2017). Modified sophorolipid derivatives developed by SyntheZyme are very effective in oil dispersion and emulsification and can be used as a dispersant (biobased surfactants).

Gas hydrate, an ice-like structure formed as a result of the reaction between natural gas and water, is an undesirable event in natural gas pipelines. Hydrate formation can lead to the shutdown of onshore and offshore operations. Rhamnolipid was quite as effective as antiagglomerant at low concentration and forms a less stable emulsion, which is advantageous for phase separation and product recovery (York and Firoozabadi 2008).

Biosurfactant is a potential replacement to chemical surfactants as it can reduce the oil viscosity, disperse the hydrocarbon, stabilize the oil emulsion, and help in the deposition of paraffin/asphalt (De Cássia et al. 2014). Advantages of using biosurfactant are their superior performance even at very low concentration when compared to chemical surfactants. Mono-rhamnolipid was highly effective at a sub-CMC level for solubilization of hydrocarbon and can be employed in surfactant-enhanced aquifer remediation (Zhong et al. 2016).

The downside of biosurfactants-enhanced bioremediation is that native microbes start utilizing biosurfactants before utilizing the contaminants. When rhamnolipids were supplemented to accelerate the degradation of pesticides in soil slurry system, biodegradation of pesticides was suppressed as the microbial inoculum, *Streptomyces* species, started utilizing the rhamnolipids (Mata-Sandoval et al. 2001).

Apart from biosurfactant-mediated bioremediation, soil washing or in situ flushing with biosurfactants is also feasible. In situ flushing with surfactants is used to treat soil and groundwater contaminated with dense nonaqueous phase liquid (DNAPL), which can retain in the polluted site for many years, if untreated. Surfactants reduce the interfacial tension between water and NAPL and increase the solubility and mobility of the pollutants. Hence, the contaminant can be recovered from the polluted site at an accelerated rate (Strbak 2000). Many technologies have been developed to automate the delivery of biosurfactants to treat the contaminated site. DO-IT (dissolved oxygen in situ) treatment developed by the ETEC LLC is used to inject biosurfactants "petrosolv" for the recovery of the contaminant from the groundwater [Advanced Bioremediation Solution ETEC].

To study the effect of biosurfactants on indigenous microbes in presence of crude oil spill, Saborimanesh and Mulligan (2015) measured the cell surface hydrophobicity of bacterial communities in presence of hydrocarbon, sophorolipids, and hydrocarbon and sophorolipids combination. Microbes are hydrophobic in presence of hydrocarbon, which is due to their tendency to interact with hydrophobic substrates. In presence of sophorolipids, cells are hydrophilic and have a limited bioavailability to utilize sophorolipids. Hydrophobicity was significantly decreased in the cell-sophorolipid-hydrocarbon system, because sophorolipids increased the bioavailability of hydrocarbons to microbes through micellar dispersion of hydrocarbons. This study showed that indigenous microbes play a significant role in hydrocarbon degradation by changing the microbial dynamics and cell surface hydrophobicity via cell surface modification (Saborimanesh and Mulligan, 2015).

A combination of food grade amphiphiles such as lecithin and tween 80 can result in smaller and stable emulsion of crude oil than Corexit 9500A and can be an effective dispersant for crude oil (Athas et al. 2014). Similar way, a more potent dispersant can be developed by blending the biosurfactants with less toxic amphiphiles. An optimized formulation of glycolipids biosurfactants such as rhamnolipids and glycolipids, sorbitol-based nonionic surfactants, and solvent ethylene glycol butyl ether exhibited high dispersion effectiveness for crude oil. The formulation also exhibited low dispersant-to-oil ratio and could diminish the environmental impact of dispersant by reducing the amount of dispersant to be added to the oil spill. In addition to this, the above formulation retained high dispersion activity at various environmental factors such as low temperature, high salinity, and high pH, and was having low aquatic toxicity as well (Song et al. 2013).

Because of their superior performance at various physicochemical conditions, biosurfactants can be incorporated into high pressure-hot water washing that is used to remove oil spills from shorelines and hard surfaces. Biosurfactants from *Pseudomonas aeruginosa* could disperse oil 2–3 times greater than water alone when applied to oil-contaminated Alaskans gravel samples (Harvey et al. 1990).

Bioaugmentation with biosurfactant-producing microbes can be used to address the challenges with biosurfactant-enhanced bioremediation. Systematic environmental molecular bioremediation technology, an approach that combines bioaugmentation and biostimulation with biosurfactants, was quite effective for ex situ treatment of oil-contaminated soil (Lin et al. 2010). Remarkable degradation of total petroleum hydrocarbon and polyaromatic hydrocarbon in multi-contaminated soil was achieved by phytoremediation supplemented with rhamnolipids (Liduino et al. 2018). Ex situ bioremediation of crude oil-contaminated soil with biosurfactants based biostimulation is shown in Figs. 15.3 and 15.4.

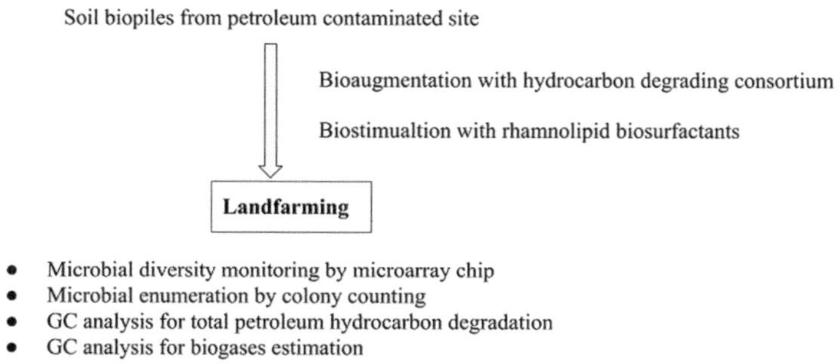

Fig. 15.3 Systematic environmental molecular bioremediation technology, a highly effective ex situ bioremediation strategy reported by Lin et al. (2010). Incorporation of bioaugmentation and biostimulation shortens the treatment time and improves the biodegradation efficiency of land farming

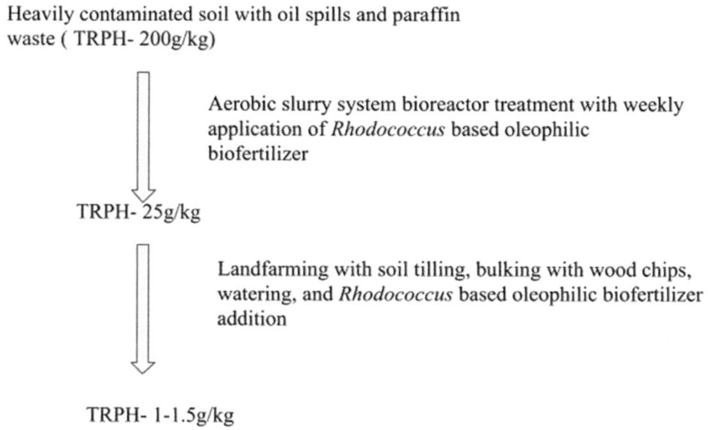

Fig. 15.4 Ex situ bioremediation of crude oil-contaminated soil using multiple approach. Here, the contamination level and treatment efficiency were monitored by measuring total recoverable petroleum hydrocarbons (TRPH). (Kuyukina et al. 2010)

Polycyclic aromatic hydrocarbons and polychlorinated biphenyl are the major organic pollutants in soil, particularly in the land exposed to emission sources such as industrial plants, oil refineries, agricultural farms, and landfills. Combustion can also be another major cause of this pollution. A recent study showed that the concentration of carcinogenic benzopyrene in agricultural and park soils of some urban areas of Havana exceeded the regulatory guidance value. Hence, periodic monitoring of organic pollutants in this kind of soil is necessary as people have direct contact with such soil (Pacheco et al. 2018).

Soil contamination with petroleum compound changes the microbial dynamics of soil to an extent that microbial community is fit to degrade the contaminant present. Many of these microbes can produce biosurfactants, which aid to emulsify the contaminant present in the soil. Some of the biosurfactant-producing genera found in the petroleum-contaminated soil are *Rhodotorula*, *Candida*, *Yarrowia*, *Geotrichum*, *Galactomyces*, and *Cystobasidium* (Yalçın et al. 2018). Apart from microbial composition, crude oil spillage alters the physicochemical properties of soil such as soil moisture content and soil permeability. Prolonged exposure of soil to petroleum hydrocarbons alter the soil wettability and induce the water repellency. In addition to this, soil water capillary water height rise and soil saturated hydraulic conductivity also decreased. Water repellency affects plant growth, changes the ecological balance, and makes the soil more prone to erosion (Roy 1999, Wei and Li 2018). Treatment with biosurfactants dispels the oil from soil particles, reduces the wettability of soil, and converts them from oil wet to water wet (Ukwungwu et al. 2017).

15.3.2 Biosurfactants for Bioremediation of Heavy Metals and Organopollutants

Pesticides and heavy metals are a big threat to the soil ecosystem. Soil washing, soil vapor extraction, thermal desorption, phytoremediation and solidification/stabilization are used to treat the pesticides and heavy metals contaminated soil. Conventional soil washing uses a combination of synthetic compounds such as SDS and EDTA. A combination of microbial-derived compounds such as rhamnolipids and citric acid effectively removed the organochlorine pesticides and heavy metals such as lindane and cadmium from contaminated soil by increasing their solubilization and desorption. These combinations are quite environmentally friendly, restore the soil ecological balance, and can cut the remediation cost by using a combination rather than individual compounds (Wan et al. 2015).

Rhamnolipids can be used as a soil bioremediation agent for the removal of heavy metals such as cadmium, nickel, lead, and zinc (Wang and Mulligan 2004, Herman et al. 1995). Sorption and desorption kinetic models revealed that rhamnolipids are compatible with various soil materials. Anionic rhamnolipids form an ionic bond with cationic cadmium ion and helps to leach out the heavy metals from the soil. Reduction of interfacial tension by rhamnolipids solubilize the cadmium ion, thus get detached from soil particles [Asci et al. 2008]. Compared to sophorolipids and lipopeptides, rhamnolipids facilitate the leaching of metals such as molybdenum, nickel, and vanadium from hazardous spent hydrosulphurozate catalyst generated by petroleum refineries (Alsaqer et al. 2018).

Rhamnolipid biosurfactant blend, JBR-425, effectively removed aged metals Zn, Cu, Pb, and Cd from the field soil deposited with metal, when compared to cationic synthetic surfactant, 1-dodecyl pyridinium chloride, and nonionic synthetic surfactant, oleyl dimethyl benzyl ammonium chloride. Remediation procedure can release metals from soil and enhance the bioavailability, which can lead to ecotoxicity of soil biota. However, treatment with JBR-425 resulted in reduced metal accumulation and increased the growth rate of two earthworm species, *Eisenia fetida* and *Lumbricus terrestris* (Slizovskiy et al. 2011).

Mobilization of heavy metals such as arsenic, copper, lead, and zinc was greatly enhanced in the presence of rhamnolipid biosurfactants. It was shown that heavy metals are incorporated into biosurfactants micelles and metal bridging might play a role in that. Hence, soil flushing with rhamnolipids can be a feasible technology to remove arsenic and other heavy metals from mine tailings (Wang and Mulligan 2009a, Wang and Mulligan 2009b).

Efficient surface-active and detergent activity of surfactin was found to be useful in the cleanup of radioactive cesium and other contaminants. Kaneka Corp., a Japanese chemical company had successfully carried out a radioactive decontamination of areas affected by the Fukushima No. 1 nuclear power plant using their biosurfactants, Kaneka Surfactin, which is composed of surfactin. Radioactive decontamination of the road using surfactin is shown in Fig. 15.5 (Sajna et al. 2015).

Biosurfactants are good additives for the phytoremediation of soil contamination. Addition of biosurfactants promotes the microbial colony formation at root surfaces and stimulates the rhizodegradation of pollutants. The plants were healthier, when compared to the addition of chemical surfactants (Al mansoori et al. 2015). Apart from rhizodegradation, rhizospheric microbes are involved in biotransformation and volatilization of organic and inorganic pollutants and biomethylation of heavy metals, which make the contaminants less toxic and more water-soluble. Since many rhizobacteria are biosurfactant producers, amendment of contaminated soil with biosurfactant-producing rhizobacteria, prior to phytoremediation, can be a promising strategy (Lal et al. 2018).

Bioaugmentation with biosurfactant-producing microbe is an effective strategy to treat persistent organic pollutants. Endosulfan, a restricted organochlorine pesticide, still used in developing countries comes under persistent organic pollutant and has a debilitating effect on humans. Biosurfactant-producing *Bordetella petrii* species could degrade α and β isomers of endosulfan up to 82%. Bioremediation of

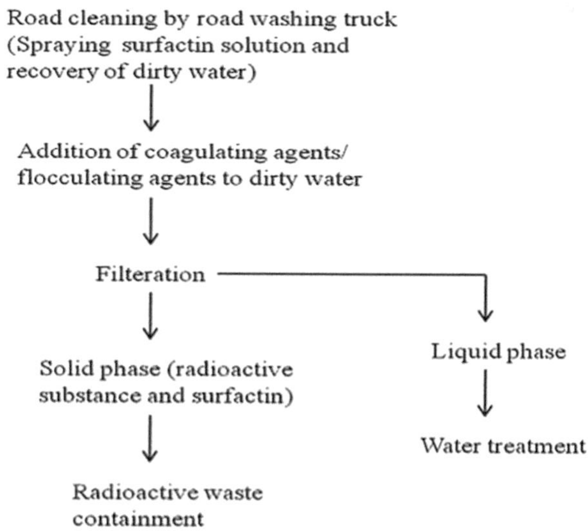

Fig. 15.5 Radioactive decontamination of roads using surfactin solution. (Sajna et al. 2015)

endosulfan-contaminated soil can be achieved by bioaugmentation with biosurfactant-producing microbes (Odukkathil and Vasudevan 2015, Odukkathil and Vasudevan 2016).

15.4 Role of Biosurfactants in Soil Health

Anthropogenic activities affect soil health and damage the ecological equilibrium supported by soil. There are physical, chemical, and biological indicators used to assess soil health [Table 15.2] (Cardoso et al. 2013). It is important to check the soil quality as soil health is the primary requirement for agriculture and environmental sustainability. Extensive farming, deforestation, and industrial effluent disposal severely affect soil health.

Surfactants are ubiquitous compounds present in the industrial formulation. Chemical surfactants used in household detergent and agricultural formulation often end up dispersed into soil and cause a negative impact on soil biota and deter soil quality. Replacing chemical surfactants with natural surfactants such as biosurfactants could be best solution to reduce the damage caused by the environmentally harmful chemicals.

15.4.1 Biosurfactants in Wastewater Treatment

Biosurfactants showed a great potential for the formulation of eco-friendly adsorbents used in wastewater treatment. A lignocellulosic biocomposite modified with natural lipopeptide biosurfactant obtained from corn steep liquor exhibited improved dye elimination and sulfate removal, when used for the treatment of winery wastewater (Perez-Ameneiro et al. 2015). The outcome can be hopefully extrapolated for the microbial-derived biosurfactants, considering the emulsification and bioadsorption properties of biosurfactants.

Dissolved air flotation (DAF) is a method of wastewater treatment to remove oily waste effluent. The efficiency of the process can be enhanced by the use of biosurfactants. Both laboratory and *in silica* analysis revealed that there was a substantial increase in efficiency of DAF to remove oil when biosurfactants were used as the collector (Rocha e Silva et al. 2018; Silva et al. 2018). Addition of biosurfactants

Table 15.2 Physical, chemical, and biological parameters to indicate soil health. (Cardoso et al. 2013)

Physical parameters	Soil texture, bulk density, porosity, and aggregate stability
Chemical parameters	Soil pH, cation exchange capacity, organic matter, and nutrient level
Biological parameters	Soil microbial activity, microbial respiration, metabolic quotient, and soil enzymes

can enhance microbial enzymatic activity and growth rate in the soil, and enhance the degradation of organic waste. Gong et al. (2017) reported that growth rate of earthworm, *Eisenia fetida* and the efficiency of vermicompost were improved on the addition of rhamnolipids (Gong et al. 2017). Biosurfactants can enhance bioenergy recovery from organic waste. Rhamnolipids and surfactin were proved to have a significant effect on hydrogen production from waste-activated sludge and organic fraction of municipal solid waste, respectively (Sharma and Melkania 2017). During the anaerobic digestion of waste-activated sludge, rhamnolipids increased the rate of acidogenesis and decreased the rate of methanogenesis, which resulted in the subsequent improved production of hydrogen within a short fermentation time (Zhou et al. 2017) [Fig. 15.6].

15.4.2 Biosurfactants in Detergent Industry

Graywater resulting from household activities contains a large number of surfactants from detergents and hygienic products. Methyl ester sulfonate, olefin sulfonates, alkyl benzene sulfonates, alkyl ether sulfates, isotridecanol ethoxylates, benzalkonium chloride, n-hexadecyl trimethyl, and ammonium chloride are the common surfactants found in the graywater. In developing countries, graywater is usually drained into the soil. Many soil properties such as soil salinity and soil pH are elevated, which leads to deterioration of soil composition and permeability. Hence, graywater should be properly disposed, and surfactants' concentration in graywater should be kept minimum (Mohamed et al. 2018). To reduce the adverse

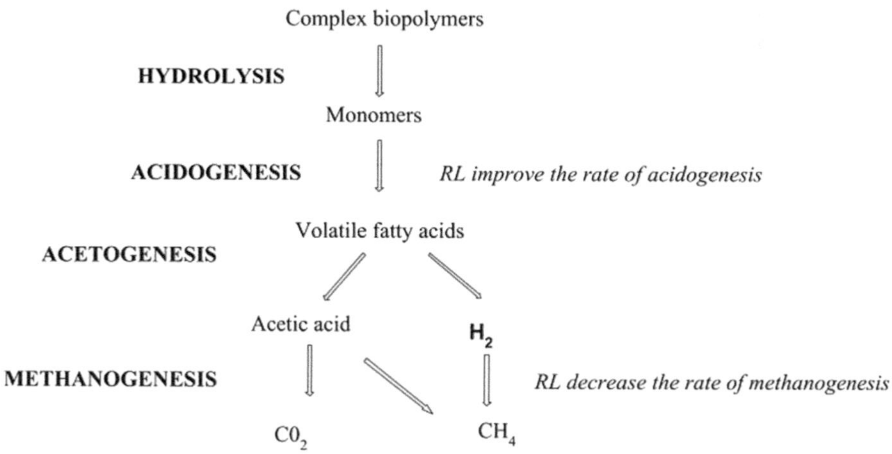

Fig. 15.6 Effect of rhamnolipids (RL) on hydrogen production from waste-activated sludge. Apart from their effect on volatile fatty acid production and conversion efficiency of methanogenesis, rhamnolipids also influence electron-proton transfer and internal resistance decrease in microbial electrolysis cell. (Zhou et al. 2017, Adapted from Bensaid et al. 2015)

effect of chemical surfactants on soil health, environmentally safe detergents containing biobased surfactants should be put to use.

Biosurfactants can be a potential replacement for chemical surfactants in detergent if they can meet large-scale production and cheap prices of synthetic surfactants. Biosurfactants can efficiently clean the stain out of soiled clothes by dispersing and solubilizing organic compounds and are compatible with enzymes used in detergents. Furthermore, they have antimicrobial and biofilm-disrupting properties (Otzen 2017). A number of dishwashing, hard surface cleaning, and laundry detergents containing sophorolipids as one of the ingredients are out in the market. Develter and Lauryssen (2010) reported the usefulness of sophorolipid in hard surface cleaning and automatic dishwashing rinse aid formulation, owing to their outstanding surface activity and low foaming properties.

A detergent formulation containing rhamnolipids as surfactant, sodium tripolyphosphate as a builder and sodium sulfate as filler had exhibited high stain removal efficiency that is comparable to commercial detergents (Bafghi and Fazaelipoor 2012). Studies showed that addition of lipopeptide biosurfactants from *Bacillus subtilis* SPB1 to commercial detergents enhanced the stain removal and wash quality (Bouassida et al. 2018). Applicability of mannosylerythritol lipids in laundry detergent formulation was explored by fabric wash analysis using biosurfactants derived from *Pseudozyma* sp. NII 08165. *Pseudozyma* biosurfactants exhibit good washing performance and are stable at high temperature and alkaline pH (Sajna et al. 2013).

Sophorolipids can be good laundry detergent additives as they possess good wetting property, emulsification index, antimicrobial activity, and clean fabric strain effectively (Joshi-Navare et al. 2013). When sophorolipids, rhamnolipids, and accell biosurfactants derived from the undisclosed yeast strain were tested to see their efficiency in removing the beef stain from cloth, in combination with either bacterial or yeast lipase enzyme, sophorolipids along with bacterial enzyme gave a satisfactory performance (Parry et al. 2012). Detergent containing a cocktail of both biosurfactants and chemical surfactants utilizing the synergistic action of sophorolipids, rhamnolipids, cellobiose lipids was shown to have enhanced oily soil detergency (Hall et al. 1995). However, a formulation containing both glycolipid biosurfactants and non-glycolipid biosurfactants in micellar phase showed an improved detergency and had been found to be suitable for all cleaning purpose, ranging from laundry detergent to hard surface cleaning. It has been noted that employing micellar phase sophorolipids is more suitable for hard surface cleaning as it possesses an efficient foam breaking activity since over-foaming of the hard surface cleaner is a disadvantage as it requires a lot of rinsing to remove the foams. Addition of micellar non-glycolipid surfactants along with sophorolipids helps to give suitable foaming properties to the hard surface cleaning formulation, as non-glycolipid biosurfactants help in the initial foaming and sophorolipids subsequently curb the foaming (Develter and Fleurackers 2010). The low foaming property of sophorolipids and their high-temperature stability can be exploited for jet washing, a washing method which uses water pressure to remove dirt from the object, and are widely used in dishwashing machine and high-tech washing machines. A mixture of lactone and acidic forms of sophorolipids exhibits better washing performance than conventionally used nonionic surfactants (Furuta et al. 2004). A formulation containing sophorolipids, cellobiose lipids, and a number of bacterial biosurfactants exhibits good flushing performance, dispersibility, foaming power, and dermatological compatibility and is found to be suitable for manual dishwashing applications (Hees and Fabry 1997).

15.4.3 Biosurfactants in Agriculture

Owing to their nontoxic and biodegradable nature, biosurfactants in agriculture can be used to achieve a sustainable environment. There are a number of applications of biosurfactants in crop protection. According to US EPA's fact sheet about rhamnolipids, rhamnolipid is an effective biofungicide against plant pathogens such as *Pythium* and *Phytophthora* species and can be used in agricultural, horticultural, and turf settings (Rhamnolipid Biosurfactant (110029) Fact Sheet). Besides being adjuvants for boosting the performance of pesticide, sophorolipids can also be used as a wetting agent, emulsifier, dispersant, and defoamer for the preparation of pesticides (Giessler-Blank et al. 2016). Sophorolipids derivatives can be effective biopesticide as their application could control plant pathogens. Among differently modified sophorolipids derivatives tested on plant pathogens, which include hydrogenated sophorolipids, sophorolipids ester derivatives, sophorolipids amide derivatives, sophorolipids biogenic amide derivatives; amide derivative shows highest antibacterial activity and sophorolipids biogenic amide derivatives show highest antifungal activity (Schofield et al. 2012).

Cellobiose lipids are more potent fungicidal agent than sophorolipids. When the antibiotic activity of both sophorolipids and cellobiose lipids were compared against *Filobasidiella neoformans* and *Candida tropicalis*, minimum inhibitory concentration (MIC) value of cellobiose lipids is much less than that of sophorolipids. Cellobiose lipids exhibit higher antifungal activity at acidic pH, while the advantage of using sophorolipids as antifungal agents is their high solubility (Kulakovskaya et al. 2014). Mode of action of cellobiose lipids in antifungal activity involves membrane permeability of target organism due to amphiphilicity of cellobiose lipids followed by membrane leakage of ATP and potassium ions (Trilisenko et al. 2012). Surfactin act as elicitor on wheat plant against *Zymoseptoria tritici* infection by stimulating both salicylic acid- and jasmonic acid-dependent signaling pathways and provided 70% protection against Septoria tritici blotch (STB) disease, caused by *Z tritici* (Le Mire et al. 2018). Biosurfactants overproducing producing *Bacillus subtilis* exhibit plant promoting trait as well (Paraszkiewicz et al. 2017). Mannosylerythritol lipids significantly reduced the infection of powdery mildew in wheat leaf (Yoshida et al. 2015).

15.5 Efficacy and Toxicity Studies of Biosurfactants

It is very necessary that efficacy and toxicity of biosurfactants should be evaluated before proceeding for their potential applications in bioremediation. Biosurfactants are relatively low toxic compounds compared to chemical surfactants. If the biosurfactant is intended for dispersant application, toxicity studies should be determined for marine test species. Toxicity was determined by LC 50 value (Median Lethal concentration), which is performed for dispersant alone and dispersant and oil mixture. LC 50 is a method that evaluates the rate of population mortality and is the concentration of a compound at which 50% population is killed in a given period of time. Marine test species such as *Phaeodactylum tricornutum*, *Oncorhynchus mykiss*, *Daphnia magna*, and *Selenastrum capricornutum* Printz are usually used for studying aquatic toxicity. However, indicator organisms vary according to the spillage site. Phytotoxicity is studied by germination index (GI), which measures the relative vegetable seed germination and relative root elongation in presence of biosurfactants (De Cássia et al. 2014). In vitro analysis of cytotoxicity and endocrine disruption ability must also be studied. Rufino et al. (2014) demonstrated the low toxicity of biosurfactants from *Candida lipolytic*a against seeds of *Brassica oleracea*, *Solanum gilo*, and *Lactuca sativa* L. and the micro-crustacean *Artemia salina*. When acute and chronic toxicities of three synthetic surfactants (PES-61, Corexit 9500, Triton X-100) and three microbiologically produced surfactants (BioEM-Glycolipid surfactant produced by *Pseudomonas aeruginosa*, emulsan, PES-51- mixture of d-limonene and a bacterial fermentation by-products) were determined and compared, biosurfactants exhibited intermediate toxicity to marine species, *Mysidopsis baha* and *Menidia beryllina* than synthetic surfactants (Edward et al. 2003). Before introducing the dispersant to oil spill site, a detailed field test is required to assess the effectiveness of biosurfactants to disperse at various environmental parameter such as temperature, pH and salinity. The effect of crude oil properties such as viscosity on dispersant activity should also be thoroughly studied.

Baffled flask test (BFT) is a standard test to study the efficacy of dispersant for the possible use in the oil spill. BFT should be performed with different variables such as temperature, oil type, mixing speed, and oil viscosity. BFT is superior and reproducible, when compared to swirl flask test, and is expected to be an official US Environmental Protection Agency's (EPA) test soon. However, detailed field trial is inevitable as laboratory-scale test does not necessarily reflect the dispersant ability in vivo (Venosa and Holder 2015). Dispersion effectiveness of Corexit 9500 on 23 crude oil measured by BFT varied from 3.4% to 93% and is higher for lighter, less viscous oil relative to heavier, more viscous oil. Furthermore, BFT revealed that dispersion effectiveness is a function of oil viscosity and gave good indication of dispersibility of oil with different variables such as mixing speed, oil type, temperature, etc. (Holder et al. 2015).

Lawniczak et al. (2013) designed a guideline for successful biosurfactant-mediated bioremediation (Fig. 15.7). Under the guideline, biocompatibility between biosurfactants, pollutants, native microbes, and plants should be taken into consid-

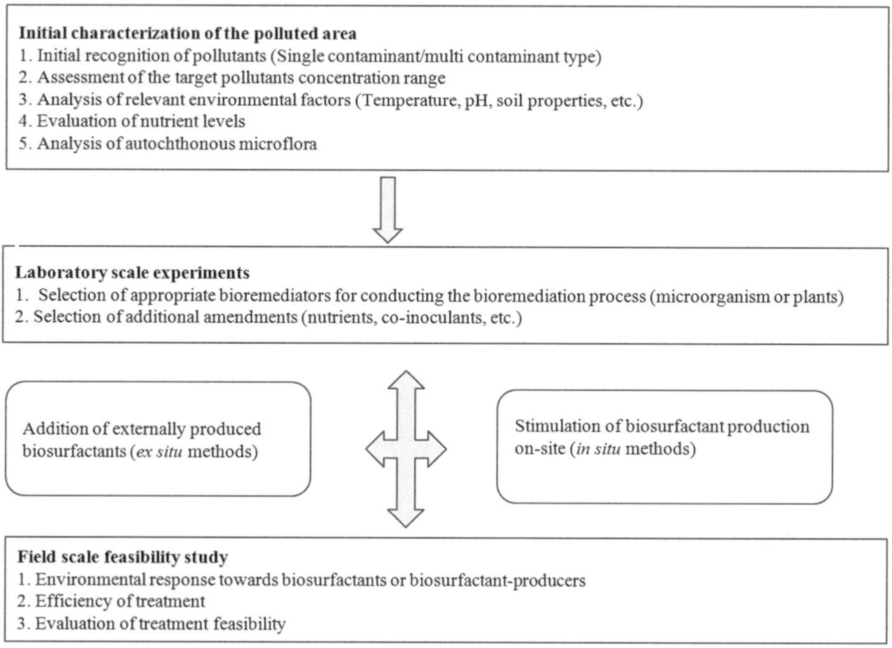

Fig. 15.7 Critical steps in biosurfactant-mediated bioremediation. (Lawniczak et al. 2013)

eration at first. Influence of native microbes on the degradation of biosurfactants should be studied in detail. An optimal concentration of biosurfactants at which an efficient biodegradation happen should be found out.

It is also imperative to perform a field study to check whether the developed technology is feasible.

15.6 Cost-Effective Production of Biosurfactants

Replacement of chemical compounds with biobased counterpart is a growing trend due to increased environmental awareness. One of the remarkable features of biosurfactant is biodegradability and environmental safety. Ideally, production of biosurfactants should be a green process from the environmental point of view, due to the utilization of renewable feedstock and no generation of hazardous by-products. However, cradle-to-grave analysis of acidic sophorolipids production by a knockout yeast for a handwash formulation revealed that use of vegetable oil and glucose as a substrate for the production of biosurfactants can contribute to much more environmental damage when compared to petroleum-derived surfactants. Hence, bioprocess development with second-generation biomass and efficient production and purification should be encouraged to reduce environmental damage (Baccile et al. 2017).

Waste cooking oil can be a cheap substrate for the production of biosurfactants. This can be a safe and environmentally sustainable solution for the disposal of waste cooking oil compared to use of waste cooking oil as an animal feed or disposal in industrial effluent which can clog the sewer in cold weather. Cultivation of *Pseudomonas* SWP-4 in medium containing waste cooking oil as sole carbon source resulted in the production of rhamnolipids at a yield of 1.9 g/L (Lan et al. 2015).

Agro-industrial waste is the abundant low-cost carbon source. Use of agro-industrial waste as a raw material for the production of value-added compounds can reduce the production cost and minimize environmental pollution (Sadh et al. 2018). Rane et al. (2017) used cheap agro-industrial waste such as molasses, orange peels extract, bagasse extract, banana peels extract and potato peels extract as a substrate for the production of biosurfactants from *Bacillus subtilis* isolate. The use of waste material as a substrate and fermentation under aseptic conditions can significantly reduce the production cost for the large-scale production of biosurfactants (Vipulanandan and Mohanty 2004). Various animal fats and tallow can be a potential substrate for biosurfactant production. Cationic biosurfactants produced by *Alcaligenes aquatilis* sp. from chicken tallow effectively remove chromium from contaminated soils (Magthalin et al. 2016). *Yarrowia lipolytica*, a biosurfactant producer, was reported to grow well in tallow derivative containing media and resulted in the production of single cell protein, microbial lipids, and lipase (Papanikolaou et al. 2007). Brewery waste is a good carbon source for the production of biosurfactants from *Bacillus subtilis* (Moshtagh et al. 2018).

Bacillus species is a potential candidate to use starchy agro-industrial waste as substrate as they are abundant in amylase enzyme. A bioprocess was developed for the simultaneous production of keratinase, amylase, and biosurfactants from a medium containing feather meal, potato peel and rapeseed cake as a carbon substrate (Bhange et al. 2016).

Downstream processing is the most expensive step of a bioprocess and account for 60-80 % of the production cost of biosurfactants. Conventional biosurfactants recovery methods such as solvent extraction have various disadvantages. In situ foam fractionation and ultrafiltration are the best choices for cost-effective continuous removal of biosurfactants. Continuous removal can result in improved yield and fermentation efficiency as product inhibition is lessened (Najmi et al. 2018).

Since high concentration of biosurfactants is usually found in the foam fraction of the fermentation broth, adsorption of biosurfactants from foam fraction is a cost-effective purification method. An integrated process foam adsorption with foam flow-through back which recirculate cell-containing collapsed foam into the bioreactor can be used for simultaneous production and recovery of rhamnolipids at high yield and purity (Anic et al. 2018). High viscosity, low dissolved oxygen, and product inhibition are the major drawbacks with large-scale production of sophorolipids. A semicontinuous sophorolipid fermentation using a novel bioreactor with dual ventilation pipes and dual sieve-plates coupled with a novel two-stage separation system resulted in a yield of 477 g/l with an improved productivity from 0.5 g g^{-1} (in the batch fermentation) to 0.6 g g^{-1} (Zhang et al. 2018).

Various recombinant organisms have been developed for the heterologous production of biosurfactants that address the challenges of biosurfactant production by natural host microbes. Pathogenicity of host-microbe and production of biosurfactants as a mixture of congeners or isomers, which aggravate the purification process and dependence on carbon sources such as vegetable oil and glucose that contribute to negative environmental impact, are the major challenges with biosurfactant production by host microbes. Development of custom-made biosurfactants and production of biosurfactants under extreme conditions can also be achieved by genetic engineering. *Pseudomonas stutzeri* Rhl was constructed for the heterologous production of rhamnolipid under anaerobic conditions (Zhao et al. 2015). A recombinant *Bacillus subtilis* was used for the production of custom-made biosurfactants called FA-Glu with applicability as a dispersant to clean up oil spills and produce biosurfactants better in medium containing glycerol than glucose (Colona et al. 2011).

Genetic-engineered *Pseudomonas putida* has been developed that utilized sustainable carbon sources such as crude glycerol and second-generation xylose and produced a set of predesigned rhamnolipid congener composition. Biosurfactants are usually synthesized as a set of congener with variation, in either the hydrophobic chain or hydrophobic moiety. The type of congener predominant in the biosurfactants determines their surface activity and other physicochemical properties. In this case, mono rhamnolipids have good foaming action, while di rhamnolipids are good emulsifiers (Tiso et al. 2017). A recombinant *Starmerella* bombicola with cytochrome P450 cyp1 gene of *Ustilago maydis* produced sophorolipids with a palmitic acid acyl chain, instead of oleic acid acyl chain (Geys et al. 2018). Hence, custom-made biosurfactants can be synthesized by playing around with biosynthetic pathways.

15.7 Conclusion

Biodegradation of pollutants is mainly achieved by the activity of indigenous microbes in the environment. It has been noted that surface-active compounds produced by microbes play an important role in the uptake of hydrophobic pollutants. Exogenous supplementation of biosurfactants improves the bioavailability of pollutants and thus accelerates the bioremediation. Considering the adverse effects of chemical surfactants, biosurfactants is the best choice to address environmental pollution. The book chapter primarily reviews the use of various biosurfactants to treat various environment contamination and improve soil sustainability. A major limitation with biosurfactant-assisted bioremediation is that biosurfactants are easily degraded than the pollutants and most microbes consume biosurfactants before utilizing the pollutants, eventually suppressing the rate of degradation. Besides the negative effects of biosurfactants addition on bioremediation, lack of consistency during scale-up experiments still questions whether biosurfactant-mediated bioremediation is a feasible technology. Hence, biosurfactant toxicity and biodegradability and substrate specificity and efficacy are the major factors to be considered for implementing biosurfactant-mediated bioremediation. Still, trials with the incorporation of biosurfactants with soil washing, in situ soil flushing, and phytoremediation were found promising. Several studies have shown that biosurfactants can be good cleaning agents owing to their amphiphilic nature and excellent performance at extreme conditions, which enable them to meet the diverse demands of the detergent industry. Using biosurfactants in detergents and cleaning agents can pave the way for environmental sustainability and preserving the human health. Being a high-value product, biosurfactants have a long way to go to meet the application in bioremediation. Use of high-titer biosurfactant-producing strains, inexpensive substrates, and a cost-effective downstream processing can make the bioprocess look appealing for remediation and sustainable technologies.

References

Açıkel YS (2011) Use of biosurfactants in the removal of heavy metal ions from soils. In: Khan M, Zaidi A, Goel R, Musarrat J (eds) Biomanagement of metal-contaminated soils. Environmental pollution, vol 20. Springer, Dordrecht, pp 183–223

Advanced Bioremediation Solution ETEC. https://www.etecllc.com/services. Accessed 7 Aug 2018

Al mansoori AF, Hasan HA, Idris M et al (2015) Potential application of a biosurfactant in phytoremediation technology for treatment of gasoline-contaminated soil. Ecol Eng 84:113–120

Almeda R, Hyatt C, Buskey EJ (2014) Toxicity of dispersant Corexit 9500A and crude oil to marine microzooplankton. Ecotoxicol Environ Saf 106:76–85. https://doi.org/10.1016/j.ecoenv.2014.04.028

Alsaqer S, Marafi M, Banat IM et al (2018) Biosurfactant-facilitated leaching of metals from spent hydrodesulphurization catalyst. J Appl Microbiol 125:1358–1369

Amanda S (2018) Bionetix Biosurf uses nature to clean oil spills on its own surf and turf. https://www.pollutionequipmentnews.com/bionetix-biosurf-uses-nature-to-clean-oil-spills-on-its-own-surf-and-turf. Accessed 7 Aug 2018

Anderson SE, Franko J, Lukomska E et al (2011) Potential immunotoxicological health effects following exposure to Corexit 9500A during cleanup of the Deepwater horizon oil spill. J Toxicol Environ Health A 74:1419–1430. https://doi.org/10.1080/15287394.2011.606797

Anic I, Apolonia I, Franco P et al (2018) Production of rhamnolipids by integrated foam adsorption in a bioreactor system. AMB Express 8:122. https://doi.org/10.1186/s13568-018-0651-y

Asci Y, Nurbas M, Acikel YS (2008) A comparative study for the sorption of Cd(II) by soils with different clay contents and mineralogy and the recovery of Cd(II) using rhamnolipid biosurfactant. J Hazard Mater 154:663–673

Aşçi Y, Nurbas M, Acikel YS (2008) Removal of zinc ions from a soil component Na-feldspar by a rhamnolipid biosurfactant. Desalination 223:361–365

Athas JC, Jun K, McCafferty C et al (2014) An effective dispersant for oil spills based on food-grade amphiphiles. Langmuir 30:9285–9294. https://doi.org/10.1021/la502312n

Baccile N, Babonneau F, Banat IM et al (2017) Development of a cradle-to-grave approach for acetylated acidic Sophorolipid biosurfactants. ACS Sustain Chem Eng 5:1186–1198

Bafghi MK, Fazaelipoor HM (2012) Application of Rhamnolipid in the formulation of a detergent. J Surfact Deterg 15:679–684

Banat IM, Franzetti A, Gandolfi I et al (2010) Microbial biosurfactants production, applications and future potential. Appl Microbiol Biotechnol 87:427–444

Barnes D, Laderrach S, Showers C (2002) Treatment of Petroleum Contaminated Soils in Cold, Wet, Remote Regions, Report No. 9E92G49, US Forest Service, Technology and

Development Program, Missoula, MT. https://www.fs.fed.us/t-d/pubs/pdfpubs/pdf02712801/pdf02712801_300dpi.pdf. Accessed 2 Aug 2018
Bensaid S, Ruggeri B, Saracco G (2015) Development of a photosynthetic microbial electrochemical cell (PMEC) reactor coupled with dark fermentation of organic wastes: medium term perspectives. Energies 8:399–429. https://doi.org/10.3390/en8010399
Bhange K, Chaturvedi V, Bhatt R (2016) Simultaneous production of detergent stable keratinolytic protease, amylase and biosurfactant by *Bacillus subtilis* PF1 using agro industrial waste. Biotechnol Rep 10:94–104
Bio-based surfactants., http://www.allied-c-s.co.jp/pdf/report12.pdf. Accessed 7 Aug 2018
Bouassida M, Fourati N, Ghazala I et al (2018) Potential application of *Bacillus subtilis* SPB1 biosurfactants in laundry detergent formulations: compatibility study with detergent ingredients and washing performance. Eng Life Sci 18:70–77. https://doi.org/10.1002/elsc.201700152
Cardoso EJBN, Vasconcellos RLF, Bini D et al (2013) Soil health: looking for suitable indicators, what should be considered to assess the effects of use and management on soil health? Sci Agric 70:274–289
Chrzanowski Ł, Wick LY, Meulenkamp R et al (2009) Rhamnolipid biosurfactants decrease the toxicity of chlorinated phenols to *Pseudomonas putida* DOT-T1E. Lett Appl Microbiol 48:756–762
Colona WJ, Marti ME, Pynn M et al (2011) Integrations of biosurfactant production into advanced biorefineries. Conference: 2011 AIChE annual meeting. Doihttps://aiche.confex.com/aiche/2011/webprogram/Paper236248.html. Accessed 12 Sept 2018
Crescenzi F, Buffagni M, Angelil D et al (2002) A new biosurfactant for use in the cleanup of oil spills on sea water environment. In: Brebbia CA (ed) Oil and Hydrocarbon Spills III. WIT Press, Southampton. https://doi.org/10.2495/OIL020231
D'Andrea MA, Reddy GK (2018) The development of Long-term adverse health effects in oil spill cleanup workers of the deepwater horizon offshore drilling rig disaster. Front Public Health 6:117. https://doi.org/10.3389/fpubh.2018.00117
Das N, Chandran P (2011, 2011) Microbial degradation of petroleum hydrocarbon contaminants: an overview. Biotechnol Res Int. https://doi.org/10.4061/2011/941810
De Cássia FSS, Almeida DG, Rufino RD et al (2014) Applications of biosurfactants in the petroleum industry and the remediation of oil spills. Int J Mol Sci 15:12523–12542
Develter D, Fleurackers S (2010) Sophorolipids and rhamnolipids. In: Johansson I, Kjellin URM (eds) Surfactants from renewable resources. Wiley, Chichester
Develter DW, Lauryssen LM (2010) Properties and industrial applications of sophorolipids. Eur J Lipid Sci Technol 112:628–638. https://doi.org/10.1002/ejlt.200900153
Dombrowski N, Donaho JA, Gutierrez T et al (2016) Reconstructing metabolic pathways of hydrocarbon-degrading bacteria from the Deepwater horizon oil spill. Nat Microbiol 17:16057. https://doi.org/10.1038/nmicrobiol.2016.57
Edward KR, Lepo JE, Lewis MA (2003) Toxicity comparison of biosurfactants and synthetic surfactants used in oil spill remediation to two estuarine species. Mar Pollut Bull 46:1309–1316
Fomina M, Gadd GM (2014) Biosorption: current perspectives on concept, definition and application. Bioresour Technol 160:3–14
Freitas BG, Brito JGM, Brasileiro PPF et al (2016) Formulation of a commercial biosurfactant for application as a dispersant of petroleum and by-products spilled in oceans. Front Microbiol 7:1646. https://doi.org/10.3389/fmicb.2016.01646
Frometa J, DeLorenzo ME, Pisarski EC et al (2017) Toxicity of oil and dispersant on the Deepwater gorgonian octocoral Swiftia exserta, with implications for the effects of the Deepwater horizon oil spill. Mar Pollut Bull 122:91–99. https://doi.org/10.1016/j.marpolbul.2017.06.009
Furuta T, Igarashi K, Hirata Y (2004) Low-foaming detergent compositions. US patent US20040171512A1
Geys R, Graevel MD, Lodens S et al (2018) Increasing uniformity of biosurfactant production in *Starmerella bombicola* via the expression of chimeric cytochrome P450s. Colloids Interf 42. https://doi.org/10.3390/colloids2040042

Giessler-Blank S, Schilling M, Thum O et al (2016) Use of sophorolipids and derivatives thereof in combination with pesticides as adjuvant/additive for plant protection and the industrial non-crop field. US patent US9351485B2

Gong X, Wei L, Yu X et al (2017) Effects of Rhamnolipid and microbial inoculants on the vermicomposting of green waste with *Eisenia fetida*. PLoS One 12:e0170820. https://doi.org/10.1371/journal.pone.0170820

Goodbody-Gringley G, Wetzel DL, Gillon D et al (2013) Toxicity of Deepwater Horizon source oil and the chemical dispersant, Corexit® 9500, to coral larvae. PLoS One 8:e45574. https://doi.org/10.1371/journal.pone.0045574

Hall PJ, Haverkamp J, Kralingen CV et al (1995) Synergistic dual-surfactant detergent composition containing sophorolipid. US patent US5417879A

Harvey S, Elashvili I, Valdes JJ et al (1990) Enhanced removal of Exxon Valdez spilled oil from Alaskan gravel by a microbial surfactant. Nat Biotechnol 8:228–230

Hees U, Fabry B (1997) Use of mixture of glyco-lipid and surfactant in hand dishwashing detergent. German patent DE19600743A1

Herman DC, Artiola JF, Miller RM (1995) Removal of cadmium, Lead, and zinc from soil by a Rhamnolipid biosurfactant. Environ Sci Technol 29:2280–2285

Holakoo L (2011) On the capability of rhamnolipids for oil spill control of surface water. Masters thesis, Concordia University, https://spectrum.library.concordia.ca/1745/. Accessed 7 Aug 2018

Holder EL, Conmy RN, Venosa AD (2015) Comparative laboratory-scale testing of dispersant effectiveness of 23 crude oils using four different testing protocols. J Environ Protect 6:628–639. https://doi.org/10.4236/jep.2015.66057

Joshi-Navare K, Khanvilkar P, Prabhune A (2013) Jatropha oil derived Sophorolipids: production and characterization as laundry detergent additive. Biochem Res Int 2013:169797. https://doi.org/10.1155/2013/169797

Khoshdast H, Abbasi H, Sam A et al (2012) Frothability and surface behavior of a rhamnolipid biosurfactant produced by *Pseudomonas aeruginosa* MA01. Biochem Eng J 60:127–134

Kleindienst S, Seidel M, Ziervogel K et al (2015) Chemical dispersants can suppress the activity of natural oil-degrading microorganisms. PNAS 112:14900–14905

Kulakovskaya E, Baskunov B, Zvonarev A (2014) The antibiotic and membrane-damaging activities of cellobiose lipids and sophorose lipids. J Oleo Sci 63:701–707

Kuyukina MS, Ivshina IB, Ritchkova MI et al (2010) Bioremediation of crude oil-contaminated soil using slurry-phase biological treatment and land farming techniques. Soil Sediment Contam 12:85–99

Lai CC, Huang YC, Wei YH et al (2009) Biosurfactant-enhanced removal of total petroleum hydrocarbons from contaminated soil. J Hazard Mater 167:609–614

Lal S, Ratna S, Said OB et al (2018) Biosurfactant and exopolysaccharide-assisted rhizobacterial technique for the remediation of heavy metal contaminated soil: an advancement in metal phytoremediation technology. Environ Technol Innov 10:243–263. https://doi.org/10.1016/j.eti.2018.02.011

Lan G, Fan Q, Liu Y et al (2015) Rhamnolipid production from waste cooking oil using *Pseudomonas* SWP-4. Biochem Eng J 101:44–54

Lawniczak L, Marecik R, Chrzanowski L (2013) Contributions of biosurfactants to natural or induced bioremediation. Appl Microbiol Biotechnol 97:2327–2339

Le Mire G, Siah A, Brisset MN et al (2018) Surfactin protects wheat against *Zymoseptoria tritici* and activates both salicylic acid- and Jasmonic acid dependent defense responses. Agriculture 8:11–23. https://doi.org/10.3390/agriculture8010011

Li S, Pi Y, Bao M et al (2015) Effect of rhamnolipid biosurfactant on solubilization of polycyclic aromatic hydrocarbons. Mar Pollut Bull 101:219–225

Liduino VS, Servulo EFC, Oliveira FJS (2018) Biosurfactant-assisted phytoremediation of multi-contaminated industrial soil using sunflower (*Helianthus annuus L.*). J Environ Sci Health A Tox Hazard Subst Environ Eng 53:609–616. https://doi.org/10.1080/10934529.2018.1429726

Lin TC, Pan PT, Cheng SS (2010) *Ex situ* bioremediation of oil-contaminated soil. J Hazard Mater 176:27–34. https://doi.org/10.1016/j.jhazmat.2009.10.080

Long X, He N, He Y et al (2017) Biosurfactant surfactin with pH-regulated emulsification activity for efficient oil separation when used as emulsifier. Bioresour Technol 241:200–206

Lovaglio RB, Jose Santos FJ, Jafelicci M Jr et al (2011) Rhamnolipid emulsifying activity and emulsion stability: pH rules. Colloids Surf B Biointerfaces 85:301–305

Madsen JK, Pihl R, Moller AH et al (2015) The anionic biosurfactant rhamnolipid does not denature industrial enzymes. Front Microbiol 6:292. https://doi.org/10.3389/fmicb.2015.00292

Magthalin CJ, Varadharajan A, Swarnalatha S et al (2016) Utilization of chicken tallow for the production of cationic biosurfactant and thereof for decontamination of Cr(III) containing soil. Procedia Environ Sci 35:895–913

Mata-Sandoval JC, Karns J, Torrents A (2001) Influence of rhamnolipids and triton X-100 on the biodegradation of three pesticides in aqueous phase and soil slurries. J Agric Food Chem 49:3296–3303

Mehta SK, Sharma S, Mehta N et al (2010) Biomimetic Amphiphiles: properties and potential use. In: Sen R (ed) Biosurfactants, advances in experimental medicine and biology, vol 672. Springer, New York, pp 102–120

Mendes A, Filgueiras L, Pinto J et al (2015) Physicochemical properties of Rhamnolipid biosurfactant from *Pseudomonas aeruginosa* PA1 to applications in microemulsions. J Biomater Nanobiotechnol 6:64–79. https://doi.org/10.4236/jbnb.2015.61007

Mohamed RM, Al-Gheethi AA, Noramira J et al (2018) Effect of detergents from laundry greywater on soil properties: a preliminary study. Appl Water Sci 8:16–24. https://doi.org/10.1007/s13201-018-0664-3

Moshtagh B, Hawboldt K, Zhang B (2018) Optimization of biosurfactant production by *Bacillus Subtilis* N3-1P using the brewery waste as the carbon source. Environ Technol. https://doi.org/10.1080/09593330.2018.1473502

Najmi Z, Ebrahimipour G, Franzetti A et al (2018) *In situ* downstream strategies for cost-effective biosurfactant recovery. Biotechnol Appl Biochem 65:523–532

National Oil and Hazardous Substances Pollution Contingency Plan., https://www.api.org/~/media/Files/Certification/ICP/ICP-Certification-Programs/1169_2017_GovRefDocs/1169_USA_40-CFR-300_Eff-04-2017.pdf. Accessed 4 Aug 2018

Odukkathil G, Vasudevan N (2015) Biodegradation of endosulfan isomers and its metabolite endosulfate by two biosurfactant producing bacterial strains of *Bordetella petrii*. J Environ Sci Health B 50:81–89

Odukkathil G, Vasudevan N (2016) Residues of endosulfan in surface and subsurface agricultural soil and its bioremediation. J Environ Manag 165:72–80

Otzen DE (2017) Biosurfactants and surfactants interacting with membranes and proteins: same but different? Biochim Biophys Acta Biomembr 1859:639–649

Pacheco S, Hilber I, Faure R et al (2018) Polycyclic aromatic hydrocarbons and polychlorinated biphenyls in urban and semi-urban soils of Havana, Cuba. J Soil Sediment. https://doi.org/10.1007/s11368-018-2137-6

Papanikolaou S, Chevalot I, Galiotou-Panayotou M et al (2007) Industrial derivative of tallow: a promising renewable substrate for microbial lipid, single-cell protein and lipase production by Yarrowia lipolytica. Electron J Biotechnol 10. https://doi.org/10.2225/vol10-issue3-fulltext-8

Paraszkiewicz K, Bernat P, Siewiera P et al (2017) Agricultural potential of rhizospheric *Bacillus subtilis* strains exhibiting varied efficiency of surfactin production. Sci Hort 225:802–809

Parry AJ, Parry NJ, Peilow AC et al (2012) Detergent compositions comprising biosurfactant and enzyme. WO patent WO2012010405A1

Passow U, Sweet J, Quigg A (2017) How the dispersant Corexit impacts the formation of sinking marine oil snow. Mar Pollut Bull 125:139–145

Penfold J, Chen M, Thomas RK et al (2011) Solution self-assembly of the sophorolipid biosurfactant and its mixture with anionic surfactant sodium dodecyl benzene sulfonate. Langmuir 27:8867–8877

Penfold J, Thomas RK, Shenc HH et al (2012) Adsorption and self-assembly of biosurfactants studied by neutron reflectivity and small angle neutron scattering: glycolipids, lipopeptides and proteins. Soft Matter 8:578–591
Perez-Ameneiro M, Vecino X, Cruz JM et al (2015) Wastewater treatment enhancement by applying a lipopeptide biosurfactant to a lignocellulosic biocomposite. Carbohyd Polym 131:186–196
Pi G, Mao L, Bao M et al (2015) Preparation of oil-in-seawater emulsions based on environmentally benign nanoparticles and biosurfactant for oil spill remediation. ACS Sustain Chem Eng 3:2686–2693
Ramachandran SD, Hodson PV, Khan CW et al (2004) Oil dispersant increases PAH uptake by fish exposed to crude oil. Ecotoxicol Environ Saf 59:300–308
Rane AN, Baikar VV, Kumar RV et al (2017) Agro-industrial wastes for production of biosurfactant by *Bacillus subtilis* ANR 88 and its application in synthesis of silver and gold nanoparticles. Front Microbiol 8:492. https://doi.org/10.3389/fmicb.2017.00492
Rhamnolipid biosurfactant (110029) Fact Sheet. https://www3.epa.gov/pesticides/chem_search/reg_actions/registration/fs_PC-110029_01-May-04.pdf. Accessed 4 Sept 2018
Rocha e Silva FCP, Rocha e Silva NMP, Luna JM et al (2018) Dissolved air flotation combined to biosurfactants: a clean and efficient alternative to treat industrial oily water. Rev Environ Sci Biotechnol 17:591–602
Rongsayamanont W, Soonglerdsongpha S, Khondee N et al (2017) Formulation of crude oil spill dispersants based on the HLD concept and using a lipopeptide biosurfactant. J Hazard Mater 334:168–177
Roy JL (1999) Soil water repellency at old crude oil spill sites. Dissertation, University of Alberta
Roz EZ, Rosenberg E (2001) Natural roles of biosurfactants. Environ Microbiol 3:229–236
Rufino RD, de Luna JM, Takaki GMC et al (2014) Characterization and properties of the biosurfactant produced by *Candida lipolytica* UCP 0988. Electron J Biotechnol 17:34–38
Saborimanesh N, Mulligan CN (2015) Effect of Sophorolipid biosurfactant on oil biodegradation by the natural oil-degrading Bacteria on the weathered biodiesel, diesel and light crude oil. J Bioremed Biodegr 6:314. https://doi.org/10.4172/2155-6199.1000314
Sadh PK, Duhan S, Duhan JS (2018) Agro-industrial wastes and their utilization using solid state fermentation: a review. Bioresour Bioprocess 5:1. https://doi.org/10.1186/s40643-017-0187-z
Sajna KV, Sukumaran RK, Jayamurthy H et al (2013) Studies on biosurfactants from *Pseudozyma* sp. NII 08165 and their potential application as laundry detergent additives. Biochem Eng J 78:85–92
Sajna KV, Höfer R, Sukumaran RK et al (2015) White biotechnology in biosurfactants. In: Pandey A, Höfer R, Taherzadeh M, Nampoothiri KM, Larroche C (eds) Industrial biorefineries and White biotechnology. Elsevier, Amsterdam, pp 499–521
Santos DK, Rufino RD, Luna JM et al (2016) Biosurfactants: multifunctional biomolecules of the 21st century. Int J Mol Sci 17(3):401. https://doi.org/10.3390/ijms17030401
Sarubbo LA, Rocha RB Jr, Luna JM et al (2015) Some aspects of heavy metals contamination remediation and role of biosurfactants. Chem Ecol 31:707–723
Saxena K, Aseri GK, Gupta AD et al (2012) Bioremediation of Xenobiotics. In: Mohee R, Mudhoo A (eds) Bioremediation and sustainability: research and applications. Wiley, Hoboken, pp 367–398
Schofield MH, Thavasi TR, Gross RA (2012) Modified sophorolipids for the inhibition of plant pathogens. WO patent WO2013052615A1
Sharma P, Melkania U (2017) Biosurfactant-enhanced hydrogen production from organic fraction of municipal solid waste using co-culture of *E. coli* and *Enterobacter aerogenes*. Bioresour Technol 243:566–572
Silva EJ, Almeida DG, Luna JM et al (2018) Use of bacterial biosurfactants as natural collectors in the dissolved air flotation process for the treatment of oily industrial effluent. Bioprocess Biosyst Eng 41:1599–1610
Slizovskiy IB, Kelsey JW, Hatzinger PB (2011) Surfactant-facilitated remediation of metal-contaminated soils: efficacy and toxicological consequences to earthworms. Environ Toxicol Chem 30:112–123. https://doi.org/10.1002/etc.357

Song D, Liang S, Zhang Q et al (2013) Development of high efficient and low toxic oil spill dispersants based on sorbitol Derivants nonionic surfactants and glycolipid biosurfactants. J Environ Protect 4:16–22. https://doi.org/10.4236/jep.2013.41B004

Song D, Liang S, Yan L et al (2016) Solubilization of polycyclic aromatic hydrocarbons by single and binary mixed Rhamnolipid-Sophorolipid biosurfactants. J Environ Qual 45:1405–1412. https://doi.org/10.2134/jeq2015.08.0443

Strbak L (2000) *In Situ* Flushing with surfactants and Cosolvents. National Network of environmental studies fellowship report for US Environmental Protection Agency. Office of Solid Waste and Emergency Response Technology Innovation Office, Washington, DC

Tiso T, Zauter R, Tulke H et al (2017) Designer rhamnolipids by reduction of congener diversity: production and characterization. Microb Cell Factories 16:225. https://doi.org/10.1186/s12934-017-0838-y

Trilisenko LV, Kulakovskaya EV, Kulakovskaya TV et al (2012) The antifungal effect of cellobiose lipid on the cells of *Saccharomyces cerevisiae* depends on carbon source. Springerplus 1:18–26. https://doi.org/10.1186/2193-1801-1-18

Ukwungwu SV, Abbas AJ, Nasr GG et al (2017) Wettability effects on Bandera gray sandstone using biosurfactants. J Eng Technol 6. http://www.joetsite.com/wp-content/uploads/2017/07/Vol.-62-47-2017.pdf

Venosa AD, Holder E (2015) Laboratory-Scale Testing of Dispersant Effectiveness of 20 Oils Using the Baffled Flask Test. http://oilspilltaskforce.org/wp-content/uploads/2015/08/Venosa-and-Holder-baffled-flask.pdf. Accessed 11 Sept 2018

Vipulanandan C, Mohanty KK (2004) Biosurfactant produced from used vegetable oil for removal of metals from wastewaters and soils. EPA project report R828598C787

Wan J, Meng D, Long T et al (2015) Simultaneous removal of Lindane, Lead and cadmium from soils by Rhamnolipids combined with citric acid. PLoS One 10:e0129978. https://doi.org/10.1371/journal.pone.0129978

Wang S, Mulligan C (2004) Rhamnolipid foam enhanced remediation of cadmium and nickel contaminated soil. Water Air Soil Pollut 157:315–330

Wang S, Mulligan CN (2009a) Arsenic mobilization from mine tailings in the presence of a biosurfactant. Appl Geochem 24:928–935

Wang S, Mulligan CN (2009b) Rhamnolipid biosurfactant-enhanced soil flushing for the removal of arsenic and heavy metals from mine tailings. Process Biochem 44:296–301

Wei Y, Li G (2018) Effect of oil Pollution on water characteristics of loessial soil. IOP Conference Series: Earth and Environmental Science, 170 032154. https://doi.org/10.1088/1755-1315/170/3/032154

White HK, Lyons SL, Harrison SJ et al (2014) Long-term persistence of dispersants following the Deepwater horizon oil spill. Environ Sci Technol Lett 1:295–299

Word JQ, Clark JR, Word LS (2015) Comparison of the acute toxicity of Corexit 9500 and household cleaning products. Hum Ecol Risk Assess 21:707–725

Yalçın HT, Ergin-Tepebaşı G, Uyar E (2018) Isolation and molecular characterization of biosurfactant producing yeasts from the soil samples contaminated with petroleum derivatives. J Basic Microbiol 58:782–792. https://doi.org/10.1002/jobm.201800126

York JD, Firoozabadi A (2008) Comparing effectiveness of Rhamnolipid biosurfactant with a quaternary ammonium salt surfactant for hydrate anti-agglomeration. J Phys Chem B 112:845–851

Yoshida S, Koitabashi M, Nakamura J et al (2015) Effects of biosurfactants, mannosylerythritol lipids, on the hydrophobicity of solid surfaces and infection behaviours of plant pathogenic fungi. J Appl Microbiol 119:215–224

Zachos E (2018) 2,400 Animals Die in Oil Spill in Colombia, https://news.nationalgeographic.com/2018/03/oil-spill-colombia-animals-killed-spd/. Accessed 3 Aug 2018

Zdziennicka A, Krawczyk J, Szymczyk K et al (2018) Macroscopic and microscopic properties of some surfactants and biosurfactants. Int J Mol Sci 19. https://doi.org/10.3390/ijms19071934

Zhang Y, Jia D, Sun W et al (2018) Semicontinuous sophorolipid fermentation using a novel bioreactor with dual ventilation pipes and dual sieve-plates coupled with a novel separation system. Microb Biotechnol 11:455–464

Zhao F, Shi R, Zhao J et al (2015) Heterologous production of *Pseudomonas aeruginosa* rhamnolipid under anaerobic conditions for microbial enhanced oil recovery. J Appl Microbiol 118:379–389

Zheng H (2012) Biological oil spilling dispersing agent and preparation method thereof. Chinese patent CN102335493A, 1 Feb 2012

Zhong H, Zhang H, Lui Z et al (2016) Sub-CMC solubilization of dodecane by rhamnolipid in saturated porous media. Sci Rep 6:33266. https://doi.org/10.1038/srep33266

Zhou A, Zhang J, Cai W et al (2017) Comparison of chemosynthetic and biological surfactants on accelerating hydrogen production from waste activated sludge in a short-cut fermentation-bioelectrochemical system. Int J Hydrog Energy 42:9044–9050

Chapter 16
Biodegradation of Non-steroidal Anti-inflammatory Drugs and Their Influence on Soil Microorganisms

Urszula Guzik and Danuta Wojcieszyńska

Abstract In recent years, particular attention has been paid to the increasing environmental pollution, including the soil environment, by pharmaceuticals. One of the most frequently detected contaminants of this kind are non-steroidal anti-inflammatory drugs (NSAIDs). As bioactive compounds, they can have a negative impact on organisms and affect the processes occurring in the environment. Therefore, it is extremely important to know the effect of these drugs on living organisms, as well as cognition of processes of their biodegradation. The presented work shows the fate of non-steroidal anti-inflammatory drugs after their release into the environment and their interactions with soil components. The processes of microbiological decomposition of NSAIDs were also discussed, and their possible negative impact on microorganisms was synthetically described. Analysis of the source data has shown that the mechanisms of NSAID toxicity on microorganisms along with their biodegradation pathways remain poorly understood.

Keywords Biodegradation · Microorganisms · Non-steroidal anti-inflammatory drugs · Soil · Toxicity

16.1 Introduction

The development of analytical tools has made it possible to analyze the problem of micro-pollutants in the soil environment. Some of the most common compounds of this type are non-steroidal anti-inflammatory drugs. This is due to their high availability because they are over-the-counter drugs. In recent years, a steady increase in both their production and consumption has been observed. Currently, there are more than 50 NSAIDs on the world market (Fokunang et al. 2018). The most popular non-steroidal anti-inflammatory drugs include ibuprofen, naproxen, diclofenac,

U. Guzik (✉) · D. Wojcieszyńska
Department of Biochemistry, Faculty of Biology and Environmental Protection, University of Silesia in Katowice, Katowice, Poland
e-mail: urszula.guzik@us.edu.pl; danuta.wojcieszynska@us.edu.pl

A. Kumar, S. Sharma (eds.), *Microbes and Enzymes in Soil Health and Bioremediation*, Microorganisms for Sustainability 16,
https://doi.org/10.1007/978-981-13-9117-0_16

ketoprofen, and paracetamol. The latter is usually included in NSAIDs although the mechanism of its action excludes it from this group of drugs.

NSAIDs work by inhibiting the conversion of arachidonic acid to prostaglandin H-PGH. Prostaglandin H is a precursor for a number of prostaglandins, which are tissue hormones (formed in damaged and inflamed tissues, which sensitizes them to other mediators, which results in the occurrence of pain and hypersensitivity in the area of damage); prostacyclins, which increase congestion in the area that is affected by the inflammatory process; and thromboxanes, which play a role in the aggregation of thrombocytes. The arachidonic acid transformation step is catalyzed by cyclic prostaglandin peroxide synthase, which is now identified with cyclooxygenase, and which includes the isoenzymes COX-1 and COX-2. Cyclooxygenase is a bifunctional enzyme that has two different enzymatic activities – cyclooxygenase and peroxidase. The conversion of arachidonic acid into prostaglandin H is performed by its cyclization to unstable 15-hydroperoxide and its double oxidation at positions 9 and 11. Owing to the peroxidase activity of cyclooxygenase, the 15-hydroxyperoxide molecule is reduced to its 15-hydroxy analog (Brune and Patrignani 2015; Osafo et al. 2017). Paracetamol, like the NSAIDs, has analgesic and antipyretic properties and suppresses prostaglandin production. However, it does not induce the adverse effects of NSAIDs on the gastrointestinal tract and has a very poor anti-inflammatory effect. Why? The answer results from the different mechanism for inhibiting cyclooxygenase by paracetamol. This probably acts as a factor that reduces the iron cation in protoporphyrin IX in the peroxidase part of the enzyme. This radical generates tyrosine radicals in the cyclooxygenase part of the enzyme, which are necessary to catalyze the oxidation reaction of arachidonic acid into 15-hydroxyperoxide (the latter is reduced to prostaglandin H in the peroxidase part of the enzyme). In the tissue that is affected by an inflammatory process, there is a high level of peroxides and these act antagonistically to paracetamol, thus inhibiting its anti-inflammatory activity (Mallet et al. 2017; Jóźwiak-Bębenista and Nowak 2014).

It is difficult to calculate the accurate global consumption of non-steroidal anti-inflammatory drugs because they are sold under various trade names and they are over-the-counter drugs. Nevertheless, in 2012, Fierce Pharma reported that diclofenac was the twelfth bestselling generic molecule globally (Fatehifar et al. 2018). The annual global production of ibuprofen is currently more than 30,000 tons (Huang et al. 2018). In turn, in Germany, about 836 tons of acetylsalicylic acid, 622 tons of paracetamol, 345 tons of ibuprofen, and 86 tons of diclofenac were consumed in 2001 (Wojcieszyńska et al. 2014). Unfortunately, the high consumption of NSAIDs has led to its occurrence in the environment. Paracetamol is one of the compounds that are most frequently detected in the aquatic environment (Abdullah et al. 2018). In natural waters, up to 10 μg/L and, in some cases, more than 65 μg/L have been reported in the USA and in the Tyne River (UK), respectively (De Gusseme et al. 2011). In turn, the concentration of paracetamol in agricultural soils is reported to be approximately 0.4 μg/kg (Liang et al. 2016). Naproxen was detected in the surface water of the Liao river in the People's Republic of China with a maximum concentration of 40.7 ng/L (Tang et al. 2014). It was also reported that the concentration of ibuprofen, naproxen, and diclofenac in the Mezquital Valley (Mexico) irrigation system was 0.22–0.38 μg/L, 2.84–6.74 μg/L, and 0.25–0.50 μg/L, respectively (Gomez-Olivan et al. 2014). Considering the numerous reports on the presence of these biologically active compounds in the environment, one can expect a negative effect of these compounds on the microorganisms that inhabit the environment.

16.2 The Fate of Non-steroidal Anti-inflammatory Drugs in the Soil

The presence of non-steroidal anti-inflammatory drugs, their metabolites, and their degradation products in the environment is primarily connected with the low removal rates in wastewater treatment plants and by the improper disposal of unused drugs (Escuder-Gilabert et al. 2018). Non-steroidal anti-inflammatory drugs can enter the soil as a result of it being irrigated with reclaimed water and/or the application of biosolids to agricultural land (Gonzales-Naranjo et al. 2013; Nowak et al. 2013; Xu et al. 2010). The fate of pharmaceuticals is determined by the filtration, sorption, and degradation processes. Moreover, the interaction of NSAIDs and other organic and inorganic compounds can result in the formation of new contaminants (Lonappan et al. 2016).

It was shown that the abiotic degradation of naproxen and diclofenac in the soil is vestigial (Caracciolo et al. 2015; Topp et al. 2008). Diclofenac (0.1 μg/g) is very stable in agricultural sterile soil, whereas in the same soil, non-autoclaving diclofenac rapidly mineralized (a half-life of fewer than 5 days) (Caracciolo et al. 2015). However, the fate of paracetamol is connected with its photosensitized transformation because this compound is highly reactive in the presence of inorganic radicals such as $OH^{\bullet}$, ${}^{\bullet}N_3$, and $CO_3^{\bullet-}$(Liang et al. 2016). Caracciolo et al. (2015) described that the half-life of ibuprofen in unsaturated and water-saturated soil was 30 and 1706 days, respectively. However, it was shown that the velocity of the degradation of NSAIDs in soil depends on the properties of the microcosms and the drug concentration. Quite a rapid degradation of ibuprofen (a half-life ranging from 0.3 to 0.9 days) was observed (Caracciolo et al. 2015). Moreover, the degradation of NSAIDs depends on environmental factors such as soil type and seasonal conditions – temperature and moisture (Cycoń et al. 2016; Nowak et al. 2013). Additionally, the apparent biodegradation rate of NSAIDs may be limited by the diffusion of oxygen and carbon dioxide between the soil and the atmosphere (Topp et al. 2008).

The speed of sorption is connected with the mobility of drugs (Martinez-Hernandez et al. 2016). The binding of NSAIDs to natural soils depends on the texture of the soil, the presence of reactive groups in the soil compounds – organic matter and mineral oxide surfaces, the exchange capacity, soil solution pH, and the retention in the soil. It was also shown that the chiral environmental soil matrices may preferentially connect some of the enantiomers of chiral pharmaceuticals (Sanganyado et al. 2017). It was also shown that acidic drugs have a lower soil sorption coefficient. Ibuprofen and naproxen have a carboxylic acid group that is ionized at a typical environmental pH (5–8) and in this form, they may conjugate and become more hydrophilic that their neutral forms. The complexes that are formed may interact with the soil organic matter and mineral surface (Ascar et al. 2017; Fent et al. 2006; Vulava et al. 2016). However, Xu et al. (2010) observed that the dissolved organic matter did not have a significant impact on the mobility of NSAIDs in the soil but that their mobility depended on the amount that passed through the soil profile (Xu et al. 2010). Moreover, Liang et al. (2016) showed that the adsorption of paracetamol depended on the type of adsorbent. There is little adsorption of paracetamol into silica, alumina, aquifer sand and sediment (Liang et al. 2016). The role of the calcium ions in the mobility of NSAIDs is unclear. Xu et al. (2010) showed that a 10 mM $CaCl_2$ solution significantly immobilized NSAIDs in the soil matrix. Conversely, the results of Graouer-Bacart et al. (2016) indicated that $CaCO_3$ increased the mobility of diclofenac significantly.

In the soil environment, reversible adsorption occurs. This may affect the bioavailability of pharmaceuticals for soil microorganisms and in this way determine the speed of degradation (Gonzales-Naranjo et al. 2013). Ascar et al. (2017) observed that the application of biosolids to soils caused a decrease in the bound NSAIDs and as a result, an increase in their biodegradation.

16.3 Mechanisms of Biodegradation of NSAIDs by Microorganisms

Most reports connected with the degradation of non-steroidal anti-inflammatory drugs concern the physicochemical methods of removing these compounds, primarily the advanced oxidation processes. However, these processes are characterized by harsh reaction conditions. Although biological methods for removing pollutants are more attractive because they are eco-friendly and inexpensive, they are not without disadvantages. The disadvantages that are most frequently mentioned are the low effectiveness of the bioremediation and the lack of microorganisms that have an increased degradation potential (Żur et al. 2018a).

The first reports concerned the biotransformation processes of NSAIDs by fungi. As early as 1975, Hart and Orr observed the transformation of paracetamol into 4-aminophenol by the *Penicillium* strain. The acetate that was released in this process was completely degraded, while 4-aminophenol accumulated in the culture medium (Hart and Orr 1975). It is known that the fungal strains that belong to the white-rot fungi (such as *Trametes* or *Phanerochaete*) and *Aspergillus* or *Cunninghamella* are also able to transform NSAIDs (Amadio et al. 2010; Aracagök et al. 2018; He and Rosazza 2003; Li et al. 2015; RatnaKumari et al. 2009; Rodarte-Morales et al. 2012). What is more, *Cunninghamella* sp. has successfully been used as a model to predict drug transformation. This microorganism has a cytochrome P-450 system that is analogous to the microsomal system of cytochrome P-450 in mammals (RatnaKumari et al. 2009; Zhong et al. 2003). It was observed that naproxen was transformed by the *Cunninghamella* species into two intermediates – desmethylnaproxen and its sulfate derivative (Zhong et al. 2003). *Cunninghamella echinulata* metabolizes paracetamol into N-acetyl-p-benzoquinoneimine (NAPQI) *via* N-hydroxylation and rearrangement (RatnaKumari et al. 2009). Amadio et al. (2010) observed the mono- and dihydroxylation of flurbiprofen by *Cunninghamella elegans*, *Cunninghamella echinulate*, and *Cunninghamella blakesleeana*. Analysis of the flurbiprofen metabolites showed that this drug was transformed into 4-hydroxyflurbiprofen, 3,4-dihydroxyflurbiprofen, and hydroxyl-methoxy-flurbiprofen. These compounds are conjugates with sulfate. This is a typical detoxification reaction that is also observed in the mammalian phase II metabolism of flurbiprofen (Amadio et al. 2010). A transformation system of naproxen that is similar to those mammalian was also observed during the removal of naproxen by *Aspergillus niger* ATCC 9142. *O*-desmethylnaproxen, 7-hydroxynaproxen, and 7-hydroxy-*O*-desmethylnaproxen as products of hydroxylation by the cytochrome P-450 system were identified (He and Rosazza 2003).

White-rot fungi are some of the most useful biocatalysts, which due to their broad specificity of enzymes such as cytochrome P-450 system or extracellular laccases and peroxidases can degrade soil pollutants (Borras et al. 2011; Marco-Urrea et al. 2009). In the last few years, promising results of NSAIDs degradation were obtained using *Trametes versicolor*. This strain is able to decompose a wide range of pharmaceuticals including non-steroidal anti-inflammatory drugs such as naproxen, ibuprofen, and ketoprofen (Borras et al. 2011; Marco-Urrea et al. 2009; Marco-Urrea et al. 2010a; Marco-Urrea et al. 2010b; Rodriguez-Rodriguez et al. 2010). During the transformation of ibuprofen, the hydroxylated derivatives 1-hydroxy ibuprofen, 2-hydroxy ibuprofen, and 1,2-dihydroxyibuprofen were detected (Marco-Urrea et al. 2009). A hydroxylation reaction was also observed during the transformation of naproxen. Cytochrome P-450 and laccase probably play a crucial role in this process (Aracagök et al. 2018; Borras et al. 2011; Marco-Urrea et al. 2010a; Rodriguez-Rodriguez et al. 2010). It was also observed 1-(6-methoxynaphthalen-2-yl)ethanone (Marco-Urrea et al. 2010a). *Trametes versicolor* also has the ability to completely biotransform 10 mg/l ketoprofen. It is likely that this compound enters a cell *via* active transport and is then converted by the cytochrome P450 enzymatic system. The main metabolite that was identified is 2-([3-hydroxy(phenyl)methyl] phenyl)-propanoic acid, and 2-[3-(4-hydroxybenzoyl)phenyl]-propanoic acid and 2-(3-benzoyl-4-hydroxyphenyl)-propanoic acid were also observed in small amounts (Marco-Urrea et al. 2010b).

Another white-rot fungus that is able to transform NSAIDs is *Phanerochaete chrysosporium*. Within 23 h, this organism completely eliminated 1 mg/L, 1 mg/L, and 1.3 mg/L of diclofenac, ibuprofen, and naproxen, respectively (Rodarte-Morales et al. 2012). Li et al. (2015) suggested that the extracellular enzymes laccases and manganese peroxidases play important roles in the removal of naproxen. The

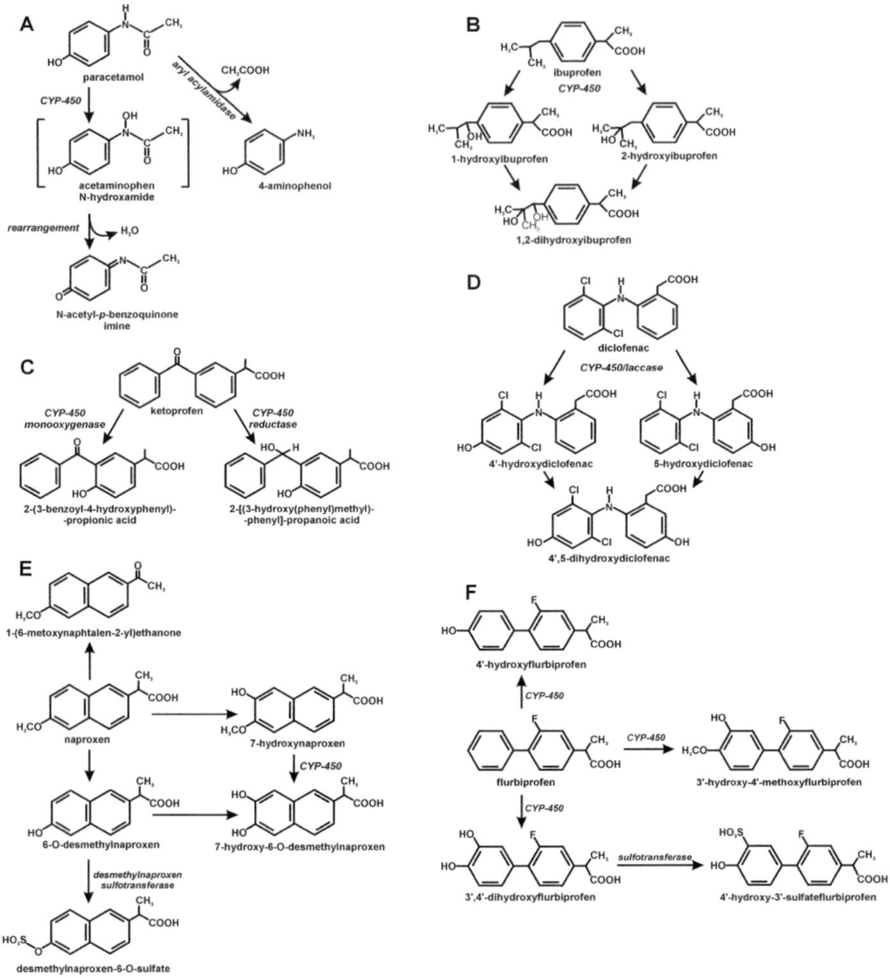

Fig. 16.1 The biodegradation pathways of non-steroidal anti-inflammatory drugs (**a** paracetamol, **b** ibuprofen, **c** ketoprofen, **d** diclofenac, **e** naproxen, **f** flurbiprofen) by fungi. (Amadio et al. 2010; Aracagök et al. 2018; Hart and Orr 1975; He and Rosazza 2003; Marco-Urrea et al. 2009; Marco-Urrea et al. 2010a, b)

biodegradation pathways of non-steroidal anti-inflammatory drugs by fungi have been presented in Fig. 16.1.

In recent years, there have been more and more reports on the degradation of non-steroidal anti-inflammatory drugs by the bacteria of the genera *Klebsiella*, *Delftia*, *Patulibacter*, *Stenotrophomonas*, *Pseudomonas*, *Labrys*, *Raoultella*, *Brevibacterium*, *Planococcus*, *Bacillus*, *Enterobacter*, *Sphingomonas*, *Pseudaminobacter*, *Ralstonia*, and *Streptomyces* (Ahn et al. 2017; Aissaoui et al. 2017; De Gusseme et al. 2011; Domaradzka et al. 2016; Fang and Zhou 2014; Górny et al. 2019; Hintner et al. 2001; Ishiyama et al. 2004; Marchlewicz et al.

2017a; Moreira et al. 2018; Murdoch and Hay 2005; Salgado et al. 2018; Stylianou et al. 2018; Wojcieszyńska et al. 2016). The best-known degradation pathway is the salicylic acid decomposition (Ahn et al. 2017; Fang and Zhou 2014; Hintner et al. 2001; Ishiyama et al. 2004; Jõesaar et al. 2017). This compound is synthesized by many plants; hence, the microorganisms developed the enzymes that are involved in its degradation in the course of evolution. Microorganisms degrade salicylic acid through three different pathways in aerobic conditions: *via* hydroxylation into catechol or gentisic acid or by cleavage into 2-oxo-3,5-heptadienedioic acid. This last is a product of the direct ring cleavage by NADH-independent salicylate 1,2-dioxygenase (Hintner et al. 2001). Degradation by gentisic acid was observed, for example, in *Ralstonia* sp. strain U2, *Ralstonia solanacearum*, *Streptomyces* sp. strain WA46, and *Pseudomonas putida strain* AK5 (Filatova et al. 2017; Fang and Zhou 2014; Ishiyama et al. 2004; Lowe-Power et al. 2016). Hydroxylation of salicylate into gentisate is catalyzed by a three-component monooxygenase – salicylate 5-hydroxylase. This enzyme is able to incorporate one atom of molecular oxygen into the structure of salicylate with a simultaneous reduction of the second oxygen atom into water (Fang and Zhou 2014; Marchlewicz et al. 2015). The cleavage of the gentisate aromatic ring leads to the formation of maleylpyruvate, which is then converted into fumarylpyruvate (Lowe-Power et al. 2016). In *Rhodococcus* sp. B4 and *Streptomyces* sp. WA46, salicylate is converted into gentisate *via* salicylyl-CoA and gentisyl-CoA. Salicylyl-CoA is the result of the activity of salicylyl-AMP ligase and probably salicylyl-CoA synthetase. AMP ligase may activate salicylate *via* the addition of AMP, and the activated adduct may be bound to the thiol group of the enzyme. It is also possible that the spontaneous conversion of salicylyl-AMP into salicylyl-CoA occurs. Salicylyl-CoA is hydroxylated by salicylyl-CoA 5-hydroxylase into gentisyl-CoA, which may undergo spontaneous hydrolysis into gentisate. This latter is converted into maleylpyruvate by gentisate 1,2-dioxygenase, which leads to the central metabolism (Ishiyama et al. 2004; Marchlewicz et al. 2015). The degradation of salicylate with catechol as a key intermediate is often observed during the decomposition of polycyclic aromatic hydrocarbons and other persistent aromatic compounds, primarily in the genus *Pseudomonas* (Ahn et al. 2017; Panov et al. 2013; Sazonova et al. 2008; Silva et al. 2007). Hydroxylation of salicylic acid into catechol is catalyzed by salicylate 1-hydroxylase. Next, the catechol that is formed may undergo two different cleavages – in an *ortho* or *meta* position. The enzymes that are engaged in these processes are catechol 1,2-dioxygenase and 2,3-dioxygenase, respectively, which are commonly present in soil microorganisms and which have been described in-depth in the literature (Guzik et al. 2013; Vaillancourt et al. 2006). The results of Kesseru et al. (2005), who observed nitrate-dependent salicylate degradation, were very surprising. Nitrate reduction led to the release of oxygen, which oxidized salicylate into catechol. The complete reduction of nitrate into molecular nitrogen is sufficient for the total oxidation of salicylate to the cleavage product of catechol – 2-hydroxymuconic semialdehyde (Kesseru et al. 2005). The biodegradation pathways of salicylic acid by bacteria have been presented in Fig. 16.2.

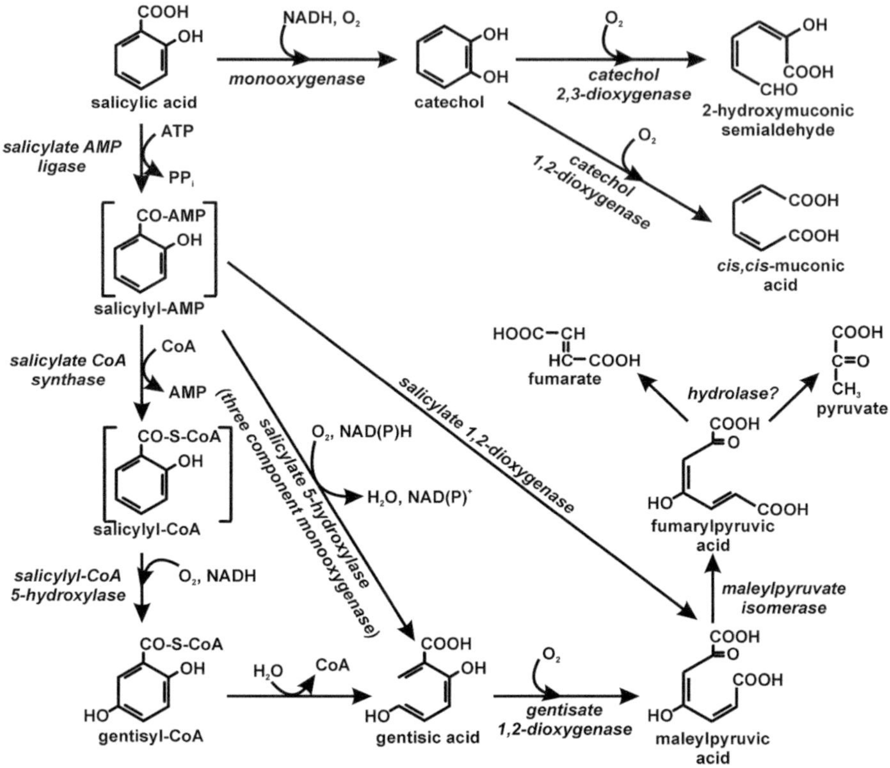

Fig. 16.2 The biodegradation pathways of salicylate by bacteria. (Fang and Zhou 2014; Hintner et al. 2001; Ishiyama et al. 2004; Kesseru et al. 2005; Lowe-Power et al. 2016)

A review of the available literature indicated that the degradation of the typically synthetic NSAIDs by bacteria is much more difficult. In recent years, however, new enzymatic pathways have been developed for these drugs. One of the best-described degradation pathways is the decomposition of ibuprofen.

The full degradation pathway of ibuprofen was described by Murdoch and Hay for the first time in 2005. The *Sphingomonas* sp. strain Ibu-2 degraded ibuprofen *via* isobutylocatechol. Isobutylocatechol is cleaved by the *meta* ring-fission enzyme to 5-formyl-2-hydroxy-7-methylocta-2,4-dienic acid and next to 2-hydroxy-5-isobutyl-hexa-2,4-dienedioic acid. The first step in the degradation of ibuprofen is it being catalyzed by ibuprofen-CoA ligase, which is encoded by *ipfF*. Moreover, an analysis of the Ibu-2 strain genome revealed the possible participation of the hydroxylation enzymes – aromatic dioxygenase, which is coded by *ipfAB*, and ferredoxin and ferredoxin reductase, which are coded by *ipfH* and *ipfI*, respectively. The enzymes that are engaged in the acidic side chain of the removal of ibuprofen are probably coded by *ipfD* and *ipfE*. These genes are similar to the genes encoding the sterol carrier protein X and the domain of the unknown function 35 (Murdoch and Hay 2013). *Meta* cleavage is also characteristic for the decomposition of ibuprofen by *Variovorax* Ibu-1. However, in this case, trihydroxy ibuprofen was observed to be the key intermediate (Murdoch and Hay 2015). The degradation pathways of ibuprofen have also been described for the gram-positive bacteria *Bacillus thuringiensis* B1(2015b) and *Patulibacter medicamentivorans* (Marchlewicz et al. 2017b; Salgado et al. 2018). *Bacillus thuringiensis* B1(2015b) is an organism that was isolated from soil from a post-industrial landfill site that had been contaminated with cyanides, heavy metals, and pesticides (dieldrin, endrin, hexachlorocyclohexanes, phenols, or hexachlorobenzene) (Marchlewicz et al. 2016). These pollutants can have an impact on the adaptation of the microflora in this area. Hence, it is not surprising that sophisticated, very diverse enzymatic systems have developed. The B1(2015b) strain is characterized by its ability to degrade both 25 mg/L ibuprofen and 6 mg/L naproxen (Marchlewicz et al. 2016). Ibuprofen is degraded into 2-hydroxyibuprofen by hydroxylation, which is then converted into 2-(4-hydroxyphenyl-) propionic acid. Similar to ibuprofen, the degradation pathway of *Sphingomonas*, the synthase acyl-CoaA activity is also observed in the next step. As a result of this activity, 1,4-hydroquinone is formed. This compound is hydroxylated by hydroquinone monooxygenase into 2-hydroxy-1,4-quinol. In contrast to the postulated *meta* cleavage of the trihydroxylated derivative by *Variovorax*, in this case, this intermediate undergoes intradiol cleavage by hydroxyquinol 1,2-dioxygenase into 3-hydroxy *cis,cis*-muconic acid (Marchlewicz et al. 2017b). *Patulibacter medicamentivorans* can degrade ibuprofen *via* two different pathways. The first is with catechols as intermediates, which leads to propionic acid and 2,4-dimethylpentanedioic acid. These compounds probably undergo the central metabolism of the cell and as a result, to the mineralization of ibuprofen. The second mechanism, which is *via* hydroxylated intermediates, is postulated by authors as being more oxidable and less toxic (Salgado et al. 2018). This conclusion is surprising because to the best of our knowledge, aromatic hydroxylated derivatives are more toxic than their parent compounds (Marchlewicz et al. 2015). Moreover, because there is no complete degradation of hydroxylated compounds that are formed, this thesis is all the more controversial. The biodegradation pathways of ibuprofen by bacteria have been presented in Fig. 16.3.

According to the WHO, the most popular monocyclic NSAID is the 4-acetaminophen called paracetamol. Liang et al. (2016) isolated and characterized the paracetamol-transforming bacteria from the soil: *Bacillus aryabhattai* strain 1-Sj-5-2-5-M, *Bacillus subtilis* strain HJ5, and *Klebsiella pneumonia* strain S001. Paracetamol oligomers and *p*-aminophenol were observed as products of the paracetamol transformation (Liang et al. 2016). The latter may transform into *p*-benzoquinone *via p*-benzoquinone imine. Degradation pathways *via p*-aminophenol as a key intermediate were also described for paracetamol degradation by *Pseudomonas moorei* KB4, *Stenotrophomonas* sp. f1, *Pseudomonas* sp. f2, and *Pseudomonas* sp. fg-2 (Żur et al. 2018b; Zhang et al. 2013). Zhang et al. (2013) proposed that the initial reaction of the decomposition of paracetamol is the release of acetate. The aminophenol that is formed is hydroxylated into hydroquinone, which subsequently is cleaved to the aliphatic acids: 2-hexenoic, succinic, malonic, oxalic, and formic acid (Zhang et al. 2013). In turn, Żur et al. (2018b) observed the

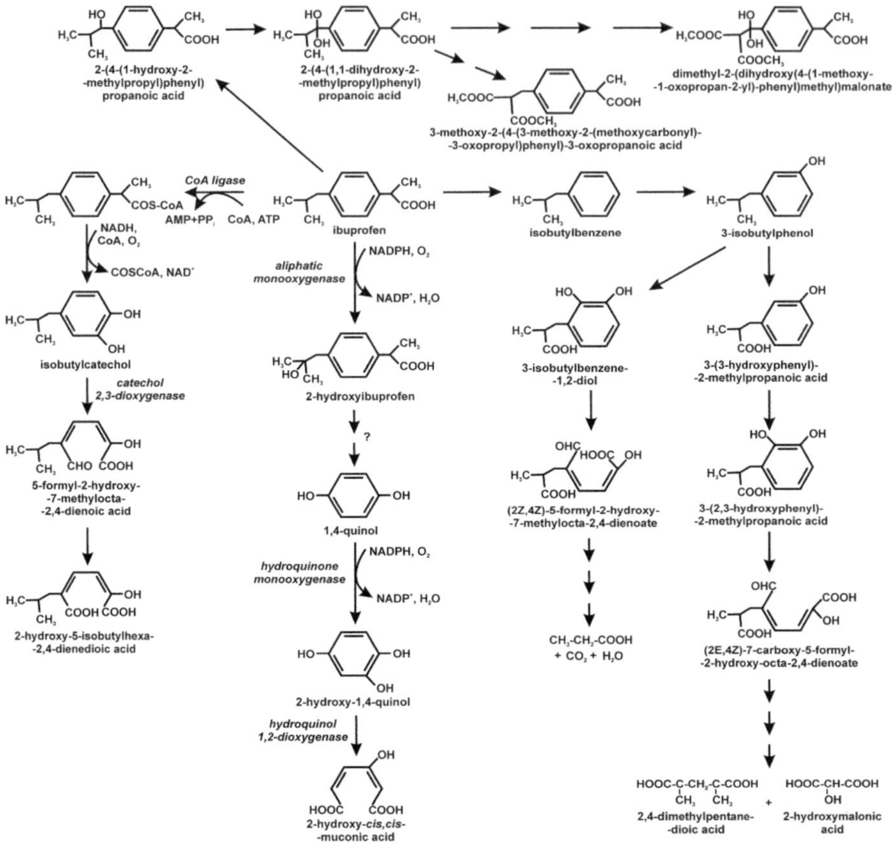

Fig. 16.3 The biodegradation pathways of ibuprofen by bacteria. (Marchlewicz et al. 2017b Murdoch and Hay 2005; Murdoch and Hay 2013; Salgado et al. 2018)

cleavage of *p*-hydroquinone by hydroquinone 1,2-dioxygenase to 4-hydroxymuconic semialdehyde. This product may be incorporated into the central metabolism and in this way be totally mineralized (Żur et al. 2018b). Li et al. (2014) analyzed the fate of paracetamol in the soil. They observed the rapid degradation of this compound and identified eight intermediates: 3-hydroxyacetaminophen, hydroquinone, 1,4-benzoquinone, N-acetyl-*p*-benzoquinone imine, *p*-acetanisidide, 4-methoxyphenol, 2-hexenoic acid, and 1,4-dimethoxybenzene. The high concentration of 2-hexenoic acid may suggest that the aromatic ring is cleaved and in this way is totally mineralized (Li et al. 2014). The biodegradation pathways of paracetamol by bacteria have been presented in Fig. 16.4.

Polycyclic non-steroidal anti-inflammatory drugs are much more difficult to degrade. In recent years, however, there have been many reports on the microbiological degradation/transformation of diclofenac. The intensification of these studies resulted from diclofenac being added to the watch list for European Union-wide

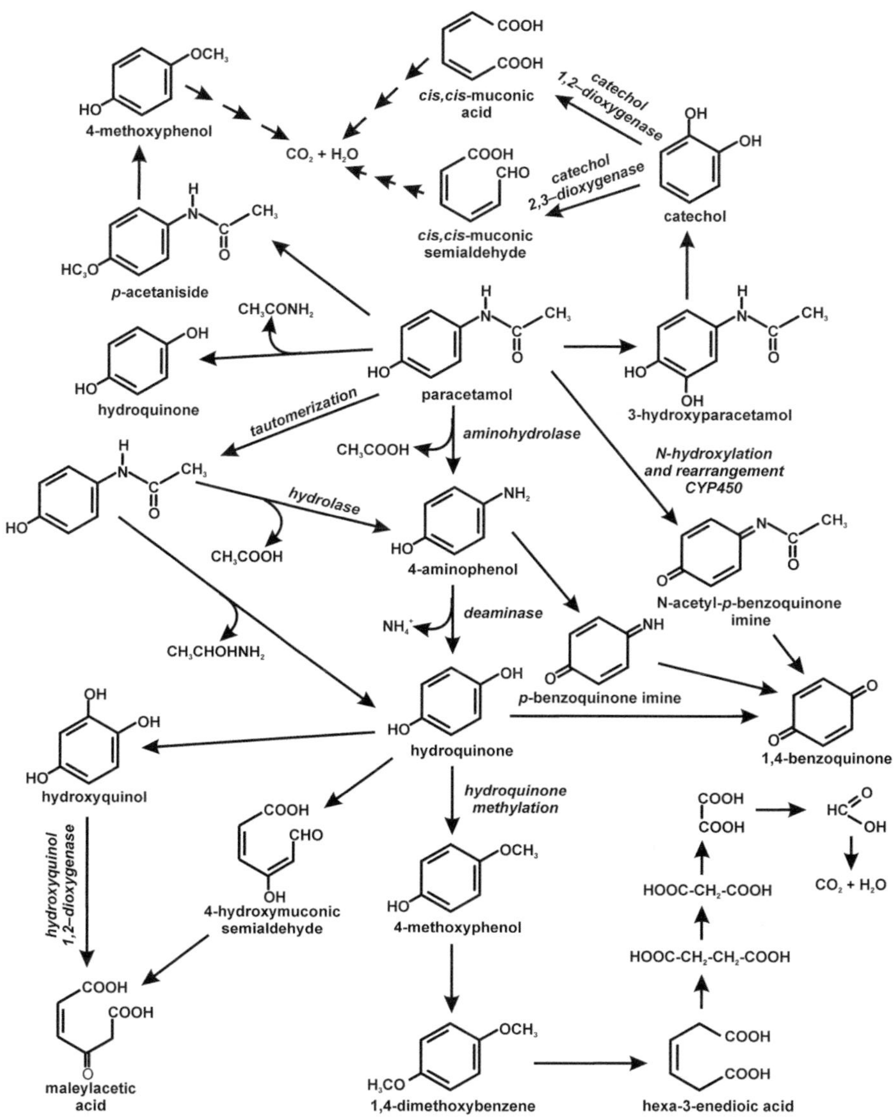

Fig. 16.4 The biodegradation pathways of paracetamol by bacteria. (Li et al. 2014; Liang et al. 2016; Wu et al. 2012; Zhang et al. 2013; Żur et al. 2018b)

monitoring in Decision 2015/495/EU (Moreira et al. 2018). These studies resulted in knowledge about a number of intermediates of the transformation of diclofenac. The first bacterial strain that was able to biotransform diclofenac was *Raoultella* sp. DD4. However, this strain only transformed 0.6 mg/L over 28 days (Domaradzka et al. 2016). The *Klebsiella* sp. KSC, *Brevibacterium* sp. D4, *Starkeya* sp. C11, *Rhizobium* sp. C12, and *Labrys portucalensis* F11 strains are characterized by a significantly better degradability (Bessa et al. 2017; Moreira et al. 2018; Stylianou et al. 2018). However, when analyzing the described products of the degradation of diclofenac, it is difficult to resist the impression that the structure of this drug most often undergoes a transformation and not complete mineralization. For example, the *Labrys portucalensis* F11 strain transforms diclofenac into a series of hydroxyl derivatives that can then be oxidized into quinone derivatives. The diclofenac side chain may be decarboxylated and the terminal hydroxyl group may be linked to sulfates. These transformations, however, do not affect the structure of the rings, whereas the amine linkage that connects the two rings is only transformed into an imine bond. The appearance of a series of derivatives with the structure of the quinone imine during the transformation of diclofenac indicates that this is the key structure in these transformations. The appearance of a peripheral dihydroxylated structure may suggest the participation of dioxygenases in the ring cleavage. However, no products of this reaction were observed (Moreira et al. 2018). Similar products were observed by Gröning et al. (2007) during the biotransformation of diclofenac by the indigenous microflora of river sediments. The strain *Enterobacter hormaechei* D15 also transformed diclofenac without affecting the structure of the aromatic rings. During decomposition, {2-[(2,6-dichlorophenyl) amino] phenyl} acetic acid and 1-(2,6-dichlorophenyl)-1,3-dihydro-2H-indol-2-one intermediates were observed (Aissaoui et al. 2017). The *Klebsiella* sp. KSC strain, which had been isolated from livestock soil, degraded/transformed high concentrations of diclofenac. Among the identified products were hydroxylated two- and one-ring derivatives. The appearance of the latter indicates a violation of the amine bond. The diclofenac side chain was decarboxylated, but no cleavage of the structure of the aromatic rings was observed (Stylianou et al. 2018).

Facey et al. (2018) proposed the complete degradation pathway of diclofenac by native soil microorganisms. The complete oxidation of diclofenac that was postulated by the authors, however, requires confirmation, because, among the intermediates that they identified, there was only a key intermediate – a carboxylated derivative of diclofenac and 2,6-dichloroaniline and carboxylated 2-hydroxyphenylacetic acid as peripheral intermediates. Previous reports clearly showed that there is a constant need to look for microorganisms that are capable of the complete mineralization of diclofenac. The biodegradation pathways of diclofenac by bacteria have been presented in Fig. 16.5.

Little is also known about the mechanism of the degradation of other popular polycyclic non-steroidal anti-inflammatory drugs, such as naproxen or ketoprofen. The naproxen-degrading strains include *Bacillus thuringiensis* B1(2015b), *Planococcus* sp. S5, and *Stenotrophomonas maltophilia* KB2. Under monosubstrate conditions, these strains only degrade naproxen partially. For this compound to be completely degraded, it is necessary to use an additional carbon source – assimilable glucose or another aromatic compound (Domaradzka et al. 2015; Górny et al. 2019; Wojcieszyńska et al. 2014). The immobilization of the *Planococcus* sp. S5 strain on a Loofah sponge affected the strain's ability to degrade higher concentrations of naproxen (Dzionek et al. 2018). The *Planococcus* sp. S5 and *Stenotrophomonas maltophilia* KB2 strains degrade naproxen *via* the trihydroxy

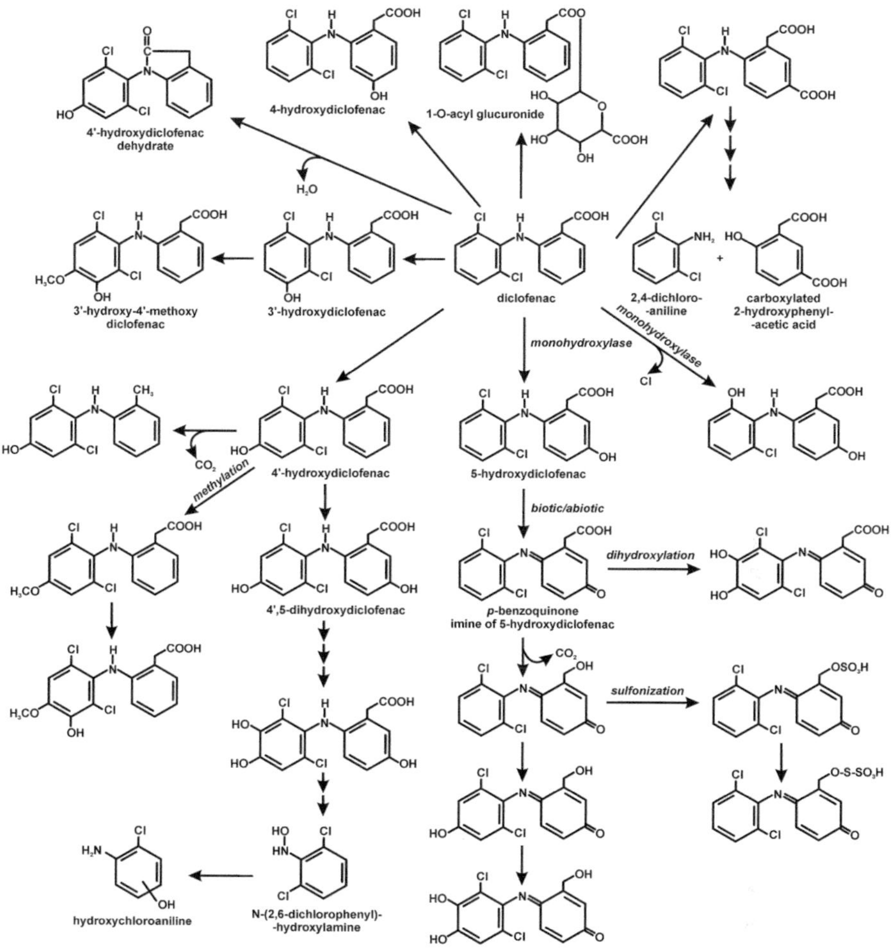

Fig. 16.5 The biodegradation pathways of diclofenac by bacteria. (Facey et al. 2018; Gröning et al. 2007; Lonappan et al. 2016; Moreira et al. 2018; Stylianou et al. 2018)

derivative which is cleaved then subsequently by hydroxyquinol 1,2-dioxygenase (Fig. 16.6) (Wojcieszyńska et al. 2014; Wojcieszyńska et al. 2016).

So far no pure strains of bacteria that are capable of transforming ketoprofen have been isolated. Quintana et al. (2005), when studying the degradation of ketoprofen by activated sludge, showed that it is transformed into two metabolites – 3-(hydroxy-carboxymethyl)hydratropic acid and 3-(keto-carboxymethyl)hydratopic acid. The presence of these metabolites indicates that ketoprofen can be broken down by the common pathway for biphenyls, biphenylethers and related compounds (Fig. 16.7) (Quintana et al. 2005).

Of the other microorganisms that are capable of transforming ketoprofen, microalgae should be mentioned. Ismail et al. (2016) showed that a consortium of

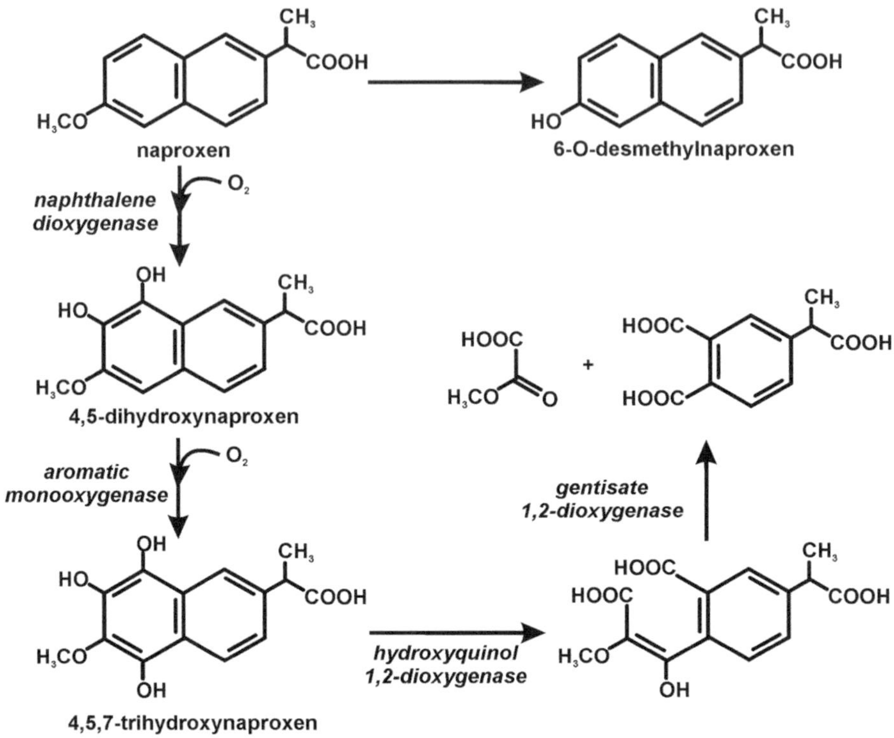

Fig. 16.6 The biodegradation pathways of naproxen by bacteria. (Wojcieszyńska et al. 2014)

microalgae degraded 5 mM ketoprofen within 7 days. The transformation products were (3-ethylphenyl)(phenyl)methanone, (3-hydroxyphenyl)(phenyl)methanone, and (3-hydroxyphenyl)(oxo)acetic acid. These compounds were less toxic than the parent compound and indicated that carboxylation and a series of oxidation/reduction reactions are the steps in ketoprofen degradation (Fig. 16.7) (Ismail et al. 2016).

The information provided clearly indicates that despite the isolation of microorganisms that are capable of degrading or transforming non-bacterial anti-inflammatory drugs, little is known about the mechanisms of the degradation of these drugs, the enzymes that are involved in these processes as well as the fact that the transformation processes occur with little efficiency.

16.4 Impact of NSAIDs on Microorganisms

Previous reports have indicated that non-steroidal anti-inflammatory drugs, which are biologically active substances, can affect organisms. In the literature reports, the effects of NSAIDs on higher organisms are increasingly being described. A remarkable example is the toxic effect of diclofenac on vultures. Because of the accidental

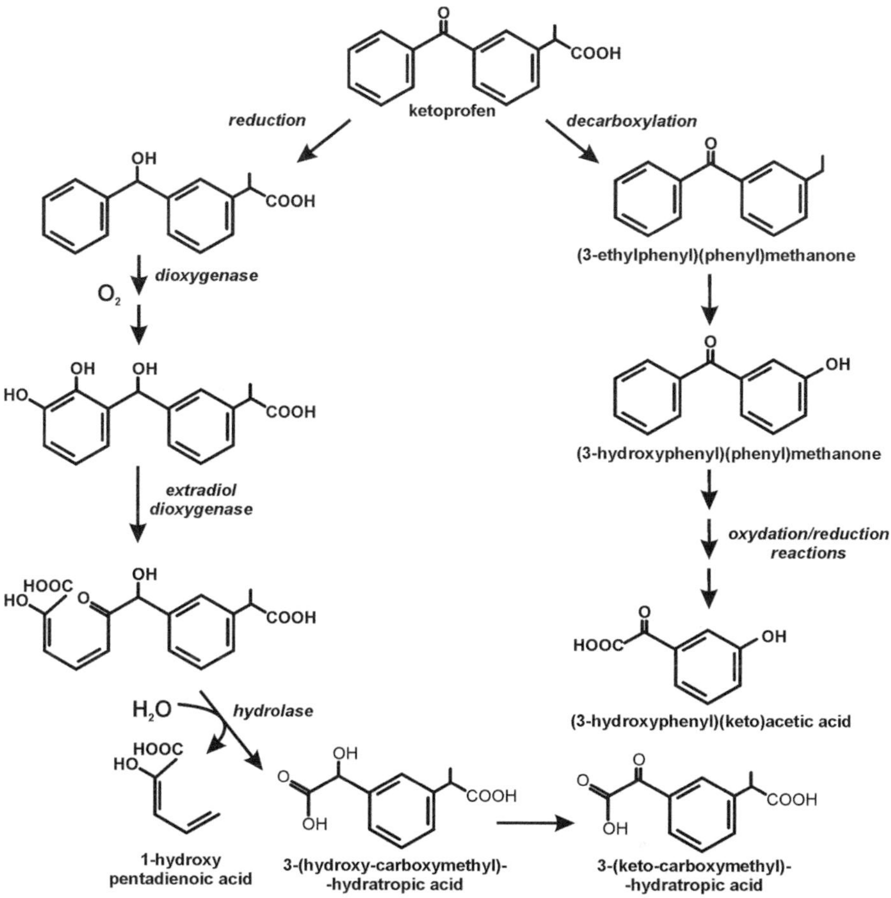

Fig. 16.7 The biodegradation pathways of ketoprofen by bacteria. (Ismail et al. 2016; Quintana et al. 2005)

poisoning by diclofenac, which resulted in renal damage, the populations of the Himalayan vulture (*Gyps himalayensis*), the white-rumped vulture (*Gyps bengalensis*), the Indian vulture (*Gyps indicus*), the slender-billed vulture (*Gyps tenuirostris*), and the red-headed vulture (*Sarcogyps calvus*) have decreased drastically (Das et al. 2011; Hla et al. 2011; Sharma and Kaushik 2017). The vultures had been exposed to diclofenac by consuming the contaminated carcasses of domestic cattle that had been treated with diclofenac shortly before their death (Das et al. 2011; Hla et al. 2011). Diclofenac might also be the cause of the reduced hatching and egg viability of the marine copepod *Gladioferens pectinatus* (Guyon et al. 2018). It was shown that ibuprofen, paracetamol, naproxen, and acetylsalicylic acid, diclofenac and their combinations induced oxidative stress in *Hyalella azteca* and *Oncorhynchus mykiss* (Gomez-Olivan et al. 2014; Żur et al. 2018a).

Little is still known about the impact of NSAIDs on microorganisms. Most of the current reports have focused on the acute toxicity that has been observed after exposure to high concentrations of these drugs (Domaradzka et al. 2016; Żur et al. 2018b; Marchlewicz et al. 2017a). However, these concentrations are not observed in the environment. This does not mean, however, that these drugs do not affect the soil microbiome. The chronic exposure of microorganisms to low concentrations of NSAIDs may cause changes in the structures of the microorganism communities in soil and might also affect the physiological state of microorganisms.

Cycoń et al. (2016) examined the impact of diclofenac, naproxen, ibuprofen, and ketoprofen on the activity of soil microorganisms during a 90-day exposure. They showed that selected NSAIDs significantly increase the microbial activity. They observed that NSAIDs at a concentration of 10 mg/kg soil affected the substrate-induced respiration, the rates of the nitrification and ammonification processes, and the soil enzyme activities – alkaline phosphatases and urease. Moreover, they observed an increase in the number of heterotrophic bacteria and fungi as a result of the evolution of the specific microorganisms that are engaged in the degradation of NSAIDs. The authors suggested that the acceleration of biochemical and microbial activity of soil after exposure to NSAIDs may cause disturbances in the soil functioning (Cycoń et al. 2016).

Grenni et al. (2013, 2014) showed that naproxen affected the structure of the natural microbial community of the River Tiber and increased *Alpha-* and *Gammaproteobacteria*. However, after the complete degradation of naproxen, the authors did not observe any significant differences between the microbial community composition in the control or in the microcosms that had been treated with the drug (Górny et al. 2019; Grenni et al. 2014). Moreover, diclofenac may affect the structural divergence of the bacterial population in active sludge. For that reason, functional changes may be observed in active sludge wastewater treatment systems (Domaradzka et al. 2016). On the other hand, Jiang et al. (2017) showed that the bacterial diversity in sequencing batch reactor after exposure to selected NSAIDs (diclofenac, diclofenac plus ibuprofen, diclofenac plus ibuprofen plus naproxen) increased. The domination of *Actinobacteria* and *Bacteroidetes* was observed, while the *Micropruina* and *Nakamurella* populations decreased (Górny et al. 2019; Jiang et al. 2017).

Diclofenac inhibited the growth of *Dunaliella tertiolecta* at concentrations of 25 mg/L and above (Santos et al. 2010). In turn, Stylianou et al. (2018) showed that this drug was toxic to *Vibrio fischeri* at a concentration of 0.1 g/L. They observed a 69% and 98% inhibition of bioluminescence after a 5- and 15-min exposure, respectively. Moreover, these authors showed a decreased toxicity of diclofenac after its biodegradation by *Klebsiella* sp. KSC. This is an indication that the intermediates that were formed during biodegradation were less toxic or hermetic (Stylianou et al. 2018). However, it was also shown that 1 g/L diclofenac caused a strong inhibition of the ability of native soil microorganisms to degrade diclofenac (Facey et al. 2018). It was also reported that after a 4-week exposure to 100 μg/L of diclofenac, a biofilm that was composed of bacteria and algae lost about 70% of its initial biomass (Caracciolo et al. 2015). On the other hand, Gomes et al. (2018) showed that exposure to the ibuprofen and diclofenac did not affect the ability of *Burkholderia cepacia* to form biofilms.

Exposure of microorganisms to diclofenac or diclofenac with ibuprofen or ibuprofen and naproxen resulted in an increase in the activity of superoxide dismutase, which suggests that NSAIDs induce oxidative stress. Moreover, a decrease in succinate dehydrogenase was observed, which is connected with damage to the mitochondrial function and to the intermediary metabolism of microorganisms. What is more, there was an increase in the production of the EPS content. This probably leads to the formation of a network structure outside the cells and protects the microorganisms against selected NSAIDs (Jiang et al. 2017).

The EC_{50} that was estimated for *Aliivibrio fischeri* was 16.31, 39.93, 47.07, and 52.64 mg/L for diclofenac, ibuprofen, naproxen, and salicylic acid, respectively. According to these values of EC_{50}, the acute toxicity alignment is diclofenac>ibuprofen>naproxen>salicylic acid. Moreover, the toxic units (TUs) that were calculated by Sprague and Ramsay's formula for diclofenac, ibuprofen, naproxen, and salicylic acid were 6.1, 2.5, 2.03, and 1.89, respectively. This indicates that these compounds should be classified as toxic for this organism. What is more, the EC_{50} and TU for combinations of these compounds were 5.13 and 19.49, respectively, which indicates a synergistic impact on *Aliivibrio fischeri* (Dökmeci et al. 2014; Żur et al. 2018a). Gonzales-Naranjo and Boltes (2014) estimated that the EC_{50} for microalgae *Pseudokirchneriella subcapitata* was 232.64 mg/L for ibuprofen. The value of this EC_{50} suggests that ibuprofen is not harmful to algae according to the criteria of EU Directive 93/67/EEC (Gonzales-Naranjo and Boltes 2014; Żur et al. 2018b). However, these authors observed a significant and measurable negative effect of ibuprofen in concentrations above 35 mg/L (Gonzales-Naranjo and Boltes 2014). Żur et al. (2018b) showed that the mean value of the microbial toxic concentration is 3435 mg/L for paracetamol. For that reason, it should be classified as a nontoxic compound. However, an evaluation of paracetamol toxicity and its metabolites using the Microtox test with bioluminescence *Vibrio fischeri* showed that a key intermediate of paracetamol – 4-aminophenol – has a significantly higher toxicity than its parent compound (Żur et al. 2018b).

Toxicity studies by Marchlewicz et al. (2017a) and Górny et al. (2019) showed a higher resistance of *Bacillus thuringiensis* B1(2015b), which is a gram-positive strain that had been isolated from the soil near the Chemical Factory "Organika-Azot" in Jaworzno (Poland), to ibuprofen and naproxen than to other microorganisms. The EC_{50} of ibuprofen on this strain was 809.3 mg/L. This value was 1.5-fold higher than the mean value of the microbial toxic concentration (545.5 mg/L) and indicates that the B1(2015b) strain is resistant to ibuprofen (Marchlewicz et al. 2017a). Moreover, it was also shown that *B. thuringiensis* B1(2015b) is extremely resistant to naproxen. The EC_{50} value of naproxen for this strain was 4.69 g/L and was threefold higher than the mean value of the toxic microbial concentration (1.66 g/L) (Górny et al. 2019). The high resistance of *B. thuringiensis* B1(2015b) to ibuprofen and naproxen is probably connected with the changes in the membrane fatty acid composition. Changes in the ratio of branched and unsaturated fatty acids were observed after the exposure of this strain to ibuprofen. In the presence of naproxen, an increased ratio of saturated and unsaturated total fatty acids was observed. In the presence of naproxen, the 16:0 iso 3OH fatty acids also appeared in the membrane of *B. thuringiensis* B1(2015b). The hydroxyl group of this compound stabilizes the structure of the membrane *via* its interaction with the membrane proteins, which may increase the tightness of bacterial membranes and may indicate an adaptive feature of the B1(2015b) strain that enables it to grow in the presence of ibuprofen and naproxen (Górny et al. 2019; Marchlewicz et al. 2017a).

Although acute toxicity tests indicated that ibuprofen is a nontoxic compound, a test of the chronic effect of ibuprofen using *Tetrahymena thermophila* showed that longer exposition of the ciliate to ibuprofen revealed the toxic effect of this compound (Marchlewicz et al. 2017a).

Wang and Gunsch (2011) showed that membrane integrity is affected by ketoprofen and naproxen. This leads to the irreversible inhibition of nitrite production in the ammonia-oxidizing bacterium *Nitrosomonas europaea* (Wang and Gunsch 2011). Moreover, COX-inhibiting NSAIDs reduced the hyphal growth, asexual and sexual reproduction, and germination in the fungi *Chaetomium globosum*, *Fusarium oxysporum*, *Fusarium solani*, *Stachybotrys atra*, *Stachybotrys chartarum*, and *Aspergillus niger* (Dalmont et al. 2017). Paracetamol may inhibit *Saccharomyces cerevisiae* but does not induce an oxidative stress response as it does in mammalian cells. The paracetamol toxicity of yeast is probably dependent on the membrane pumps that are associated with the ergosterol biosynthesis pathway (Dalmont et al. 2017; Srikanth et al. 2005). The inhibition of DNA synthesis by diclofenac at 50–100 mg/L may be the reason for the inhibition of the growth of gram-positive and gram-negative bacteria (Caracciolo et al. 2015; Domaradzka et al. 2016).

The inhibitory effect of naproxen on algae was both concentration- and time-dependent. The complete inhibition of the growth of *Cymbella* sp. and *Scenedesmus quadricauda* at 100 mg/L naproxen was observed. However, *Cymbella* sp. was more tolerant to naproxen than *S. quadricauda* because the diatom frustules played a protective role in preventing naproxen from entering the algae cell. Moreover, it was also observed that the content of chlorophyll *a* and carotenoid were significantly reduced at 100 mg/L of naproxen. This may be caused by the fact that carotenoids can deactivate excited chlorophyll in order to scavenge the accumulated reactive oxygen species in a chloroplast. Exposure to 1 mg/L of naproxen caused a significant increase in superoxide dismutase (SOD). A high concentration led to a decrease in the activity of SOD. In turn, the catalase activity was significantly increased in the presence of naproxen (Ding et al. 2017).

Not only can the parent compounds have a negative impact on microorganisms, but also on their metabolites. The degradation of NSAIDs has been shown to lead to more toxic intermediates. For example, the isobutylbenzene, 3-isobutylphenol, 4-isobutylbenzene-1,2-diol, and 2-hydroxy-5-isobutyl-hexa-2,4-dienoic acid that were observed during the degradation of ibuprofen showed, in both acute and chronic toxicity tests, higher toxicity from ibuprofen (Salgado et al. 2018). Moreover, the metabolites of paracetamol – benzoquinone and N-acetyl-*p*-benzoquinone imine (NAPQI) – are considered to be more toxic than their parent compound (Li et al. 2014).

Non-steroidal anti-inflammatory drugs present in the environment can affect microorganisms in different ways depending on the environmental conditions. This is mainly due to their availability to microorganisms. The literature reports show that NSAIDs may have a negative effect on nitrification processes, can cause oxidative stress in microorganisms, as well as affect the state of the cell membrane. There are microorganisms that have the ability to transform/biodegrade these bioactive compounds. NSAIDs with the monocyclic structure are usually fully biodegradable, while polycyclic NSAIDs are usually only biotransformed to their hydroxyl derivatives. The analysis of the literature data indicates a continuous need for further research that will allow filling numerous gaps in the current knowledge on the fate of NSAIDs in the soil environment.

Acknowledgments This research was funded by the National Science Centre, Poland (grant number 2018/29/B/NZ9/00424).

References

Abdullah Q, Edrees W, Al-Kaf AG et al (2018) Biodegradation of paracetamol by native bacterial strains isolated from Yemeni pharmaceutical wastewater plant in Sana'a. Chron Pharma Sci 2:512–522

Ahn E, Choi KY, Kang BS et al (2017) Salicylate degradation by a cold-adapted *Pseudomonas* sp. Ann Microbiol 67:417–424

Aissaoui S, Ouled-Haddar H, Sifour M et al (2017) Metabolic and co-metabolic transformation of diclofenac by *Enterobacter hormaechei* D15 isolated from activated sludge. Curr Microbiol 74:381–388

Amadio J, Gordon K, Murphy CD et al (2010) Biotransformation of flurbiprofen by *Cunninghamella* species. Appl Environ Microbiol 76:6299–6303

Aracagök YD, Göker H, Cihangir N (2018) Biodegradation of diclofenac with fungal strains. Arch Environ Prot 44:55–62

Ascar L, Ahmuda I, Morales N et al (2017) Mobility of nonsteroidal anti-inflammatory drugs in soils with and without amendment of biosolid. J Chil Chem Soc 62:3593–3596

Bessa VS, Moreira IS, Tiritan ME et al (2017) Enrichment of bacterial strains for the biodegradation of diclofenac and carbamazepine from activated sludge. Int Biodeterior Biodegradation 120:135–142

Borras E, Llorens-Blanch G, Rodriguez-Rodriguez CE et al (2011) Soil colonization by *Trametes versicolor* grown on lignocellulosic materials: substrate selection and naproxen degradation. Int Biodeterior Biodegradation 65:846–852

Brune K, Patrignani P (2015) New insights into the use of currently available non-steroidal anti-inflammatory drugs. J Pain Res 8:105–118

Caracciolo AB, Topp E, Grenni P (2015) Pharmaceuticals in the environment: biodegradation and effects on natural microbial communities. A review. J Pharm Biomed Anal 106:25–36

Cycoń M, Borymski S, Żołnierczyk B et al (2016) Variable effects of non-steroidal anti-inflammatory drugs (NSAIDs) on selected biochemical processes mediated by soil microorganisms. Front Microbiol 7:1969. https://doi.org/10.3389/fmicb.2016.01969

Dalmont K, Biles CL, Konsure H et al (2017) Nonsteroidal anti-inflammatory drugs (NSAIDS) inhibit the growth and reproduction of *Chaetomium globosum* and other fungi associated with water-damaged buildings. Mycopathologia 182:1025–1036

Das D, Cuthbert RJ, Jakati RD et al (2011) Diclofenac is toxic to the Himalayan Vulture *Gyps himalayensis*. Bird Conserv Int 21:72–75

De Gusseme B, Vanhaecke L, Verstraete W et al (2011) Degradation of acetaminophen by *Delftia tsuruhatensis* and *Pseudomonas aeruginosa* in a membrane bioreactor. Water Res 45:1829–1837

Ding T, Lin K, Yang B et al (2017) Biodegradation of naproxen by freshwater algae *Cymbella* sp. and *Scenedesmus quadricauda* and the comparative toxicity. Bioresour Technol 238:164–173

Dökmeci AH, Dökmeci I, Ibar H (2014) The determination of single and mixture toxicity at high concentrations of some acidic pharmaceuticals via *Aliivibrio fischeri*. Environ Process 1:95–103

Domaradzka D, Guzik U, Hupert-Kocurek K et al (2015) Cometabolic degradation of naproxen by *Planococcus* sp. strain S5. Water Air Soil Pollut 226:297. https://doi.org/10.1007/s11270-015-2564-6

Domaradzka D, Guzik U, Hupert-Kocurek K et al (2016) Toxicity of diclofenac and its biotransformation by *Raoultella* sp. DD4. Pol J Environ Stud 5:2211–2216

Dzionek A, Wojcieszyńska D, Hupert-Kocurek K et al (2018) Immobilization of *Planococcus* sp. S5 strain on the loofah sponge and its application in naproxen removal. Catalysts 8:176. https://doi.org/10.3391/catal8050176

Escuder-Gilabert L, Martin-Biosca Y, Perez-Baeza M et al (2018) Direct chromatographic study of the enantioselective biodegradation of ibuprofen and ketoprofen by an activated sludge. J Chromat A 1568:140–148

Facey SJ, Nebel BA, Kontny L et al (2018) Rapid and complete degradation of diclofenac by native soil microorganisms. Environ Technol Innov 10:55–61

Fang T, Zhou NY (2014) Purification and characterization of salicylate 5-hydroxylase, a three-component monooxygenase from *Ralstonia* sp. strain U2. Appl Microbiol Biotechnol 98:671–679

Fatehifar M, Borghei SM, Ekhlasinia A (2018) Application of moving bed biofilm reactor in the removal of pharmaceutical compounds (diclofenac and ibuprofen). J Environ Chem Eng 6:5530–5535

Fent K, Weston AA, Caminada D (2006) Ecotoxicology of human pharmaceuticals. Aquat Toxicol 76:122–159

Filatova IY, Kazakov AS, Muzafarov EN et al (2017) Protein SgpR of *Pseudomonas putida* strain AK5 is a LysR-type regulator of salicylate degradation through gentisate. FEMS Microbiol Lett 364:1–7

Fokunang CN, Fokunang ET, Frederick K et al (2018) Overview of non-steroidal anti-inflammatory drugs (nsaids) in resource limited countries. MOJ Toxicol 4:5–13

Gomes IB, Madureira D, Simoes LC et al (2018) The effects of pharmaceutical and personal care products on the behavior of *Burkholderia cepacia* isolated from drinking water. Int Biodeter Biodegrad. https://doi.org/10.1016/j.ibiod.2018.03.018

Gomez-Olivan LM, Neri-Cruz G-MM et al (2014) Binary mixtures of diclofenac with paracetamol, ibuprofen, naproxen, and acetylsalicylic acid and these pharmaceuticals in isolated form induce oxidative stress on *Hyalella azteca*. Environ Monit Asses 186:7259–7271

Gonzales-Naranjo V, Boltes K (2014) Toxicity of ibuprofen and perfluorooctanoic acid for risk assessment of mixtures in aquatic and terrestrial environments. Int J Environ Sci Technol 11:1743–1750

Gonzales-Naranjo V, Boltes K, Biel M (2013) Mobility of ibuprofen, a persistent active drug, in soils irrigated with reclaimed water. Plant Soil Environ 59:68–73

Górny D, Guzik U, Hupert-Kocurek K et al (2019) Naproxen ecotoxicity and biodegradation by *Bacillus thuringiensis* B1(2015b) strain. Ecotoxicol Environ Saf 167:505–512

Graouer-Bacart M, Sayen S, Guillon E (2016) Adsorption and co-adsorption of diclofenac and Cu(II) on calcareous soils. Ecotoxicol Environ Saf 124:386–392

Grenni P, Patrolecco L, Ademollo N et al (2013) Degradation of gemfibrozil and naproxen in a river water ecosystem. Microchem J 107:158–164

Grenni P, Patrolecco L, Ademollo N et al (2014) Capability of the natural microbial community in a river water ecosystem to degrade the drug naproxen. Environ Sci Pollut Res 21:13470–13479

Gröning J, Held C, Garten C et al (2007) Transformation of diclofenac by the indigenous microflora of river sediments and identification of a major intermediate. Chemosphere 69:509–516

Guyon A, Smith KF, Charry MP et al (2018) Effects of chronic exposure to benzophenone and diclofenac on DNA methylation levels and reproductive success in a marine copepod. J Xenobiotics 8:7674. https://doi.org/10.4081/xeno.2018.7674

Guzik U, Hupert-Kocurek K, Wojcieszyńska D (2013) Intradiol dioxygenases – the key enzymes in xenobiotics degradation. Chapter 7. In: Chamy R, Rosenkranz F (eds) Biodegradation of hazardous and special products. InTech Rijeka, Rijeka, pp 129–153, ISBN 978-953-51-1155-9

Hart A, Orr DLJ (1975) The degradation of paracetamol (4-hydroxyacetanilide) and other substituted acetanilides by a *Penicillium* species. Antonie Van Leeuwenhoek 41:239–247

He A, Rosazza JPN (2003) Microbial transformation of S-naproxen by *Aspergillus niger* ATCC 9142. Pharmazie 58:420–423

Hintner JP, Lechner C, Riegert U et al (2001) Direct ring fission of salicylate by a salicylate 1,2-dioxygenase activity form *Pseudaminobacter salicylatoxidans*. J Bacteriol 183:6936–6942

Hla H, Shwe NM, Htun TW et al (2011) Historical and current status of vultures in Myanmar. Bird Conserv Int 21:376–387

Huang CY, Fu LH, Sung MH et al (2018) Ibuprofen biodegradation by hospital, municipal, and distillery activated sludges. Environ Technol 5:1–10

Ishiyama D, Vujaklija D, Davies J (2004) Novel pathway of salicylate degradation by *Streptomyces* sp. strain WA46. Appl Environ Microbiol 70:1297–1306

Ismail MM, Essam TM, Ragab YM et al (2016) Biodegradation of ketoprofen using a microalgal-bacterial consortium. Biotechnol Lett 38:1493–1502

Jiang C, Geng J, Hu H et al (2017) Impact of selected non-steroidal anti-inflammatory pharmaceuticals on microbial community assembly and activity in sequencing batch reactors. PLoS One 12(6):e0179236. https://doi.org/10.1371/journal.pone.0179236

Jõessar M, Viggor S, Heinaru E et al (2017) Strategy of *Pseudomonas pseudoalcaligenes* C70 for effective degradation of phenol and salicylate. PLoS One 12(3):e0173180. https://doi.org/10.1371/journal.pone.0173180

Jóźwiak-Bębenista M, Nowak JZ (2014) Paracetamol: mechanism of action, applications and safety concern. Acta Pol Pharm 71:11–23

Kesseru P, Kiss I, Bihari Z et al (2005) Nitrate-dependent salicylate degradation by *Pseudomonas butanovora* under anaerobic conditions. Bioresour Technol 96:779–784

Li J, Ye Q, Gan J (2014) Degradation and transformation products of acetaminophen in soil. Water Res 49:44–52

Li X, Toledo RA, Wang S et al (2015) Removal of carbamazepine and naproxen by immobilized *Phanerochaete chrysosporium* under non-sterile condition. New Biotechnol 32:282–289

Liang C, Lan Z, Zhang X et al (2016) Mechanism for the primary transformation of acetaminophen in a soil/water. Water Res 98:215–224

Lonappan L, Brar SK, Das RK et al (2016) Diclofenac and its transformation products: environmental occurrence and toxicity – a review. Environ Int 96:127–138

Lowe-Power TM, Jacobs JM, Ailloud F et al (2016) Degradation of the plant defense signal salicylic acid protects *Ralstonia solanacearum* from toxicity and enhances virulence on tobacco. mBio 7(3):e00656–e00616. https://doi.org/10.1128/mBio.00656-15

Mallet C, Eschalier A, Daulhac L (2017) Paracetamol: update on its analgesic mechanism of action. Pain Relief, Cecilia Maldonado, IntechOpen. https://doi.org/10.5772/66649

Marchlewicz A, Guzik U, Wojcieszyńska D (2015) Over-the-counter monocyclic non-steroidal anti-inflammatory drugs in environment – sources, risks, biodegradation. Water Air Soil Pollut 226:355. https://doi.org/10.1007/s11270-015-2622-0

Marchlewicz A, Domaradzka D, Guzik U et al (2016) *Bacillus thuringiensis* B1(2015b) is a gram-positive bacteria able to degrade naproxen and ibuprofen. Water Air Soil Pollut 227:197

Marchlewicz A, Guzik U, Hupert-Kocurek K et al (2017a) Toxicity and biodegradation of ibuprofen by *Bacillus thuringiensis* B1(2015b). Environ SciPollut Res 24:7572–7584
Marchlewicz A, Guzik U, Smułek W et al (2017b) Exploring the degradation of ibuprofen by *Bacillus thuringiensis* B1(2015b): the new pathway and factors affecting degradation. Molecules 22:1676. https://doi.org/10.3390/molecules22101676
Marco-Urrea E, Perez-Trujillo M, Vicent T et al (2009) Ability of white-rot fungi to remove selected pharmaceuticals and identification of degradation products of ibuprofen by *Trametes versicolor*. Chemosphere 74:765–772
Marco-Urrea E, Perez-Trujillo M, Blanquez P et al (2010a) Biodegradation of the analgesic naproxen by *Trametes versicolor* and identification of intermediates using HPLC-DAD-MS and NMR. Bioresour Technol 101:2159–2166
Marco-Urrea E, Pérez-Trujillo M, Cruz-Morató C et al (2010b) White-rot fungus-mediated degradation of the analgesic ketoprofen and identification of intermediates by HPLC–DAD–MS and NMR. Chemosphere 78:474–481
Martinez-Hernandez V, Meffe R, Lopez SH et al (2016) The role of sorption and biodegradation in the removal of acetaminophen, carbamazepine, caffeine, naproxen, and sulfamethoxazole during soil contact: a kinetics study. Sci Total Environ 559:232–241
Moreira IS, Bessa VS, Murgolo S et al (2018) Biodegradation of diclofenac by the bacterial strain *Labrys portucalensis* F11. Ecotoxicol Environ Saf 152:104–113
Murdoch RW, Hay AG (2005) Formation of catechols via removal of acid side chains from ibuprofen and related aromatic acids. Appl Environ Microbiol 71:6121–6125
Murdoch RW, Hay AG (2013) Genetic and chemical characterization of ibuprofen degradation by *Sphingomonas* Ibu-2. Microbiology 159:621–632
Murdoch RW, Hay AG (2015) The biotransformation of ibuprofen to trihydroxyibuprofen in activated sludge and by *Variovorax* Ibu-1. Biodegradation 26:105–113
Nowak KM, Girardi C, Miltner A et al (2013) Contribution of microorganisms to non-extractable residue formation during biodegradation of ibuprofen in soil. Sci Total Environ 445–446:377–384
Osafo N, Agyare C, Obiri DD et al (2017) Mechanism of action of nonsteroidal anti-inflammatory drugs. Nonsteroidal Anti-Inflammatory Drugs, AGA Al-kaf, IntechOpen. https://doi.org/10.5772/68090
Panov AV, Volkova OV, Puntus IF et al (2013) *scpA*, a new salicylate hydroxylase gene localized in salicylate/caprolactam degradation plasmids. MolBiol 47:105–111
Quintana JB, Weiss S, Reemtsma T (2005) Pathways and metabolites of microbial degradation of selected acidic pharmaceutical and their occurrence in municipal wastewater treated by a membrane bioreactor. Water Res 39:2654–2664
RatnaKumari T, Vidyavathi M, Asha S et al (2009) Biotransformation of paracetamol by *Cunninghamella echinulate*. J Pharm Res 8:1–5
Rodarte-Morales AI, Feijoo G, Moreira MT et al (2012) Biotransformation of three pharmaceutical active compounds by the fungus *Phanerochaete chrysosporium* in a fedbatch stirred reactor under air and oxygen supply. Biodegradation 23:145–156
Rodriguez-Rodriguez CE, Marco-Urrea E, Caminal G (2010) Degradation of naproxen and carbamazepine in spiked sludge by slurry and solid-phase *Trametes versicolor* systems. Bioresour Technol 101:2259–2266
Salgado R, Brito D, Noronha JP et al (2018) Metabolite identification of ibuprofen biodegradation by *Patulibacter medicamentivorans* under aerobic conditions. Environ Technol 27:1–16
Sanganyado E, Lu Z, Fu Q et al (2017) Chiral pharmaceuticals: a review on their environmental occurrence and fate processes. Water Res 124:527–542
Santos LHMLM, Araujo AN, Fachini A et al (2010) Ecotoxicological aspects related to the presence of pharmaceuticals in the aquatic environment. J Hazard Mater 175:45–95
Sazonova OI, Izmalkova TY, Kosheleva IA et al (2008) Salicylate degradation by *Pseudomonas putida* strains not involving the "classical" *nah2* operon. Microbiology 77:710–716

Sharma K, Kaushik G (2017) NSAIDS in the environment: from emerging problem to green solution. J Pharm Pharmacol 2:1077

Silva TR, Valdman E, Valdman B et al (2007) Salicylic acid degradation from aqueous solutions using *Pseudomonas fluorescens* HK44: parameters studies and application tools. Braz J Microbiol 38:39–44

Srikanth CV, Chakraborti AK, Bachhawat AK (2005) Acetaminophen toxicity and resistance in the yeast *Saccharomyces cerevisiae*. Microbiology 151:99–111

Stylianou K, Hapeshi E, Vasquez MI et al (2018) Diclofenac biodegradation by newly isolated *Klebsiella* sp. KSC: microbial intermediates and ecotoxicological assessment. J Environ Chem Eng 6:3242–3248

Tang Y, Li XM, Xu ZC et al (2014) Removal of naproxen and bezafibrate by activated sludge under aerobic conditions: kinetics and effect of substrates. Biotechnol Appl Biochem 61:333–341

Topp E, Hendel JG, Lapen DR et al (2008) Fate of the nonsteroidal anti-inflammatory drug naproxen in agricultural soil receiving liquid municipal biosolids. Environ Toxicol Chem 27:2005–2010

Vaillancourt FH, Bolin JT, Eltis LD (2006) The ins and outs of ring-cleaving dioxygenases. Crit Rev Biochem MolBiol 41:241–267

Vulava VM, Cory WC, Murphey VL et al (2016) Sorption, photodegradation, and chemical transformation of naproxen and ibuprofen in soils and water. Sci Total Environ 565:1063–1070

Wang S, Gunsch CK (2011) Effects of selected pharmaceutically active compounds on the ammonia oxidizing bacterium *Nitrosomonas europaea*. Chemosphere 82:565–572

Wojcieszyńska D, Domaradzka D, Hupert-Kocurek K et al (2014) Bacterial degradation of naproxen – undisclosed pollutant in the environment. J Environ Manag 145:157–161

Wojcieszyńska D, Domaradzka D, Hupert-Kocurek K et al (2016) Enzymes involved in naproxen degradation by *Planococcus* sp S5. Pol J Microbiol 65:177–182

Wu S, Zhang L, Chen J (2012) Paracetamol in the environment and its degradation by microorganisms. Appl Microbiol Biotechnol 96:875–884

Xu J, Wu L, Chen W et al (2010) Leaching potential od nonsteroidal anti-inflammatory drugs in soils. Environ Toxicol Chem 29:800–807

Zhang L, Hu J, Zhu R et al (2013) Degradation of paracetamol by pure bacterial cultures and their microbial consortium. Appl Microbiol Biotechnol 97:3687–3698

Zhong D-F, Sun L, Liu L et al (2003) Microbial transformation of naproxen by *Cunninghamella* species. Acta Pharmacol Sin 24:442–447

Żur J, Piński A, Marchlewicz A et al (2018a) Organic micropollutants paracetamol and ibuprofen – toxicity, biodegradation, and genetic background of their utilization by bacteria. Environ Sci Pollut Res 25:21498–21524

Żur J, Wojcieszyńska D, Hupert-Kocurek K et al (2018b) Paracetamol – toxicity and microbial utilization. *Pseudomonas moorei* KB4 as a case study for exploring degradation pathway. Chemosphere 206:192–202

MIX
Papier aus verantwortungsvollen Quellen
Paper from responsible sources
FSC® C105338

If you have any concerns about our products,
you can contact us on
ProductSafety@springernature.com

In case Publisher is established outside the EU,
the EU authorized representative is:
Springer Nature Customer Service Center GmbH
Europaplatz 3, 69115 Heidelberg, Germany

Printed by Libri Plureos GmbH
in Hamburg, Germany